Dodge & Plymouth Vans Automotive Repair Manual

by Rob Maddox and John H Haynes

Member of the Guild of Motoring Writers

Models covered:

All Full-size Dodge and Plymouth Vans

1971 through 1999

(9D6 - 30065)

(349)

ABCDE
FGHIJ

Haynes Publishing Group
Sparkford Nr Yeovil
Somerset BA22 7JJ England

Haynes North America, Inc
861 Lawrence Drive
Newbury Park
California 91320 USA

About this manual

Its purpose

The purpose of this manual is to help you get the best value from your vehicle. It can do so in several ways. It can help you decide what work must be done, even if you choose to have it done by a dealer service department or a repair shop; it provides information and procedures for routine maintenance and servicing; and it offers diagnostic and repair procedures to follow when trouble occurs.

We hope you use the manual to tackle the work yourself. For many simpler jobs, doing it yourself may be quicker than arranging an appointment to get the vehicle into a shop and making the trips to leave it and pick it up. More importantly, a lot of money can be saved by avoiding the expense the shop must pass on to you to cover its labor and overhead costs. An added benefit is the sense of satisfaction and accomplishment that you feel after doing the job yourself.

Using the manual

The manual is divided into Chapters. Each Chapter is divided into numbered Sections, which are headed in bold type between horizontal lines. Each Section consists of consecutively numbered paragraphs.

At the beginning of each numbered Section you will be referred to any illustrations which apply to the procedures in that Section. The reference numbers used in illustration captions pinpoint the pertinent Section and the Step within that Section. That is, illustration 3.2 means the illustration refers to Section 3 and Step (or paragraph) 2 within that Section.

Procedures, once described in the text, are not normally repeated. When it's necessary to refer to another Chapter, the reference will be given as Chapter and Section number. Cross references given without use of the word "Chapter" apply to Sections and/or paragraphs in the same Chapter. For example, "see Section 8" means in the same Chapter.

References to the left or right side of the vehicle assume you are sitting in the driver's seat, facing forward.

Even though we have prepared this manual with extreme care, neither the publisher nor the author can accept responsibility for any errors in, or omissions from, the information given.

NOTE

A **Note** provides information necessary to properly complete a procedure or information which will make the procedure easier to understand.

CAUTION

A **Caution** provides a special procedure or special steps which must be taken while completing the procedure where the Caution is found. Not heeding a Caution can result in damage to the assembly being worked on.

WARNING

A **Warning** provides a special procedure or special steps which must be taken while completing the procedure where the Warning is found. Not heeding a Warning can result in personal injury.

Acknowledgements

We are grateful to the Chrysler Corporation for providing technical information and certain illustrations. Technical writers who contributed to this project include Jay Storer, Mike Stubblefield and Larry Warren.

© **Haynes North America, Inc. 1996, 1998, 2000**

With permission from J.H. Haynes & Co. Ltd.

A book in the Haynes Automotive Repair Manual Series

Printed in the U.S.A.

ISBN 1 56392 378 5

Library of Congress Control Number 00-102051

Contents

Introduction to the Dodge and Plymouth vans

Dodge and Plymouth Full-size vans feature conventional chassis layout, with the engine mounted at the front and the power being transmitted through either a manual or automatic transmission to a driveshaft and solid rear axle. Transmissions used are a three-, four- or five-speed manual and three- or four-speed automatic.

The front suspension features independent coil spring type front suspension and solid axle and leaf springs at the rear.

All models are equipped with power assisted front drum or disc and either drum or disc rear brakes. An Anti-lock Braking System (ABS) is used on some later models.

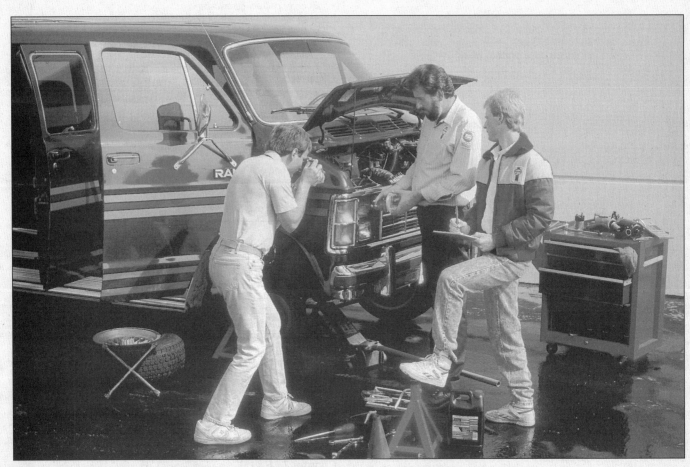

Haynes photographer, mechanic and author with 1986 Dodge Ram Van

Vehicle identification numbers

Modifications are a continuing and unpublicized process in vehicle manufacturing. Since spare parts manuals and lists are compiled on a numerical basis, the individual vehicle numbers are essential to correctly identify the component required.

Vehicle Identification Number (VIN)

This very important identification number is stamped on a plate attached to the left door or pillar (earlier models) or side of the dashboard just inside the windshield on the driver's side of the vehicle (later models) **(see illustrations)**. The VIN also appears on the Vehicle Certificate of Title and Registration. It contains information such as where and when the vehicle was manufactured, the model year and the body style.

VIN year and engine codes

Two particularly important pieces of information located in the VIN are the model year and engine codes. On models through 1980, counting from the left, the engine code is the fourth digit and the model year code is the fifth digit. On 1981 and later models, the 8th digit is the engine code and the 10th digit is the model year code. After obtaining the engine code from the VIN, use the accompanying chart to identify which engine was installed in your vehicle at the factory **(see illustration)**. On earlier models, it's possible

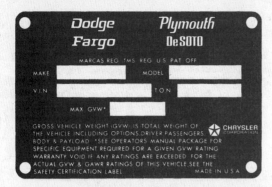

On earlier models the Vehicle Identification Number (VIN) is found on the vehicle identification plate on the driver's side door or door post

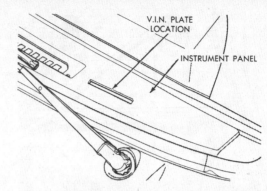

On later models the VIN is visible from the outside of the vehicle, through the driver's side of the windshield

the original engine may have been "swapped" for a different engine; see *Engine sequence number* below to determine if this has happened.

Equipment identification plate

This number-stamped metal plate is located on inside of the hood. It contains valuable information concerning the production of the vehicle as well as information about the way in which the vehicle is equipped. This plate is especially useful for matching the color and type of paint during repair work.

Safety Certification label

The Safety Certification label is affixed to the left front door pillar. The plate contains

the name of the manufacturer, the month and year of production, the Gross Vehicle Weight Rating (GVWR) and the certification statement.

Engine identification number

The engine ID number on the six-cylinder engine is stamped into the rear of the block, just below the cylinder head and in front of the bellhousing **(see illustration)**. On the earlier small-block V8 engine, the number is on the front of the left cylinder bank, just below the cylinder head **(see illustration)**. On the later small block V8 and V6 engines the number is a pad located next to the right engine mount **(see illustration)**. On big-block 440 and 400 engines, the pad is on the right (passenger's side) of the distributor.

Example (1971 through 1980 models):

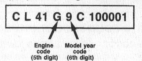

Example (1981 and later models):

Model Year Codes

1 = 1971	B = 1981	M = 1991
2 = 1972	C = 1982	N = 1992
3 = 1973	D = 1983	P = 1993
4 = 1974	E = 1984	R = 1994
5 = 1975	F = 1985	S = 1995
6 = 1976	G = 1986	T = 1996
7 = 1977	H = 1987	V = 1997
8 = 1978	J = 1988	W = 1998
9 = 1979	K = 1989	X = 1999
0 = 1980	L = 1990	

Engine and model year codes

A	360 CID V8 EFI engine (1994)
B	198 CID six-cylinder engine one-barrel carburetor (1971 through 1974) 225 CID six-cylinder engine one-barrel carburetor (1974 through 1980)
C	225 CID six-cylinder engine two-barrel carburetor (1974 through 1980)
D	440 CID V8 engine one-barrel carburetor (1974 through 1978)
E	318 CID V8 engine (1 barrel carburetor) (1971 through 1980)
F	360 CID V8 engine (1974 through 1980)
G	318 CID V8 engine two-barrel carburetor (1980)
H	225 CID six-cylinder engine one-barrel carburetor (1983 through 1987) 225 CID six-cylinder engine, (1 barrel carburetor) (1985 through 1987)
I	360 CID V8 engine (California) (1983 through 1987)
J	400 CID V8 (1974 through 1978) 225 CID six-cylinder engine, one-barrel carburetor (1983)
N	225 CID six-cylinder engine one-barrel carburetor (1980)
P	318 CID V8, four-barrel carburetor (1980) 318 CID V8, two-barrel carburetor) (1981 and 1982)
M	318 CID V8, four-barrel carburetor (1981 and 1982)
R	318 CID V8 engine four-barrel carburetor (1982)
S	360 CID V8 high-performance engine (1980) 360 CID V8 engine two-barrel carburetor (1981 and 1982)
T	360 CID V8 engine (1974 through 1982) 318 CID V8 engine two-barrel carburetor (1983 through 1987)
U	360 CID V8 engine (1980 through 1982) 318 CID V8 engine four-barrel carburetor (1983 through 1987) 360 CID V8 high performance (1981 through 1983)
V	360 CID V8 engine two-barrel carburetor (1983 through 1989)
W	360 CID V8 engine (Federal) (1983 through 1991)
X	238 CID V6 EFI engine (1989 through 1999)
Y	318 CID V8 EFI engine (1989 through 1999)
Z	360 CID V8 MPI engine (1990 through 1999)
5	5.9 liter V8 EFI engine (heavy duty) (1990 through 1993)

Engine identification codes for the models covered in this manual

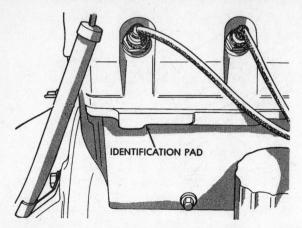

On the inline six-cylinder engine, the engine number is located on the identification pad below the right rear corner of the cylinder head

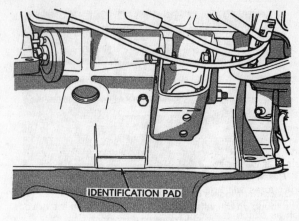

On later small block V8 and V6 engines the identification number is on a pad near the engine mount

Engine sequence number

In addition to the engine identification number, a sequence number, which may be helpful when buying replacement parts, is used. On six-cylinder engines, the number is directly below the engine identification number. On V8 engines, the number is on the right side of the block, just above the oil pan **(see illustration)**. If the original engine is still installed in your vehicle, the last six digits of this number will match the last six digits of the VIN number.

Transmission identification number

The ID number on manual transmissions is located on a pad on the right side of the case **(see illustration)**. On automatic transmissions, the number is stamped on the transmission case near the fluid dipstick tube **(see illustration)**.

Axle identification numbers

The identification number is located on a tag attached to the differential cover **(see illustration)**.

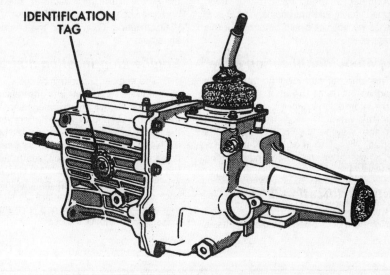

Typical manual transmission identification number location

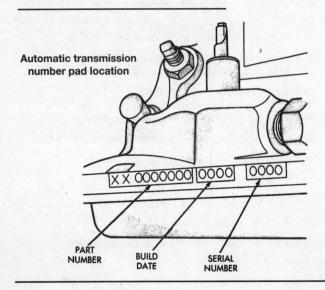

Automatic transmission number pad location

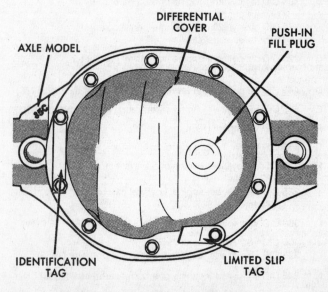

Typical axle identification tag location

Buying parts

Replacement parts are available from many sources, which generally fall into one of two categories - authorized dealer parts departments and independent retail auto parts stores. Our advice concerning these parts is as follows:

Retail auto parts stores: Good auto parts stores will stock frequently needed components which wear out relatively fast, such as clutch components, exhaust systems, brake parts, tune-up parts, etc. These stores often supply new or reconditioned parts on an exchange basis, which can save a considerable amount of money. Discount auto parts stores are often very good places to buy materials and parts needed for general vehicle maintenance such as oil, grease, filters, spark plugs, belts, touch-up paint, bulbs, etc. They also usually sell tools and general accessories, have convenient hours, charge lower prices and can often be found not far from home.

Authorized dealer parts department: This is the best source for parts which are unique to the vehicle and not generally available elsewhere (such as major engine parts, transmission parts, trim pieces, etc.).

Warranty information: If the vehicle is still covered under warranty, be sure that any replacement parts purchased - regardless of the source - do not invalidate the warranty!

To be sure of obtaining the correct parts, have engine and chassis numbers available and, if possible, take the old parts along for positive identification.

Maintenance techniques, tools and working facilities

Maintenance techniques

There are a number of techniques involved in maintenance and repair that will be referred to throughout this manual. Application of these techniques will enable the home mechanic to be more efficient, better organized and capable of performing the various tasks properly, which will ensure that the repair job is thorough and complete.

Fasteners

Fasteners are nuts, bolts, studs and screws used to hold two or more parts together. There are a few things to keep in mind when working with fasteners. Almost all of them use a locking device of some type, either a lockwasher, locknut, locking tab or thread adhesive. All threaded fasteners should be clean and straight, with undamaged threads and undamaged corners on the hex head where the wrench fits. Develop the habit of replacing all damaged nuts and bolts with new ones. Special locknuts with nylon or fiber inserts can only be used once. If they are removed, they lose their locking ability and must be replaced with new ones.

Rusted nuts and bolts should be treated with a penetrating fluid to ease removal and prevent breakage. Some mechanics use turpentine in a spout-type oil can, which works quite well. After applying the rust penetrant, let it work for a few minutes before trying to loosen the nut or bolt. Badly rusted fasteners may have to be chiseled or sawed off or removed with a special nut breaker, available at tool stores.

If a bolt or stud breaks off in an assembly, it can be drilled and removed with a special tool commonly available for this purpose. Most automotive machine shops can perform this task, as well as other repair procedures, such as the repair of threaded holes that have been stripped out.

Flat washers and lockwashers, when removed from an assembly, should always be replaced exactly as removed. Replace any damaged washers with new ones. Never use a lockwasher on any soft metal surface (such as aluminum), thin sheet metal or plastic.

Grade 1 or 2   Grade 5   Grade 8

Bolt strength marking (standard/SAE/USS; bottom - metric)

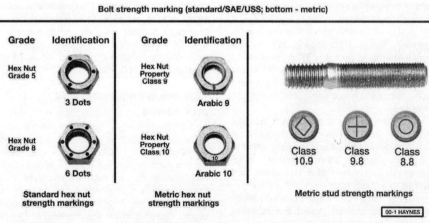

Grade	Identification
Hex Nut Grade 5	3 Dots
Hex Nut Grade 8	6 Dots

Standard hex nut strength markings

Grade	Identification
Hex Nut Property Class 9	Arabic 9
Hex Nut Property Class 10	Arabic 10

Metric hex nut strength markings

Class 10.9   Class 9.8   Class 8.8

Metric stud strength markings

00-1 HAYNES

Fastener sizes

For a number of reasons, automobile manufacturers are making wider and wider use of metric fasteners. Therefore, it is important to be able to tell the difference between standard (sometimes called U.S. or SAE) and metric hardware, since they cannot be interchanged.

All bolts, whether standard or metric, are sized according to diameter, thread pitch and length. For example, a standard 1/2 - 13 x 1 bolt is 1/2 inch in diameter, has 13 threads per inch and is 1 inch long. An M12 - 1.75 x 25 metric bolt is 12 mm in diameter, has a thread pitch of 1.75 mm (the distance between threads) and is 25 mm long. The two bolts are nearly identical, and easily confused, but they are not interchangeable.

In addition to the differences in diameter, thread pitch and length, metric and standard bolts can also be distinguished by examining the bolt heads. To begin with, the distance across the flats on a standard bolt head is measured in inches, while the same dimension on a metric bolt is sized in millimeters (the same is true for nuts). As a result, a standard wrench should not be used on a metric bolt and a metric wrench should not be used on a standard bolt. Also, most standard bolts have slashes radiating out from the center of the head to denote the grade or strength of the bolt, which is an indication of the amount of torque that can be applied to it. The greater the number of slashes, the greater the strength of the bolt. Grades 0 through 5 are commonly used on automobiles. Metric bolts have a property class (grade) number, rather than a slash, molded into their heads to indicate bolt strength. In this case, the higher the number, the stronger the bolt. Property class numbers 8.8, 9.8 and 10.9 are commonly used on automobiles.

Strength markings can also be used to distinguish standard hex nuts from metric hex nuts. Many standard nuts have dots stamped into one side, while metric nuts are marked with a number. The greater the number of dots, or the higher the number, the greater the strength of the nut.

Metric studs are also marked on their ends according to property class (grade). Larger studs are numbered (the same as metric bolts), while smaller studs carry a geometric code to denote grade.

It should be noted that many fasteners, especially Grades 0 through 2, have no distinguishing marks on them. When such is the case, the only way to determine whether it is standard or metric is to measure the thread pitch or compare it to a known fastener of the same size.

Standard fasteners are often referred to as SAE, as opposed to metric. However, it should be noted that SAE technically refers to a non-metric fine thread fastener only. Coarse thread non-metric fasteners are referred to as USS sizes.

Since fasteners of the same size (both standard and metric) may have different

Metric thread sizes

	Ft-lbs	Nm
M-6	6 to 9	9 to 12
M-8	14 to 21	19 to 28
M-10	28 to 40	38 to 54
M-12	50 to 71	68 to 96
M-14	80 to 140	109 to 154

Pipe thread sizes

	Ft-lbs	Nm
1/8	5 to 8	7 to 10
1/4	12 to 18	17 to 24
3/8	22 to 33	30 to 44
1/2	25 to 35	34 to 47

U.S. thread sizes

	Ft-lbs	Nm
1/4 - 20	6 to 9	9 to 12
5/16 - 18	12 to 18	17 to 24
5/16 - 24	14 to 20	19 to 27
3/8 - 16	22 to 32	30 to 43
3/8 - 24	27 to 38	37 to 51
7/16 - 14	40 to 55	55 to 74
7/16 - 20	40 to 60	55 to 81
1/2 - 13	55 to 80	75 to 108

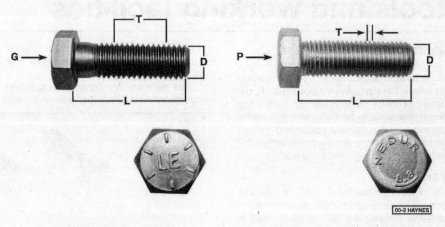

Standard (SAE and USS) bolt dimensions/grade marks

G   Grade marks (bolt strength)
L   Length (in inches)
T   Thread pitch (number of threads per inch)
D   Nominal diameter (in inches)

Metric bolt dimensions/grade marks

P   Property class (bolt strength)
L   Length (in millimeters)
T   Thread pitch (distance between threads in millimeters)
D   Diameter

strength ratings, be sure to reinstall any bolts, studs or nuts removed from your vehicle in their original locations. Also, when replacing a fastener with a new one, make sure that the new one has a strength rating equal to or greater than the original.

Tightening sequences and procedures

Most threaded fasteners should be tightened to a specific torque value (torque is the twisting force applied to a threaded component such as a nut or bolt). Overtightening the fastener can weaken it and cause it to break, while undertightening can cause it to eventually come loose. Bolts, screws and studs, depending on the material they are

made of and their thread diameters, have specific torque values, many of which are noted in the Specifications at the beginning of each Chapter. Be sure to follow the torque recommendations closely. For fasteners not assigned a specific torque, a general torque value chart is presented here as a guide. These torque values are for dry (unlubricated) fasteners threaded into steel or cast iron (not aluminum). As was previously mentioned, the size and grade of a fastener determine the amount of torque that can safely be applied to it. The figures listed here are approximate for Grade 2 and Grade 3 fasteners. Higher grades can tolerate higher torque values.

Fasteners laid out in a pattern, such as cylinder head bolts, oil pan bolts, differential

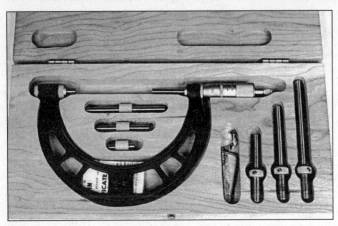

Micrometer set

Dial indicator set

cover bolts, etc., must be loosened or tightened in sequence to avoid warping the component. This sequence will normally be shown in the appropriate Chapter. If a specific pattern is not given, the following procedures can be used to prevent warping.

Initially, the bolts or nuts should be assembled finger-tight only. Next, they should be tightened one full turn each, in a criss-cross or diagonal pattern. After each one has been tightened one full turn, return to the first one and tighten them all one-half turn, following the same pattern. Finally, tighten each of them one-quarter turn at a time until each fastener has been tightened to the proper torque. To loosen and remove the fasteners, the procedure would be reversed.

Component disassembly

Component disassembly should be done with care and purpose to help ensure that the parts go back together properly. Always keep track of the sequence in which parts are removed. Make note of special characteristics or marks on parts that can be installed more than one way, such as a grooved thrust washer on a shaft. It is a good idea to lay the disassembled parts out on a clean surface in the order that they were removed. It may also be helpful to make sketches or take instant photos of components before removal.

When removing fasteners from a component, keep track of their locations. Sometimes threading a bolt back in a part, or putting the washers and nut back on a stud, can prevent mix-ups later. If nuts and bolts cannot be returned to their original locations, they should be kept in a compartmented box or a series of small boxes. A cupcake or muffin tin is ideal for this purpose, since each cavity can hold the bolts and nuts from a particular area (i.e. oil pan bolts, valve cover bolts, engine mount bolts, etc.). A pan of this type is especially helpful when working on assemblies with very small parts, such as the carburetor, alternator, valve train or interior dash and trim pieces. The cavities can be marked with paint or tape to identify the contents.

Whenever wiring looms, harnesses or connectors are separated, it is a good idea to identify the two halves with numbered pieces of masking tape so they can be easily reconnected.

Gasket sealing surfaces

Throughout any vehicle, gaskets are used to seal the mating surfaces between two parts and keep lubricants, fluids, vacuum or pressure contained in an assembly.

Many times these gaskets are coated with a liquid or paste-type gasket sealing compound before assembly. Age, heat and pressure can sometimes cause the two parts to stick together so tightly that they are very difficult to separate. Often, the assembly can be loosened by striking it with a soft-face hammer near the mating surfaces. A regular hammer can be used if a block of wood is placed between the hammer and the part. Do not hammer on cast parts or parts that could be easily damaged. With any particularly stubborn part, always recheck to make sure that every fastener has been removed.

Avoid using a screwdriver or bar to pry apart an assembly, as they can easily mar the gasket sealing surfaces of the parts, which must remain smooth. If prying is absolutely necessary, use an old broom handle, but keep in mind that extra clean up will be necessary if the wood splinters.

After the parts are separated, the old gasket must be carefully scraped off and the gasket surfaces cleaned. Stubborn gasket material can be soaked with rust penetrant or treated with a special chemical to soften it so it can be easily scraped off. A scraper can be fashioned from a piece of copper tubing by flattening and sharpening one end. Copper is recommended because it is usually softer than the surfaces to be scraped, which reduces the chance of gouging the part. Some gaskets can be removed with a wire brush, but regardless of the method used, the mating surfaces must be left clean and smooth. If for some reason the gasket surface is gouged, then a gasket sealer thick enough to fill scratches will have to be used during reassembly of the components. For most applications, a non-drying (or semi-drying) gasket sealer should be used.

Hose removal tips

Warning: *If the vehicle is equipped with air conditioning, do not disconnect any of the A/C hoses without first having the system depressurized by a dealer service department or a service station.*

Hose removal precautions closely parallel gasket removal precautions. Avoid scratching or gouging the surface that the hose mates against or the connection may leak. This is especially true for radiator hoses. Because of various chemical reactions, the rubber in hoses can bond itself to the metal spigot that the hose fits over. To remove a hose, first loosen the hose clamps that secure it to the spigot. Then, with slip-joint pliers, grab the hose at the clamp and rotate it around the spigot. Work it back and forth until it is completely free, then pull it off. Silicone or other lubricants will ease removal if they can be applied between the hose and the outside of the spigot. Apply the same lubricant to the inside of the hose and the outside of the spigot to simplify installation.

As a last resort (and if the hose is to be replaced with a new one anyway), the rubber can be slit with a knife and the hose peeled from the spigot. If this must be done, be careful that the metal connection is not damaged.

If a hose clamp is broken or damaged, do not reuse it. Wire-type clamps usually weaken with age, so it is a good idea to replace them with screw-type clamps whenever a hose is removed.

Tools

A selection of good tools is a basic requirement for anyone who plans to maintain and repair his or her own vehicle. For the owner who has few tools, the initial investment might seem high, but when compared to the spiraling costs of professional auto maintenance and repair, it is a wise one.

To help the owner decide which tools are needed to perform the tasks detailed in this manual, the following tool lists are offered: *Maintenance and minor repair, Repair/overhaul* and *Special.*

The newcomer to practical mechanics

Dial caliper

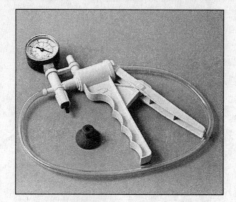

Hand-operated vacuum pump

Timing light

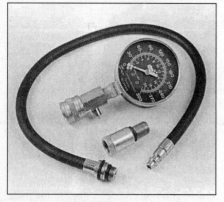

Compression gauge with spark plug hole adapter

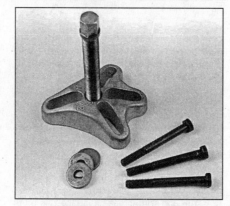

Damper/steering wheel puller

General purpose puller

Hydraulic lifter removal tool

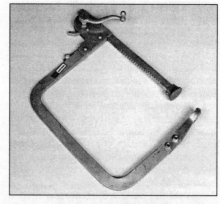

Valve spring compressor

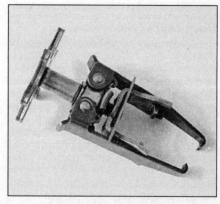

Valve spring compressor

Ridge reamer

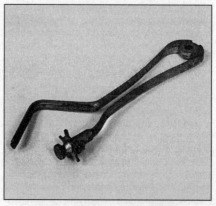

Piston ring groove cleaning tool

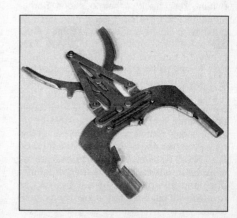

Ring removal/installation tool

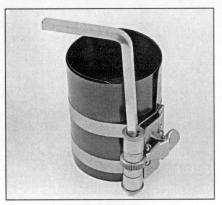

Ring compressor

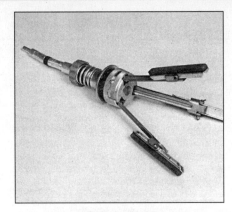

Cylinder hone

Brake hold-down spring tool

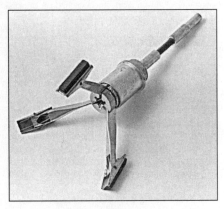

Brake cylinder hone

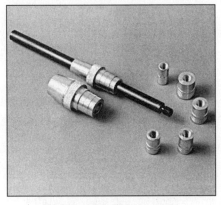

Clutch plate alignment tool

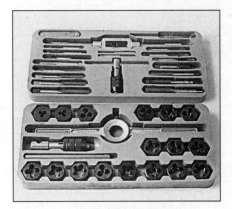

Tap and die set

should start off with the *maintenance and minor repair* tool kit, which is adequate for the simpler jobs performed on a vehicle. Then, as confidence and experience grow, the owner can tackle more difficult tasks, buying additional tools as they are needed. Eventually the basic kit will be expanded into the *repair and overhaul* tool set. Over a period of time, the experienced do-it-yourselfer will assemble a tool set complete enough for most repair and overhaul procedures and will add tools from the special category when it is felt that the expense is justified by the frequency of use.

Maintenance and minor repair tool kit

The tools in this list should be considered the minimum required for performance of routine maintenance, servicing and minor repair work. We recommend the purchase of combination wrenches (box-end and open-end combined in one wrench). While more expensive than open end wrenches, they offer the advantages of both types of wrench.

Combination wrench set (1/4-inch to 1 inch or 6 mm to 19 mm)
Adjustable wrench, 8 inch
Spark plug wrench with rubber insert
Spark plug gap adjusting tool
Feeler gauge set
Brake bleeder wrench
Standard screwdriver (5/16-inch x 6 inch)

Phillips screwdriver (No. 2 x 6 inch)
Combination pliers - 6 inch
Hacksaw and assortment of blades
Tire pressure gauge
Grease gun
Oil can
Fine emery cloth
Wire brush
Battery post and cable cleaning tool
Oil filter wrench
Funnel (medium size)
Safety goggles
Jackstands (2)
Drain pan

Note: *If basic tune-ups are going to be part of routine maintenance, it will be necessary to purchase a good quality stroboscopic timing light and combination tachometer/dwell meter. Although they are included in the list of special tools, it is mentioned here because they are absolutely necessary for tuning most vehicles properly.*

Repair and overhaul tool set

These tools are essential for anyone who plans to perform major repairs and are in addition to those in the maintenance and minor repair tool kit. Included is a comprehensive set of sockets which, though expensive, are invaluable because of their versatility, especially when various extensions and drives are available. We recommend the 1/2-inch drive over the 3/8-inch drive. Although the larger drive is bulky and more expensive,

it has the capacity of accepting a very wide range of large sockets. Ideally, however, the mechanic should have a 3/8-inch drive set and a 1/2-inch drive set.

Socket set(s)
Reversible ratchet
Extension - 10 inch
Universal joint
Torque wrench (same size drive as sockets)
Ball peen hammer - 8 ounce
Soft-face hammer (plastic/rubber)
Standard screwdriver (1/4-inch x 6 inch)
Standard screwdriver (stubby - 5/16-inch)
Phillips screwdriver (No. 3 x 8 inch)
Phillips screwdriver (stubby - No. 2)
Pliers - vise grip
Pliers - lineman's
Pliers - needle nose
Pliers - snap-ring (internal and external)
Cold chisel - 1/2-inch
Scribe
Scraper (made from flattened copper tubing)
Centerpunch
Pin punches (1/16, 1/8, 3/16-inch)
Steel rule/straightedge - 12 inch
Allen wrench set (1/8 to 3/8-inch or 4 mm to 10 mm)
A selection of files
Wire brush (large)
Jackstands (second set)
Jack (scissor or hydraulic type)

Note: *Another tool which is often useful is an electric drill with a chuck capacity of 3/8-inch and a set of good quality drill bits.*

Special tools

The tools in this list include those which are not used regularly, are expensive to buy, or which need to be used in accordance with their manufacturer's instructions. Unless these tools will be used frequently, it is not very economical to purchase many of them. A consideration would be to split the cost and use between yourself and a friend or friends. In addition, most of these tools can be obtained from a tool rental shop on a temporary basis.

This list primarily contains only those tools and instruments widely available to the public, and not those special tools produced by the vehicle manufacturer for distribution to dealer service departments. Occasionally, references to the manufacturer's special tools are included in the text of this manual. Generally, an alternative method of doing the job without the special tool is offered. However, sometimes there is no alternative to their use. Where this is the case, and the tool cannot be purchased or borrowed, the work should be turned over to the dealer service department or an automotive repair shop.

> *Valve spring compressor*
> *Piston ring groove cleaning tool*
> *Piston ring compressor*
> *Piston ring installation tool*
> *Cylinder compression gauge*
> *Cylinder ridge reamer*
> *Cylinder surfacing hone*
> *Cylinder bore gauge*
> *Micrometers and/or dial calipers*
> *Hydraulic lifter removal tool*
> *Balljoint separator*
> *Universal-type puller*
> *Impact screwdriver*
> *Dial indicator set*
> *Stroboscopic timing light (inductive pick-up)*
> *Hand operated vacuum/pressure pump*
> *Tachometer/dwell meter*
> *Universal electrical multimeter*
> *Cable hoist*
> *Brake spring removal and installation tools*
> *Floor jack*

Buying tools

For the do-it-yourselfer who is just starting to get involved in vehicle maintenance and repair, there are a number of options available when purchasing tools. If maintenance and minor repair is the extent of the work to be done, the purchase of individual tools is satisfactory. If, on the other hand, extensive work is planned, it would be a good idea to purchase a modest tool set from one of the large retail chain stores. A set can usually be bought at a substantial savings over the individual tool prices, and they often come with a tool box. As additional tools are

needed, add-on sets, individual tools and a larger tool box can be purchased to expand the tool selection. Building a tool set gradually allows the cost of the tools to be spread over a longer period of time and gives the mechanic the freedom to choose only those tools that will actually be used.

Tool stores will often be the only source of some of the special tools that are needed, but regardless of where tools are bought, try to avoid cheap ones, especially when buying screwdrivers and sockets, because they won't last very long. The expense involved in replacing cheap tools will eventually be greater than the initial cost of quality tools.

Care and maintenance of tools

Good tools are expensive, so it makes sense to treat them with respect. Keep them clean and in usable condition and store them properly when not in use. Always wipe off any dirt, grease or metal chips before putting them away. Never leave tools lying around in the work area. Upon completion of a job, always check closely under the hood for tools that may have been left there so they won't get lost during a test drive.

Some tools, such as screwdrivers, pliers, wrenches and sockets, can be hung on a panel mounted on the garage or workshop wall, while others should be kept in a tool box or tray. Measuring instruments, gauges, meters, etc. must be carefully stored where they cannot be damaged by weather or impact from other tools.

When tools are used with care and stored properly, they will last a very long time. Even with the best of care, though, tools will wear out if used frequently. When a tool is damaged or worn out, replace it. Subsequent jobs will be safer and more enjoyable if you do.

How to repair damaged threads

Sometimes, the internal threads of a nut or bolt hole can become stripped, usually from overtightening. Stripping threads is an all-too-common occurrence, especially when working with aluminum parts, because aluminum is so soft that it easily strips out.

Usually, external or internal threads are only partially stripped. After they've been cleaned up with a tap or die, they'll still work. Sometimes, however, threads are badly damaged. When this happens, you've got three choices:

1) *Drill and tap the hole to the next suitable oversize and install a larger diameter bolt, screw or stud.*
2) *Drill and tap the hole to accept a threaded plug, then drill and tap the plug to the original screw size. You can also buy a plug already threaded to the original size. Then you simply drill a hole to the specified size, then run the threaded plug into the hole with a bolt and jam*

nut. Once the plug is fully seated, remove the jam nut and bolt.
3) *The third method uses a patented thread repair kit like Heli-Coil or Slimsert. These easy-to-use kits are designed to repair damaged threads in straight-through holes and blind holes. Both are available as kits which can handle a variety of sizes and thread patterns. Drill the hole, then tap it with the special included tap. Install the Heli-Coil and the hole is back to its original diameter and thread pitch.*

Regardless of which method you use, be sure to proceed calmly and carefully. A little impatience or carelessness during one of these relatively simple procedures can ruin your whole day's work and cost you a bundle if you wreck an expensive part.

Working facilities

Not to be overlooked when discussing tools is the workshop. If anything more than routine maintenance is to be carried out, some sort of suitable work area is essential.

It is understood, and appreciated, that many home mechanics do not have a good workshop or garage available, and end up removing an engine or doing major repairs outside. It is recommended, however, that the overhaul or repair be completed under the cover of a roof.

A clean, flat workbench or table of comfortable working height is an absolute necessity. The workbench should be equipped with a vise that has a jaw opening of at least four inches.

As mentioned previously, some clean, dry storage space is also required for tools, as well as the lubricants, fluids, cleaning solvents, etc. which soon become necessary.

Sometimes waste oil and fluids, drained from the engine or cooling system during normal maintenance or repairs, present a disposal problem. To avoid pouring them on the ground or into a sewage system, pour the used fluids into large containers, seal them with caps and take them to an authorized disposal site or recycling center. Plastic jugs, such as old antifreeze containers, are ideal for this purpose.

Always keep a supply of old newspapers and clean rags available. Old towels are excellent for mopping up spills. Many mechanics use rolls of paper towels for most work because they are readily available and disposable. To help keep the area under the vehicle clean, a large cardboard box can be cut open and flattened to protect the garage or shop floor.

Whenever working over a painted surface, such as when leaning over a fender to service something under the hood, always cover it with an old blanket or bedspread to protect the finish. Vinyl covered pads, made especially for this purpose, are available at auto parts stores.

Jacking and towing

Jacking

The jack supplied with the vehicle should only be used for raising the vehicle when changing a tire or placing jackstands under the frame. NEVER work under the vehicle or start the engine when the vehicle supported only by a jack.

The vehicle should be parked on level ground with the wheels blocked, the parking brake applied and the transmission in Park (automatic) or Reverse (manual). If the vehicle is parked alongside the roadway, or in any other hazardous situation, turn on the emergency hazard flashers. If a tire is to be changed, loosen the lug nuts one-half turn before raising off the ground.

Place the jack under the vehicle in the indicated position (see illustration). Operate the jack with a slow, smooth motion until the wheel is raised off the ground. Remove the lug nuts, pull off the wheel, install the spare and thread the lug nuts back on with the beveled side facing in. Tighten the lug nuts snugly, lower the vehicle until some weight is on the wheel, tighten them completely in a criss-cross pattern and remove the jack. Note that some spare tires are designed for temporary use only - don't exceed the recommended speed, mileage or other restriction instructions accompanying the spare.

Towing

Equipment specifically designed for towing should be used and attached to the main structural members of the vehicle. Optional tow hooks may be attached to the frame at both ends of the vehicle; they are intended for emergency use only, for rescuing a stranded vehicle. Do not use the tow hooks for highway towing. Stand clear when using tow straps or chains, they may break causing serious injury.

Safety is a major consideration when towing and all applicable state and local laws must be obeyed. In addition to a tow bar, a safety chain must be used for all towing.

These vehicles may be towed with four wheels on the ground for a distance of 15 miles or less, as long as the speed doesn't exceed 30 mph. If the vehicle has to be towed more than 15 miles, place the rear wheels on a towing dolly.

If any vehicle is to be towed with the front wheels on the ground and the rear wheels raised, the ignition key must be turned to the OFF position to unlock the steering column and a steering wheel clamping device designed for towing must be used or damage to the steering column lock may occur.

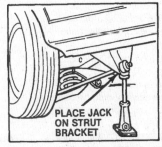 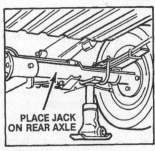

Front and rear jacking points

Booster battery (jump) starting

Observe these precautions when using a booster battery to start a vehicle:

a) *Before connecting the booster battery, make sure the ignition switch is in the Off position.*
b) *Turn off the lights, heater and other electrical loads.*
c) *Your eyes should be shielded. Safety goggles are a good idea.*
d) *Make sure the booster battery is the same voltage as the dead one in the vehicle.*
e) *The two vehicles MUST NOT TOUCH each other!*
f) *Make sure the transaxle is in Neutral (manual) or Park (automatic).*
g) *If the booster battery is not a maintenance-free type, remove the vent caps and lay a cloth over the vent holes.*

Connect the red jumper cable to the positive (+) terminals of each battery **(see illustration)**.

Connect one end of the black jumper cable to the negative (-) terminal of the booster battery. The other end of this cable should be connected to a good ground on the vehicle to be started, such as a bolt or bracket on the body.

Start the engine using the booster battery, then, with the engine running at idle speed, disconnect the jumper cables in the reverse order of connection.

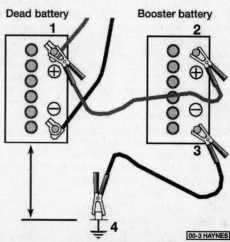

Make the booster battery cable connections in the numerical order shown (note that the negative cable of the booster battery is NOT attached to the negative terminal of the dead battery)

Automotive chemicals and lubricants

A number of automotive chemicals and lubricants are available for use during vehicle maintenance and repair. They include a wide variety of products ranging from cleaning solvents and degreasers to lubricants and protective sprays for rubber, plastic and vinyl.

Cleaners

Carburetor cleaner and choke cleaner is a strong solvent for gum, varnish and carbon. Most carburetor cleaners leave a dry-type lubricant film which will not harden or gum up. Because of this film it is not recommended for use on electrical components.

Brake system cleaner is used to remove grease and brake fluid from the brake system, where clean surfaces are absolutely necessary. It leaves no residue and often eliminates brake squeal caused by contaminants.

Electrical cleaner removes oxidation, corrosion and carbon deposits from electrical contacts, restoring full current flow. It can also be used to clean spark plugs, carburetor jets, voltage regulators and other parts where an oil-free surface is desired.

Demoisturants remove water and moisture from electrical components such as alternators, voltage regulators, electrical connectors and fuse blocks. They are non-conductive, non-corrosive and non-flammable.

Degreasers are heavy-duty solvents used to remove grease from the outside of the engine and from chassis components. They can be sprayed or brushed on and, depending on the type, are rinsed off either with water or solvent.

Lubricants

Motor oil is the lubricant formulated for use in engines. It normally contains a wide variety of additives to prevent corrosion and reduce foaming and wear. Motor oil comes in various weights (viscosity ratings) from 5 to 80. The recommended weight of the oil depends on the season, temperature and the demands on the engine. Light oil is used in cold climates and under light load conditions. Heavy oil is used in hot climates and where high loads are encountered. Multi-viscosity oils are designed to have characteristics of both light and heavy oils and are available in a number of weights from 5W-20 to 20W-50.

Gear oil is designed to be used in differentials, manual transmissions and other areas where high-temperature lubrication is required.

Chassis and wheel bearing grease is a heavy grease used where increased loads and friction are encountered, such as for wheel bearings, balljoints, tie-rod ends and universal joints.

High-temperature wheel bearing grease is designed to withstand the extreme temperatures encountered by wheel bearings in disc brake equipped vehicles. It usually contains molybdenum disulfide (moly), which is a dry-type lubricant.

White grease is a heavy grease for metal-to-metal applications where water is a problem. White grease stays soft under both low and high temperatures (usually from -100 to +190-degrees F), and will not wash off or dilute in the presence of water.

Assembly lube is a special extreme pressure lubricant, usually containing moly, used to lubricate high-load parts (such as main and rod bearings and cam lobes) for initial start-up of a new engine. The assembly lube lubricates the parts without being squeezed out or washed away until the engine oiling system begins to function.

Silicone lubricants are used to protect rubber, plastic, vinyl and nylon parts.

Graphite lubricants are used where oils cannot be used due to contamination problems, such as in locks. The dry graphite will lubricate metal parts while remaining uncontaminated by dirt, water, oil or acids. It is electrically conductive and will not foul electrical contacts in locks such as the ignition switch.

Moly penetrants loosen and lubricate frozen, rusted and corroded fasteners and prevent future rusting or freezing.

Heat-sink grease is a special electrically non-conductive grease that is used for mounting electronic ignition modules where it is essential that heat is transferred away from the module.

Sealants

RTV sealant is one of the most widely used gasket compounds. Made from silicone, RTV is air curing, it seals, bonds, waterproofs, fills surface irregularities, remains flexible, doesn't shrink, is relatively easy to remove, and is used as a supplementary sealer with almost all low and medium temperature gaskets.

Anaerobic sealant is much like RTV in that it can be used either to seal gaskets or to form gaskets by itself. It remains flexible, is solvent resistant and fills surface imperfections. The difference between an anaerobic sealant and an RTV-type sealant is in the curing. RTV cures when exposed to air, while an anaerobic sealant cures only in the absence of air. This means that an anaerobic sealant cures only after the assembly of parts, sealing them together.

Thread and pipe sealant is used for sealing hydraulic and pneumatic fittings and vacuum lines. It is usually made from a Teflon compound, and comes in a spray, a paint-on liquid and as a wrap-around tape.

Chemicals

Anti-seize compound prevents seizing, galling, cold welding, rust and corrosion in fasteners. High-temperature ant-seize, usually made with copper and graphite lubricants, is used for exhaust system and exhaust manifold bolts.

Anaerobic locking compounds are used to keep fasteners from vibrating or working loose and cure only after installation, in the absence of air. Medium strength locking compound is used for small nuts, bolts and screws that may be removed later. High-strength locking compound is for large nuts, bolts and studs which aren't removed on a regular basis.

Oil additives range from viscosity index improvers to chemical treatments that claim to reduce internal engine friction. It should be noted that most oil manufacturers caution against using additives with their oils.

Gas additives perform several functions, depending on their chemical makeup. They usually contain solvents that help dissolve gum and varnish that build up on carburetor, fuel injection and intake parts. They also serve to break down carbon deposits that form on the inside surfaces of the combustion chambers. Some additives contain upper cylinder lubricants for valves and piston rings, and others contain chemicals to remove condensation from the gas tank.

Miscellaneous

Brake fluid is specially formulated hydraulic fluid that can withstand the heat and pressure encountered in brake systems. Care must be taken so this fluid does not come in contact with painted surfaces or plastics. An opened container should always be resealed to prevent contamination by water or dirt.

Weatherstrip adhesive is used to bond weatherstripping around doors, windows and trunk lids. It is sometimes used to attach trim pieces.

Undercoating is a petroleum-based, tar-like substance that is designed to protect metal surfaces on the underside of the vehicle from corrosion. It also acts as a sound-deadening agent by insulating the bottom of the vehicle.

Waxes and polishes are used to help protect painted and plated surfaces from the weather. Different types of paint may require the use of different types of wax and polish. Some polishes utilize a chemical or abrasive cleaner to help remove the top layer of oxidized (dull) paint on older vehicles. In recent years many non-wax polishes that contain a wide variety of chemicals such as polymers and silicones have been introduced. These non-wax polishes are usually easier to apply and last longer than conventional waxes and polishes.

Conversion factors

Length (distance)
Inches (in)	X 25.4	= Millimetres (mm)	X 0.0394	= Inches (in)
Feet (ft)	X 0.305	= Metres (m)	X 3.281	= Feet (ft)
Miles	X 1.609	= Kilometres (km)	X 0.621	= Miles

Volume (capacity)
Cubic inches (cu in; in³)	X 16.387	= Cubic centimetres (cc; cm³)	X 0.061	= Cubic inches (cu in; in³)
Imperial pints (Imp pt)	X 0.568	= Litres (l)	X 1.76	= Imperial pints (Imp pt)
Imperial quarts (Imp qt)	X 1.137	= Litres (l)	X 0.88	= Imperial quarts (Imp qt)
Imperial quarts (Imp qt)	X 1.201	= US quarts (US qt)	X 0.833	= Imperial quarts (Imp qt)
US quarts (US qt)	X 0.946	= Litres (l)	X 1.057	= US quarts (US qt)
Imperial gallons (Imp gal)	X 4.546	= Litres (l)	X 0.22	= Imperial gallons (Imp gal)
Imperial gallons (Imp gal)	X 1.201	= US gallons (US gal)	X 0.833	= Imperial gallons (Imp gal)
US gallons (US gal)	X 3.785	= Litres (l)	X 0.264	= US gallons (US gal)

Mass (weight)
Ounces (oz)	X 28.35	= Grams (g)	X 0.035	= Ounces (oz)
Pounds (lb)	X 0.454	= Kilograms (kg)	X 2.205	= Pounds (lb)

Force
Ounces-force (ozf; oz)	X 0.278	= Newtons (N)	X 3.6	= Ounces-force (ozf; oz)
Pounds-force (lbf; lb)	X 4.448	= Newtons (N)	X 0.225	= Pounds-force (lbf; lb)
Newtons (N)	X 0.1	= Kilograms-force (kgf; kg)	X 9.81	= Newtons (N)

Pressure
Pounds-force per square inch (psi; lbf/in²; lb/in²)	X 0.070	= Kilograms-force per square centimetre (kgf/cm²; kg/cm²)	X 14.223	= Pounds-force per square inch (psi; lbf/in²; lb/in²)
Pounds-force per square inch (psi; lbf/in²; lb/in²)	X 0.068	= Atmospheres (atm)	X 14.696	= Pounds-force per square inch (psi; lbf/in²; lb/in²)
Pounds-force per square inch (psi; lbf/in²; lb/in²)	X 0.069	= Bars	X 14.5	= Pounds-force per square inch (psi; lbf/in²; lb/in²)
Pounds-force per square inch (psi; lbf/in²; lb/in²)	X 6.895	= Kilopascals (kPa)	X 0.145	= Pounds-force per square inch (psi; lbf/in²; lb/in²)
Kilopascals (kPa)	X 0.01	= Kilograms-force per square centimetre (kgf/cm²; kg/cm²)	X 98.1	= Kilopascals (kPa)

Torque (moment of force)
Pounds-force inches (lbf in; lb in)	X 1.152	= Kilograms-force centimetre (kgf cm; kg cm)	X 0.868	= Pounds-force inches (lbf in; lb in)
Pounds-force inches (lbf in; lb in)	X 0.113	= Newton metres (Nm)	X 8.85	= Pounds-force inches (lbf in; lb in)
Pounds-force inches (lbf in; lb in)	X 0.083	= Pounds-force feet (lbf ft; lb ft)	X 12	= Pounds-force inches (lbf in; lb in)
Pounds-force feet (lbf ft; lb ft)	X 0.138	= Kilograms-force metres (kgf m; kg m)	X 7.233	= Pounds-force feet (lbf ft; lb ft)
Pounds-force feet (lbf ft; lb ft)	X 1.356	= Newton metres (Nm)	X 0.738	= Pounds-force feet (lbf ft; lb ft)
Newton metres (Nm)	X 0.102	= Kilograms-force metres (kgf m; kg m)	X 9.804	= Newton metres (Nm)

Vacuum
Inches mercury (in. Hg)	X 3.377	= Kilopascals (kPa)	X 0.2961	= Inches mercury
Inches mercury (in. Hg)	X 25.4	= Millimeters mercury (mm Hg)	X 0.0394	= Inches mercury

Power
Horsepower (hp)	X 745.7	= Watts (W)	X 0.0013	= Horsepower (hp)

Velocity (speed)
Miles per hour (miles/hr; mph)	X 1.609	= Kilometres per hour (km/hr; kph)	X 0.621	= Miles per hour (miles/hr; mph)

Fuel consumption*
Miles per gallon, Imperial (mpg)	X 0.354	= Kilometres per litre (km/l)	X 2.825	= Miles per gallon, Imperial (mpg)
Miles per gallon, US (mpg)	X 0.425	= Kilometres per litre (km/l)	X 2.352	= Miles per gallon, US (mpg)

Temperature
Degrees Fahrenheit = (°C x 1.8) + 32   Degrees Celsius (Degrees Centigrade; °C) = (°F - 32) x 0.56

*It is common practice to convert from miles per gallon (mpg) to litres/100 kilometres (l/100km), where mpg (Imperial) x l/100 km = 282 and mpg (US) x l/100 km = 235

Safety first!

Regardless of how enthusiastic you may be about getting on with the job at hand, take the time to ensure that your safety is not jeopardized. A moment's lack of attention can result in an accident, as can failure to observe certain simple safety precautions. The possibility of an accident will always exist, and the following points should not be considered a comprehensive list of all dangers. Rather, they are intended to make you aware of the risks and to encourage a safety conscious approach to all work you carry out on your vehicle.

Essential DOs and DON'Ts

DON'T rely on a jack when working under the vehicle. Always use approved jackstands to support the weight of the vehicle and place them under the recommended lift or support points.

DON'T attempt to loosen extremely tight fasteners (i.e. wheel lug nuts) while the vehicle is on a jack - it may fall.

DON'T start the engine without first making sure that the transmission is in Neutral (or Park where applicable) and the parking brake is set.

DON'T remove the radiator cap from a hot cooling system - let it cool or cover it with a cloth and release the pressure gradually.

DON'T attempt to drain the engine oil until you are sure it has cooled to the point that it will not burn you.

DON'T touch any part of the engine or exhaust system until it has cooled sufficiently to avoid burns.

DON'T siphon toxic liquids such as gasoline, antifreeze and brake fluid by mouth, or allow them to remain on your skin.

DON'T inhale brake lining dust - it is potentially hazardous (see Asbestos below).

DON'T allow spilled oil or grease to remain on the floor - wipe it up before someone slips on it.

DON'T use loose fitting wrenches or other tools which may slip and cause injury.

DON'T push on wrenches when loosening or tightening nuts or bolts. Always try to pull the wrench toward you. If the situation calls for pushing the wrench away, push with an open hand to avoid scraped knuckles if the wrench should slip.

DON'T attempt to lift a heavy component alone - get someone to help you.

DON'T rush or take unsafe shortcuts to finish a job.

DON'T allow children or animals in or around the vehicle while you are working on it.

DO wear eye protection when using power tools such as a drill, sander, bench grinder, etc. and when working under a vehicle.

DO keep loose clothing and long hair well out of the way of moving parts.

DO make sure that any hoist used has a safe working load rating adequate for the job.

DO get someone to check on you periodically when working alone on a vehicle.

DO carry out work in a logical sequence and make sure that everything is correctly assembled and tightened.

DO keep chemicals and fluids tightly capped and out of the reach of children and pets.

DO remember that your vehicle's safety affects that of yourself and others. If in doubt on any point, get professional advice.

Asbestos

Certain friction, insulating, sealing, and other products - such as brake linings, brake bands, clutch linings, torque converters, gaskets, etc. - may contain asbestos. Extreme care must be taken to avoid inhalation of dust from such products, since it is hazardous to health. If in doubt, assume that they do contain asbestos.

Fire

Remember at all times that gasoline is highly flammable. Never smoke or have any kind of open flame around when working on a vehicle. But the risk does not end there. A spark caused by an electrical short circuit, by two metal surfaces contacting each other, or even by static electricity built up in your body under certain conditions, can ignite gasoline vapors, which in a confined space are highly explosive. Do not, under any circumstances, use gasoline for cleaning parts. Use an approved safety solvent.

Always disconnect the battery ground (-) cable at the battery before working on any part of the fuel system or electrical system. Never risk spilling fuel on a hot engine or exhaust component. It is strongly recommended that a fire extinguisher suitable for use on fuel and electrical fires be kept handy in the garage or workshop at all times. Never try to extinguish a fuel or electrical fire with water.

Fumes

Certain fumes are highly toxic and can quickly cause unconsciousness and even death if inhaled to any extent. Gasoline vapor falls into this category, as do the vapors from some cleaning solvents. Any draining or pouring of such volatile fluids should be done in a well ventilated area.

When using cleaning fluids and solvents, read the instructions on the container carefully. Never use materials from unmarked containers.

Never run the engine in an enclosed space, such as a garage. Exhaust fumes contain carbon monoxide, which is extremely poisonous. If you need to run the engine, always do so in the open air, or at least have the rear of the vehicle outside the work area.

If you are fortunate enough to have the use of an inspection pit, never drain or pour gasoline and never run the engine while the vehicle is over the pit. The fumes, being heavier than air, will concentrate in the pit with possibly lethal results.

The battery

Never create a spark or allow a bare light bulb near a battery. They normally give off a certain amount of hydrogen gas, which is highly explosive.

Always disconnect the battery ground (-) cable at the battery before working on the fuel or electrical systems.

If possible, loosen the filler caps or cover when charging the battery from an external source (this does not apply to sealed or maintenance-free batteries). Do not charge at an excessive rate or the battery may burst.

Take care when adding water to a non maintenance-free battery and when carrying a battery. The electrolyte, even when diluted, is very corrosive and should not be allowed to contact clothing or skin.

Always wear eye protection when cleaning the battery to prevent the caustic deposits from entering your eyes.

Household current

When using an electric power tool, inspection light, etc., which operates on household current, always make sure that the tool is correctly connected to its plug and that, where necessary, it is properly grounded. Do not use such items in damp conditions and, again, do not create a spark or apply excessive heat in the vicinity of fuel or fuel vapor.

Secondary ignition system voltage

A severe electric shock can result from touching certain parts of the ignition system (such as the spark plug wires) when the engine is running or being cranked, particularly if components are damp or the insulation is defective. In the case of an electronic ignition system, the secondary system voltage is much higher and could prove fatal.

Troubleshooting

Contents

This Section provides an easy reference guide to the more common problems that may occur during the operation of your vehicle. Various symptoms and their probable causes are grouped under headings denoting components or systems, such as Engine, Cooling system, etc. They also refer to the Chapter and/or Section that deals with the problem.

Remember that successful troubleshooting isn't a mysterious 'black art' practiced only by professional mechanics, it's simply the result of knowledge combined with an intelligent, systematic approach to a problem. Always use a process of elimination starting, with the simplest solution and working through to the most complex - and never overlook the obvious. Anyone can run the gas tank dry or leave the lights on overnight, so don't assume that you're exempt from such oversights.

Finally, always establish a clear idea why a problem has occurred and take steps to ensure that it doesn't happen again. If the electrical system fails because of a poor connection, check all other connections in the system to make sure they don't fail as well. If a particular fuse continues to blow, find out why - don't just go on replacing fuses. Remember, failure of a small component can often be indicative of potential failure or incorrect functioning of a more important component or system.

Engine and performance

1 Engine will not rotate when attempting to start

1   Battery terminal connections loose or corroded. Check the cable terminals at the battery; tighten cable clamp and/or clean off corrosion as necessary (see Chapter 1).
2   Battery discharged or faulty. If the cable ends are clean and tight on the battery posts, turn the key to the On position and switch on the headlights or windshield wipers. If they won't run, the battery is discharged.
3   Automatic transmission not engaged in park (P) or Neutral (N).
4   Broken, loose or disconnected wires in the starting circuit. Inspect all wires and connectors at the battery, starter solenoid and ignition switch (in steering column).
5   Starter motor pinion jammed in flywheel ring gear. If manual transmission, place transmission in gear and rock the vehicle to manually turn the engine. Remove starter (Chapter 5) and inspect pinion and flywheel (Chapter 2) at earliest convenience.
6   Starter solenoid faulty (Chapter 5).
7   Starter relay faulty (Chapter 5)
8   Starter motor faulty (Chapter 5).
9   Neutral start switch faulty (automatic transmission models) (Chapter 7B).
10   Ignition switch faulty (Chapter 12).
11   Engine seized. Try to turn the crankshaft with a large socket and breaker bar on the pulley bolt.

2 Engine rotates but will not start

1   Fuel tank empty.
2   Fault in carburetor or fuel injection system (Chapter 4).
3   Battery discharged (engine rotates slowly). Check the operation of electrical components as described in previous Section.
4   Battery terminal connections loose or corroded. See previous Section.
5   Choke not operating properly (Chapter 1).
6   Water inside the distributor cap (particularly in foul weather). Remove the cap and dry it (see Chapter 1).
7   Fouled spark plugs or bad spark-plug wires (Chapter 1).

8   Faulty distributor components. Check the cap and rotor (Chapter 1).
9   Fuel not reaching the carburetor. With the engine off, remove the air cleaner top plate, open the choke plate by hand and have an assistant press the accelerator while you look down the carburetor bore - if you can see small streams of gasoline squirting into the carburetor, fuel is reaching the carburetor; otherwise, check for a clogged fuel filter or lines and a defective fuel pump. Also make sure the tank vent lines aren't clogged (Chapter 4).
10   Faulty distributor pick-up coil or ignition module (Chapter 5).
11   Burned or otherwise damaged ignition points (1971 and 1972 models) (Chapter 1).
12   Low cylinder compression. Check as described in Chapter 2.
13   Severe vacuum leak. Check for a loose carburetor or intake manifold or a disconnected PCV or power brake booster vacuum hose.
14   Water in fuel. Drain tank and fill with new fuel.
15   Defective ignition coil (Chapter 5).
16   Broken, loose or disconnected wires in the starting circuit (see previous Section).
17   Loose distributor (changing ignition timing). Turn the distributor body as necessary to start the engine, then adjust the ignition timing as soon as possible (Chapter 1).
18   Broken, loose or disconnected wires at the ignition coil or faulty coil (Chapter 5).
19   Timing chain failure or wear affecting valve timing (Chapter 2).

3 Starter motor operates without turning engine

1   Starter pinion sticking. Remove the starter (Chapter 5) and Inspect.
2   Starter pinion or flywheel/driveplate teeth worn or broken. Remove the inspection cover and inspect.

4 Engine hard to start when cold

1   Battery discharged or low. Check as described in Chapter 1.
2   Fuel not reaching the carburetor or fuel

injection. Check the fuel filter, lines and fuel pump (Chapters 1 and 4).
3   Choke inoperative (Chapters 1 and 4).
4   Defective spark plugs (Chapter 1).

5 Engine hard to start when hot

1   Air filter dirty (Chapter 1).
2   Fuel not reaching the carburetor or fuel injection (see Section 4). On carbureted models, check for a vapor lock situation, brought about by clogged fuel tank vent lines.
3   Bad engine ground connection.
4   Choke sticking (Chapter 1).
5   Defective pick-up coil in the distributor (Chapter 5).
6   Float level too high (Chapter 4).

6 Starter motor noisy or engages roughly

1   Pinion or flywheel/driveplate teeth worn or broken. Remove the inspection cover on the left side of the engine and inspect.
2   Starter motor mounting bolts loose or missing.

7 Engine starts but stops immediately

1   Loose or damaged wire harness connections at distributor, coil or alternator.
2   If the engine stalls immediately after the ignition key is turned away from Start, check the ignition ballast resistor (Chapter 5).
3   Intake manifold vacuum leaks. Make sure all mounting bolts/nuts are tight and all vacuum hoses connected to the manifold are attached properly and in good condition.
4   Insufficient fuel flow (see Chapter 4).

8 Engine 'lopes' while idling or idles erratically

1   Vacuum leaks. Check the mounting bolts at the intake manifold and carburetor throttle body (see Chapter 1) for tightness. Make sure that all vacuum hoses are connected and in good condition. Use a length of fuel hose held against your ear to listen for

vacuum leaks while the engine is running. A hissing sound will be heard. A soapy water solution will also detect leaks - spray it around suspected vacuum leak areas (for example, the intake manifold-to-cylinder head mating point, the base of the carburetor, the base of the EGR valve, etc). If the idle becomes smooth when the solution is sprayed at a particular point, you've found the area of the vacuum leak.

2 Leaking EGR valve or plugged PCV valve (see Chapters 1 and 6).

3 Air filter clogged (Chapter 1).

4 Fuel pump not delivering sufficient fuel to the carburetor or fuel injection system (Chapter 4).

5 Leaking head gasket. Perform a cylinder compression check (Chapter 2).

6 Timing chain worn (Chapter 2).

7 Camshaft lobes worn (Chapter 2).

8 Valves burned or otherwise leaking (Chapter 2).

9 Ignition timing out of adjustment (Chapter 1).

10 Defective ignition points or loose distributor shaft, causing dwell to change (1971 and 1972 models) (Chapter 1).

11 Faulty spark plugs, spark plug wires, distributor cap or rotor (Chapter 1).

12 Thermostatic air cleaner not operating properly (Chapter 1).

13 Choke not operating properly (Chapters 1 and 4).

14 Carburetor or fuel injection system dirty, clogged or out of adjustment. Check the float level (Chapter 4).

15 Idle speed out of adjustment (Chapter 1).

9 Engine misses at idle speed

1 Spark plugs fouled, faulty or not gapped properly (Chapter 1).

2 Faulty spark plug wires (Chapter 1).

3 Wet or damaged distributor components (Chapter 1).

4 Short circuits in ignition, coil or spark-plug wires.

5 Sticking or faulty EGR valve (see Chapters 1 and 6).

6 Clogged fuel filter and/or foreign matter in fuel. Remove the fuel filter (Chapter 1) and inspect.

7 Vacuum leaks at intake manifold or hose connections. Check as described in Section 8 of *Troubleshooting*.

8 Defective ignition points or loose distributor shaft, causing dwell to change (1971 and 1972 models) (Chapter 1).

9 Incorrect idle speed (Chapter 4) or idle mixture (Chapter 4).

10 Incorrect ignition timing (Chapter 1).

11 Low or uneven cylinder compression. Check as described in Chapter 2.

12 Choke not operating properly (Chapter 1).

10 Excessively high idle speed

1 Vacuum leak. Check as described in Section 8 of *Troubleshooting*.

2 Idle speed incorrectly adjusted (Chapter 1).

3 Sticking throttle linkage (Chapter 4).

4 Choke opened excessively at idle (Chapter 4).

11 Battery will not hold a charge

1 Alternator drivebelt defective or not adjusted properly (Chapter 1).

2 Battery cables loose or corroded (Chapter 1).

3 Alternator not charging properly or voltage regulator faulty (Chapter 5).

4 Loose, broken or faulty wires in the charging circuit (Chapter 5).

5 Short circuit causing a continuous drain on the battery.

6 Battery defective internally.

12 Ammeter registers a discharge or alternator light stays on

1 Fault in alternator, voltage regulator or charging circuit (Chapter 5).

2 Alternator drivebelt defective or not properly adjusted (Chapter 1).

3 Loose, broken or faulty wires in the charging circuit (Chapter 5). Sometimes, the connections to the ammeter, behind the instrument cluster, become loose, causing a discharge condition. Bypass the ammeter with a heavy-gauge jumper wire to check for this.

13 Ammeter does not function (needle never moves)

1 Fault in the instrument cluster printed circuit or dash wiring (Chapter 12).

2 Short circuit that bypasses the ammeter.

3 Faulty ammeter.

14 Engine misses throughout driving speed range

1 Fuel filter clogged and/or impurities in the fuel system. Check fuel filter (Chapter 1) or clean system (Chapter 4).

2 Fouled, faulty or incorrectly gapped spark plugs (Chapter 1).

3 Incorrect ignition timing (Chapter 1).

4 Cracked distributor cap, disconnected distributor wires or damaged distributor components (Chapter 1).

5 Defective spark plug wires (Chapter 1).

6 Defective ignition points (early models)

or loose distributor shaft, causing dwell to change (see Chapter 1).

7 Low or uneven cylinder compression pressures. Check as described in Chapter 2.

8 Weak or faulty ignition coil or loose connections at the coil (Chapter 5).

9 Weak or faulty ignition system (Chapter 5).

10 Vacuum leaks (see Section 8 of *Troubleshooting*).

11 Dirty or clogged carburetor or fuel injection system (Chapter 4).

12 Leaky EGR valve (Chapter 6).

13 Carburetor or fuel injection system out of adjustment (Chapter 4).

15 Hesitation or stumble during acceleration

1 Ignition timing incorrect (Chapter 1).

2 Faulty spark-plug wires, distributor cap or ignition coil (Chapters 1 and 5).

3 Dirty or clogged carburetor or fuel injection system (Chapter 4).

4 Low fuel pressure. Check for proper operation of the fuel pump and for restrictions in the fuel filter and lines (Chapter 4).

5 Carburetor or fuel injection system out of adjustment (Chapter 4).

16 Engine stalls

1 Idle speed incorrect (Chapter 4).

2 Fuel filter clogged and/or water and impurities in the fuel system (Chapter 1).

3 Choke not operating properly (Chapter 1).

4 Damaged or wet distributor cap and wires.

5 Emissions system components faulty (Chapter 6).

6 Faulty or incorrectly gapped spark plugs (Chapter 1). Also check the spark plug wires (Chapter 1).

7 Vacuum leak at the carburetor or fuel injection system, intake manifold or vacuum hoses. Check as described in Section 8 of *Troubleshooting*.

17 Engine lacks power

1 Incorrect ignition timing (Chapter 1).

2 Excessive play in distributor shaft, causing dwell to change (1971 and 1972 models) (Chapter 1). At the same time, check for faulty distributor cap, wires, etc. (Chapter 1).

3 Faulty or incorrectly gapped spark plugs (Chapter 1).

4 Air filter dirty (Chapter 1).

5 Faulty ignition coil (Chapter 5).

6 Incorrectly adjusted throttle cable (not allowing throttle plates to open completely) (Chapter 4).

7 Choke not opening completely (Chapter 1).

8   Brakes binding (Chapters 1 and 10).
9   Automatic transmission fluid level incorrect, causing slippage (Chapter 1).
10   Clutch slipping (Chapter 8).
11   Fuel filter clogged and/or impurities in the fuel system (Chapters 1 and 4).
12   EGR system not functioning properly (Chapters 1 and 6).
13   Use of sub-standard fuel. Fill tank with proper octane fuel.
14   Low or uneven cylinder compression pressures. Check as described in Chapter 2.
15   Vacuum leak at the carburetor or fuel injection system or intake manifold (check as described in Section 8 of *Troubleshooting*).
16   Dirty or clogged carburetor jets or malfunctioning choke (Chapters 1 and 4).

18   Engine backfires

1   EGR system not functioning properly (Chapter 6).
2   Ignition timing incorrect (Chapter 1).
3   Thermostatic air cleaner system not operating properly (Chapter 6).
4   Vacuum leak (refer to Section 8 of *Troubleshooting*).
5   Damaged valve springs or sticking or burned valves - a vacuum-gauge check will often reveal this problem (Chapter 2C).
6   Carburetor float level out of adjustment (Chapter 4).

19   Engine surges while holding accelerator steady

1   Intake air (vacuum) leak (see Section 8 of *Troubleshooting*).
2   Fuel pump not working properly (Chapter 4).

20   Pinging or knocking engine sounds when engine is under load

1   Incorrect grade of fuel. Fill tank with fuel of the proper octane rating.
2   Ignition timing incorrect (Chapter 1).
3   Carbon build-up in combustion chambers. Remove cylinder heads and clean combustion chambers.
4   Incorrect spark plugs (Chapter 1).

21   Engine diesels (continues to run) after being turned off

1   Idle speed too high (Chapter 4).
2   Ignition timing incorrect (Chapter 1).
3   Incorrect spark plug heat range (Chapter 1).
4   Intake air (vacuum) leak (refer to Section 8 of *Troubleshooting*).
5   Carbon build-up in combustion chambers. Remove cylinder heads and clean combustion chambers.
6   Leaking fuel injectors.
7   Valves sticking (Chapter 2).
8   EGR system not operating properly (Chapter 6).
9   Fuel shut-off system not operating properly (Chapter 6).
10   Check for causes of overheating (Section 27).

22   Low oil pressure

1   Improper grade of oil (viscosity too low) - if you're using the correct grade, try a higher grade to raise the pressure.
2   Oil pump worn or damaged (Chapter 2).
3   Engine overheating (refer to Section 27).
4   Clogged oil filter (Chapter 1).
5   Clogged oil strainer (Chapter 2).
6   Oil pressure gauge not working properly (Chapter 2).

23   Excessive oil consumption

1   Loose oil pan drain plug.
2   Loose bolts or damaged oil pan gasket (Chapter 2).
3   Loose bolts or damaged front cover gasket (Chapter 2).
4   Front or rear crankshaft oil seal leaking (Chapter 2).
5   Loose bolts or damaged valve cover gasket (Chapter 2).
6   Loose oil filter (Chapter 1).
7   Loose or damaged oil pressure switch (Chapter 2).
8   Pistons and cylinders excessively worn (Chapter 2).
9   Piston rings not installed correctly on pistons (Chapter 2).
10   Worn or damaged piston rings (Chapter 2).
11   Intake and/or exhaust valve oil seals worn or damaged (Chapter 2).
12   Valve stem oil seals missing or damaged.
13   Worn or damaged valves/guides (Chapter 2).

24   Excessive fuel consumption

1   Dirty or clogged air filter element (Chapter 1).
2   Incorrect ignition timing (Chapter 1).
3   Incorrect idle speed (Chapter 4).
4   Idle mixture not correctly set (Chapter 4).
5   Low tire pressure or incorrect tire size (Chapter 10).
6   Fuel leakage. Check all connections, lines and components in the fuel system (Chapters 1 and 4).
7   Choke sticking closed (Chapter 1).
8   Dirty or clogged carburetor jets or fuel injectors (Chapter 4).

25   Fuel odor

1   Fuel leakage. Check all connections, lines and components in the fuel system (Chapter 4).
2   Fuel tank overfilled. Fill only to automatic shut-off.
3   Charcoal canister filter in Evaporative Emissions Control system clogged (Chapter 1).
4   Vapor leaks from Evaporative Emissions Control system lines (Chapter 6).

26   Miscellaneous engine noises

1   A strong dull noise that becomes more rapid as the engine accelerates indicates worn or damaged crankshaft bearings or an unevenly worn crankshaft. To pinpoint the trouble spot, remove the spark plug wire from one plug at a time and crank the engine over. If the noise stops, the cylinder with the removed plug wire indicates the problem area. Replace the bearing and/or service or replace the crankshaft (Chapter 2).
2   A similar (yet slightly higher pitched) noise to the crankshaft knocking described in the previous paragraph, that becomes more rapid as the engine accelerates, indicates worn or damaged connecting rod bearings (Chapter 2). The procedure for locating the problem cylinder is the same as described in Paragraph 1.
3   An overlapping metallic noise that increases in intensity as the engine speed increases, yet diminishes as the engine warms up indicates abnormal piston and cylinder wear (Chapter 2). To locate the problem cylinder, use the procedure described in Paragraph 1.
4   A rapid clicking noise that becomes faster as the engine accelerates indicates a worn piston pin or piston pin hole. This sound will happen each time the piston hits the highest and lowest points in the stroke (Chapter 2). The procedure for locating the problem piston is described in Paragraph 1.
5   A metallic clicking or groaning noise coming from the water pump indicates worn or damaged water pump bearings or pump. Replace the water pump with a new one (Chapter 3).
6   A rapid tapping sound or clicking sound that becomes faster as the engine speed increases indicates "valve tapping." This can be identified by holding one end of a section of hose to your ear and placing the other end at different spots along the valve cover. The point where the sound is loudest indicates the problem valve. If the pushrod and rocker arm components are in good shape, you likely have a collapsed valve lifter. Changing the engine oil and adding a high-viscosity oil treatment (such as STP) will sometimes cure a stuck lifter problem. If the problem persists, the lifters, pushrods and rocker arms must be removed for inspection (see Chapter 2).

7   A steady metallic rattling or rapping sound coming from the area of the timing chain cover indicates a worn, damaged or out-of-adjustment timing chain. Service or replace the chain and related components (Chapter 2).

Cooling system

27   Overheating

1   Insufficient coolant in system (Chapter 1).
2   Drivebelt defective or not adjusted properly (Chapter 1).
3   Radiator core blocked by corrosion or radiator grille dirty and restricted (Chapter 3).
4   Thermostat faulty (Chapter 3).
5   Fan not functioning properly (Chapter 3).
6   Radiator cap not maintaining proper pressure. Have cap pressure tested by gas station or repair shop.
7   Ignition timing incorrect (Chapter 1).
8   Defective water pump (Chapter 3).
9   Improper grade of engine oil.
10   Inaccurate temperature gauge (Chapter 12).

28   Overcooling

1   Thermostat faulty, not installed or of too low a temperature (Chapter 3).
2   Inaccurate temperature gauge (Chapter 12).

29   External coolant leakage

1   Deteriorated or damaged hoses. Loose clamps at hose connections (Chapter 1).
2   Water pump seals defective. If this is the case, water will drip from the weep hole in the water pump body (Chapter 3).
3   Leakage from radiator core or header tank. This will require the radiator to be professionally repaired (see Chapter 3 for removal procedures).
4   Engine drain plugs or water jacket freeze plugs leaking (see Chapters 1 and 2C).
5   Leak from coolant temperature switch (Chapter 3).
6   Leak from damaged gaskets or small cracks (Chapter 2).
7   Damaged head gasket. This can be verified by checking the condition of the engine oil, as noted in Section 30.

30   Internal coolant leakage

Note: *Internal coolant leaks can usually be detected by examining the oil. Check the dipstick and inside the valve cover for water deposits and an oil consistency like that of a milkshake.*

1   Leaking cylinder head gasket. Have the system pressure tested or remove the cylinder head (Chapter 2) and inspect.
2   Cracked cylinder bore or cylinder head. Dismantle engine and inspect (Chapter 2).
3   Loose cylinder head bolts (tighten as described in Chapter 2).

31   Abnormal coolant loss

1   Overfilling system (Chapter 1).
2   Coolant boiling away due to overheating (see causes in Section 27).
3   Internal or external leakage (see Sections 29 and 30).
4   Faulty radiator cap. Have the cap pressure tested.
5   Cooling system being pressurized by engine compression. This could be due to a cracked head or block or leaking head gasket(s).

32   Poor coolant circulation

1   Inoperative water pump. A quick test is to pinch the top radiator hose closed with your hand while the engine is idling, then release it. You should feel a surge of coolant if the pump is working properly (Chapter 3).
2   Restriction in the cooling system. Drain, flush and refill the system (Chapter 1). If necessary, remove the radiator (Chapter 3) and have it reverse flushed or professionally cleaned.
3   Loose water pump drivebelt (Chapter 1).
4   Thermostat sticking (Chapter 3).
5   Insufficient coolant (Chapter 1).

33   Corrosion

1   Excessive impurities in the water. Soft, clean water is recommended. Distilled or rainwater is satisfactory.
2   Insufficient antifreeze solution (refer to Chapter 1 for the proper ratio of water to antifreeze).
3   Infrequent flushing and draining of system. Regular flushing of the cooling system should be carried out at the specified intervals as described in (Chapter 1).

Clutch

Note: *All clutch related service information is located in Chapter 8, unless otherwise noted.*

34   Fails to release (pedal pressed to the floor - shift lever does not move freely in and out of Reverse)

1   Clutch contaminated with oil. Remove clutch plate and inspect.

2   Clutch plate warped, distorted or otherwise damaged.
3   Diaphragm spring fatigued. Remove clutch cover/pressure plate assembly and inspect.
4   Insufficient pedal stroke. Check and adjust as necessary.
5   Lack of grease on pilot bushing.

35   Clutch slips (engine speed increases with no increase in vehicle speed)

1   Worn or oil soaked clutch plate.
2   Clutch plate not broken in. It may take 30 or 40 normal starts for a new clutch to seat.
3   Diaphragm spring weak or damaged. Remove clutch cover/pressure plate assembly and inspect.
4   Clutch adjusted too tight (see Chapter 1).
5   Flywheel warped (Chapter 2).

36   Grabbing (chattering) as clutch is engaged

1   Oil on clutch plate. Remove and inspect. Repair any leaks.
2   Worn or loose engine or transmission mounts. They may move slightly when clutch is released. Inspect mounts and bolts.
3   Worn splines on transmission input shaft. Remove clutch components and inspect.
4   Warped pressure plate or flywheel. Remove clutch components and inspect.
5   Diaphragm spring fatigued. Remove clutch cover/pressure plate assembly and inspect.
6   Clutch linings hardened or warped.
7   Clutch lining rivets loose.

37   Squeal or rumble with clutch engaged (pedal released)

1   Improper pedal adjustment. Adjust pedal freeplay (Chapter 1).
2   Release bearing binding on transmission shaft. Remove clutch components and check bearing. Remove any burrs or other damage on the shaft.
3   Pilot bushing worn or damaged.
4   Clutch rivets loose.
5   Clutch plate cracked.
6   Fatigued clutch plate torsion springs. Replace clutch plate.

38   Squeal or rumble with clutch disengaged (pedal depressed)

1   Worn or damaged release bearing.
2   Worn or broken pressure plate diaphragm fingers.

39 Clutch pedal stays on floor when disengaged

Binding linkage or release bearing. Inspect linkage or remove clutch components as necessary.

Manual transmission

Note: *All manual transmission service information is located in Chapter 7, unless otherwise noted.*

40 Noisy in Neutral with engine running

1   Input shaft bearing worn.
2   Damaged main drive gear bearing.
3   Insufficient transmission lubricant (see Chapter 1).
4   Transmission lubricant in poor condition. Drain and fill with proper grade lubricant. Check old lubricant for water and debris (Chapter 1).
5   Noise can be caused by variations in engine torque. Change the idle speed and see if noise disappears.

41 Noisy in all gears

1   Any of the above causes, and/or:
2   Worn or damaged output gear bearings or shaft.

42 Noisy in one particular gear

1   Worn, damaged or chipped gear teeth.
2   Worn or damaged synchronizer.

43 Slips out of gear

1   Transmission loose on clutch housing.
2   Stiff shift lever seal.
3   Shift linkage binding.
4   Broken or loose input gear bearing retainer.
5   Dirt between clutch lever and engine housing.
6   Worn linkage.
7   Damaged or worn check balls, fork rod ball grooves or check springs.
8   Worn mainshaft or countershaft bearings.
9   Loose engine mounts (Chapter 2).
10 Excessive gear endplay.
11 Worn synchronizers.

44 Oil leaks

1   Excessive amount of lubricant in trans-

mission (see Chapter 1 for correct checking procedures). Drain lubricant as required.
2   Rear oil seal or speedometer oil seal damaged.
3   To pinpoint a leak, first remove all built-up dirt and grime from the transmission. Degreasing agents and/or steam cleaning will achieve this. With the underside clean, drive the vehicle at low speeds so the air flow will not blow the leak far from its source. Raise the vehicle and determine where the leak is located.

45 Difficulty engaging gears

1   Clutch not releasing completely.
2   Loose or damaged shift linkage. Make a thorough inspection, replacing parts as necessary.
3   Insufficient transmission lubricant (Chapter 1).
4   Transmission lubricant in poor condition. Drain and fill with proper grade lubricant. Check lubricant for water and debris (Chapter 1).
5   Worn or damaged shift rod.
6   Sticking or jamming gears.

46 Noise occurs while shifting gears

1   Check for proper operation of the clutch (Chapter 8).
2   Faulty synchronizer assemblies. Measure baulk ring-to-gear clearance. Also, check for wear or damage to baulk rings or any parts of the synchromesh assemblies.

Automatic transmission

Note: *Due to the complexity of the automatic transmission, it's difficult for the home mechanic to properly diagnose and service. For problems other than the following, the vehicle should be taken to a reputable mechanic.*

47 Fluid leakage

Note: *For further leak diagnosis and seal replacement information, refer to Chapter 7B.*
1   Automatic transmission fluid is a deep red color, and fluid leaks should not be confused with engine oil which can easily be blown by air flow to the transmission.
2   To pinpoint a leak, first remove all built-up dirt and grime from the transmission. Degreasing agents and/or steam cleaning will achieve this. With the underside clean, drive the vehicle at low speeds so the air flow will not blow the leak far from its source. Raise the vehicle and determine where the leak is located. Common areas of leakage are:

a) *Fluid pan*: *tighten mounting bolts and/or replace pan gasket as necessary (Chapter 1).*
b) *Rear extension*: *tighten bolts and/or replace oil seal as necessary.*
c) *Filler pipe*: *replace the rubber oil seal where pipe enters transmission case.*
d) *Transmission oil lines*: *tighten fittings where lines enter transmission case and/or replace lines.*
e) *Vent pipe*: *transmission overfilled and/or water in fluid (see checking procedures, Chapter 1).*
f) *Speedometer connector*: *replace the O-ring where speedometer cable enters transmission case.*

48 General shift mechanism problems

Chapter 7 deals with checking and adjusting the shift linkage on automatic transmissions. Common problems which may be caused by out of adjustment linkage are:

a) *Engine starting in gears other than P (park) or N (Neutral).*
b) *Indicator pointing to a gear other than the one actually engaged.*
c) *Vehicle moves with transmission in P (Park) position.*

49 Transmission will not downshift with the accelerator pedal pressed to the floor

Chapter 7 deals with adjusting the throttle rod to enable the transmission to downshift properly.

50 Engine will start in gears other than Park or Neutral

Chapter 7 deals with adjusting the Neutral start switch installed on automatic transmissions.

51 Transmission slips, shifts rough, is noisy or has no drive in forward or Reverse gears

1   There are many probable causes for the above problems, but the home mechanic should concern himself only with one possibility; fluid level.
2   Before taking the vehicle to a shop, check the fluid level and condition as described in Chapter 1. Add fluid, if necessary, or change the fluid and filter if needed. If problems persist, have a professional diagnose the transmission.

Driveshaft

Note: *Refer to Chapter 8, unless otherwise specified, for service information.*

52 Leaks at front of driveshaft

Defective transmission rear seal and/or bushing. See Chapter 7B for replacement procedure. As this is done, check the splined yoke for burrs or roughness that could damage the new seal. Remove burrs with a fine file or whetstone.

53 Knock or clunk when transmission is under initial load (just after transmission is put into gear)

1   Loose or disconnected rear suspension components. Check all mounting bolts and bushings (Chapters 7 and 10).
2   Loose driveshaft bolts. Inspect all bolts and nuts and tighten them securely.
3   Worn or damaged universal joint bearings (Chapter 8).
4   Worn sleeve yoke and mainshaft spline.

54 Metallic grating sound consistent with vehicle speed

Pronounced wear in the universal joint bearings. Replace U-joints or driveshafts, as necessary.

55 Vibration

Note: *Before blaming the driveshaft, make sure the tires are perfectly balanced and perform the following test.*
1   Install a tachometer inside the vehicle to monitor engine speed as the vehicle is driven. Drive the vehicle and note the engine speed at which the vibration (roughness) is most pronounced. Now shift the transmission to a different gear and bring the engine speed to the same point.
2   If the vibration occurs at the same engine speed (rpm) regardless of which gear the transmission is in, the driveshaft is NOT at fault since the driveshaft speed varies.
3   If the vibration decreases or is eliminated when the transmission is in a different gear at the same engine speed, refer to the following probable causes.
4   Bent or dented driveshaft. Inspect and replace as necessary.
5   Undercoating or built-up dirt, etc. on the driveshaft. Clean the shaft thoroughly.
6   Worn universal joint bearings. Replace the U-joints or driveshaft as necessary.
7   Driveshaft and/or companion flange out of balance. Check for missing weights on the shaft. Remove driveshaft and reinstall

180-degrees from original position, then recheck. Have the driveshaft balanced if problem persists.
8   Loose driveshaft mounting bolts/nuts.
9   Defective center bearing, if so equipped.
10   Worn transmission rear bushing (Chapter 7).

56 Scraping noise

Look for scrape marks around the driveshaft that will help you locate the source of the scraping.

57 Whining or whistling noise

Defective center bearing, if so equipped.

Rear axle and differential

Note: *For differential servicing information, refer to Chapter 8, unless otherwise specified.*

58 Noise - same when in drive as when vehicle is coasting

1   Road noise. No corrective action available.
2   Tire noise. Inspect tires and check tire pressures (Chapter 1).
3   Front wheel bearings loose, worn or damaged (Chapter 1).
4   Insufficient differential oil (Chapter 1).
5   Defective differential.

59 Knocking sound when starting or shifting gears

Defective or incorrectly adjusted differential.

60 Noise when turning

Defective differential.

61 Vibration

See probable causes under Driveshaft. Proceed under the guidelines listed for the driveshaft. If the problem persists, check the rear wheel bearings by raising the rear of the vehicle and spinning the wheels by hand. Listen for evidence of rough (noisy) bearings. Remove and inspect (Chapter 8).

62 Oil leaks

1   Pinion oil seal damaged (Chapter 8).
2   Axleshaft oil seals damaged (Chapter 8).
3   Differential cover leaking. Tighten mounting bolts or replace the gasket as required.

4   Loose filler plug on differential (Chapter 1).
5   Clogged or damaged breather on differential.

Brakes

Note: *Before assuming a brake problem exists, make sure the tires are in good condition and inflated properly, the front end alignment is correct and the vehicle is not loaded with weight in an unequal manner. All service procedures for the brakes are included in Chapter 9, unless otherwise noted.*

63 Vehicle pulls to one side during braking

1   Defective, damaged or oil contaminated brake pad or shoe on one side. Inspect as described in Chapter 1. Refer to Chapter 9 if replacement is required.
2   Excessive wear of brake pad/shoe material or disc/drum on one side. Inspect and repair as necessary.
3   Loose or disconnected front suspension components. Inspect and tighten all bolts securely (Chapters 1 and 10).
4   Defective caliper or wheel cylinder assembly. Remove the caliper or wheel cylinder and inspect for a stuck piston or damage.
5   Brake pad to rotor adjustment needed. Inspect automatic adjusting mechanism for proper operation.
6   Scored or out-of-round rotor.
7   Loose caliper mounting bolts.
8   Incorrect wheel bearing adjustment.
9   On models with front drum brakes, the automatic adjusters may not be functioning correctly (see Chapter 9).

64 Noise (high-pitched squeal)

Note: *Squeal coming from front disc brakes is common and sometimes hard to fix. Anti-squeal compounds that are applied to the back of the brake pads often help. These are available from auto parts stores. Also available are anti-squeal shims that help as well. Semi-metallic brake pads last longer than conventional organic linings, but they also contribute to squeal and cause more rotor wear. Also make sure the caliper mounting bolts are tight and all the caliper sliding surfaces are well lubricated (see Chapter 9).*
1   Front brake pads worn out. Replace pads with new ones immediately!
2   Glazed or contaminated pads.
3   Dirty or scored rotor.
4   Bent support plate.

65 Excessive brake pedal travel

1   Partial brake system failure. Inspect entire system (Chapter 1) and correct as required.

2   Insufficient fluid in master cylinder. Check (Chapter 1) and add fluid - bleed system if necessary.
3   Air in system. Bleed system.
4   Excessive lateral rotor play.
5   Brakes out of adjustment. Check the operation of the automatic adjusters.
6   Defective proportioning valve. Replace valve and bleed system.

66   Brake pedal feels spongy when depressed

1   Air in brake lines. Bleed the brake system.
2   Deteriorated rubber brake hoses. Inspect all system hoses and lines. Replace parts as necessary.
3   Master cylinder mounting nuts loose. Inspect master cylinder bolts (nuts) and tighten them securely.
4   Master cylinder faulty.
5   Incorrect shoe or pad clearance.
6   Defective check valve. Replace valve and bleed system.
7   Clogged reservoir cap vent hole.
8   Deformed rubber brake lines.
9   Soft or swollen caliper seals.
10 Poor quality brake fluid. Bleed entire system and fill with new approved fluid.

67   Excessive effort required to stop vehicle

1   Power brake booster not operating properly.
2   Excessively worn linings or pads. Check and replace if necessary.
3   One or more caliper pistons seized or sticking. Inspect and rebuild as required.
4   Brake pads or linings contaminated with oil or grease. Inspect and replace as required.
5   New pads or linings installed and not yet seated. It'll take a while for the new material to seat against the rotor or drum.
6   Worn or damaged master cylinder or caliper assemblies. Check particularly for frozen pistons.
7   Also see causes listed under Section 66.

68   Pedal travels to the floor with little resistance

Little or no fluid in the master cylinder reservoir caused by leaking caliper piston(s) or loose, damaged or disconnected brake lines. Inspect entire system and repair as necessary.

69   Brake pedal pulsates during brake application

Note: *Brake pedal pulsation during operation of the Anti-Lock Brake System (ABS) is normal.*

1   Wheel bearings damaged, worn or out of adjustment (Chapter 1).
2   Caliper not sliding properly due to improper installation or obstructions. Remove and inspect.
3   Rotor not within specifications. Remove the rotor and check for excessive lateral runout and parallelism. Have the rotors resurfaced or replace them with new ones. Also make sure that all rotors are the same thickness.
4   Out of round rear brake drums. Remove the drums and have them turned or replace them with new ones.

70   Brakes drag (indicated by sluggish engine performance or wheels being very hot after driving)

1   Output rod adjustment incorrect at the brake pedal or between the brake booster and master cylinder.
2   Obstructed master cylinder compensator. Disassemble master cylinder and clean.
3   Master cylinder piston seized in bore. Overhaul master cylinder.
4   Caliper assembly in need of overhaul.
5   Brake pads or shoes worn out.
6   Piston cups in master cylinder or caliper assembly deformed. Overhaul master cylinder.
7   Rotor not within specifications (Section 69).
8   Parking brake assembly will not release.
9   Clogged brake lines.
10 Wheel bearings out of adjustment (Chapter 1).
11 Brake pedal height improperly adjusted.
12 Wheel cylinder needs overhaul.
13 Improper shoe-to-drum clearance. Adjust as necessary.

71   Rear brakes lock up under light brake application

1   Tire pressures too high.
2   Tires excessively worn (Chapter 1).
3   Defective power brake booster.

72   Rear brakes lock up under heavy brake application

1   Tire pressures too high.
2   Tires excessively worn (Chapter 1).
3   Front brake pads or shoes contaminated with oil, mud or water. Replace the pads.
4   Front brake pads or shoes excessively worn.
5   Defective master cylinder or caliper assembly.

Suspension and steering

Note: *All service procedures for the suspension and steering systems are included in Chapter 10, unless otherwise noted.*

73   Vehicle pulls to one side

1   Tire pressures uneven (Chapter 1).
2   Defective tire (Chapter 1).
3   Excessive wear in suspension or steering components (Chapters 1 and 10).
4   Front end alignment incorrect.
5   Front brakes dragging. Inspect as described in Section 71.
6   Wheel bearings improperly adjusted (Chapter 1).
7   Wheel lug nuts loose.

74   Shimmy, shake or vibration

1   Tire or wheel out of balance or out of round. Have them balanced on the vehicle.
2   Loose, worn or out of adjustment wheel bearings (Chapter 1).
3   Shock absorbers and/or suspension components worn or damaged (see Chapter 10).

75   Excessive pitching and/or rolling around corners or during braking

1   Defective shock absorbers. Replace as a set.
2   Broken or weak leaf springs and/or suspension components.
3   Worn or damaged stabilizer bar or bushings.

76   Wandering or general instability

1   Improper tire pressures.
2   Incorrect front end alignment.
3   Worn or damaged steering linkage or suspension components.
4   Improperly adjusted steering gear.
5   Out-of-balance wheels.
6   Loose wheel lug nuts.
7   Worn rear shock absorbers.
8   Fatigued or damaged rear leaf springs.

77   Excessively stiff steering

1   Lack of fluid in the power steering fluid reservoir, where appropriate (Chapter 1).
2   Incorrect tire pressures (Chapter 1).
3   Lack of lubrication at balljoints (Chapter 1).
4   Front end out of alignment.
5   Steering gear out of adjustment or lacking lubrication.
6   Improperly adjusted wheel bearings.

7 Worn or damaged steering gear.
8 Interference of steering column with turn signal switch.
9 Low tire pressures.
10 Worn or damaged balljoints.
11 Worn or damaged steering linkage.
12 See also Section 76.

78 Excessive play in steering

1 Loose wheel bearings (Chapter 1).
2 Excessive wear in suspension bushings (Chapter 1).
3 Steering gear improperly adjusted.
4 Incorrect front end alignment.
5 Steering gear mounting bolts loose.
6 Worn steering linkage.

79 Lack of power assistance

1 Steering pump drivebelt faulty or not adjusted properly (Chapter 1).
2 Fluid level low (Chapter 1).
3 Hoses or pipes restricting the flow. Inspect and replace parts as necessary.
4 Air in power steering system. Bleed system.
5 Defective power steering pump.

80 Steering wheel fails to return to straight-ahead position

1 Incorrect front end alignment.
2 Tire pressures low.
3 Steering gears improperly engaged.
4 Steering column out of alignment.
5 Worn or damaged balljoint.

6 Worn or damaged steering linkage.
7 Improperly lubricated idler arm.
8 Insufficient oil in steering gear.
9 Lack of fluid in power steering pump.

81 Steering effort not the same in both directions (power system)

1 Leaks in steering gear.
2 Clogged fluid passage in steering gear.

82 Noisy power steering pump

1 Insufficient oil in pump.
2 Clogged hoses or oil filter in pump.
3 Loose pulley.
4 Improperly adjusted drivebelt (Chapter 1).
5 Defective pump.

83 Miscellaneous noises

1 Improper tire pressures.
2 Insufficiently lubricated balljoint or steering linkage.
3 Loose or worn steering gear, steering linkage or suspension components.
4 Defective shock absorber.
5 Defective wheel bearing.
6 Worn or damaged suspension bushings.
7 Damaged leaf spring.
8 Loose wheel lug nuts.
9 Worn or damaged rear axleshaft spline.
10 Worn or damaged rear shock absorber mounting bushing.
11 Incorrect rear axle endplay.
12 See also causes of noises at the rear axle and driveshaft.

84 Excessive tire wear (not specific to one area)

1 Incorrect tire pressures.
2 Tires out of balance. Have them balanced on the vehicle.
3 Wheels damaged. Inspect and replace as necessary.
4 Suspension or steering components worn (Chapter 1).

85 Excessive tire wear on outside edge

1 Incorrect tire pressure.
2 Excessive speed in turns.
3 Front end alignment incorrect (excessive toe-in).

86 Excessive tire wear on inside edge

1 Incorrect tire pressure.
2 Front end alignment incorrect (toe-out).
3 Loose or damaged steering components (Chapter 1).

87 Tire tread worn in one place

1 Tires out of balance. Have them balanced on the vehicle.
2 Damaged or buckled wheel. Inspect and replace if necessary.
3 Defective tire.

Chapter 1
Tune-up and routine maintenance

Contents

Specifications

Recommended lubricants and fluids

Note: *Listed here are the manufacturers recommendations at the time this manual was written. Manufacturers occasionally upgrade their fluid and lubricant specifications, so check with your auto parts store for current recommendations.*

Engine oil type .... API grade "certified for gasoline engines"
Engine oil viscosity .... See accompanying chart

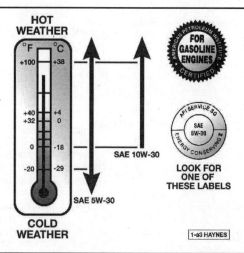

Engine oil viscosity chart - for best fuel economy and cold starting, select the lowest SAE viscosity grade for the expected temperature range

Recommended lubricants and fluids (continued)

Automatic transmission fluid type
    1971 through 1978 .... Dexron type automatic transmission fluid
    1979 through 1985
        Without lockup torque converter .... Dexron type automatic transmission fluid
        With lockup torque converter .... Mopar ATF +3 Type 7176 automatic transmission fluid or equivalent
    1986 and later .... Mopar ATF +3 Type 7176 automatic transmission fluid or equivalent
Manual transmission lubricant type........................ Dexron II or Mercon automatic transmission fluid*
Differential lubricant type .... SAE 80W-90 GL-5 gear lubricant
Limited slip differential .... Add Chrysler Friction Modifier No. 4318060, or equivalent,
                                         to the specified lubricant
Brake fluid type .... DOT 3 brake fluid
Power steering fluid .... Chrysler power steering fluid or equivalent
Manual steering gear lubricant type .... SAE 90 GL-5 hypoid gear lubricant
Chassis grease type .... NLGI no. 2 EP chassis grease
Front wheel bearing grease .... NLGI no. 2 EP high-temperature wheel bearing grease

* **Note:** *All manual transmissions covered by this manual came from the factory filled with Dexron ATF; however, a worn or noisy transmission may function better with SAE 75W-90 GL-5 gear lubricant. If switching to gear lubricant, be sure all ATF is drained from the transmission before refilling.*

Capacities

Engine oil (with filter change - approximate) .... 5 qts
Automatic transmission (approximate)
    When draining pan and replacing filter .... 4 qts
    When filling a new or rebuilt transmission from dry .... 8 qts
Manual transmission .... 2 qts

Ignition system

Spark plug type
    Inline six-cylinder engine
        1974 and earlier .... Champion no. RN14YC
        1975 and later .... Champion no. RV17YC
    Small-block V8 engine - 318 CID (5.2L) and 360 CID (5.9L)
        1971 and 1972
            With two-barrel carburetor .... Champion no. RN14YC
            With four-barrel carburetor .... Champion no. RN12YC
        1973 through 1985 (all) .... Champion no. RN14YC
        1986 and later (all).. Champion no. RN12YC
    V6 engine - 238 CID (3.9L) (all)............................. Champion no. RN12YC
    Big-block V8 engine - 400 CID (6.6L) and 440 CID (7.2L)
        1977 and 1978 .... Champion no. RJ12YC
Spark plug gap (all engines, all years) .... 0.035 inch
Ignition timing .... Refer to the *Vehicle Emissions Control Information* label in the engine compartment

Cylinder location and distributor rotation diagram - inline six-cylinder engine

The blackened terminal shown on the distributor cap indicates the Number One spark plug wire position

318, 360 engines   400, 440 engines

Cylinder location and distributor rotation diagram - V8 engine

Firing order
    Inline six-cylinder engine.. 1-5-3-6-2-4
    V6 engine... 1-6-5-4-3-2
    V8 engine... 1-8-4-3-6-5-7-2

General

Valve clearance - engine hot (1980 and earlier inline six-cylinder engines)
    Intake .... 0.010 inch
    Exhaust .... 0.020 inch
Clutch pedal freeplay.. 1 inch
Disc brake pad lining thickness (minimum) .... 5/16 inch
Drum brake shoe lining thickness (minimum) .... 1/16 inch

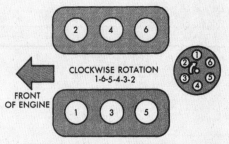

CLOCKWISE ROTATION
1-6-5-4-3-2

FRONT OF ENGINE

Cylinder location and distributor rotation diagram - V6 engine

Automatic transmission band adjustment

1971 through 1980
    Kickdown band
        Six-cylinder engine .... Tighten to 72 in-lbs, back off 2-1/2 turns
        V8 engines .... Tighten to 72 in-lbs, back off 2 turns
    Low-reverse band .... Tighten to 72 in-lbs, back off 2 turns
1981 through 1991
    Kickdown band .... Tighten to 72 in-lbs, back off 2-1/2 turns
    Low-reverse band
        A-904, A-998, A-999, A-500 .... Tighten to 72 in-lbs, back off 4 turns
        A-727, A-518.. Tighten to 72 in-lbs, back off 2 turns

1992 through 1994
  Kickdown band .... Tighten to 72 in-lbs, back off 2-1/2 turns
  Low-reverse band
    32RH, 42RH .... Tighten to 72 in-lbs, back off 4 turns
    36RH, 37RH, 46RH .... Tighten to 72 in-lbs, back off 2 turns
1995 and later
  Kickdown band
    32RH .... Tighten to 72 in-lbs, back off 2-1/4 turns
    36RH, 46RH, 46RE .... Tighten to 72 in-lbs, back off 2-7/8 turns
  Low-reverse band
    32RH, 42RH .... Tighten to 72 in-lbs, back off 4 turns
    36RH, 46RH, 46RE .... Tighten to 72 in-lbs, back off 2 turns

Torque specifications

Ft-lbs (unless otherwise indicated)

Automatic transmission band adjusting screw locknut...........................	30
Automatic transmission pan bolts	96 in-lbs
Carburetor mounting nuts/bolts	96 in-lbs
Differential cover bolts	
1971 through 1997..	10 to 20
1998 and later..	30
EGR valve bolts	17
Spark plugs	
Inline six-cylinder engine	120 in-lbs
V6 and V8 engines	30
Engine oil drain plug	20
Oxygen sensor	20
Wheel lug nuts	85

Engine compartment components (V8 engine model)

1 Oil filler cap
2 Brake fluid reservoir
3 Power brake booster
4 Automatic transmission fluid dipstick
5 Engine oil dipstick
6 Radiator cap
7 Coolant reservoir
8 Windshield washer fluid reservoir
9 Battery

Engine compartment underside components

1	Lower radiator hose	4	Brake hose	7	Automatic transmission fluid pan
2	Timing marks	5	Evaporative Emissions Control canister	8	Lower balljoint
3	Power steering pump	6	Oil pan drain plug		

Typical V8 engine components (engine cover removed)

1	PCV valve	3	Crankcase inlet filter	5	Oil pressure sending unit
2	Air cleaner housing	4	Distributor	6	Ignition coil

1 Maintenance Schedule

Every 250 miles or weekly, whichever comes first

Check the engine oil level (Section 4)
Check the engine coolant level (Section 4)
Check the windshield washer fluid level (Section 4)
Check the brake and clutch fluid levels (Section 4)
Check the tires and tire pressures (Section 5)

Every 3000 miles or 3 months, whichever comes first

All items listed above, plus . . .
Check the automatic transmission fluid level (Section 6)
Check the power steering fluid level (Section 7)
Change the engine oil and filter (Section 8)
Check and service the battery (Section 9)
Check the cooling system (Section 10)
Inspect and replace, if necessary, all underhood hoses (Section 11)
Inspect and replace, if necessary, the windshield wiper blades (Section 12)
Inspect the suspension and steering components (Section 13)
Inspect the exhaust system (Section 14)
Check the clutch pedal freeplay (Section 15)
Check the manual transmission lubricant (Section 16)
Check the differential lubricant level (Section 17)

Every 7500 miles or 6 months, whichever comes first

Rotate the tires (Section 18)
Check the brakes (Section 19) *
Inspect the fuel system (Section 20)
Check the carburetor mounting bolt/nut torque (Section 21)
Check the throttle linkage (Section 22)
Check the engine drivebelts (Section 23)
Check the seat belts (Section 24)
Check the neutral start switch (Section 25)

Every 15,000 miles or 12 months, whichever comes first

All items listed above, plus . . .
Lubricate the chassis components (Section 26) *
Check and replace if necessary, the ignition points (1971 and 1972 models only) (Section 27)
Check and adjust if necessary, the valve clearances (1980 and earlier six-cylinder models) (Section 28)

Every 30,000 miles or 24 months, whichever comes first

All items listed above, plus . . .
Replace the air filter (Section 29)
Change the automatic transmission fluid and filter (Section 30)**
Adjust the automatic transmission bands (Section 31)**
Change the manual transmission lubricant (Section 32)
Change the differential lubricant (Section 33)
Check and repack the front wheel bearings (Section 34)*
Service the cooling system (drain, flush and refill) (Section 35)
Check the Positive Crankcase Ventilation (PCV) system (Section 36)
Check the evaporative emissions control system (Section 37)
Check the Exhaust Gas Recirculation (EGR) system (Section 38)
Check the thermostatically-controlled air cleaner (Section 39)
Check the crankcase inlet filter (Section 40)
Replace the fuel filter (Section 41)
Check and clean, if necessary, the exhaust manifold heat control valve (Section 42)
Clean the carburetor choke shaft (Section 43)
Replace the spark plugs (Section 44)
Inspect the spark plug wires, distributor cap and rotor (Section 45)
Check and adjust, if necessary, the idle speed (Section 46)
Check and adjust, if necessary, the ignition timing (Section 47)

Every 52,500 miles

Replace the oxygen sensor (1985 through 1987 models) (Section 48)

Every 82,500 miles

Replace the oxygen sensor (1988 and later models) (Section 48)

*This item is affected by "severe" operating conditions, as described below. If the vehicle is operated under severe conditions, perform all maintenance indicated with an asterisk (*) at 3000 mile/three-month intervals. Severe conditions exist if you mainly operate the vehicle . . .
in dusty areas
towing a trailer
idling for extended periods and/or driving at low speeds when outside temperatures remain below freezing and most trips are less than four miles long
**If operated under one or more of the following conditions, change the automatic transmission fluid and adjust the bands every 15,000 miles:
in heavy city traffic where the outside temperature regularly reaches 90-degrees F or higher
in hilly or mountainous terrain
frequent trailer pulling

Typical rear underside components

1	*Exhaust pipe*	4	*Leaf spring*	7	*Differential check/fill plug*
2	*Fuel tank*	5	*Parking brake cable adjuster*	8	*Parking brake adjuster*
3	*Flexible brake hose*	6	*Rear universal joint*		

2   Introduction

This Chapter is designed to help the home mechanic maintain Chrysler rear-wheel drive models with the goals of maximum performance, economy, safety and reliability in mind.

Included is a master maintenance schedule, followed by procedures dealing specifically with each item on the schedule. Visual checks, adjustments, component replacement and other helpful items are included. Refer to the accompanying illustrations of the engine compartment and the underside of the vehicle for the locations of various components.

Servicing your vehicle in accordance with the mileage/time maintenance schedule and the step-by-step procedures will result in a planned maintenance program that should produce a long and reliable service life. Keep in mind that it's a comprehensive plan, so maintaining some items but not others at the specified intervals will not produce the same results.

As you service your vehicle, you will discover that many of the procedures can - and should - be grouped together because of the nature of the particular procedure you're performing or because of the close proximity of two otherwise unrelated components to one another.

For example, if the vehicle is raised for chassis lubrication, you should inspect the exhaust, suspension, steering and fuel systems while you're under the vehicle. When you're rotating the tires, it makes good sense to check the brakes since the wheels are already removed. Finally, let's suppose you have to borrow or rent a torque wrench. Even if you only need it to tighten the spark plugs, you might as well check the torque of as many critical fasteners as time allows.

The first step in this maintenance program is to prepare yourself before the actual work begins. Read through all the procedures you're planning to do, then gather up all the parts and tools needed. If it looks like you might run into problems during a particular job, seek advice from a mechanic or an experienced do-it-yourselfer.

The maintenance intervals are based on the assumption the vehicle owner will be doing the maintenance or service work, as opposed to having a dealer service department do the work. Although the time/mileage intervals are loosely based on factory recommendations, most have been shortened to ensure, for example, that such items as lubricants and fluids are checked/changed at intervals that promote maximum engine/driveline service life. Also, subject to the preference of the individual owner interested in keeping the vehicle in peak condition at all times, and with the vehicle's ultimate resale in mind, many of the maintenance procedures may be performed more often than recommended in the following schedule. We encourage such owner initiative.

When the vehicle is new it should be serviced initially by a factory authorized dealer service department to protect the factory warranty. In many cases the initial maintenance check is done at no cost to the owner (check with your dealer service department for additional information).

3   Tune-up general information

The term tune-up is used in this manual to represent a combination of individual operations rather than one specific procedure.

If, from the time the vehicle is new, the routine maintenance schedule is followed closely and frequent checks are made of fluid levels and high wear items, as suggested throughout this manual, the engine will be kept in relatively good running condition and the need for additional work will be minimized.

More likely than not, however, there will be times when the engine is running poorly due to lack of regular maintenance. This is even more likely if a used vehicle, which has not received regular and frequent maintenance checks, is purchased. In such cases, an engine tune-up will be needed outside of the regular routine maintenance intervals.

The first step in any tune-up or diagnostic procedure to help correct a poor running engine is a cylinder compression check. A compression check (see Chapter 2, Part C) will help determine the condition of internal engine components and should be used as a guide for tune-up and repair procedures. If, for instance, the compression check indicates serious internal engine wear, a conventional tune-up won't improve the performance of the engine and would be a waste of time and money. Because of its importance, the compression check should be done by someone with the right equipment and the knowledge to use it properly.

The following procedures are those most often needed to bring a generally poor running engine back into a proper state of tune.

Minor tune-up

Check all engine related fluids (Section 4)
Clean, inspect and test the battery (Section 9)
Check the cooling system (Section 10)
Check all underhood hoses (Section 11)
Check and adjust the drivebelts (Section 23)
Replace the ignition points (1971 and 1972 models only) (Section 27)
Check the air filter (Section 29)
Check the PCV valve (Section 36)
Replace the spark plugs (Section 44)
Inspect the spark plug and coil wires (Section 45)
Inspect the distributor cap and rotor (Section 45)
Check and adjust the idle speed (Section 46)
Check and adjust the ignition timing (Section 47)

Major tune-up

All items listed under Minor tune-up, plus . . .

Check the fuel system (Section 20)
Replace the air filter (Section 29)
Check the EGR system (Section 38)
Replace the spark plug wires (Section 45)
Replace the distributor cap and rotor (Section 45)
Check the ignition system (Chapter 5)
Check the charging system (Chapter 5)

---

4   Fluid level checks (every 250 miles or weekly)

Note: *The following are fluid level checks to be done on a 250 mile or weekly basis. Additional fluid level checks can be found in specific maintenance procedures which follow. Regardless of intervals, be alert to fluid leaks under the vehicle which would indicate a fault to be corrected immediately.*

1   Fluids are an essential part of the lubrication, cooling, brake, clutch and windshield washer systems. Because the fluids gradually become depleted and/or contaminated during normal operation of the vehicle, they must be periodically replenished. See *Recommended lubricants and fluids* at the beginning of this Chapter before adding fluid to any of the following components. **Note**: *The vehicle must be on level ground when fluid levels are checked.*

4.2 The engine oil dipstick (arrow) is located at the front of the engine

Engine oil

Refer to illustrations 4.2, 4.4 and 4.6

2   The engine oil level is checked with a dipstick that extends through a tube and into the oil pan at the bottom of the engine **(see illustration)**.

3   The oil level should be checked before the vehicle has been driven, or about 15 minutes after the engine has been shut off. If the oil is checked immediately after driving the vehicle, some of the oil will remain in the upper engine components, resulting in an inaccurate reading on the dipstick.

4   Pull the dipstick out of the tube and wipe all the oil from the end with a clean rag or paper towel. Insert the clean dipstick all the way back into the tube, then pull it out again. Note the oil at the end of the dipstick. Add oil as necessary to keep the level between the ADD and FULL marks on the dipstick **(see illustration)**.

5   Do not overfill the engine by adding too much oil since this may result in oil fouled spark plugs, oil leaks or oil seal failures.

6   Oil is added to the engine after pulling the cap from the oil fill tube **(see illustration)**. A funnel may help to reduce spills.

7   Checking the oil level is an important preventive maintenance step. A consistently low oil level indicates oil leakage through damaged seals, defective gaskets or past worn rings or valve guides. If the oil looks

4.6 Pull the oil filler cap straight off to add oil to the engine - always make sure the area around the opening is clean before removing the cap to prevent dirt from contaminating the engine

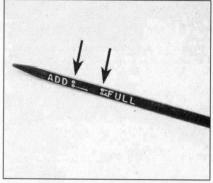

4.4 The oil level must be maintained between the marks at all times - it takes one quart of oil to raise the level from the ADD to the FULL mark

milky or has water droplets in it, the cylinder head gasket(s) may be blown or the head(s) or block may be cracked. The engine should be checked immediately. Whenever you check the oil level, slide your thumb and index finger up the dipstick before wiping off the oil. If you see small dirt or metal particles clinging to the dipstick, the oil should be changed (see Section 8).

Engine coolant

Refer to illustration 4.8

Warning: *Do not allow antifreeze to come in contact with your skin or painted surfaces of the vehicle. Flush contaminated areas immediately with plenty of water. Don't store new coolant or leave old coolant lying around where it's accessible to children or pets - they're attracted by its sweet smell. Ingestion of even a small amount of coolant can be fatal! Wipe up garage floor and drip pan coolant spills immediately. Keep antifreeze containers covered and repair leaks in the cooling system as soon as they are noted.*

8   Most vehicles covered by this manual are equipped with a pressurized coolant recovery system. A white plastic coolant reservoir located in the engine compartment is connected by a hose to the radiator filler neck **(see illustration)**. If the engine overheats,

4.8 The coolant reservoir is located in the left front corner of the engine compartment on most models - keep the level between the MAX and COOL marks on the side of the reservoir

1

4.14 This windshield washer fluid reservoir can be easily confused with the coolant reservoir if you're not careful - remember that the coolant reservoir is connected to the radiator, just below the radiator cap, by a small-diameter rubber hose

coolant escapes through a valve in the radiator cap and travels through the hose into the reservoir. As the engine cools, the coolant is automatically drawn back into the cooling system to maintain the correct level. **Warning:** *Do not remove the radiator cap to check the coolant level when the engine is warm.*

9   The coolant level in the reservoir should be checked regularly. The level in the reservoir varies with the temperature of the engine. When the engine is cold, the coolant level should be at or slightly above the MIN mark on the reservoir. Once the engine has warmed up, the level should be at or near the MAX mark. If it isn't, allow the engine to cool, then remove the cap from the reservoir and add a 50/50 mixture of ethylene glycol-based antifreeze and water.

10   Drive the vehicle and recheck the coolant level. If only a small amount of coolant is required to bring the system up to the proper level, water can be used. However, repeated additions of water will dilute the antifreeze and water solution. In order to maintain the proper ratio of antifreeze and water, always top up the coolant level with the correct mixture. An empty plastic milk jug or bleach bottle makes

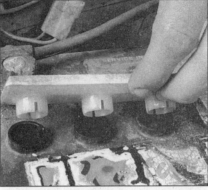

4.17 Remove the cell caps to check the electrolyte level in the battery - if the level is low, add distilled water only

an excellent container for mixing coolant. Do not use rust inhibitors or additives.

11   If the coolant level drops consistently, there may be a leak in the system. Inspect the radiator, hoses, filler cap, drain plugs and water pump (see Section 10). If no leaks are noted, have the radiator cap pressure tested by a service station.

12   If you have to remove the radiator cap, wait until the engine has cooled, then wrap a thick cloth around the cap and turn it to the first stop. If coolant or steam escapes, let the engine cool down longer, then remove the cap.

13   Check the condition of the coolant as well. It should be relatively clear. If it's brown or rust colored, the system should be drained, flushed and refilled. Even if the coolant appears to be normal, the corrosion inhibitors wear out, so it must be replaced at the specified intervals.

Windshield washer fluid

Refer to illustration 4.14

14   Fluid for the windshield washer system is located in a plastic reservoir in the engine compartment **(see illustration)**.

15   In milder climates, plain water can be used in the reservoir, but it should be kept no more than 2/3 full to allow for expansion if the

water freezes. In colder climates, use windshield washer system antifreeze, available at any auto parts store, to lower the freezing point of the fluid. Mix the antifreeze with water in accordance with the manufacturer's directions on the container. **Caution:** *Don't use cooling system antifreeze - it will damage the vehicle's paint.*

16   To help prevent icing in cold weather, warm the windshield with the defroster before using the washer.

Battery electrolyte

Refer to illustration 4.17

17   Most vehicles with which this manual is concerned are equipped with a battery which is permanently sealed (except for vent holes) and has no filler caps. Water doesn't have to be added to these batteries at any time. If a maintenance-type battery is installed, the caps on the top of the battery should be removed periodically to check for a low electrolyte level **(see illustration)**. This check is most critical during the warm summer months.

Brake and clutch fluid

Refer to illustrations 4.19a and 4.19b

18   The brake master cylinder is mounted on the upper left of the engine compartment firewall. The clutch master cylinder used on later model manual transmission models is mounted next to the brake master cylinder.

19   The fluid inside can be checked after removing the cover or cap **(see illustrations)**. Be sure to wipe the top of the reservoir cover with a clean rag to prevent contamination of the brake and/or clutch system before removing the cover.

20   When adding fluid, pour it carefully into the reservoir to avoid spilling it on surrounding painted surfaces or onto the radiator grille. Be sure the specified fluid is used, since mixing different types of brake fluid can cause damage to the system. See *Recommended lubricants and fluids* at the front of this Chapter or your owner's manual. **Warning:** *Brake fluid can harm your eyes and damage painted surfaces, so use extreme caution*

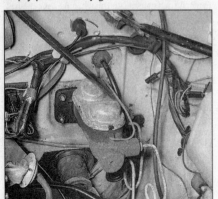

4.19a On earlier models, use a screwdriver to detach the spring retainer, then remove the cover to check the brake fluid level - it should be about 1/4-inch from the top (but not completely full)

14.19b The brake fluid level on later models is easily checked after unscrewing the reservoir caps - the level should be even with the ring at the base of the filler neck

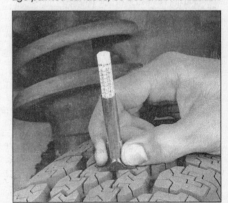

5.2 Use a tire tread depth indicator to monitor tire wear - they are available at auto parts stores and service stations and cost very little

UNDERINFLATION

CUPPING

Cupping may be caused by:
- Underinflation and/or mechanical irregularities such as out-of-balance condition of wheel and/or tire, and bent or damaged wheel.
- Loose or worn steering tie-rod or steering idler arm.
- Loose, damaged or worn front suspension parts.

5.3 Perform a periodic examination of your tires, looking for any of these conditions - see a tire shop or service station if any of the irregularities are present

OVERINFLATION

INCORRECT TOE-IN OR EXTREME CAMBER

FEATHERING DUE TO MISALIGNMENT

when handling or pouring it. Do not use brake fluid that has been standing open or is more than one year old. Brake fluid absorbs moisture from the air. Excess moisture can cause a dangerous loss of brake performance.

21 At this time, the fluid and master cylinder can be inspected for contamination. The system should be drained and refilled if deposits, dirt particles or water droplets are seen in the fluid.

22 After filling the reservoir to the proper level, make sure the cover or cap is on tight to prevent fluid leakage.

23 The brake fluid level in the master cylinder will drop slightly as the pads at the front wheels wear down during normal operation. If the master cylinder requires repeated additions to keep it at the proper level, it's an indication of leakage in the brake system, which should be corrected immediately. Check all brake lines and connections (see Section 19 for more information).

24 If, upon checking the master cylinder fluid level, you discover one or both reservoirs empty or nearly empty, the brake system should be bled (Chapter 9).

5 Tire and tire pressure checks (every 250 miles or weekly)

Refer to illustrations 5.2, 5.3, 5.4a, 5.4b and 5.8

1 Periodic inspection of the tires may spare you the inconvenience of being stranded with a flat tire. It can also provide you with vital information regarding possible problems in the steering and suspension systems before major damage occurs.

2 The original tires on this vehicle are equipped with 1/2-inch wear bands that will appear when tread depth reaches 1/16-inch, but they don't appear until the tires are worn out. Tread wear can be monitored with a simple, inexpensive device known as a tread depth indicator **(see illustration)**.

3 Note any abnormal tread wear **(see illustration)**. Tread pattern irregularities such as cupping, flat spots and more wear on one side than the other are indications of front end alignment and/or balance problems. If any of these conditions are noted, take the vehicle to a tire shop or service station to correct the problem.

4 Look closely for cuts, punctures and embedded nails or tacks. Sometimes a tire will hold air pressure for a short time or leak down very slowly after a nail has embedded itself in the tread. If a slow leak persists,

check the valve stem core to make sure it's tight **(see illustration)**. Examine the tread for an object that may have embedded itself in the tire or for a "plug" that may have begun to leak (radial tire punctures are repaired with a plug that's installed in a puncture). If a puncture is suspected, it can be easily verified by spraying a solution of soapy water onto the puncture area **(see illustration)**. The soapy solution will bubble if there's a leak. Unless the puncture is unusually large, a tire shop or service station can usually repair the tire.

5 Carefully inspect the inner sidewall of each tire for evidence of brake fluid leakage. If you see any, inspect the brakes immediately.

6 Correct air pressure adds miles to the lifespan of the tires, improves mileage and enhances overall ride quality. Tire pressure

5.4a If a tire loses air on a steady basis, check the valve stem core first to make sure it's snug (special inexpensive wrenches are commonly available at auto parts stores)

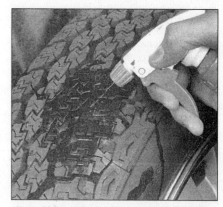

5.4b If the valve stem core is tight, raise the corner of the vehicle with the low tire and spray a soapy water solution onto the tread as the tire is turned slowly - leaks will cause small bubbles to appear

5.8 To extend the life of the tires, check the air pressure at least once a week with an accurate gauge (don't forget the spare!)

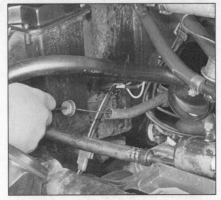

6.3 The automatic transmission dipstick (arrow) is located at the front of the engine compartment, on the right (passenger's) side

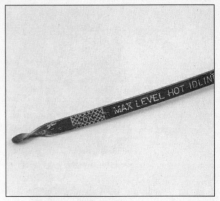

6.6 Check the fluid with the transmission at normal operating temperature - the level should be kept in the OK range

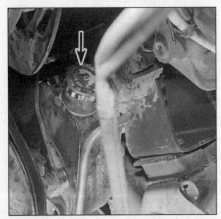

7.2 The power steering fluid reservoir is located near the front of the engine, down low - turn the cap (arrow) counterclockwise to remove it

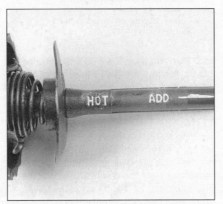

7.6a The power steering fluid dipstick has marks on both sides so the fluid level can be checked when hot . . .

7.6b . . . or cold

cannot be accurately estimated by looking at a tire, especially if it's a radial. A tire pressure gauge is essential. Keep an accurate gauge in the vehicle. The pressure gauges attached to the nozzles of air hoses at gas stations are often inaccurate.

7   Always check tire pressure when the tires are cold. Cold, in this case, means the vehicle has not been driven over a mile in the three hours preceding a tire pressure check. A pressure rise of four to eight pounds is not uncommon once the tires are warm.

8   Unscrew the valve cap protruding from the wheel or hubcap and push the gauge firmly onto the valve stem **(see illustration)**. Note the reading on the gauge and compare the figure to the recommended tire pressure shown on the placard on the driver's side door pillar. Be sure to reinstall the valve cap to keep dirt and moisture out of the valve stem mechanism. Check all four tires and, if necessary, add enough air to bring them up to the recommended pressure.

9   Don't forget to keep the spare tire inflated to the specified pressure (refer to your owner's manual or the tire sidewall).

6   Automatic transmission fluid level check (every 250 miles or weekly)

Refer to illustrations 6.3 and 6.6

1   The automatic transmission fluid level should be carefully maintained. Low fluid level can lead to slipping or loss of drive, while overfilling can cause foaming and loss of fluid.

2   With the parking brake set, start the engine, then move the shift lever through all the gear ranges, ending in Neutral. The fluid level must be checked with the vehicle level and the engine running at idle. **Note:** *Incorrect fluid level readings will result if the vehicle has just been driven at high speeds for an extended period, in hot weather in city traffic, or if it has been pulling a trailer. If any of these conditions apply, wait until the fluid has cooled (about 30 minutes).*

3   With the transmission at normal operating temperature, remove the dipstick from the filler tube. The dipstick is located at the front of the engine compartment on the passenger's side **(see illustration)**.

4   Carefully touch the fluid at the end of the dipstick to determine if it's warm or hot. Wipe

the fluid from the dipstick with a clean rag and push it back into the filler tube until the cap seats.

5   Pull the dipstick out again and note the fluid level.

6   If the fluid felt warm, the level should be between the two dimples **(see illustration)**. If it felt hot, the level should be in the crosshatched area, near the MAX line. If additional fluid is required, add it directly into the tube using a funnel. It takes about one pint to raise the level from the bottom of the crosshatched area to the MAX line with a hot transmission, so add the fluid a little at a time and keep checking the level until it's correct.

7   The condition of the fluid should also be checked along with the level. If the fluid at the end of the dipstick is a dark reddish-brown color, or if it smells burned, it should be changed. If you are in doubt about the condition of the fluid, purchase some new fluid and compare the two for color and smell.

7   Power steering fluid level check (every 3000 miles or 3 months)

Refer to illustrations 7.2, 7.6a and 7.6b

1   Unlike manual steering, the power steering system relies on fluid which may, over a period of time, require replenishing.

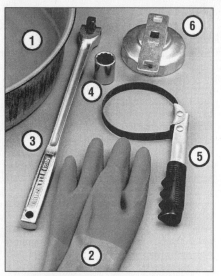

8.3 These tools are required when changing the engine oil and filter

1   **Drain pan** - It should be fairly shallow in depth, but wide to prevent spills
2   **Rubber gloves** - When removing the drain plug and filter, you will get oil on your hands (the gloves will prevent burns)
3   **Breaker bar** - Sometimes the oil drain plug is tight, and a long breaker bar is needed to loosen it
4   **Socket** – To be used with the breaker bar or a ratchet (must be the correct size to fit the drain plug - six-point preferred)
5   **Filter wrench** - This is a metal band-type wrench, which requires clearance around the filter to be effective
6   **Filter wrench** - This type fits on the bottom of the filter and can be turned with a ratchet or breaker bar (different-size wrenches are available for different types of filters)

2   The fluid reservoir for the power steering pump is located on the pump body at the front of the engine **(see illustration)**.
3   For the check, the front wheels should be pointed straight ahead and the engine should be off.
4   Use a clean rag to wipe off the reservoir cap and the area around the cap. This will help prevent any foreign matter from entering the reservoir during the check.
5   Twist off the cap and check the temperature of the fluid at the end of the dipstick with your finger.
6   Wipe off the fluid with a clean rag, reinsert the dipstick, then withdraw it and read the fluid level. The fluid should be at the proper level, depending on whether it was checked hot or cold **(see illustrations)**. Never allow the fluid level to drop below the lower mark on the dipstick.
7   If additional fluid is required, pour the specified type directly into the reservoir, using a funnel to prevent spills.
8   If the reservoir requires frequent fluid

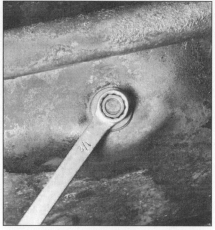

8.9 Use a box-end wrench or socket to remove the oil drain plug (this prevents rounding off the corners of the plug hex)

additions, all power steering hoses, hose connections and the power steering pump should be carefully checked for leaks.

---

8   Engine oil and filter change (every 3000 miles or 3 months)

Refer to illustrations 8.3, 8.9, 8.14 and 8.18

1   Frequent oil changes are the most important preventive maintenance procedures that can be done by the home mechanic. As engine oil ages, it becomes diluted and contaminated, which leads to premature engine wear.
2   Although some sources recommend oil filter changes every other oil change, we feel that the minimal cost of an oil filter and the relative ease with which it is installed dictate that a new filter be installed every time the oil is changed.
3   Gather together all necessary tools and materials before beginning this procedure **(see illustration)**.
4   You should have plenty of clean rags and newspapers handy to mop up any spills. Access to the underside of the vehicle may be improved if the vehicle can be lifted on a hoist, driven onto ramps or supported by jackstands. **Warning:** *Do not work under a vehicle which is supported only by a bumper, hydraulic or scissors-type jack.*
5   If this is your first oil change, get under the vehicle and familiarize yourself with the locations of the oil drain plug and the oil filter. The engine and exhaust components will be warm during the actual work, so note how they are situated to avoid touching them when working under the vehicle.
6   Warm the engine to normal operating temperature. If the new oil or any tools are needed, use this warm-up time to gather everything necessary for the job. The correct type of oil for your application can be found in *Recommended lubricants and fluids* at the beginning of this Chapter.
7   With the engine oil warm (warm engine

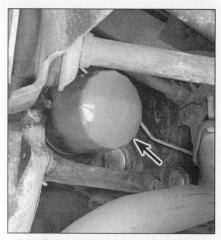

8.14 Since the oil filter is on very tight, you'll need a special wrench for removal - DO NOT use the wrench to tighten the new filter. This photo shows the filter on a small-block V8 engine (the filter is at the right rear of the engine). On six-cylinder and big-block V8 engines, the filter is also on the right (passenger's) side, but near the front of the engine.

oil will drain better and more built-up sludge will be removed with it), raise and support the vehicle. Make sure it's safely supported!
8   Move all necessary tools, rags and newspapers under the vehicle. Set the drain pan under the drain plug. Keep in mind that the oil will initially flow from the pan with some force; position the pan accordingly.
9   Being careful not to touch any of the hot exhaust components, use a wrench to remove the drain plug near the bottom of the oil pan **(see illustration)**. Depending on how hot the oil is, you may want to wear gloves while unscrewing the plug the final few turns.
10   Allow the old oil to drain into the pan. It may be necessary to move the pan as the oil flow slows to a trickle.
11   After all the oil has drained, wipe off the drain plug with a clean rag. Small metal particles may cling to the plug and would immediately contaminate the new oil.
12   Clean the area around the drain plug opening and reinstall the plug. Tighten the plug securely with the wrench. If a torque wrench is available, use it to tighten the plug to the torque listed in this Chapter's Specifications.
13   Move the drain pan into position under the oil filter.
14   Use the filter wrench to loosen the oil filter **(see illustration)**. Chain or metal-band filter wrenches may distort the filter canister, but it doesn't matter, since the filter will be discarded anyway.
15   Completely unscrew the old filter. Be careful; it's full of oil. Empty the oil inside the filter into the drain pan, then lower the filter (V8 and V6 engines) or lift the filter out from above (inline six-cylinder engines).
16   Compare the old filter with the new one to make sure they're the same type.

8.18 Lubricate the oil filter gasket with clean engine oil before installing the filter on the engine

17   Use a clean rag to remove all oil, dirt and sludge from the area where the oil filter mounts to the engine. Check the old filter to make sure the rubber gasket isn't stuck to the engine. If the gasket is stuck to the engine (use a flashlight if necessary), remove it.

18   Apply a light coat of clean oil to the rubber gasket on the new oil filter **(see illustration)**.

19   Attach the new filter to the engine, following the tightening directions printed on the filter canister or packing box. Most filter manufacturers recommend against using a filter wrench due to the possibility of overtightening and damage to the seal.

20   Remove all tools, rags, etc. from under the vehicle, being careful not to spill the oil in the drain pan, then lower the vehicle.

21   Move to the engine compartment and locate the oil filler cap.

22   Pour the fresh oil through the filler opening. A funnel may be helpful.

23   Pour four quarts of fresh oil into the engine. Wait a few minutes to allow the oil to drain into the pan, then check the level on the oil dipstick (see Section 4 if necessary). If the oil level is above the ADD mark, start the engine and allow the new oil to circulate.

24   Run the engine for only about a minute and then shut it off. Immediately look under the vehicle and check for leaks at the oil pan drain plug and around the oil filter. If either is leaking, tighten with a bit more force.

25   With the new oil circulated and the filter now completely full, recheck the level on the dipstick and add more oil as necessary.

26   During the first few trips after an oil change, make it a point to check frequently for leaks and proper oil level.

27   The old oil drained from the engine cannot be reused in its present state and should be disposed of. Oil reclamation centers, auto repair shops and gas stations will normally accept the oil, which can be refined and used again. After the oil has cooled it can be drained into a container (capped plastic jugs, topped bottles, milk cartons, etc.) for transport to one of these disposal sites. Don't dispose of the oil by pouring it on the ground or down a drain!

9   Battery check, maintenance and charging (every 7500 miles or 6 months)

Refer to illustrations 9.1, 9.6a, 9.6b, 9.7a and 9.7b

Warning: *Certain precautions must be followed when checking and servicing the battery. Hydrogen gas, which is highly flammable, is always present in the battery cells, so keep lighted tobacco and all other open flames and sparks away from the battery. The electrolyte inside the battery is actually dilute sulfuric acid, which will cause injury if splashed on your skin or in your eyes. It will also ruin clothes and painted surfaces. When removing the battery cables, always detach the negative cable first and hook it up last!*

1   A routine preventive maintenance program for the battery in your vehicle is the only way to ensure quick and reliable starts. But before performing any battery maintenance, make sure that you have the proper equipment necessary to work safely around the battery **(see illustration)**.

2   There are also several precautions that should be taken whenever battery maintenance is performed. Before servicing the battery, always turn the engine and all accessories off and disconnect the cable from the negative terminal of the battery.

3   The battery produces hydrogen gas, which is both flammable and explosive. Never create a spark, smoke or light a match around the battery. Always charge the battery in a ventilated area.

4   Electrolyte contains poisonous and corrosive sulfuric acid. Do not allow it to get in your eyes, on your skin on your clothes. Never ingest it. Wear protective safety glasses when working near the battery. Keep children away from the battery.

5   Note the external condition of the battery. If the positive terminal and cable clamp on your vehicle's battery is equipped with a rubber protector, make sure that it's not torn or damaged. It should completely cover the terminal. Look for any corroded or loose connections, cracks in the case or cover or loose hold-down clamps. Also check the entire length of each cable for cracks and frayed conductors.

6   If corrosion, which looks like white, fluffy deposits **(see illustration)** is evident, particularly around the terminals, the battery should be removed for cleaning. Loosen the cable clamp bolts with a wrench, being careful to remove the ground cable first, and slide them off the terminals **(see illustration)**. Then disconnect the hold-down clamp bolt and nut, remove the clamp and lift the battery from the engine compartment.

9.1 Tools and materials required for battery maintenance

1   *Face shield/safety goggles - When removing corrosion with a brush, the acidic particles can easily fly up into your eyes*

2   *Baking soda - A solution of baking soda and water can be used to neutralize corrosion*

3   *Petroleum jelly - A layer of this on the battery posts will help prevent corrosion*

4   *Battery post/cable cleaner - This wire brush cleaning tool will remove all traces of corrosion from the battery posts and cable clamps*

5   *Treated felt washers - Placing one of these on each post, directly under the cable clamps, will help prevent corrosion*

6   *Puller - Sometimes the cable clamps are very difficult to pull off the posts, even after the nut/bolt has been completely loosened. This tool pulls the clamp straight up and off the post without damage*

7   *Battery post/cable cleaner - Here is another cleaning tool which is a slightly different version of Number 4 above, but it does the same thing*

8   *Rubber gloves - Another safety item to consider when servicing the battery; remember that's acid inside the battery!*

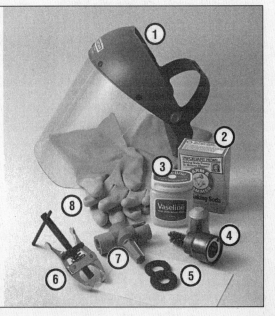

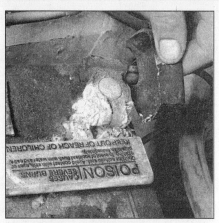

9.6a Battery terminal corrosion usually appears as light, fluffy powder

9.6b Removing the cable from a battery post with a wrench - sometimes special battery pliers are required for this procedure if corrosion has caused deterioration of the nut hex (always remove the ground cable first and hook it up last!)

9.7a When cleaning the cable clamps, all corrosion must be removed (the inside of the clamp is tapered to match the taper on the post, so don't remove too much material)

7   Clean the cable clamps thoroughly with a battery brush or a terminal cleaner and a solution of warm water and baking soda **(see illustration)**. Wash the terminals and the top of the battery case with the same solution but make sure that the solution doesn't get into the battery. When cleaning the cables, terminals and battery top, wear safety goggles and rubber gloves to prevent any solution from coming in contact with your eyes or hands. Wear old clothes too - even diluted, sulfuric acid splashed onto clothes will burn holes in them. If the terminals have been extensively corroded, clean them up with a terminal cleaner **(see illustration)**. Thoroughly wash all cleaned areas with plain water.

8   Make sure that the battery tray is in good condition and the hold-down clamp bolts are tight. If the battery is removed from the tray, make sure no parts remain in the bottom of the tray when the battery is reinstalled. When reinstalling the hold-down clamp bolts, do not overtighten them.

9   Any metal parts of the vehicle damaged by corrosion should be covered with a zinc-based primer, then painted.

10   Information on removing and installing the battery can be found in Chapter 5. Information on jump starting can be found at the front of this manual. For more detailed battery checking procedures, refer to the *Haynes Automotive Electrical Manual*.

Charging

Warning: *When batteries are being charged, hydrogen gas, which is very explosive and flammable, is produced. Do not smoke or allow open flames near a charging or a recently charged battery. Wear eye protection when near the battery during charging. Also, make sure the charger is unplugged before connecting or disconnecting the battery from the charger.*
Note: *The manufacturer recommends the battery be removed from the vehicle for charging because the gas that escapes during this procedure can damage the paint or interior, depending on the model. Fast charging with the battery cables connected can result in damage to the electrical system.*

11   Slow-rate charging is the best way to restore a battery that's discharged to the point where it will not start the engine. It's also a good way to maintain the battery charge in a vehicle that's only driven a few miles between starts. Maintaining the battery charge is particularly important in the winter when the battery must work harder to start the engine and electrical accessories that drain the battery are in greater use.

12   It's best to use a one or two-amp battery charger (sometimes called a "trickle" charger). They are the safest and put the least strain on the battery. They are also the least expensive. For a faster charge, you can use a higher amperage charger, but don't use one rated more than 1/10th the amp/hour rating of the battery. Rapid boost charges that claim to restore the power of the battery in one to two hours are hardest on the battery and can damage batteries not in good condition. This type of charging should only be used in emergency situations.

13   The average time necessary to charge a battery should be listed in the instructions that come with the charger. As a general rule, a trickle charger will charge a battery in 12 to 16 hours.

14   Remove all the cell caps (if equipped) and cover the holes with a clean cloth to prevent spattering electrolyte. Disconnect the negative battery cable and hook the battery charger cable clamps up to the battery posts (positive to positive, negative to negative), then plug in the charger. Make sure it is set at 12-volts if it has a selector switch.

15   If you're using a charger with a rate higher than two amps, check the battery regularly during charging to make sure it doesn't overheat. If you're using a trickle charger, you can safely let the battery charge overnight after you've checked it regularly for the first couple of hours.

16   If the battery has removable cell caps, measure the specific gravity with a hydrome-

9.7b Regardless of the type of tool used on the battery posts, a clean, shiny surface should be the result

ter every hour during the last few hours of the charging cycle. Hydrometers are available inexpensively from auto parts stores - follow the instructions that come with the hydrometer. Consider the battery charged when there's no change in the specific gravity reading for two hours and the electrolyte in the cells is gassing (bubbling) freely. The specific gravity reading from each cell should be very close to the others. If not, the battery probably has a bad cell(s).

17   Some batteries with sealed tops have built-in hydrometers on the top that indicate the state of charge by the color displayed in the hydrometer window. Normally, a bright-colored hydrometer indicates a full charge and a dark hydrometer indicates the battery still needs charging.

18   If the battery has a sealed top and no built-in hydrometer, you can hook up a digital voltmeter across the battery terminals to check the charge. A fully charged battery should read 12.6 volts or higher.

19   Further information on the battery and jump starting can be found in Chapter 5 and at the front of this manual.

Check for a chafed area that could fail prematurely.

Check for a soft area indicating the hose has deteriorated inside.

Overtightening the clamp on a hardened hose will damage the hose and cause a leak.

Check each hose for swelling and oil-soaked ends. Cracks and breaks can be located by squeezing the hose.

10.4 Hoses, like drivebelts, have a habit of failing at the worst possible time - to prevent the inconvenience of a blown radiator or heater hose, inspect them carefully as shown here

10   Cooling system check (every 7500 miles or 6 months)

Refer to illustration 10.4

1   Many major engine failures can be attributed to a faulty cooling system. If the vehicle is equipped with an automatic transmission, the cooling system also cools the transmission fluid and thus plays an important role in prolonging transmission life.

2   The cooling system should be checked with the engine cold. Do this before the vehicle is driven for the day or after it has been shut off for at least three hours.

3   Remove the radiator cap by turning it to the left until it reaches a stop. If you hear a hissing sound (indicating there is still pressure in the system), wait until this stops. Now press down on the cap with the palm of your hand and continue turning to the left until the cap can be removed. Thoroughly clean the cap,

inside and out, with clean water. Also clean the filler neck on the radiator. All traces of corrosion should be removed. The coolant inside the radiator should be relatively transparent. If it is rust colored, the system should be drained and refilled (see Section 35). If the coolant level is not up to the top, add additional antifreeze/coolant mixture (see Section 4).

4   Carefully check the large upper and lower radiator hoses along with the smaller diameter heater hoses which run from the engine to the firewall. Inspect each hose along its entire length, replacing any hose which is cracked, swollen or shows signs of deterioration. Cracks may become more apparent if the hose is squeezed **(see illustration)**. Regardless of condition, it's a good idea to replace hoses with new ones every two years.

5   Make sure all hose connections are tight. A leak in the cooling system will usually show up as white or rust colored deposits on the areas adjoining the leak. If wire-type clamps are used at the ends of the hoses, it may be a good idea to replace them with more secure screw-type clamps.

6   Use compressed air or a soft brush to remove bugs, leaves, etc. from the front of the radiator or air conditioning condenser. Be careful not to damage the delicate cooling fins or cut yourself on them.

7   Every other inspection, or at the first indication of cooling system problems, have the cap and system pressure tested. If you don't have a pressure tester, most gas stations and repair shops will do this for a minimal charge.

11   Underhood hose check and replacement (every 7500 miles or 6 months)

Refer to illustration 11.1

General

1   **Caution:** *Replacement of air conditioning hoses must be left to a dealer service department or air conditioning shop that has the equipment to depressurize the system safely. Never remove air conditioning components or hoses* **(see illustration)** *until the system has been depressurized.*

2   High temperatures in the engine compartment can cause the deterioration of the rubber and plastic hoses used for engine, accessory and emission systems operation. Periodic inspection should be made for cracks, loose clamps, material hardening and leaks. Information specific to the cooling system hoses can be found in Section 10.

3   Some, but not all, hoses are secured to the fittings with clamps. Where clamps are used, check to be sure they haven't lost their tension, allowing the hose to leak. If clamps aren't used, make sure the hose has not expanded and/or hardened where it slips over the fitting, allowing it to leak.

Vacuum hoses

4   It's quite common for vacuum hoses,

especially those in the emissions system, to be color coded or identified by colored stripes molded into them. Various systems require hoses with different wall thicknesses, collapse resistance and temperature resistance. When replacing hoses, be sure the new ones are made of the same material.

5   Often the only effective way to check a hose is to remove it completely from the vehicle. If more than one hose is removed, be sure to label the hoses and fittings to ensure correct installation.

6   When checking vacuum hoses, be sure to include any plastic T-fittings in the check. Inspect the fittings for cracks and the hose where it fits over the fitting for distortion, which could cause leakage.

7   A small piece of vacuum hose (1/4-inch inside diameter) can be used as a stethoscope to detect vacuum leaks. Hold one end of the hose to your ear and probe around vacuum hoses and fittings, listening for the "hissing" sound characteristic of a vacuum leak. **Warning:** *When probing with the vacuum hose stethoscope, be very careful not to come into contact with moving engine components such as the drivebelt, cooling fan, etc.*

Fuel hose

Warning: *There are certain precautions which must be taken when inspecting or servicing fuel system components. Work in a well ventilated area and do not allow open flames (cigarettes, appliance pilot lights, etc.) or bare light bulbs near the work area. Mop up any spills immediately and do not store fuel soaked rags where they could ignite. On fuel-injected models, the fuel system is under high pressure, so if any fuel lines are to be disconnected, the pressure in the system must be relieved first (see Chapter 4 for more information).*

8   Check all rubber fuel lines for deterioration and chafing. Check especially for cracks in areas where the hose bends and just before fittings, such as where a hose attaches to the fuel filter.

11.1 Air conditioning hoses are identified by the metal tubes used at connections (arrow) - DO NOT disconnect or accidentally damage the air conditioning hoses (the system is under high pressure)

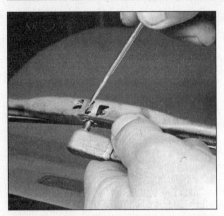

12.6 Use a small screwdriver to pry the release lever up (on some models, you will have to press down on a tab)

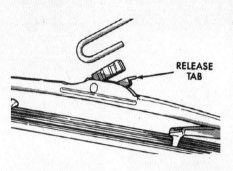

12.7 Disengage the release tab holding the wiper blade holding the wiper arm to the wiper arm

12.8a On some models, you'll have to squeeze the tabs on the end of the blade, then slide the element out of the blade

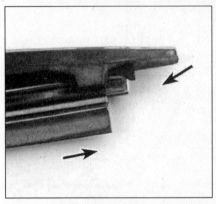

12.8b On other models, lift the end of the element up over the tang on the end of the blade, then slide the element out of the blade

9 High quality fuel line must be used for fuel line replacement. Never, under any circumstances, use unreinforced vacuum line, clear plastic tubing or water hose for fuel lines. On fuel-injected models, use hose made specifically for high-pressure fuel injection systems.

10 Spring-type clamps are commonly used on fuel lines. These clamps often lose their tension over a period of time, and can be "sprung" during removal. Replace all spring-type clamps with screw clamps whenever a hose is replaced.

Metal lines

11 Sections of metal line are often used for fuel line between the fuel pump and carburetor. Check carefully to be sure the line has not been bent or crimped and that cracks have not started in the line.

12 If a section of metal fuel line must be replaced, only seamless steel tubing should be used, since copper and aluminum tubing don't have the strength necessary to withstand normal engine vibration.

13 Check the metal brake lines where they enter the master cylinder and brake proportioning unit (if used) for cracks in the lines or loose fittings. Any sign of brake fluid leakage calls for an immediate thorough inspection of the brake system.

12 Wiper blade inspection and replacement (every 7500 miles or 6 months)

Refer to illustrations 12.6, 12.7, 12.8a and 12.8b

1 The windshield wiper and blade assembly should be inspected periodically for damage, loose components and cracked or worn blade elements.

2 Road film can build up on the wiper blades and affect their efficiency, so they should be washed regularly with a mild detergent solution.

3 The action of the wiping mechanism can loosen the bolts, nuts and fasteners, so they should be checked and tightened, as neces-

sary, at the same time the wiper blades are checked.

4 If the wiper blade elements (sometimes called inserts) are cracked, worn or warped, they should be replaced with new ones.

5 Pull the wiper blade/arm assembly away from the glass.

6 On 1997 and earlier models, insert a small screwdriver into the release slot and pry up or depress the tab (depending on model) to detach the blade assembly. Slide it off the wiper arm and over the retaining stud **(see illustration)**.

7 On 1998 and later models, disengage the release tab holding the wiper blade to the wiper arm **(see illustration)**.

8 On some models, squeeze the tabs on the end, pull the blade out of the bridge claw, then slide the element out of the blade assembly bridge **(see illustration)**. On other models, lift the end of the assembly up and over the tang in the end of the bridge **(see illustration)**.

9 Compare the new element with the old for length, design, etc.

10 Slide the new element into place. It will automatically lock at the correct location.

11 Reinstall the blade assembly on the arm, wet the windshield and check for proper operation.

13 Suspension and steering check (every 7500 miles or 6 months)

Refer to illustration 13.4

1 Indications of a fault in these systems are excessive play in the steering wheel before the front wheels react, excessive sway around corners, body movement over rough roads or binding at some point as the steering wheel is turned.

2 Raise the front of the vehicle periodically and visually check the suspension and steering components for wear. Because of the work to be done, make sure the vehicle cannot fall from the stands.

3 Check the wheel bearings. Do this by spinning the front wheels. Listen for any abnormal noises and watch to make sure the wheel spins true (doesn't wobble). Grab the

top and bottom of the tire and pull in-and-out on it. Notice any movement which would indicate a loose wheel bearing assembly. If the bearings are suspect, refer to Section 34 and Chapter 10 for more information.

4 From under the vehicle, check for loose bolts, broken or disconnected parts and deteriorated rubber bushings on all suspension and steering components **(see illustration)**. Look for fluid leaking from the steering

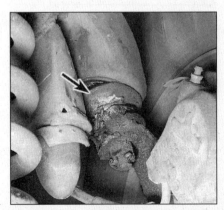

13.4 Inspect the balljoint boots for leaking grease - this one has a split boot and should be replaced

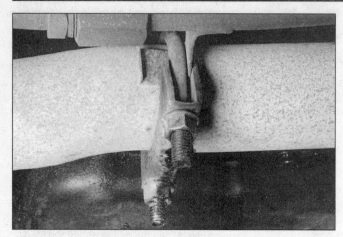

14.4a Check the exhaust system clamp nuts to make sure they are tight

14.4b Inspect the rubber mounts (1) for damage, deterioration and sagging. Also check for rusted holes in the exhaust pipe and muffler (2); this pipe should be repaired or replaced

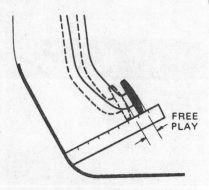

15.1 To determine the clutch pedal freeplay, depress the pedal, stop the moment clutch resistance is felt, then measure the distance (freeplay)

gear assembly. Check the power steering hoses, belts and connections for leaks. Inspect the shock absorbers for fluid leaks, indicating the need for replacement.

5   Have an assistant turn the steering wheel from side-to-side and check the steering components for free movement, chafing and binding. If the steering doesn't react with the movement of the steering wheel, try to determine where the slack is located.

14   Exhaust system check (every 7500 miles or 6 months)

Refer to illustrations 14.4a and 14.4b

1   With the engine cold (at least three hours after the vehicle has been driven), check the complete exhaust system from the manifold to the end of the tailpipe. Be careful around the catalytic converter, which may be hot even after three hours. The inspection should be done with the vehicle on a hoist to permit unrestricted access. If a hoist isn't available, raise the vehicle and support it securely on jackstands.

2   Check the exhaust pipes and connections for signs of leakage and/or corrosion

indicating a potential failure. Make sure that all brackets and hangers are in good condition and tight.

3   Inspect the underside of the body for holes, corrosion, open seams, etc. which may allow exhaust gasses to enter the passenger compartment. Seal all body openings with silicone sealant or body putty.

4   Rattles and other noises can often be traced to the exhaust system, especially the hangers, mounts and heat shields **(see illustrations)**. Try to move the pipes, mufflers and catalytic converter. If the components can come in contact with the body or suspension parts, secure the exhaust system with new brackets and hangers.

15   Clutch pedal freeplay check and adjustment (1987 and earlier models) (every 7500 miles or 6 months)

Refer to illustration 15.1

1   Press down lightly on the clutch pedal and, with a ruler, measure the distance that it moves freely before the clutch resistance is felt **(see illustration)**. The freeplay should be within the specified limits. If it isn't, it must be adjusted.

2   On all models, turn the self-locking adjusting nut on the clutch fork rod to achieve the specified freeplay at the pedal **(see Chapter 8, illustration 3.4)**.

16   Manual transmission lubricant level check (every 75 00 miles or 6 months)

Refer to illustration 16.2

1   The manual transmission has a filler plug which must be removed to check the lubricant level. If the vehicle is raised to gain access to the plug, be sure to support it safely on jack-

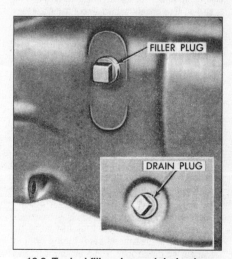

16.2 Typical filler plug and drain plug locations on manual transmissions - use an open-end wrench to unscrew the plugs

stands - DO NOT crawl under a vehicle which is supported only by a jack! Be sure the vehicle is level or the check may be inaccurate.

2   Using an open-end wrench, unscrew the plug from the transmission **(see illustration)** and use your little finger to reach inside the housing to feel the lubricant level. The level should be at or near the bottom of the plug hole.

3   If it isn't, add the recommended lubricant through the plug hole with a syringe or squeeze bottle.

4   Install and tighten the plug and check for leaks after the first few miles of driving.

17   Differential lubricant level check (every 7500 miles or 6 months)

Refer to illustrations 17.2a, 17.2b and 17.3

1   The rear differential has a threaded metal or rubber press-in filler plug which must be removed to check the lubricant level. If the

17.2a On some models, the inspection plug is removed by unscrewing it from the differential case

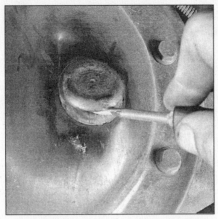

17.2b On later models, use a small screwdriver to pry out the rubber plug

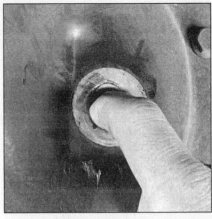

17.3 Use your finger as a dipstick to check the lubricant level

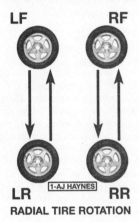

RADIAL TIRE ROTATION

18.2 The recommended tire rotation pattern for these vehicles

19.5a You will find an inspection hole like this in most calipers - placing a ruler across the hole should enable you to determine the thickness of remaining pad material

19.5b Some models don't have an inspection hole, but you can check the pad thickness by looking at the end of the caliper

vehicle is raised to gain access to the plug, be sure to support it safely on jackstands - DO NOT crawl under the vehicle when it's supported only by the jack. Be sure the vehicle is level or the check may not be accurate.

2   Remove the plug from the filler hole in the differential housing or cover (see illustrations).

3   The lubricant level should be at the bottom of the filler hole (see illustration). If not, use a syringe to add the recommended lubricant until it just starts to run out of the opening. On some models a tag is located in the area of the plug which gives information regarding lubricant type, particularly on models equipped with Sure Grip.

4   Install the plug securely into the filler hole.

18 Tire rotation (every 7500 miles or 6 months)

Refer to illustration 18.2

1   The tires should be rotated at the specified intervals and whenever uneven wear is noticed.

2   Refer to the accompanying illustration for the preferred tire rotation pattern.

3   Refer to the information in *Jacking and towing* at the front of this manual for the proper procedures to follow when raising the vehicle and changing a tire. If the brakes are to be checked, don't apply the parking brake as stated. Make sure the tires are blocked to prevent the vehicle from rolling as it's raised.

4   Preferably, the entire vehicle should be raised at the same time. This can be done on a hoist or by jacking up each corner and then lowering the vehicle onto jackstands placed under the frame rails. Always use four jackstands and make sure the vehicle is safely supported.

5   After rotation, check and adjust the tire pressures as necessary and be sure to check the lug nut tightness.

19 Brake check (every 7500 miles or 6 months)

Warning: *Brake system dust may contain asbestos, which is hazardous to your health. DO NOT blow it out with compressed air, inhale it or use gasoline or solvents to remove it. Use brake system cleaner only.*

Note: *For detailed photographs of the brake system, refer to Chapter 9.*

1   In addition to the specified intervals, the brakes should be inspected every time the wheels are removed or whenever a defect is suspected.

2   To check the brakes, raise the vehicle and place it securely on jackstands. Remove the wheels (see *Jacking and towing* at the front of the manual, if necessary).

Disc brakes

Refer to illustrations 19.5a and 19.5b

3   Disc brakes are used on the front wheels on most models. Extensive damage to the disc can occur if the pads are not replaced when needed.

4   The disc brake calipers, which contain the pads, are visible with the wheels removed. There is an outer pad and an inner pad in each caliper. All pads should be inspected.

5   Each caliper has a "window" or opening to inspect the pads. Check the thickness of the pad lining by looking into the caliper at each end and down through the inspection window at the top of the housing (see illustrations). If the pad material has worn to about 5/16-inch or less, the pads should be replaced.

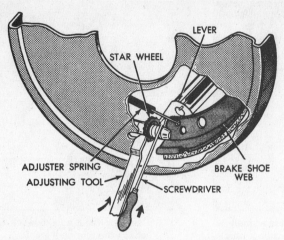

19.11 If the brake drum won't pull off the hub, use a thin screwdriver to push the lever away, then use an adjusting tool or another screwdriver to back off the star wheel

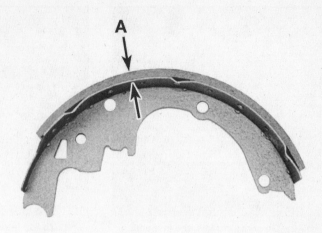

19.13 If the lining is bonded to the brake shoe, measure the lining thickness from the outer surface to the metal shoe, as shown here; if the lining is riveted to the shoe, measure from the lining outer surface to the rivet head

6   If you're unsure about the exact thickness of the remaining lining material, remove the pads for further inspection or replacement (refer to Chapter 9).

7   Before installing the wheels, check for leakage and/or damage (cracks, splitting, etc.) around the brake hose connections. Replace the hose or fittings as necessary, referring to Chapter 9.

8   Check the condition of the disc. Look for score marks, deep scratches and burned spots. If these conditions exist, the hub/disc assembly should be removed for servicing (see Section 34).

Drum brakes

Refer to illustrations 19.11, 19.13 and 19.15

9   On front brakes, remove the drum/hub assembly as described in Section 34. On rear brakes, remove the drum by pulling it off the axle and brake assembly. If this proves difficult, make sure the parking brake is released, then squirt penetrating oil around the center hub areas. Allow the oil to soak in and try to pull the drum off again.

10   If the drum still cannot be pulled off, the parking brake lever will have to be lifted slightly off its stop. This is done by first removing the small plug from the backing plate.

11   With the plug removed, insert a thin screwdriver and lift the adjusting lever off the star wheel, then use an adjusting tool or screwdriver to back off the star wheel several turns **(see illustration)**. This will move the brake shoes away from the drum. If the drum still won't pull off, tap around its inner circumference with a soft-face hammer.

12   With the drum removed, do not touch any brake dust (see the Warning at the beginning of this Section).

13   Note the thickness of the lining material on both the front and rear brake shoes. If the material has worn away to within 1/16-inch of the recessed rivets or metal backing, the shoes should be replaced **(see illustration)**. The shoes should also be replaced if they're cracked, glazed (shiny surface) or contaminated with brake fluid.

14   Make sure that all the brake assembly springs are connected and in good condition.

15   Check the brake components for any signs of fluid leakage. Carefully pry back the rubber cups on the wheel cylinders located at the top of the brake shoes **(see illustration)**. Any leakage is an indication that the wheel cylinders should be overhauled immediately (see Chapter 9). Also check brake hoses and connections for signs of leakage.

16   Wipe the inside of the drum with a clean rag and brake system cleaner or denatured alcohol. Again, be careful not to breathe the dangerous asbestos dust.

17   Check the inside of the drum for cracks, score marks, deep scratches and hard spots, which will appear as small discolorations. If these imperfections cannot be removed with fine emery cloth, the drum must be taken to a machine shop equipped to resurface the drums.

18   If after the inspection process all parts are in good working condition, reinstall the brake drum.

19   Install the wheels and lower the vehicle.

Parking brake

20   The parking brake is operated by a hand lever or a foot pedal and locks the rear brake system. The easiest, and perhaps most obvious method of periodically checking the operation of the parking brake assembly is to stop the vehicle on a steep hill with the parking brake set and the transmission in Neutral. If the parking brake cannot prevent the vehicle from rolling, adjust it (see Chapter 9).

20   Fuel system check (every 7500 miles or 6 months)

Refer to illustration 20.5

Warning: *Gasoline is extremely flammable, so take extra precautions when working on any part of the fuel system. Don't smoke or*

19.15 To check for wheel cylinder leakage, use a small screwdriver to pry the boot away from the cylinder

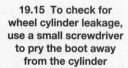

20.5 Check the fuel tank fill connections, as well as all hoses connected to the tank, for cracks or leaks. Leaks can often be found by looking for clean, damp areas around the hoses

21.4 There are usually four carburetor mounting nuts/bolts (arrows) on V8 models, but many six-cylinder models only have two

1

allow open flames or bare light bulbs in or near the work area, and don't work in a garage where a natural gas-type appliance (such as a water heater or clothes dryer) with a pilot light is present. Since gasoline is car-cinogenic, wear latex gloves when there's a possibility of being exposed to fuel, and, if you spill fuel on your skin, rinse it off immedi-ately with soap and water. Have a Class B fire extinguisher on hand.

1   The fuel tank is located under the rear of the vehicle.

2   The fuel system is most easily checked with the vehicle raised on a hoist so the com-ponents underneath the vehicle are readily visible and accessible.

3   If the smell of gasoline is noticed while driving or after the vehicle has been in the sun, the system should be thoroughly inspected immediately.

4   Remove the gas tank cap and check for damage, corrosion and an unbroken sealing imprint on the gasket. Replace the cap with a new one if necessary.

5   With the vehicle raised, check the fuel tank and filler neck for punctures, cracks and other damage (see illustration). The connec-tion between the filler neck and the tank is especially critical. Sometimes a rubber filler neck will leak due to loose clamps or deterio-rated rubber; problems a home mechanic can usually rectify. Warning: Do not, under any cir-cumstances, try to repair a fuel tank yourself (except rubber components). A welding torch or any open flame can easily cause the fuel vapors to explode if the proper precautions are not taken!

6   Carefully check all rubber hoses and metal lines leading away from the fuel tank. Look for loose connections, deteriorated hoses, crimped lines and other damage. Fol-low the lines to the front of the vehicle, care-fully inspecting them all the way. Repair or replace damaged sections as necessary.

7   If a fuel odor is still evident after the inspection, refer to Section 37.

21 Carburetor mounting bolt/nut torque check

Refer to illustration 21.4

1   The carburetor is attached to the top of the intake manifold by bolts and/or nuts. These fasteners can sometimes work loose from vibration and temperature changes dur-ing normal engine operation and cause a vac-uum leak.

2   If you suspect that a vacuum leak exists at the bottom of the carburetor, obtain a length of hose. Start the engine and place one end of the hose next to your ear as you probe around the base with the other end. You will hear a hissing sound if a leak exists (be careful of hot or moving engine compo-nents). See Chapter 2C, Section 6 for more information on finding vacuum leaks.

3   Remove the air cleaner assembly (see Chapter 4), tagging each hose to be discon-nected with a piece of numbered tape to make reassembly easier.

4   Locate the mounting bolts at the base of the carburetor. Decide what special tools or adapters will be necessary, if any, to tighten the fasteners (see illustration).

5   Tighten the bolts or nuts to the torque listed in this Chapter's Specifications. Don't overtighten them, as the threads could strip.

6   If, after the bolts are properly tightened, a vacuum leak still exists, the carburetor must be removed and a new gasket installed. See Chapter 4 for more information.

7   After tightening the fasteners, reinstall the air cleaner and return all hoses to their original positions.

22 Throttle linkage inspection (every 7500 miles or 6 months)

1   Inspect the throttle linkage for damage, missing parts, binding or interference when the accelerator pedal is depressed.

2   Lubricate the various linkage pivot points with chassis grease. Do not lubricate carburetor or throttle body linkage balljoint type pivots, throttle control cables or trans-mission control cables.

23 Drivebelt check, adjustment and replacement (every 7500 miles or 6 months)

Refer to illustrations 23.3, 23.4 and 23.6

1   The drivebelts are located at the front of the engine and play an important role in the overall operation of the vehicle and its com-ponents. Due to their function and material make-up, the belts are prone to failure after a period of time and should be inspected and adjusted periodically to prevent major engine damage.

2   The number of belts used on a particular vehicle depends on the accessories installed. Drivebelts are used to turn the alternator, power steering pump, water pump and air conditioning compressor. Depending on the pulley arrangement, more than one of these components may be driven by a single belt. On later models a single, self-adjusting ser-pentine drivebelt is used to drive all of the components.

Check

Refer to illustrations 23.3, 23.4 and 23.5

3   With the engine off, open the hood and locate the various belts at the front of the engine. Using your fingers (and a flashlight, if necessary), move along the belts checking for cracks and separation of the belt plies. Also check for fraying and glazing, which gives the belt a shiny appearance (see illus-

STREAKED SIDEWALL

FRAYING

CRACKS

SEPARATION

GLAZING

OIL SOAKED

TENSILE BREAK

23.3 Here are some of the more common problems associated with drivebelts (check the belts very carefully to prevent an untimely breakdown)

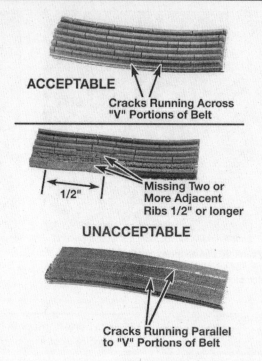

ACCEPTABLE

Cracks Running Across "V" Portions of Belt

1/2"

Missing Two or More Adjacent Ribs 1/2" or longer

UNACCEPTABLE

Cracks Running Parallel to "V" Portions of Belt

23.4 Small cracks in the underside of a V-ribbed belt are acceptable - lengthwise cracks, or missing pieces that cause the belt to make noise, are cause for replacement

tration). Both sides of each belt should be inspected, which means you will have to twist the belt to check the underside.

4   If your vehicle is equipped with a serpentine drivebelt, check the ribs on the underside of the belt. They should all be the same depth, with none of the surface uneven **(see illustration)**.

5   The tension of each V-belt is checked by pushing on the belt at a distance halfway between the pulleys. Push firmly with your thumb and see how much the belt moves (deflects) **(see illustration)**. A rule of thumb is that if the distance from pulley center-to-pulley center is between 7 and 11 inches, the belt should deflect 1/4-inch. If the belt travels between pulleys spaced 12 to 16 inches apart, the belt should deflect 1/2-inch. **Note:** *On serpentine belts, the belt tension is controlled by a tensioner and the belt does not require periodic adjustment.*

Adjustment

Refer to illustration 23.6

Note: *Models with a serpentine drivebelt require no adjustment - belt tension is maintained by a spring-loaded pulley.*

6   If it is necessary to adjust the belt tension, either to make the belt tighter or looser, it is done by moving the belt-driven accessory on the bracket. For each component there is an adjusting bolt and a pivot bolt. Both bolts must be loosened slightly to enable you to move the component **(see illustration)**.

7   After the two bolts have been loosened, move the component away from the engine to tighten the belt or toward the engine to loosen the belt. Hold the accessory in posi-

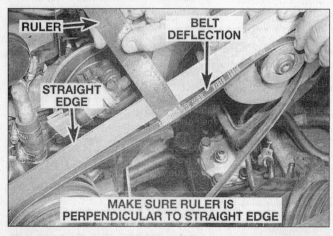

23.5 Measuring drivebelt deflection with a straightedge and ruler

23.6 With the pivot bolt and adjusting bolt (arrow) loose, pivot the alternator up to increase the drivebelt tension, then tighten the pivot bolt and adjusting bolts

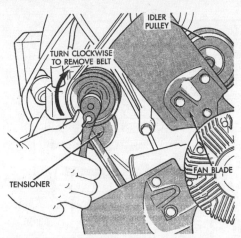

23.13a Rotate the tensioner clockwise to release tension and remove the serpentine drivebelt

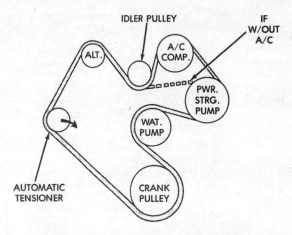

23.13b Serpentine drivebelt routing details

tion and check the belt tension. If it is correct, tighten the two bolts until just snug, then recheck the tension. If the tension is all right, tighten the bolts.

8 To adjust the alternator drivebelt on some later models, loosen the pivot bolt and locknut, then turn the adjusting bolt to tension the belt. After the belt is tensioned, tighten the pivot bolt and locknut.

9 It will often be necessary to use some sort of prybar to move the accessory while the belt is adjusted. If this must be done to gain the proper leverage, be very careful not to damage the component being moved or the part being pried against.

Replacement

Conventional drivebelt

10 To replace a belt, follow the above procedures for drivebelt adjustment but slip the belt off the pulleys and remove it. Since belts tend to wear out more or less at the same time, it's a good idea to replace all of them at the same time. Mark each belt and the corresponding pulley grooves so the replacement belts can be installed properly.

11 Take the old belts with you when purchasing new ones in order to make a direct comparison for length, width and design.

12 Adjust the belts as described earlier in this Section.

Serpentine drivebelt

Refer to illustrations 23.13a and 23.13b

Note: *If the indicator mark on the automatic tensioner is outside the operating range, the belt must be replaced.*

13 Use a ratchet and a socket on the tensioner pulley bolt to rotate the tensioner clockwise and release the tension on the belt **(see illustration)**. Remove the belt and carefully release the tensioner. Route the new belt over the pulleys, again rotating the tensioner to allow the belt to be installed, then release the tensioner. Make sure the belt is routed properly because improper routing can cause the water pump to rotate backwards, causing overheat-

ing **(see illustration)**. The belt routing diagram is usually located in the engine compartment.

24 Seat belt check (every 7500 miles or 6 months)

1 Check the seat belts, buckles, latch plates and guide loops for any obvious damage or signs of wear.

2 On later models, make sure the seat belt reminder light comes on when the key is turned on.

3 The seat belts are designed to lock up during a sudden stop or impact, yet allow free movement during normal driving. The retractors should hold the belt against your chest while driving and rewind the belt when the buckle is unlatched.

4 If any of the above checks reveal problems with the seat belt system, replace parts as necessary.

25 Neutral start switch check (models with automatic transmission) (every 7500 miles or 6 months)

Warning: *During the following checks there is a chance the vehicle could lunge forward, possibly causing damage or injuries. Allow plenty of room around the vehicle, apply the parking brake firmly and depress the regular brake pedal during the checks.*

1 The neutral start switch prevents the engine from being cranked unless the gear selector is in Park or Neutral.

2 Try to start the vehicle in each gear. The engine should crank only in Park or Neutral. If it cranks in any other gear, the neutral start switch is faulty or in need of adjustment (see Chapter 7 Part B).

3 Make sure the steering column lock allows the key to go into the Lock position only when the shift lever is in Park.

4 The ignition key should come out only in the Lock position.

26 Chassis lubrication (every 15,000 miles or 12 months)

Refer to illustrations 26.1, 26.2 and 26.6

1 Refer to *Recommended lubricants and fluids* at the front of this Chapter to obtain the necessary grease, etc. You'll also need a grease gun **(see illustration)**. Occasionally

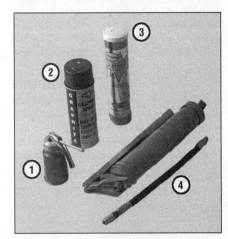

26.1 Materials required for chassis and body lubrication

1 **Engine oil** - *Light engine oil in a can like this can be used for door and hood hinges*

2 **Graphite spray** - *Used to lubricate lock cylinders*

3 **Grease** - *Grease, in a variety of types and weights, is available for use in a grease gun. Check the Specifications for your requirements*

4 **Grease gun** - *A common grease gun, shown here with a detachable hose and nozzle, is needed for chassis lubrication. After use, clean it thoroughly!*

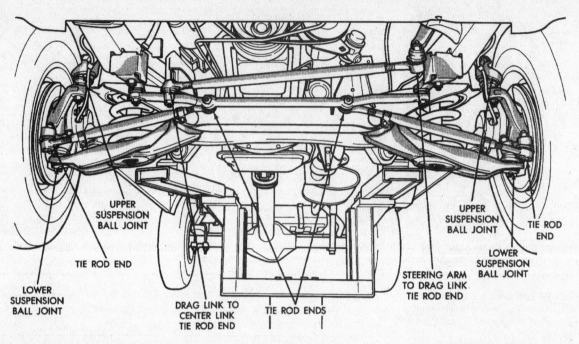

26.2 Here are the typical grease fitting locations - the suspension balljoints usually require more grease than the steering linkage components

plugs will be installed rather than grease fittings. If so, grease fittings will have to be purchased and installed.

2   Look under the vehicle and locate the grease fittings **(see illustration)**. Some models with manual transmissions have grease fittings on the shift linkage which should be lubricated whenever the chassis is lubricated.

3   For easier access under the vehicle, raise it with a jack and place jackstands under the frame. Make sure it's safely supported by the stands. If the wheels are to be removed at this interval for tire rotation or brake inspection, loosen the lug nuts slightly while the vehicle is still on the ground.

4   Before beginning, force a little grease out of the nozzle to remove any dirt from the end

26.6 Pump the grease into each suspension balljoint fitting until the rubber seal is firm to the touch; don't pump in too much grease or you'll rupture the seal!

of the gun. Wipe the nozzle clean with a rag.

5   With the grease gun and plenty of clean rags, crawl under the vehicle and begin lubricating the components.

6   Wipe one of the grease fitting nipples clean and push the nozzle firmly over it. Pump the gun until the component is completely lubricated. On balljoints, stop pumping when the rubber seal is firm to the touch **(see illustration)**. Do not pump too much grease into the fitting as it could rupture the seal. For all other suspension and steering components, continue pumping grease into the fitting until it oozes out of the joint between the two components. If it escapes around the grease gun nozzle, the nipple is clogged or the nozzle is not completely seated on the fitting. Resecure the gun nozzle to the fitting and try again. If necessary, replace the fitting with a new one.

7   Wipe the excess grease from the components and the grease fitting. Repeat the procedure for the remaining fittings.

8   Also clean and lubricate the parking brake cable, along with the cable guides and levers. This can be done by smearing some of the chassis grease onto the cable and its related parts with your fingers.

---

27   Ignition point check and replacement (1971 and 1972 models) (every 15,000 miles or 12 months)

Refer to illustrations 27.1 and 27.16
Caution: *The ignition key should be in the OFF position during this procedure.*

1   The ignition points must be replaced at regular intervals on vehicles not equipped with electronic ignition. Occasionally, the rubbing block on the points will wear sufficiently to require readjustment. Several special tools are required to replace the points **(see illustration)**.

2   After removing the distributor cap and rotor, the points and condenser are plainly visible. The points may be examined by gently prying them open to reveal the condition of the contact surfaces. If they are rough, pitted or dirty, they should be replaced. **Caution:** *The following procedure requires the removal and installation of small screws which can easily fall into the distributor. To retrieve them, the distributor would have to be removed and disassembled. Use a magnetic or spring-loaded screwdriver and exercise caution.*

3   To replace the condenser, which should be replaced along with the points, remove the screw that secures the condenser to the point plate. Loosen the nut or screw retaining the condenser wire and the primary wire to the point assembly. Remove the condenser and mounting bracket.

4   Remove the point assembly mounting screw.

5   Attach the new point assembly to the distributor plate with the mounting screw.

6   Install the new condenser and tighten the mounting screw.

7   Attach the primary ignition wire and the condenser lead to the point assembly. Make sure that the forked connectors for the primary ignition and condenser leads do not touch the distributor plate or any other grounded surface.

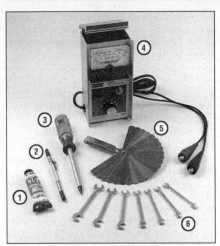

27.1 Tools and materials needed for ignition point replacement and dwell angle adjustment

1 **Distributor cam lube** - Sometimes this special lubricant comes with the new points; however, its a good idea to buy a tube and have it on hand

2 **Screw starter** - This tool has special claws which hold the screw securely as it is started, which helps prevent accidental dropping of the screw

3 **Magnetic screwdriver** - Serves the same purpose as 2 above. If you do not have one of these special screwdrivers, you risk dropping the point mounting screws down into the distributor body

4 **Dwell meter** - A dwell meter is the only accurate way to determine the point setting (gap). Connect the meter according to the instructions supplied with it

5 **Blade-type feeler gauges** - These are required to set the initial point gap (space between the points when they are open)

6 **Ignition wrenches** - These special wrenches are made to work within the tight confines of the distributor. Specifically, they are needed to loosen the nut/bolt which secures the leads to the points

8 Two adjusting methods are available for setting the points. The first and most effective method involves an instrument called a dwell meter.

9 Connect one dwell meter lead to the primary ignition terminal of the point assembly or to the distributor terminal of the coil. Connect the other lead of the dwell meter to an engine ground. Some dwell meters may have different connecting instructions, so always follow the instrument manufacturer's directions.

10 Have an assistant crank the engine over or use a remote starter and crank the engine with the ignition switch turned to On.

11 Note the dwell reading on the meter and compare it to the Specifications found at the front of this chapter or on the engine tune-up

27.16 If a dwell meter is not available, a feeler gauge can be used to adjust the point gap (which is equivalent to setting the dwell angle); however, you must be very precise when using this method, and, even then, you probably won't get the accuracy of using a dwell meter (dwell meters are quite inexpensive)

decal. If the dwell reading is incorrect, adjust it by first loosening the point assembly mounting screw a small amount.

12 Move the point assembly plate with a screwdriver inserted into the slot provided next to the points. Closing the gap on the points will increase the dwell reading, while opening the gap will decrease the dwell.

13 Tighten the screw after the correct reading is obtained and recheck the setting.

14 If a dwell meter is unavailable, a feeler gauge can be used to set the points.

15 Have a helper crank the engine in short bursts until the rubbing block of the points rests on a high point of the cam assembly. The rubbing block must be exactly on the apex of one of the cam lobes for correct point adjustment. It may be necessary to rotate the front pulley of the engine with a socket and breaker bar to position the cam lobe exactly.

16 Measure the gap between the contact points with the correct size feeler gauge **(see illustration)**. If the gap is incorrect, loosen the mounting screw and move the point assembly until the correct gap is achieved.

17 Tighten the screw and check the gap one more time.

18 Apply a small amount of distributor cam lube to the point cam on the distributor shaft.

28 Valve clearance check and adjustment (1980 and earlier six-cylinder models) (every 15,000 miles or 12 months)

Refer to illustration 28.4
Note: *Although the factory-recommended procedure requires the engine to be running while the valve clearances are checked/adjusted, the following procedure is easier, safer and less*

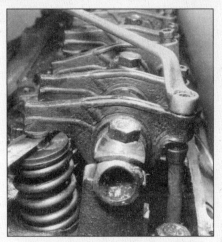

28.4 Be sure to use the correct feeler gauge when adjusting the valves - intake and exhaust valves require different clearances

messy and will produce the same results.

1 Make sure the engine is at normal operating temperature before beginning this procedure.

2 Refer to Chapter 2A and remove the valve cover, then position the number one piston at TDC on the compression stroke (the procedure is in Chapter 2, Part C).

3 Slip the specified size feeler gauge between the number one cylinder intake valve stem and the rocker arm (the intake valves are directly opposite the intake manifold runners, while the exhaust valves are opposite the exhaust manifold runners). It should fit between them with a slight amount of drag if the clearance is correct.

4 If the feeler gauge will not fit, or if it fits loosely, turn the adjusting screw with a box-end wrench until you can feel a slight drag on the feeler gauge as it is withdrawn **(see illustration)**. **Note:** *The rocker arm adjusting screws have interference threads and no locknut, so always use the correct size box-end wrench to turn them (a six-point wrench is preferred).*

5 Repeat the procedure for the number one cylinder exhaust valve. Be sure to use the correct size feeler gauge - intake and exhaust valves require different clearances (see the Specifications at the front of this chapter).

6 Turn the crankshaft 120-degrees to position the next piston in the firing order at TDC on the compression stroke. The ignition rotor in the distributor will be pointing at the appropriate terminal in the cap. Refer to the Specifications at the front of this chapter for the firing order.

7 Adjust the valves, then repeat the procedure to position the next piston in the firing order at TDC.

8 Continue working through the firing order until all valves for all cylinders have been adjusted, then reinstall the rocker arm cover.

9 Start the engine and check for oil leaks at the valve cover-to-cylinder head junction.

1

29.2 Remove the air cleaner
wing nut (arrow)

29.4 Lift the element out of the
air cleaner housing

29   Air filter replacement (every 30,000 miles or 24 months)

Refer to illustrations 29.2 and 29.4

1   At the specified intervals, the air filter element should be replaced with a new one.
2   The filter is located on top of, or adjacent to the carburetor or throttle body, and is replaced by unscrewing the wing nut from the top of the filter housing and lifting off the cover **(see illustration)**. **Note:** *The engine cover must be removed first.*
3   While the top plate is off, be careful not to drop anything down into the carburetor or throttle body.
4   Lift the air filter element out of the housing **(see illustration)** and wipe out the inside of the air cleaner housing with a clean rag.
5   Place the new filter element in the air cleaner housing. Make sure it seats properly in the bottom of the housing.
6   Install the top plate and tighten the wing nut securely.

30   Automatic transmission fluid and filter change (every 30,000 miles or 24 months)

Refer to illustrations 30.6, 30.9, 30.11a and 30.11b

1   At the specified intervals, the transmission fluid should be drained and replaced. Since the fluid will remain hot long after driving, perform this procedure only after the engine has cooled down completely. The manufacturer also recommends adjusting the transmission bands at this time, since this procedure requires removing the fluid pan (see Section 31).
2   Before beginning work, purchase the specified transmission fluid (see *Recommended lubricants and fluids* at the front of this Chapter) and a new filter.
3   Other tools necessary for this job include a floor jack, jackstands to support the vehicle in a raised position, a drain pan capable of holding at least eight pints, newspapers and clean rags.

4   Raise the vehicle and support it securely on jackstands.
5   Place the drain pan underneath the transmission pan. Remove the rear and side pan mounting bolts, but only loosen the front pan bolts approximately four turns.
6   Carefully pry the transmission pan loose with a screwdriver, allowing the fluid to drain **(see illustration)**.
7   Remove the remaining bolts, pan and gasket. Carefully clean the gasket surface of the transmission to remove all traces of the old gasket and sealant.
8   Drain the fluid from the transmission pan, clean the pan with solvent and dry it with compressed air, if available.
9   Remove the filter from the valve body inside the transmission **(see illustration)**. Use a gasket scraper to remove any traces of old gasket material that remain on the valve body. **Note:** *Be very careful not to gouge the delicate aluminum gasket surface on the valve body.*
10   Install a new gasket and filter. On many replacement filters, the gasket is attached to the filter to simplify installation.
11   Make sure the gasket surface on the transmission pan is clean, then install a new gasket on the pan **(see illustration)**. Put the pan in place against the transmission and, working around the pan, tighten each bolt a little at a time to the torque listed in this Chapter's Specifications **(see illustration)**.
12   Lower the vehicle and add approximately seven pints of the specified type of automatic transmission fluid through the filler tube (see Section 6).
13   With the transmission in Park and the parking brake set, run the engine at a fast idle, but don't race it.
14   Move the gear selector through each range and back to Park. Check the fluid level. It will probably be low. Add enough fluid to bring the level between the two dimples on the dipstick.
15   Check under the vehicle for leaks during the first few trips. Check the fluid level again when the transmission is hot (see Section 6).

30.6 With the front bolts in place but loose, pull the rear of the
pan down to drain the fluid

30.9 Remove the three filter screws and lower the filter from the
transmission valve body - some more fluid will drain when the
filter is removed

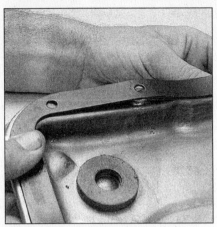

30.11a Place a new gasket in position on the pan and install two bolts to hold it in place - some later gaskets, like the one shown here, have two holes slightly smaller than the others so the bolts can grip the gasket securely

30.11b Hold the pan in place and install all of the bolts snugly before tightening them fully

31 Automatic transmission band adjustment (every 30,000 miles or 24 months)

Refer to illustrations 31.2 and 31.7

1   The transmission bands should be adjusted at the specified interval when the transmission fluid and filter are being replaced (see Section 30).

Kickdown band

2   The kickdown band adjusting screw is located on the left side of the transmission **(see illustration)**.
3   Raise the front of the vehicle and support it securely on jackstands.
4   Loosen the adjusting screw locknut approximately five turns, then loosen the adjusting screw a few turns. Make sure the adjusting screw turns freely, with no binding; lubricate it with penetrating oil if necessary.
5   Tighten the adjusting screw to the torque listed in the Specifications section at

the beginning of this Chapter, then back it off the number of turns listed in the Specifications. Hold the screw from turning, then tighten the locknut securely.

Low-Reverse (rear) band

6   To gain access to the Low-Reverse band, the fluid pan must be removed (see Section 30).
7   Loosen the adjusting screw locknut and back it off four turns **(see illustration)**. Make sure the screw turns freely in the lever.
8   Tighten the adjusting screw to the torque listed in the Specifications section at the beginning of this Chapter, then back it off the number of turns listed in the Specifications. Hold the screw from turning, then tighten the locknut securely.
9   Install the transmission fluid pan.

32 Manual transmission lubricant change (every 30,000 miles or 24 months)

1   Raise the vehicle and support it securely on jackstands.

2   Move a drain pan, rags, newspapers and wrenches under the transmission.
3   Remove the transmission drain plug at the bottom of the case and allow the lubricant to drain into the pan **(see illustration 16.2)**.
4   After the lubricant has drained completely, reinstall the plug and tighten it securely.
5   Remove the fill plug from the side of the transmission case. Using a hand pump, syringe or funnel, fill the transmission with the specified lubricant until it begins to leak out through the hole. Reinstall the fill plug and tighten it securely.
6   Lower the vehicle.
7   Drive the vehicle for a short distance, then check the drain and fill plugs for leakage.

33 Differential lubricant change (every 30,000 miles or 24 months)

All models

Refer to illustration 33.3

1   This procedure should be performed after the vehicle has been driven so the lubricant will be warm and therefore will flow out of the differential more easily.
2   Raise the vehicle and support it securely on jackstands. If the differential has a bolt-on cover at the rear, it is usually easiest to remove the cover to drain the lubricant (which will also allow you to inspect the differential). If there's no bolt-on pan, look for a drain plug at the bottom of the differential housing. If there's not a drain plug and no cover, you'll have to remove the lubricant through the filler plug hole with a suction pump. If you'll be draining the lubricant by removing the cover or a drain plug, move a drain pan, rags, newspapers and wrenches under the vehicle.
3   Remove the filler plug from the differential (see Section 17). If a suction pump is being used, insert the flexible hose. Work the hose down to the bottom of the differential

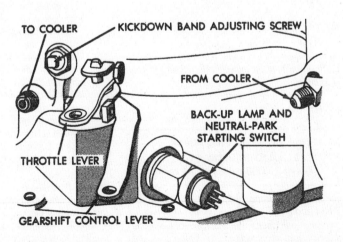

31.2 Location of the kickdown band adjusting screw

TO COOLER

KICKDOWN BAND ADJUSTING SCREW

FROM COOLER

BACK-UP LAMP AND NEUTRAL-PARK STARTING SWITCH

THROTTLE LEVER

GEARSHIFT CONTROL LEVER

31.7 Loosen the Low-Reverse band locknut (arrow)

33.3 This is the easiest way to remove the lubricant: Work the end of the hose to the bottom of the differential housing and draw out the old lubricant with a suction pump. Since this differential has a bolt-on cover, the preferred method is to unbolt and remove the cover, as shown in the following illustrations

33.4c After the lubricant has drained, remove the bolts and the cover

housing and pump the lubricant out **(see illustration)**. If you'll be draining the lubricant through a drain plug, remove the plug and allow the lubricant to drain into the pan, then reinstall the drain plug.

Models where the rear cover must be removed

Refer to illustrations 33.4a, 33.4b, 33.4c and 33.6

4   If the differential is being drained by removing the cover plate, remove the bolts on the lower half of the plate **(see illustration)**. Loosen the bolts on the upper half and use them to keep the cover loosely attached **(see illustration)**. Allow the oil to drain into the pan, then completely remove the cover **(see illustration)**.

5   Using a lint-free rag, clean the inside of the cover and the accessible areas of the differential housing. As this is done, check for chipped gears and metal particles in the lubricant, indicating that the differential should be more thoroughly inspected and/or repaired.

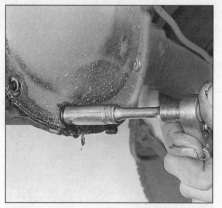

33.4a Remove the bolts from the lower edge of the cover . . .

33.6 Carefully scrape the old gasket material off to ensure a leak-free seal

6   Thoroughly clean the gasket mating surfaces of the differential housing and the cover plate. Use a gasket scraper or putty knife to remove all traces of the old gasket **(see illustration)**.
7   Apply a thin layer of RTV sealant to the cover flange, then press a new gasket into position on the cover. Make sure the bolt holes align properly.
8   Place the cover on the differential housing and install the bolts. Tighten the bolts securely.

All models

9   Use a hand pump, syringe or funnel to fill the differential housing with the specified lubricant until it's level with the bottom of the plug hole.
10   Install the filler plug and make sure it is secure.

34 Front wheel bearing check, repack and adjustment (every 30,000 miles or 24 months)

Refer to illustrations 34.1, 34.6, 34.7, 34.8, 34.11 and 34.15

1   In most cases the front wheel bearings will not need servicing until the brake pads are

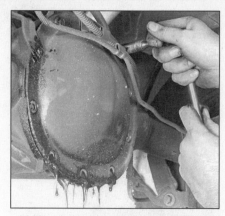

33.4b . . . then loosen the top bolts and let the lubricant drain out

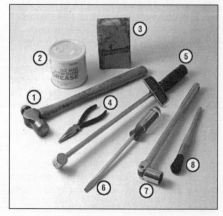

34.1 Tools and materials needed for front wheel bearing maintenance

1   *Hammer - A common hammer will do just fine*
2   *Grease - High-temperature grease that is formulated for front wheel bearings should be used*
3   *Wood block - If you have a scrap piece of 2x4, it can be used to drive the new seal into the hub*
4   *Needle-nose pliers - Used to straighten and remove the cotter pin in the spindle*
5   *Torque wrench - This is very important in this procedure; if the bearing is too tight, the wheel won't turn freely - if it's too loose, the wheel will "wobble" on the spindle. Either way, it could mean extensive damage*
6   *Screwdriver - Used to remove the seal from the hub (a long screwdriver is preferred)*
7   *Socket/breaker bar - Needed to loosen the nut on the spindle if it's extremely tight*
8   *Brush - Together with some clean solvent, this will be used to remove old grease from the hub and spindle*

changed. However, the bearings should be checked whenever the front of the vehicle is raised for any reason. Several items, including a torque wrench and special grease, are required for this procedure **(see illustration)**.

2   With the vehicle securely supported on jackstands, spin each wheel and check for noise, rolling resistance and freeplay.

3   Grasp the top of each tire with one hand and the bottom with the other. Move the wheel in-and-out on the spindle. If there's any noticeable movement, the bearings should be checked and then repacked with grease or replaced if necessary.

4   Remove the wheel.

5   Remove the brake caliper (see Chapter 9) and hang it out of the way on a piece of wire. A wood block can be slid between the brake pads to keep them separated, if necessary.

6   Pry the dust cap out of the hub using a screwdriver or hammer and chisel (see illustration).

7   Straighten the bent ends of the cotter pin, then pull the cotter pin out of the nut lock (see illustration). Discard the cotter pin and use a new one during reassembly.

8   Remove the nut lock, nut and washer from the end of the spindle (see illustration).

9   Pull the hub/disc assembly out slightly, then push it back into its original position. This should force the outer bearing off the spindle enough so it can be removed.

10   Pull the hub/disc assembly off the spindle.

11   Use a screwdriver to pry the seal out of the rear of the hub (see illustration). As this is done, note how the seal is installed.

12   Remove the inner wheel bearing from the hub.

13   Use solvent to remove all traces of the old grease from the bearings, hub and spindle. A small brush may prove helpful; however make sure no bristles from the brush embed themselves inside the bearing rollers. Allow the parts to air dry.

14   Carefully inspect the bearings for cracks, heat discoloration, worn rollers, etc. Check the bearing races inside the hub for wear and damage. If the bearing races are defective, the hubs should be taken to a machine shop with the facilities to remove the old races and press new ones in. Note that the bearings and races come as matched sets and old bearings should never be installed on new races.

15   Use high-temperature front wheel bearing grease to pack the bearings. Work the grease completely into the bearings, forcing it between the rollers, cone and cage from the back side (see illustration).

16   Apply a thin coat of grease to the spindle at the outer bearing seat, inner bearing seat, shoulder and seal seat.

17   Put a small quantity of grease inboard of each bearing race inside the hub. Using your finger, form a dam at these points to provide extra grease availability and to keep thinned grease from flowing out of the bearing.

18   Place the grease-packed inner bearing into the rear of the hub and put a little more grease outboard of the bearing.

19   Place a new seal over the inner bearing and tap the seal evenly into place with a hammer and blunt punch until it's flush with the hub.

34.6 Dislodge the dust cap by working around the outer circumference with a hammer and chisel

34.7 Remove the cotter pin and discard it - use a new one when the hub is reinstalled

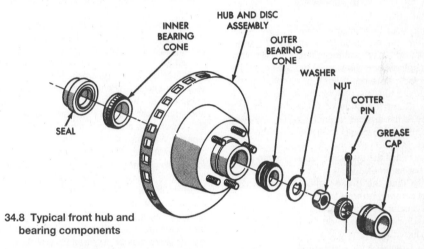

34.8 Typical front hub and bearing components

34.11 Use a large screwdriver to pry the grease seal out of the rear of the hub

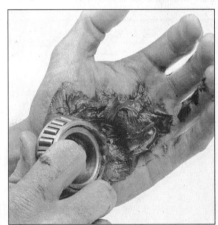

34.15 Work the grease completely into the bearing rollers - if you don't like getting greasy, special bearing packing tools that work with a common grease gun are available inexpensively from auto parts stores

20   Carefully place the hub assembly onto the spindle and push the grease-packed outer bearing into position.

21   Install the washer and spindle nut. Tighten the nut only slightly (no more than 12 ft-lbs of torque).

22   Spin the hub in a forward direction while tightening the spindle nut to approximately 20 ft-lbs to seat the bearings and remove any grease or burrs which could cause excessive bearing play later.

23   Loosen the spindle nut 1/4-turn, then

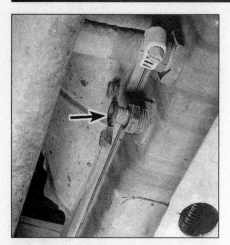

35.3 The drain fitting (arrow) is located at the bottom of the radiator - turn this "wing bolt" counterclockwise to drain the coolant

using your hand (not a wrench of any kind), tighten the nut until it's snug. Install the nut lock and a new cotter pin through the hole in the spindle and the slots in the nut lock. If the nut lock slots don't line up, remove the nut lock and turn it slightly until they do.

24   Bend the ends of the cotter pin until they're flat against the nut. Cut off any extra length which could interfere with the dust cap.

25   Install the dust cap, tapping it into place with a hammer.

26   Place the brake caliper near the disc and carefully remove the wood spacer. Install the caliper (see Chapter 9).

27   Install the wheel on the hub and tighten the lug nuts.

28   Grasp the top and bottom of the tire and

check the bearings in the manner described earlier in this Section.

29   Lower the vehicle.

35   Cooling system servicing (draining, flushing and refilling) (every 30,000 miles or 24 months)

Refer to illustrations 35.3 and 35.4

Warning: *Do not allow antifreeze to come in contact with your skin or painted surfaces of the vehicle. Rinse off spills immediately with plenty of water. Antifreeze is highly toxic if ingested. Never leave antifreeze lying around in an open container or in puddles on the floor; children and pets are attracted by it's sweet smell and may drink it. Check with local authorities about disposing of used antifreeze. Many communities have collection centers which will see that antifreeze is disposed of safely.*

1   Periodically, the cooling system should be drained, flushed and refilled to replenish the antifreeze mixture and prevent formation of rust and corrosion, which can impair the performance of the cooling system and cause engine damage. When the cooling system is serviced, all hoses and the radiator cap should be checked and replaced if necessary.

2   Apply the parking brake and block the wheels. **Warning:** *If the vehicle has just been driven, wait several hours to allow the engine to cool down before beginning this procedure.*

3   Move a large container under the radiator drain to catch the coolant. Attach a 3/8-inch diameter hose to the drain fitting (if possible) to direct the coolant into the container, then open the drain fitting **(see illustration)** (a pair of pliers may be required to turn it).

4   After coolant stops flowing out of the coolant reservoir, remove the radiator cap, allow the radiator to drain, then move the container under the engine block drain plugs - there's one on each side of the block on V8 engines and one on the right side only on six-cylinder engines **(see illustration)**. Remove the plugs and allow the coolant in the block to drain. **Note:** *Frequently, the coolant will not drain from the block after the plug is removed. This is due to a rust layer that has built up behind the plug. Insert a Phillips screwdriver into the hole to break the rust barrier.*

5   While the coolant is draining, check the condition of the radiator hoses, heater hoses and clamps (refer to Section 11 if necessary).

6   Replace any damaged clamps or hoses.

7   Once the system is completely drained, flush the radiator with fresh water from a garden hose until it runs clear at the drain. The flushing action of the water will remove sediments from the radiator but will not remove rust and scale from the engine and cooling tube surfaces.

8   These deposits can be removed with a chemical cleaner. Follow the procedure outlined in the manufacturer's instructions. If the radiator is severely corroded, damaged or leaking, it should be removed (see Chapter 3) and taken to a radiator repair shop.

9   Remove the overflow hose from the coolant reservoir and flush the reservoir with clean water, then reconnect the hose.

10   Close and tighten the radiator drain fitting. Install and tighten the block drain plugs.

11   Place the heater temperature control in the maximum heat position.

12   Slowly add new coolant (a 50/50 mixture of water and antifreeze) to the radiator until it's full. Add coolant to the reservoir up to the

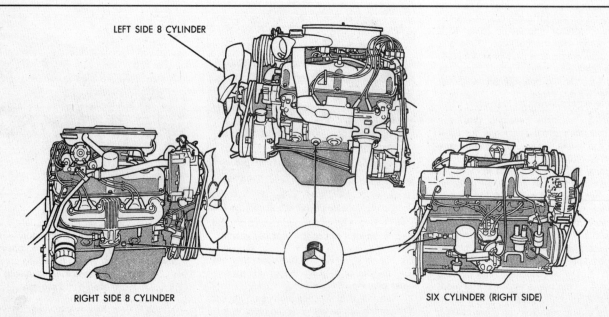

LEFT SIDE 8 CYLINDER

RIGHT SIDE 8 CYLINDER   SIX CYLINDER (RIGHT SIDE)

35.4 Cylinder block drain plug locations - the V8 engine shown is a small-block, but V6 and big-block V8 engines also have a drain plug on each side of the block

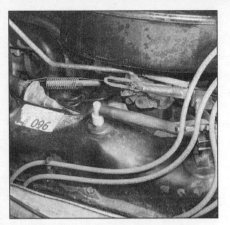

36.2 The PCV valve fits into the valve cover

37.2 The evaporative emissions canister is located underneath the body, behind the right front wheel - check the hoses and connections (arrows) for damage

38.2 Look under the EGR valve housing (arrow) to make sure the stem moves

lower mark.

13   Leave the radiator cap off and run the engine in a well-ventilated area until the thermostat opens (coolant will begin flowing through the radiator and the upper radiator hose will become hot).

14   Turn the engine off and let it cool. Add more coolant mixture to bring the level back up to the lip on the radiator filler neck.

15   Squeeze the upper radiator hose to expel air, then add more coolant mixture if necessary. Replace the radiator cap.

16   Start the engine, allow it to reach normal operating temperature and check for leaks.

36   Positive Crankcase Ventilation (PCV) valve check and replacement (every 30,000 miles or 24 months)

Refer to illustration 36.2

1   The PCV valve is located in the valve cover.

2   With the engine idling at normal operating temperature, pull the valve (with hose attached) from the rubber grommet in the cover **(see illustration)**.

3   Place your finger over the valve opening. If there's no vacuum at the valve, check for a plugged hose, manifold port, or the valve itself. Replace any plugged or deteriorated hoses.

4   Turn off the engine and shake the PCV valve, listening for a rattle. If the valve doesn't rattle, replace it with a new one.

5   To replace the valve, pull it from the end of the hose, noting its installed position.

6   When purchasing a replacement PCV valve, make sure it's for your particular vehicle and engine size. Compare the old valve with the new one to make sure they're the same.

7   Push the valve into the end of the hose until it's seated.

8   Inspect the rubber grommet for damage and hardening. Replace it with a new one if necessary.

9   Push the PCV valve and hose securely into position.

37   Evaporative emissions control system check (every 30,000 miles or 24 months)

Refer to illustration 37.2

1   The function of the evaporative emissions control system is to draw fuel vapors from the gas tank and fuel system, store them in a charcoal canister and route them to the intake manifold during normal engine operation.

2   The most common symptom of a fault in the evaporative emissions system is a strong fuel odor in the engine compartment. If a fuel odor is detected, inspect the charcoal canister, located in the engine compartment or under the vehicle **(see illustration)**. Check the canister and all hoses for damage and deterioration.

3   The evaporative emissions control system is explained in more detail in Chapter 6.

38   Exhaust Gas Recirculation (EGR) system check (every 30,000 miles or 24 months)

Refer to illustration 38.2

1   The EGR valve is usually located on the intake manifold, adjacent to the carburetor or throttle body. The EGR valve is operated by engine vacuum. Most of the time when a problem develops in the emissions system, it's due to a stuck or corroded EGR valve or a damaged or leaking hose or vacuum diaphragm.

2   Warm the engine up to operating temperature and allow it to idle with the transmission in Neutral. Accelerate the engine abruptly up to about 2000 rpm and look under the valve housing to make sure the movement marker groove in the stem moves **(see illustration)**.

3   If the stem doesn't move, check all the vacuum hoses for leaks or damage. If the hoses are not damaged, check the EGR valve operation. Disconnect the vacuum hose, connect a hand-held vacuum pump and apply 10 inches Hg of vacuum. If the valve stem still does not move, the valve itself is faulty and should be replaced.

4   If the valve stem moves, clamp the hose and make sure the valve stays open (diaphragm holds vacuum) 30 seconds or more. If it does not, the diaphragm is faulty and the valve should be replaced.

5   If necessary, reset the Emissions Maintenance Reminder light (see Section 49).

6   Refer to Chapter 6 for more information on the EGR system.

39   Thermostatic air cleaner check (every 30,000 miles or 24 months)

Refer to illustration 39.3

1   These models are equipped with a thermostatically controlled air cleaner which draws air to the carburetor/throttle body from different locations, depending upon engine temperature.

2   This is a visual check. If access is limited, a small mirror may have to be used.

3   Open the hood and locate the damper door inside the air cleaner assembly **(see illustration)**.

39.3 The damper door (arrow) is located in the air cleaner housing snorkel

40.3 Lubricate the filter by pouring clean engine oil into the large opening until it drains out the smaller opening

4   If there is a flexible air duct attached to the end of the air cleaner, disconnect it at the air cleaner. This will enable you to see the damper inside.

5   The check should be done when the engine is cold. Start the engine and look at the damper, which should move to a closed position. With the damper closed, air cannot enter through the air cleaner opening, but instead enters the air cleaner through a flexible duct attached to a heat stove on the exhaust manifold.

6   As the engine warms up to operating temperature, the damper should open to allow air through the air cleaner opening. Depending on outside temperature, this may take 10 to 15 minutes. To speed up this check, you can reconnect the air duct, drive the vehicle, then check to see if the damper is completely open.

7   If the thermo-controlled air cleaner is not operating properly, see Chapter 6 for more information.

40   Crankcase inlet filter cleaning (every 30,000 miles or 24 months)

Refer to illustration 40.3

1   Disconnect the hose and pull the crankcase inlet filter out of the valve cover.

2   Wash the inside of the filter with solvent.

3   Lubricate the filter by pouring clean engine oil into the large opening and allowing it to drain out through the smaller (inlet) opening **(see illustration)**.

4   Reinstall the filter in the valve cover and connect the hose.

41   Fuel filter replacement (every 30,000 miles or 24 months)

Warning: *Gasoline is extremely flammable, so take extra precautions when working on any part of the fuel system. Don't smoke or allow open flames or bare light bulbs in or near the work area, and don't work in a garage where a natural gas-type appliance (such as a water heater or clothes dryer) with a pilot light is present. Since gasoline is carcinogenic, wear latex gloves when there's a possibility of being exposed to fuel, and, if you spill fuel on your skin, rinse it off immediately with soap and water. Have a Class B fire extinguisher on hand.*

Carbureted models

Refer to illustration 41.1

1   The fuel filter is mounted in the fuel line, between the carburetor and the fuel pump **(see illustration)**.

2   Place rags or newspapers under the filter to absorb any spilled fuel. If the original fuel filter is still installed, use wire cutters to cut the Keystone clamps securing the fuel and vapor return hoses to the filter, then detach the hoses from the filter. Loosen aftermarket screw-type clamps with a screwdriver or the correct-size six-point socket.

3   Installation is the reverse of removal. Use new screw-type clamps to replace the Keystone clamps. Make sure the vapor return line is routed correctly.

Fuel-injected models

Refer to illustration 41.7

Note: *This procedure does not apply to 1993 and later 5.9L engines or 1994 and later V6 and 5.2L V8 engines. Fuel filter replacement on these models is not a routine maintenance item. The procedure can be found in Chapter 4B, Section 12.*

4   The fuel filter is located on the right-side frame rail near the engine compartment.

5   Relieve the fuel system pressure as described in Chapter 4.

6   Raise the vehicle and support it securely on jackstands.

7   Place shop towels around and under the filter. Disconnect the fuel lines, remove the mounting bolts and lower the filter from the vehicle. **(see illustration)**.

8   Installation is the reverse of removal. Use new clamps when connecting the hoses to the filter.

9   Start the engine and check for leaks.

42   Exhaust manifold heat control valve - check and servicing (every 30,000 miles or 24 months)

Warning: *The heat control valve, as well as the exhaust manifold and pipe, are very hot. Never touch these parts with bare hands until the engine has cooled completely. If you must touch exhaust system parts when they're hot, wear heavy, heat-resistant gloves, such as welding gloves.*

1   The manifold heat control valve is located on the exhaust manifold.

Check

2   Allow the engine to warm up to normal operating temperature, then have an assis-

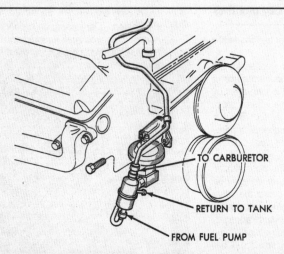

41.1 Typical fuel filter location (V8 engine shown, inline six-cylinder engine similar)

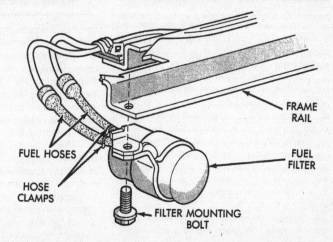

41.7 Fuel filter mounting details - fuel-injected models

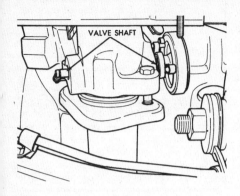

42.4 Spray the solvent at both ends of the valve shaft, let it penetrate, then work the valve back and forth to loosen any deposits which will cause binding

43.3 The choke plate (arrow) is located at the top of the carburetor air horn (shown in the open position)

43.9 Work the choke shaft back-and-forth while spraying carburetor cleaner into the contact areas

1

tant accelerate the engine from idle. The valve counterweight should move counterclockwise when the engine accelerates.

3 If it doesn't, deposits have probably built up on the valve shaft which inhibit its movement.

Servicing

Refer to illustration 42.4

4 With the engine cold, spray MOPAR Manifold Heat Control Valve solvent, or equivalent, to each end of the shaft **(see illustration)**.

5 Allow the solvent to soak in for a few minutes, then work the valve shaft back and forth by hand until it moves freely.

43 Carburetor choke check and shaft cleaning (every 30,000 miles or 24 months)

Refer to illustrations 43.3 and 43.9

1 The choke operates when the engine is cold, so this check can only be performed before the vehicle has been started for the day.

2 Remove the engine cover and remove the top plate of the air cleaner assembly (see Section 29). If any vacuum hoses must be disconnected, make sure you tag the hoses for reinstallation in their original positions. Place the top plate and wing nut aside, out of the way of moving engine components.

3 Look at the center of the air cleaner housing. You will notice a flat plate at the carburetor opening **(see illustration)**.

4 Press the accelerator pedal to the floor. The plate should close completely. Start the engine while you watch the plate at the carburetor. **Warning:** *Do not position your face directly over the carburetor, as the engine could backfire, causing serious burns.* When the engine starts, the choke plate should open slightly.

5 Allow the engine to continue running at

an idle speed. As the engine warms up to operating temperature, the plate should slowly open, allowing more air to enter through the top of the carburetor.

6 After a few minutes, the choke plate should be fully open to the vertical position. Lightly tap the accelerator to make sure the fast-idle cam disengages.

7 You will notice that the engine speed corresponds with the plate opening. With the plate fully closed, the engine should run at a fast idle speed. As the plate opens and the throttle is moved to disengage the fast idle cam, the engine speed will decrease.

8 Refer to Chapter 4 for specific information on adjusting and servicing the choke components.

9 At the specified intervals, or when the choke shaft binds, apply MOPAR Combustion Chamber Conditioner 4318001 or equivalent to the choke shaft and fast idle cam contact surfaces to remove deposits and ensure free movement **(see illustration)**.

44 Spark plug replacement (every 30,000 miles or 24 months)

Refer to illustrations 44.2, 44.5a, 44.5b, 44.6 and 44.10

1 Remove the engine cover and label each spark plug wire to ensure proper installation.

2 In most cases, the tools necessary for spark plug replacement include a spark plug socket which fits onto a ratchet (spark plug sockets are padded inside to prevent damage to the porcelain insulators on the new plugs), various extensions and a gap gauge to check and adjust the gaps on the new plugs **(see illustration)**. A special plug wire removal tool is available for separating the wire boots from the spark plugs, but it isn't absolutely necessary. A torque wrench should be used to tighten the new plugs.

3 The best approach when replacing the spark plugs is to purchase the new ones in advance, adjust them to the proper gap and replace them one at a time. When buying the

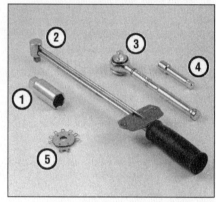

44.2 Tools required for changing spark plugs

1 *Spark plug socket - This will have special padding inside to protect the spark plug's porcelain insulator*
2 *Torque wrench - Although not mandatory, using this tool is the best way to ensure the plugs are tightened properly*
3 *Ratchet - Standard hand tool to fit the spark plug socket*
4 *Extension - Depending on model and accessories, you may need special extensions and universal joints to reach one or more of the plugs*
5 *Spark plug gap gauge - This gauge for checking the gap comes in a variety of styles. Make sure the gap for your engine is included*

new spark plugs, be sure to obtain the correct plug type for your particular engine. This information can be found on the *Emission Control Information* label located under the hood, in the factory owner's manual and the Specifications at the front of this Chapter. If differences exist between the plug specified on the emissions label and in the owner's manual, assume that the emissions label is correct.

4 Allow the engine to cool completely before attempting to remove any of the

44.5a Spark plug manufacturers recommend using a wire-type gauge when checking the gap - if the wire does not slide between the electrodes with a slight drag, adjustment is required

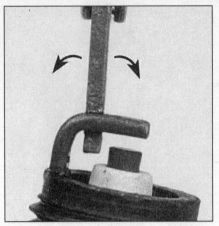

44.5b To change the gap, bend the side electrode only, as indicated by the arrows, and be very careful not to crack or chip the porcelain insulator surrounding the center electrode

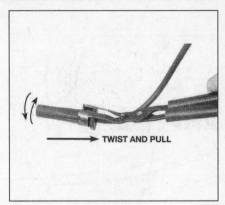

44.6 When removing the spark plug wires, pull only on the boot and twist it back-and-forth

plugs. While you're waiting for the engine to cool, check the new plugs for defects and adjust the gaps.

5   The gap is checked by inserting the proper-thickness gauge between the electrodes at the tip of the plug **(see illustration)**. The gap between the electrodes should be the same as the one specified on the *Emissions Control Information* label or in this Chapter's Specifications. The wire should just slide between the electrodes with a slight amount of drag. If the gap is incorrect, use the adjuster on the gauge body to bend the curved side electrode slightly until the proper gap is obtained **(see illustration)**. If the side electrode is not exactly over the center electrode, bend it with the adjuster until it is. Check for cracks in the porcelain insulator (if any are found, the plug should not be used).

6   With the engine cool, remove the spark plug wire from one spark plug. Pull only on the boot at the end of the wire - do not pull on the wire. A plug wire removal tool should be used if available **(see illustration)**.

7   If compressed air is available, use it to blow any dirt or foreign material away from the spark plug hole. A common bicycle pump will also work. The idea here is to eliminate the possibility of debris falling into the cylinder as the spark plug is removed.

8   Place the spark plug socket over the plug and remove it from the engine by turning it in a counterclockwise direction. On six-cylinder engines, also remove the stamped-steel spark plug cup from the cylinder head and replace the rubber oil seal on installation.

9   Compare the spark plug to those shown in the color photos inside the back cover of this manual to get an indication of the general running condition of the engine.

10   Thread one of the new plugs into the hole until you can no longer turn it with your fingers, then tighten it with a torque wrench (if available) or the ratchet. It might be a good idea to slip a short length of rubber hose over the end of the plug to use as a tool to thread

it into place **(see illustration)**. The hose will grip the plug well enough to turn it, but will start to slip if the plug begins to cross-thread in the hole - this will prevent damaged threads and the accompanying repair costs.

11   Before pushing the spark plug wire onto the end of the plug, inspect it following the procedures outlined in Section 45.

12   Attach the plug wire to the new spark plug, again using a twisting motion on the boot until it's seated on the spark plug.

13   Repeat the procedure for the remaining spark plugs, replacing them one at a time to prevent mixing up the spark plug wires.

45 Spark plug wires, distributor cap and rotor - check and replacement (every 30,000 miles or 24 months)

Refer to illustrations 45.10 and 45.13

1   The spark plug wires should be checked

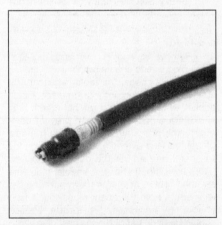

44.10 A length of snug-fitting rubber hose will save time and prevent damaged threads when installing the spark plugs

at the recommended intervals and whenever new spark plugs are installed in the engine.

2   The wires should be inspected one at a time to prevent mixing up the order, which is essential for proper engine operation. On some models it will be necessary to remove the two screws and detach the distributor splash shield for access.

3   Disconnect the plug wire from one spark plug. To do this, grab the rubber boot, twist slightly and pull the wire free. Do not pull on the wire itself, only on the rubber boot.

4   Check inside the boot for corrosion, which will look like a white crusty powder. Push the wire and boot back onto the end of the spark plug. It should be a tight fit on the plug. If it isn't, remove the wire and use a pair of pliers to carefully crimp the metal connector inside the boot until it fits securely on the end of the spark plug.

5   Using a clean rag, wipe the entire length of the wire to remove any built-up dirt and grease. Once the wire is clean, check for holes, burned areas, cracks and other damage. Don't bend the wire excessively or the conductor inside might break.

6   Disconnect the wire from the distributor cap. Pull the wire straight out of the cap. Pull only on the rubber boot during removal.

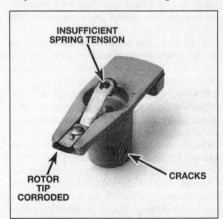

45.10 The ignition rotor should be checked for wear and corrosion as indicated here (if in doubt about its condition, buy a new one)

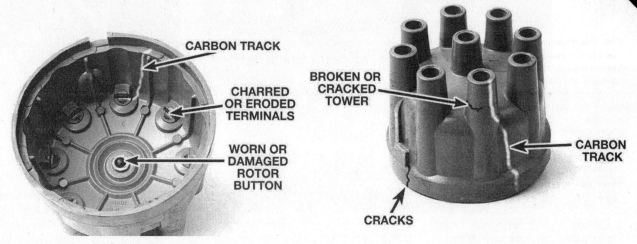

45.13 Shown here are some of the common defects to look for when inspecting the distributor cap (if in doubt about its condition, install a new one)

1

Check for corrosion and a tight fit in the same manner as the spark plug end. Reattach the wire to the distributor cap.

7   Check the remaining spark plug wires one at a time, making sure they are securely fastened at the distributor and the spark plug when the check is complete.

8   If new spark plug wires are required, purchase a new set for your specific engine model. Wire sets are available pre-cut, with the rubber boots already installed. Remove and replace the wires one at a time to avoid mix-ups in the firing order. The wire routing is extremely important, so be sure to note exactly how each wire is situated before removing it.

9   Release the distributor cap clips. Pull up on the cap, with the wires attached, to separate it from the distributor, then position it to one side.

10   The rotor is now visible on the end of the distributor shaft. Check it carefully for cracks and carbon tracks. Make sure the center terminal spring tension is adequate and look for corrosion and wear on the rotor tip **(see illustration)**. If in doubt about its condition, replace it with a new one.

11   If replacement is required, detach the rotor from the shaft and install a new one. The rotor is a press fit on the shaft and can be pried or pulled off.

12   The rotor is indexed to the shaft so it can only be installed one way. It has an internal key that must line up with a slot in the end of the shaft (or vice versa).

13   Check the distributor cap for carbon tracks, cracks and other damage. Closely examine the terminals on the inside of the cap for excessive corrosion and damage **(see illustration)**. Slight deposits are normal. Again, if in doubt about the condition of the cap, replace it with a new one. Be sure to apply a small dab of silicone dielectric grease to each terminal before installing the cap. Also, make sure the carbon brush (center terminal) is correctly installed in the cap - a wide

gap between the brush and rotor will result in rotor burn-through and/or damage to the distributor cap.

14   To replace the cap, simply separate it from the distributor and transfer the spark plug wires, one at a time, to the new cap. Be very careful not to mix up the wires!

15   Reattach the cap to the distributor, then reposition the clips or tighten the screw latches to hold it in place.

46   Idle speed check and adjustment (every 30,000 miles or 24 months)

This procedure is covered under *Carburetor adjustments* in Chapter 4.

47   Ignition timing check and adjustment (every 30,000 miles or 24 months)

Refer to illustrations 47.2 and 47.5

1   All vehicles are equipped with an *Emissions Control Information* label inside the engine compartment. The label contains important ignition timing specifications and the proper timing procedure for your specific vehicle. If the information on the emissions label is different from the information included in this Section, follow the procedure on the label. **Note:** *Ignition timing on 1995 and later models isn't adjustable.*

2   At the specified intervals, or when the distributor has been removed, the ignition timing must be checked and adjusted if necessary. Tools required for this procedure include an inductive pick-up timing light, a tachometer, a distributor wrench and, in some cases, a means of plugging vacuum hoses **(see illustration)**.

3   Before you check the timing, make sure the idle speed is correct (see Chapter 4) and

the engine is at normal operating temperature.

4   With the engine off, connect a timing light in accordance with the manufacturer's instructions. Usually, the light must be connected to the battery and the number one spark plug wire in some fashion. The number one spark plug is the one at the very front on six-cylinder engines and the one at the front of the left (driver's side) of the engine on V8 engines. Attach the timing light lead near the plug.

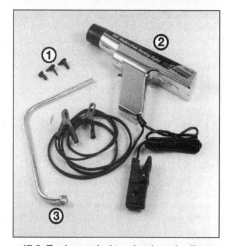

47.2 Tools needed to check and adjust the ignition timing

1   *Vacuum plugs - Vacuum hoses will, in most cases, have to be disconnected and plugged. Molded plugs in various shapes and sizes are available for this*

2   *Inductive pick-up timing light - Flashes a bright concentrated beam of light when the number one spark plug fires. Connect the leads according to the instructions supplied with the light*

3   *Distributor wrench - On some models, the hold-down bolt for the distributor is difficult to reach and turn with a conventional wrench or socket. A special wrench like this must be used*

...g scale is ...r of the damper and is marked to indicate the number of degrees before and after TDC - the receptacle, used on some later models, is where dealer mechanics can hook up a special magnetic probe to monitor ignition timing (small-block V8 engine shown)

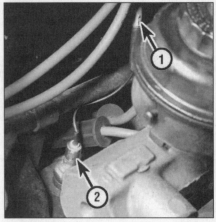

48.2 Unplug the connector (1), then unscrew the oxygen sensor (2)

5   Locate the numbered timing scale on the front cover of the engine (see illustration). On small-block V8 engines, it's cast into the front cover on the left (driver's side) of the engine. On six-cylinder and big-block V8 engines, the scale is a stamped-steel tab on the right (passenger's side) of the engine front cover.
6   Locate the notched groove across the crankshaft pulley or the flywheel. It may be necessary to have an assistant temporarily turn the ignition on and off in short bursts without starting the engine in order to bring the groove into a position where it can easily be cleaned and marked. **Warning:** *Stay clear of all moving engine components when the engine is turned over in this manner.*
7   Use white soap-stone, chalk or paint to mark the groove in the pulley or flywheel. Also, put a mark on the timing scale corresponding to the number of degrees specified on the *Vehicle Emission Control Information* label in the engine compartment.
8   Aim the timing light at the marks, again being careful not to come into contact with moving parts. The marks made should appear stationary. If the marks are in alignment, the timing is correct. If the marks are not aligned, turn off the engine.
9   Loosen the hold-down bolt or nut at the base of the distributor. Loosen the bolt/nut only slightly, just enough to turn the distributor (see Chapter 5).
10   Now restart the engine and turn the distributor very slowly until the timing marks are aligned.
11   Shut off the engine and tighten the distributor bolt/nut, being careful not to move the distributor.

12   Start the engine and recheck the timing to make sure the marks are still in alignment.
13   Disconnect the timing light and tachometer and reconnect any components which were disconnected for this procedure.

48   Oxygen sensor replacement (Electronic Fuel Control system-equipped models) (every 52,000 miles or 82,500 miles, depending on model year)

Refer to illustration 48.2

1   The oxygen sensor must be replaced at the specified interval (see the *Maintenance schedule* at the front of this Chapter). The sensor is threaded into the exhaust pipe or manifold. It may be necessary on some models to raise the front of the vehicle and support it securely on jackstands for access to the underside of the engine compartment.
2   Disconnect the oxygen sensor wire by unplugging the connector (see illustration).
3   Unscrew the sensor with a box-end wrench.
4   Use a tap to clean the threads in the exhaust pipe or manifold.
5   If the old sensor is to be reinstalled, apply anti-seize compound to the threads. New sensors will already have the anti-seize compound on the threads.
6   Install the sensor, tighten it to the torque listed in this Chapter's Specifications and plug in the connector.
7   If necessary, reset the Emissions Maintenance Reminder light (see Section 49).

49   Emissions Maintenance Reminder (EMR) light resetting (1980 through 1988 models)

General information

1   1980 through 1988 models are equipped with an Emissions Maintenance Reminder light that comes on at certain predetermined intervals. This light is sometimes marked *CHECK EGR* or *MAINT REQD* and doesn't indicate a problem - it simply alerts the driver that it's time to check the EGR system and/or replace the oxygen sensor (see the *Maintenance schedule* at the beginning of this Chapter).

Light resetting

Note: *Beginning with the 1989 model year, a special Chrysler DRB-II tester must be used to reset the EMR light.*
2   Find the Emissions Maintenance Reminder module - it's a small plastic box located under the instrument panel, usually to the left of the steering column. Remove the module from its bracket (if so equipped).
3   On 1987 and earlier models, remove the 9-volt battery from the module.
4   On all models, insert a thin screwdriver or rod into the hole in the module and depress the switch contacts with the ignition key in the On position.
5   On 1987 and earlier models, install a new 9-volt battery in the module.
6   Reinstall the module to its bracket.

Chapter 2 Part A
Inline six-cylinder engine

Contents

Specifications

General

Cylinder numbers (front-to-rear)	1-2-3-4-5-6
Displacement	225 cubic inches
Bore	3.40 inches
Stroke	4.125 inches
Distributor rotation	Clockwise, as viewed from top
Firing order	1-5-3-6-2-4
Minimum compression	100 psi
Maximum compression variation between cylinders	25 psi

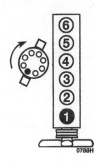

The blackened terminal shown on the distributor cap indicates the Number One spark plug wire position

Cylinder location and distributor rotation

Camshaft

Bearing journal diameter	
No. 1	1.998 to 1.999 inches
No. 2	1.982 to 1.983 inches
No. 3	1.967 to 1.968 inches
No. 4	1.951 to 1.952 inches
Bearing oil clearance	0.001 to 0.003 inch

Oil pump

Minimum oil pressure at curb idle speed	10 psi
Normal operating oil pressure	30 to 70 psi at 2000 rpm
Oil pump cover surface variation	0.0015 inch maximum
Outer rotor thickness	
1980 and earlier	0.649 inch minimum
1981 and later	0.825 inch minimum
Outer rotor diameter	2.469 inches minimum
Inner rotor thickness	
1980 and earlier	0.649 inch minimum
1981 and later	0.825 inch minimum
Rotor-to-cover clearance	0.004 inch maximum
Outer rotor clearance	0.014 inch maximum
Rotor tip clearance	0.010 inch maximum

Valve lash/clearance - solid lifters (1980 and earlier models)

Intake	0.010 inch
Exhaust	0.020 inch

Valve lift

1980 and earlier	
Intake	0.406 inch
Exhaust	0.414 inch
1981 and later	
Intake	0.378 inch
Exhaust	0.378 inch

Torque specifications*

	Ft-lbs (unless otherwise indicated)
Camshaft sprocket bolt	50
Cooling fan-to-water pump bolts or nuts	200 in-lbs
Crankshaft rear main seal retainer bolts	30
Cylinder head bolts	70
Engine mounts	
Bracket-to-engine bolts	50
Mount-to-bracket nuts	75
Exhaust manifold-to-exhaust pipe bolts/nuts	35
Flywheel/driveplate-to-crankshaft bolts	55
Intake-to-exhaust manifold	
Stud nut	25
Bolts	21
Manifold assembly-to-cylinder head nuts	120 in-lbs
Oil pan screws	200 in-lbs
Oil pump attaching bolts	200 in-lbs
Oil pump cover bolts	95 in-lbs
Oil filter attaching stud	120 in-lbs
Oil pressure gauge sending unit	60 in-lbs
Rocker arm shaft bolts	24
Timing chain cover bolts	200 in-lbs
Valve cover bolts	40 in-lbs

*Note: *Refer to Part C for additional specifications*

1   General information

This part of Chapter 2 is devoted to in-vehicle repair procedures for the 225-cubic inch inline six-cylinder engine, which was used in Dodge trucks for many years, until supplanted by the V6 engine in 1988 (see Part B of this Chapter). All information concerning engine removal and installation and engine block and cylinder head overhaul can be found in Part C of this Chapter.

Since the repair procedures included in this Part are based on the assumption that the engine is still installed in the vehicle, if they are being used during a complete engine overhaul (with the engine already out of the vehicle and on a stand) many of the steps included here will not apply.

The specifications included in this Part of Chapter 2 apply only to the procedures found here. The specifications necessary for rebuilding the block and cylinder heads are included in Part 2C.

2   Repair operations possible with the engine in the vehicle

Many major repair operations can be accomplished without removing the engine from the vehicle.

Clean the engine compartment and the exterior of the engine with some type of pressure washer before any work is done, making sure to keep the carburetor and distributor covered with plastic-wrap to protect them from water. A clean engine will make the job easier and will help keep dirt out of the internal areas of the engine.

Almost all work to the engine will require the removal of the engine cover. This is done by adjusting both front seats as far back as possible, unbolting the two brackets at the rear of the cover (some models also have a third bracket on the driver's side of the cover), and releasing the two forward latches (see Chapter 2 Part B, **illustrations 2.3a and 2.3b**).

If oil or coolant leaks develop, indicating a need for gasket or seal replacement, the repairs can generally be made with the engine in the vehicle. The oil pan gasket, the cylinder head gaskets, intake and exhaust manifold gaskets, timing cover gaskets and the crankshaft oil seals are accessible with the engine in place.

Exterior engine components, such as the water pump, the starter motor, the alternator, the distributor, oil pump and the carburetor as well as the intake and exhaust manifolds, can be removed for repair with the engine in place.

Since the cylinder head can be removed without pulling the engine, valve component servicing can also be accomplished with the engine in the vehicle.

Replacement of, repairs to or inspection of the timing chain and sprockets and the oil pump pick-up tube with the screen are all possible with the engine in place.

In extreme cases caused by a lack of necessary equipment, repair or replacement of piston rings, pistons, connecting rods and rod bearings is possible with the engine in the vehicle. However, this practice is not recommended because of the cleaning and preparation work that must be done to the components involved.

3   Top Dead Center (TDC) for number one piston - locating

Refer to illustration 3.6

1   Top Dead Center (TDC) is the highest point in the cylinder that each piston reaches as it travels up-and-down when the crankshaft turns. Each piston reaches TDC on the compression stroke and again on the exhaust stroke, but TDC generally refers to piston position on the compression stroke. The timing marks are referenced to the number one piston at TDC.

2   Positioning the pistons at TDC is an essential part of many procedures such as camshaft removal, timing chain replacement and distributor removal.

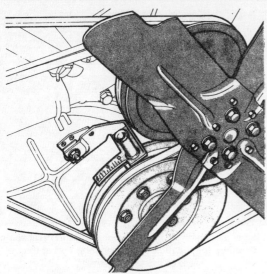

3.6 Align the mark on the vibration damper with the 0 mark on the timing plate

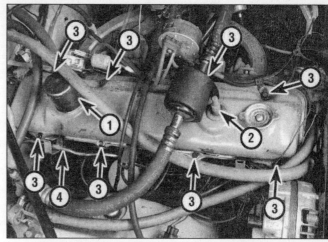

4.3 Disconnect anything attached to the valve cover

1 PCV breather (lift to   3 Valve cover bolts (two are
   remove)   hidden by hoses and
2 PCV valve (lift to remove)   wiring in this photo)

3   To bring any piston to TDC, the crankshaft must be turned using one of the methods outlined below. When looking at the front of the engine, normal crankshaft rotation is clockwise. **Warning:** *Before beginning this procedure, be sure to place the transmission in Neutral and disable the ignition system by disconnecting the coil wire from the distributor cap and grounding it on the engine block.*

a) *The preferred method is to turn the crankshaft with a large socket and breaker bar attached to the large bolt that is threaded into the front of the crankshaft.* **Note:** *There is no bolt in the front of the crankshaft on many six-cylinder engines, so you will have to get a 3/4 inch x 16 threads-per-inch bolt that is about 1-1/2 or 2-inches long.*

b) *A remote-starter switch, which may save some time, can also be used. Attach the switch leads to the S (switch) and B (battery) terminals on the starter solenoid, keeping the wires and your hands away from the fan and drive belts. Once the piston is close to TDC, use a socket and breaker bar as described in the previous paragraph.*

c) *If an assistant is available to turn the ignition switch to the Start position in short bursts, you can get the piston close to TDC without a remote starter switch. Use a socket and breaker bar as described in Paragraph (a) to complete the procedure.*

4   Scribe or paint a small mark on the distributor body directly below the number one spark plug wire terminal in the distributor cap.
5   Remove the engine cover (see Section 2). Remove the distributor cap as described in Chapter 1.
6   Turn the crankshaft (see Step 3 above) until the line on the vibration damper is aligned with the zero or "TDC" mark on the timing indicator **(see illustration)**.

7   The rotor should now be pointing directly at the mark on the distributor base. If it is 180-degrees off, the piston is at TDC on the exhaust stroke.
8   If the rotor is 180-degrees off, turn the crankshaft one complete turn (360-degrees) clockwise. The rotor should now be pointing at the mark. When the rotor is pointing at the number-one spark plug wire terminal in the distributor cap (which is indicated by the mark on the distributor body) and the timing marks are aligned, the number one piston is at TDC on the compression stroke.
9   After the number one piston has been positioned at TDC on the compression stroke, TDC for any of the remaining cylinders is located by turning the engine over in the normal direction of rotation until the rotor is pointing to the spark plug wire terminal in the distributor cap (you will have to install and remove the distributor cap as you are turning the engine over) for the piston you wish to bring to TDC.

4   Valve cover - removal and installation

Removal

Refer to illustration 4.3

1   Disconnect the cable from the negative battery terminal and remove the engine cover (see Section 2).
2   Remove the air cleaner assembly (see Chapter 4).
3   Remove the breather tube, PCV valve and hoses, vacuum line and electrical wiring clips from the valve cover **(see illustration)**.
4   Remove the valve cover mounting bolts **(see illustration 4.3)**.
5   Remove the valve cover while lifting up gently on the air conditioning hose, and moving the valve cover up and then forward until it can be removed. **Note:** *If the cover is stuck*

to the head, bump the cover with a block of wood and a hammer to release it. If it still will not come loose, try to slip a flexible putty knife between the head and cover to break the seal. Don't pry at the cover-to-head joint, as damage to the sealing surface and cover flange will result and oil leaks will develop.

Installation

6   The mating surfaces of the cylinder head and valve cover must be perfectly clean when the cover is installed. Use a gasket scraper to remove all traces of sealant or old gasket, then wipe the mating surfaces with a cloth saturated with lacquer thinner or acetone. If there is sealant or oil on the mating surfaces when the cover is installed, oil leaks may develop. **Note:** *Gasket removal solvents are available from auto parts stores and may prove helpful.*
7   Inspect the valve cover gasket flange for distortion. Carefully straighten it with a wooden block and the rounded end of a ball-peen hammer, if necessary.
8   Make sure the threaded holes in the head are clean. If necessary, run a tap into them to remove corrosion or sealant and restore damaged threads.
9   Mate the new gaskets to the cover before the cover is installed. Apply a thin coat of RTV sealant to the cover flange, then position the gasket inside the cover lip and allow the sealant to set up so the gasket adheres to the cover (if the sealant is not allowed to set, the gasket may fall out of the cover as it is installed on the engine).
10   Carefully position the cover on the head and install the bolts.
11   Tighten the bolts to the torque listed in this Chapter's Specifications.
12   The remaining installation steps are the reverse of removal. Allow twenty minutes for the RTV to cure before starting the engine.
13   Start the engine and check carefully for oil leaks as the engine warms up.

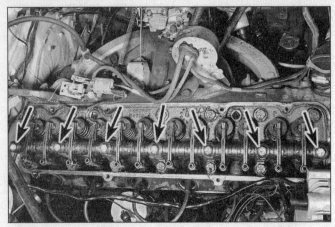

5.2 Remove the rocker arm shaft bolts and curved washers (arrows), working from the end bolts towards the center, loosening them a little at a time

5.3 Lift the shaft assembly from the head without disassembling it; notice the location of the one long bolt at the rear of the shaft (arrow)

5   Rocker arms and shaft - removal, inspection and installation

Removal

Refer to illustrations 5.2 and 5.3

1   Remove the valve cover and gasket (see Section 4).

2   Loosen the rocker shaft bolts 1/2-turn at a time, moving from bolt to bolt **(see illustration)**.

3   When all the bolts are loose, remove the rocker arm shaft as an assembly with the bolts still installed **(see illustration)**. The bolts will keep the assembly together.

4   Remove the pushrods and store them in a cardboard box with numbered holes so they can be reinstalled in their original locations.

Inspection

Refer to illustrations 5.5 and 5.6

5   Once the shaft assembly is removed, disassemble the components for cleaning and inspection **(see illustration)**. **Caution:** *All parts MUST be reinstalled in their original locations. Store or lay the parts out in order so they can be put back on the rocker shaft in their original positions. One strategy is to tie a heavy piece of wire (such as a straightened*

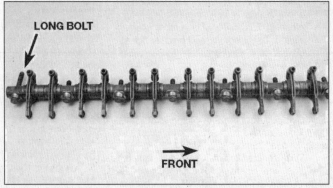

5.5 Disassemble the components for cleaning and inspection, being very careful to keep everything in order so all the parts can be reassembled in exactly the same locations

coat hanger) around a short wood dowel or similar object. Label the dowel to indicate the front of the engine, then remove each rocker arm and spacer from the engine, starting at the front of the engine, and slide them over the wire (the dowel will keep them from falling off at the other end).

6   Inspect the rocker arms (especially the surfaces where they contact the valve stems and rocker arms) and rocker arm shaft for signs of excessive wear, galling or damage. If any of these conditions exist, replace the

rocker arm and shaft assembly with new parts. **Note:** *Normal wear will appear as shiny areas where there is metal-to-metal contact* **(see illustration)** *and, unless there is not excessive wear, scoring, damage or overheating, the parts are reusable.*

7   Inspect the pushrods to see if they are bent, worn or cracked. Make sure the cup at the top of each rocker arm is not loose. Roll each pushrod across a piece of plate glass - if it wobbles, it's bent. If any of these conditions exist, replace the pushrods with new

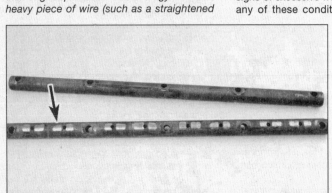

5.6 Inspect all the surfaces of the rocker arm shaft, looking carefully at metal-to-metal contact areas (arrow)

5.12 The one long bolt goes in the rear position

6.7 Once the spring is depressed, the keepers can be removed with a small magnet or a needle-nose pliers (a magnet is preferred to prevent dropping the keepers)

6.16 Be sure to install the seals on the correct valves

1	*Exhaust valve seal*	
	2	*Intake valve seal*

2A

ones. **Note:** *When purchasing replacement pushrods, always bring one of the old ones with you and compare it for exact length with the new ones.*

Installation

Refer to illustration 5.12

8   Reassemble the rocker shaft, rocker arms, spacers, retainers and bolts, applying moly-base grease or engine assembly lube to the wear areas on the rocker-arm shaft. **Caution:** *Make sure the oil holes in the rocker shaft are located correctly so the rocker assemblies are being oiled properly* **(see illustration 5.5)**. *Also, make sure there is a spacer between intake and exhaust rocker arms for the same cylinder.*

9   Apply moly-base grease or engine assembly lube to the tops of the valve stems.

10   Install the pushrods in the correct positions, applying moly-base grease or engine assembly lube to the rounded lower end and the cupped upper end.

11   Install the rocker arm shaft with the correct end facing the front of the engine.

12   Install the rocker shaft bolts (the long bolt is installed at the rear) **(see illustration)**.

13   Tighten each bolt a little at a time, working from bolt to bolt, until the torque listed in this Chapter's Specifications is obtained. Be careful not to bend the pushrods. Refer to Chapter 1 for valve adjustment procedure on 1980 and earlier models with mechanical lifters.

14   Attach a new valve cover gasket to the cover. Notice that there are tabs provided in the cover to retain the gasket. It may be necessary to apply RTV sealant to the corners of the cover to retain the gasket there.

15   The remainder of the installation procedure is the reverse of removal.

16   Start the engine and run it until it reaches normal operating temperature, then make sure that there are no leaks.

6   Valve springs, retainers and seals - replacement

Refer to illustrations 6.7, 6.16, 6.18 and 6.20
Note: *Broken valve springs and defective valve stem seals can be replaced without removing the cylinder head. Two special tools and a compressed air source are normally required to perform this operation, so read through this Section carefully and rent or buy the tools before beginning the job.*

1   Remove the valve cover (see Section 4).

2   Remove the spark plugs (see Chapter 1).

3   Turn the crankshaft until the number one piston is at top dead center on the compression stroke (see Section 3).

4   Remove the rocker arm shaft (see Section 5). If you intend to use a lever-type spring compressor, remove the rocker arms and reinstall the shaft.

5   Thread an adapter into the spark plug hole and connect an air hose from a compressed air source to it. Most auto parts stores can supply the air hose adapter. **Note:** *Many cylinder compression gauges utilize a screw-in fitting that may work with your air hose quick-disconnect fitting.*

6   Apply compressed air to the cylinder. The valves should be held in place by the air pressure. If the valve faces or seats are in poor condition, leaks may prevent the air pressure from retaining the valves - valve reconditioning is indicated.

7   Stuff shop rags into the cylinder head holes around the valves to prevent parts and tools from falling into the engine, then use a valve spring compressor to compress the spring. Remove the keepers with small needle-nose pliers or a magnet **(see illustration)**. **Note:** *Several different types of tools are available for compressing the valve springs with the head in place. One type, shown here,*

grips the lower spring coils and presses on the retainer as the knob is turned, while the lever-type utilizes the rocker arm shaft for leverage. Both types work very well, although the lever type is usually less expensive.

8   Remove the valve spring and retainer.

9   Remove the old valve stem seals.

10   If there are any valve spring shims that have been used, to get the correct assembled height, note where and how many shims are used at each valve and then remove the shims. **Caution:** *The shims and valve springs must be reinstalled in their original locations for correct reassembly.*

11   Wrap a rubber band or tape around the top of the valve stem so the valve won't fall into the combustion chamber, then release the air pressure.

12   Inspect the valve stem for damage. Rotate the valve in the guide and check the end for eccentric movement, which would indicate that the valve is bent.

13   Move the valve up-and-down in the guide and make sure it does not bind. If the valve stem binds, either the valve is bent or the guide is damaged. In either case, the head will have to be removed for repair.

14   Reapply air pressure to the cylinder to retain the valve in the closed position, then remove the tape or rubber band from the valve stem.

15   If shims were used on any of the valves, place them back in their original positions. This will keep the same valve spring installed height as prior to disassembly.

16   If you're working on an exhaust valve, install the new shield on the valve stem and push it down to the top of the valve guide **(see illustration)**.

17   If you're working on an intake valve, push a new valve stem seal down over the valve guide, using the valve stem as a guide, but don't force the seal against the top of the guide.

18   Install the spring and retainer or rotator in position over the valve **(see illustration)**. **Caution:** *Different-length springs are used on the intake and exhaust valves - accidentally installing an intake spring on an exhaust valve can cause major engine damage!*

19   Compress the valve spring assembly only enough to install the keepers in the valve stem.

20   Position the keepers in the valve stem groove. Apply a small dab of grease to the inside of each keeper to hold it in place if necessary **(see illustration)**. Remove the pressure from the spring tool and make sure the keepers are seated.

21   Disconnect the air hose and remove the adapter from the spark plug hole.

22   Repeat the above procedure on the remaining cylinders, following the firing order sequence. Bring each piston to top dead center on the compression stroke before applying air pressure.

23   Reinstall the rocker arm assemblies and the valve covers.

24   Start the engine, then check for oil leaks and unusual sounds coming from the valve cover area.

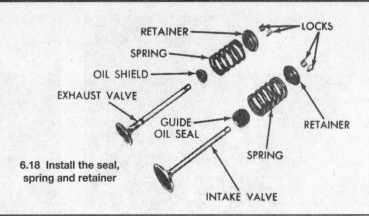

6.18 Install the seal, spring and retainer

---

7   Intake and exhaust manifolds - removal and installation

Refer to illustrations 7.5, 7.6 and 7.17

Removal

1   Disconnect the cable from the negative battery terminal and remove the engine cover (see Section 2).

2   Remove the air cleaner, vacuum lines and heat stove flexible connector (see Chapter 4).

3   Disconnect the throttle linkage (see Chapter 4).

4   If the manifold is going to be replaced, remove the carburetor, EGR and divorced choke assembly (see Chapter 4) so that it can be installed on the new manifold.

5   Disconnect the exhaust pipe at the exhaust manifold **(see illustration)**.

6   Remove the nuts attaching the intake/exhaust manifold assembly to the cylinder head in the reverse order of the tightening sequence **(see illustration 7.17)** and detach the manifold assembly **(see illustration)**.

7   Apply some penetrating oil and let it sit for a few minutes, then remove the three vertical bolts (or stud nuts) to separate the intake manifold from the exhaust manifold. The bolts are in the center of the intake manifold, around the carburetor mounting point.

8   Use a gasket scraper to remove all traces of sealant and old gasket material, then wipe the mating surfaces with a cloth saturated with lacquer thinner or acetone. If there is old sealant or oil on the mating surfaces when the manifolds are installed, vacuum or exhaust leaks may develop. Don't forget to clean the intake-to-exhaust manifold mating surfaces. **Note:** *Aerosol gasket removal solvents are available from auto parts stores and may prove helpful.*

9   Check the gasket surfaces for warpage with a straightedge and feeler gauges. They must not be warped more than 0.009 inch per foot of manifold length. Check the manifolds carefully for cracks and distortion - it's common to find cracks in these long exhaust manifolds. Replace the manifolds if any cracks or distortion are found.

Installation

10   Use new gaskets when rejoining and installing the manifold assembly.

11   Install, but **do not tighten**, the bolts attaching the intake manifold to the exhaust manifold.

12   Install the manifold assembly and tighten the fasteners finger-tight only! Washers spanning the intake and exhaust manifold flanges must be flat - replace any that are distorted by previous over-tightening. Install the steel conical washers with the cup side facing the nut. Install the brass washers with the flat sides facing the manifold. Install the nuts with the cone-shaped side facing the washer.

13   Carefully tighten all fasteners to only 10 in-lbs; do not over-tighten them!

14   Tighten the inboard intake-to-exhaust manifold bolts or nuts to 10 ft-lbs.

15   Tighten the outboard intake-to-exhaust manifold bolts or nuts to 10 ft-lbs.

16   Repeat the procedure, tightening the intake-to-exhaust manifold bolts or nuts to the torque listed in this Chapter's Specifications.

17   Tighten the manifold assembly-to-cylinder head nuts to the torque listed in this Chapter's Specifications, following the proper sequence **(see illustration)**. Work up

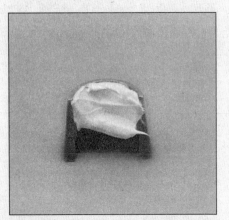

6.20 Apply a small dab of grease to each keeper as shown here before installation - it'll hold them in place on the valve stem as the spring is released

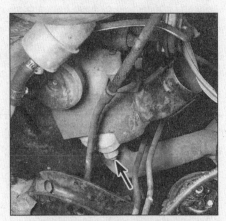

7.5 Remove the two bolts and nuts to disconnect the exhaust pipe at the exhaust manifold (arrow); the second nut is on the opposite side of the manifold

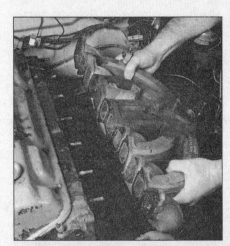

7.6 The intake/exhaust manifolds come off the engine as a single unit

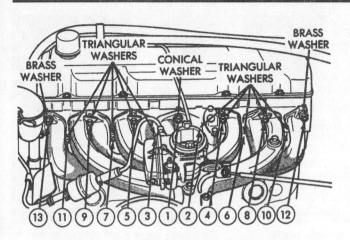

7.17 Intake/exhaust manifold nut TIGHTENING sequence - reverse this sequence when loosening the bolts. Be sure to note the locations of the special conical and triangular washers when removing them

8.5 Use a bolt-type puller to remove the vibration damper

2A

to the final torque in three steps and DO NOT OVER-TIGHTEN THEM!

18   The rest of the installation procedure is the reverse of removal.

8   Crankshaft front oil seal - replacement

Removal

Refer to illustrations 8.5, 8.6a, 8.6b and 8.8

Note: *On all models the crankshaft front oil seal can be replaced once the timing chain cover is removed. However, on the 1974 and earlier models, the seal MUST be replaced from inside the cover (cover must be removed), while on later models (1975 through 1987), the seal can be replaced by carefully prying it out of the timing chain cover while it is still bolted to the engine block.*

1   Drain the cooling system (see Chapter 1).

2   Unbolt the fan assembly and remove the fan and shroud (see Chapter 3).

3   Remove the radiator (see Chapter 3).

4   Unbolt and remove the drivebelt pulley from the vibration damper.

5   Remove the vibration damper assembly from the front end of the crankshaft. A special puller, which can be found at your local parts store, will be required for this operation **(see illustration)**.

1974 and earlier models

6   Remove the timing chain cover **(see illustrations)**. **Note:** *This procedure can be* accomplished by removing the front oil pan bolts, that connect the pan and front cover, and loosening the remaining oil pan bolts so the pan can be slightly lowered. It can be extremely difficult on some models, however, to do this without removal of the oil pan. If necessary, remove the oil pan (see Section 13).

7   After the front cover is removed, remove all traces of gasket sealant from the cover-to-block mating surfaces with a scraper, then wipe the surfaces clean with a rag soaked in lacquer thinner or acetone. Remove the old rubber end seal from the front cover. Support the cover on top of two blocks of wood and drive the seal out from the front with a hammer and punch. **Caution:** *Be careful not to scratch, gouge or distort the area that the seal fits into or a leak will develop.*

8.6a Remove all the timing chain cover bolts (arrows) - don't forget the oil pan-to-cover bolts that are removed from below

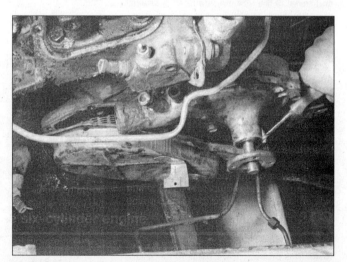

8.6b Separate the front cover from the block by prying it enough to break the gasket's hold, but be careful not gouge or scratch the gasket surface on either part or a leak could develop - also note that there are two dowel pins that align the cover with the block, so the cover should be removed as straight as possible to prevent distorting the dowel holes in the cover

1975 and later models

8   If the seal is being removed while the cover is still attached to the engine block, carefully pry the seal out of the cover with a seal puller or a large screwdriver **(see illustration)**. **Caution:** *Be careful not to scratch, gouge or distort the area that the seal fits into or a leak will develop.*

9   Clean the bore to remove any old seal material and corrosion. Wipe the surface clean with a rag soaked in lacquer thinner or acetone.

Installation

Refer to illustration 8.10

10   After either method of seal removal is complete and the new seal is ready to be installed, coat the metal outside diameter of the seal with RTV sealant. Position the seal on the cover so the seal lip (end with the spring) will face IN, toward the engine. Drive the seal into the bore with a large socket and hammer until it's completely seated **(see illustration)**. Select a socket that's the same outside diameter as the seal and make sure the new seal is pressed into place until it bottoms against the cover flange. **Note:** *The seal on 1974 models is driven in from the back side of the cover. On 1975 and later models, the seal is driven in from the front.*

11   If the front cover was removed, place a new cover-to-block gasket and rubber end seal on the front cover, attaching the gasket to the cover with RTV sealant. Install the front cover, tightening the cover-to-block bolts to the torque listed in this Chapter's Specifications. Apply a bead of RTV sealant just to the rear of the rubber timing cover end seal, to make sure there's a good seal at this problem sealing area, then install and tighten the oil pan bolts to the torque listed in this Chapter's Specifications.

12   Lubricate the seal lips with engine oil and reinstall the vibration damper. A special threaded tool, available at most auto parts stores, is normally used to install the vibration damper. However, if care is taken, a large wooden block and a hammer can be used to seat the damper. Take special care to ensure the Woodruff key and keyway in the damper line up before pressing the damper into place. The remainder of installation is the reverse of removal.

9   Timing chain, cover and sprockets - removal, inspection and installation

Timing chain slack check (cover on engine)

Note: *This procedure will allow you to check the amount of slack in the timing chain without removing the timing chain cover from the engine.*

1   Disconnect the negative battery cable from the battery and remove the engine cover (see Section 2).

8.8 On 1975 and later models, pry the old seal out with a seal puller or screwdriver - the front cover can remain attached to the block

8.10 Drive the new seal into place with a socket and hammer

2   Place the engine at TDC (Top Dead Center) (see Section 3).

3   Keeping the piston on the compression stroke, place the number one piston about 30-degrees before TDC (BTDC).

4   Remove the distributor cap (see Chapter 1).

5   You must rotate the crankshaft very slowly by hand for accuracy on this check, so get a 1/2-inch-drive breaker bar, extension and correct-size socket to use on the large bolt in the center of the crankshaft pulley. **Note:** *Not all models are equipped with a bolt in the center of the pulley, but all crankshafts are threaded so a bolt of the correct size and pitch could be installed.*

6   Turn the crankshaft clockwise until the number one piston is at TDC. This will take up the slack on the left side of the timing chain.

7   Mark the position of the distributor rotor on the distributor housing.

8   Slowly turn the crankshaft counterclockwise until the slightest movement is seen at the distributor rotor. Stop and note how far

the number one piston has moved away from the TDC mark by looking at the ignition timing marks.

9   If the mark has moved more than 10-degrees, the timing chain is probably worn excessively. Remove the timing chain cover for a more accurate check.

Cover removal

10   Remove the vibration damper (see Section 8).

11   Loosen the front oil pan bolts, then remove the timing cover bolts **(see illustration 8.6a)**. **Note:** *This procedure can be done by removing the front oil pan bolts and loosening the remaining oil pan bolts so the oil pan can be slightly lowered. It can be extremely difficult on some models, however, to do this without removing the oil pan to properly reseal the gasket areas. If necessary, remove the oil pan.*

12   Detach the timing chain cover and gasket (see Section 8).

13   Remove the oil slinger from the end of the crankshaft, if equipped.

9.14 Attach a torque wrench, as shown, and measure the amount of movement to determine the need for timing chain replacement

Timing chain and gear inspection (cover removed from engine)

Refer to illustrations 9.14 and 9.15

14   Attach a socket and torque wrench to the camshaft sprocket bolt **(see illustration)** and apply force in the normal direction of crankshaft rotation (30 ft-lbs if the cylinder head is still in position, complete with rocker arms, or 15 ft-lbs if the cylinder head has been removed). Don't allow the crankshaft to rotate. If necessary, jam the crankshaft sprocket or flywheel so that it can't move. Using a ruler, note the amount of movement of the chain. If it exceeds 3/16-inch, a new timing chain will be required. **Note:** *Whenever a new timing chain is required, the entire set (chain, camshaft and crankshaft sprockets) must be replaced as an assembly.*

15   Inspect the camshaft gear for damage or wear **(see illustration)**. The camshaft gear on early models may have been steel, but most cam gears will be an aluminum gear with a nylon coating on the teeth. This nylon coating may be cracked or breaking off in small pieces. These pieces tend to end up in the oil pan and may eventually plug the oil pump pick-up screen. If the pieces have come off the camshaft gear, the oil pan should be removed to properly clean or replace the oil pump pick-up screen (see Section 13).

16   Inspect the crankshaft gear for damage or wear. The crankshaft gear is a steel gear, but the teeth can be grooved or worn enough to cause a poor meshing of the gear and the chain.

Chain removal

Refer to illustration 9.17

17   Be sure the timing marks are aligned **(see illustration)**. Hold a prybar through one of the camshaft sprocket holes to prevent the camshaft from turning. Remove the bolt from the camshaft sprocket and withdraw the camshaft sprocket and chain. The chain can be disengaged from the teeth of the crankshaft sprocket.

18   The sprocket on the crankshaft can be removed with a two or three jaw puller, but be careful not to damage the threads in the end of the crankshaft. **Note:** *If the timing chain cover oil seal has been leaking, refer to Section 8 and install a new one.*

Installation

19   Installation is the reverse of removal. Make sure the timing marks on the gears are properly aligned **(see illustration 9.17)**. Note the metal camshaft gear that replaces the original aluminum/plastic combination (this is preferred for long life). Refer to Section 8 to reinstall the front cover. Be sure to apply a 1/8-inch bead of RTV-type sealant just to the rear of the timing cover rubber seal (always use a new timing chain cover lower seal). Also use a new timing cover gasket.

20   A special threaded tool, available at

9.15 Inspect the timing chain and both gears for wear or damage

1   *Camshaft gear (original-style aluminum gear with nylon teeth)*
2   *Crankshaft gear*

most auto parts stores, is normally used to install the vibration damper. However, if care is taken, a large wooden block and a hammer can be used to seat the damper. Take special care to ensure the Woodruff key and keyway in the damper line up before pressing the damper into place.

---

10   Camshaft and lifters - removal, inspection and installation

Camshaft lobe wear check

1   To determine the extent of cam lobe wear, the valve lift should be checked prior to camshaft removal. Refer to Section 4 and remove the valve cover.

2   Position the number one piston at TDC on the compression stroke (see Section 3).

3   Beginning with the number one cylinder valves, mount a dial indicator on the engine and position the plunger against the top surface of the first rocker arm. The plunger should be directly above and in line with the valve, not the pushrod.

4   Zero the dial indicator, then very slowly turn the crankshaft in the normal direction of rotation until the indicator needle stops and begins to move in the opposite direction. The point at which it stops indicates maximum valve lift.

5   Record this figure for future reference, then reposition the piston at TDC on the compression stroke.

6   Move the dial indicator to the remaining number one cylinder rocker arm and repeat the check. Be sure to record the results for each valve.

7   Repeat the check for the remaining valves. Since each piston must be at TDC on the compression stroke for this procedure, work from cylinder-to-cylinder following the firing order sequence (see Chapter 1), i.e. after checking number 1, turn the engine until the rotor points at number 5.

9.17 The timing marks must be aligned as shown here when installing the timing chain and sprockets

8   After the check is complete, compare the result to this Chapter's Specifications. If valve lift is less than specified, cam lobe wear has probably occurred and a new camshaft should be installed.

Removal

9   Refer to the appropriate Sections and remove the rocker arms and shafts, the pushrods, the cylinder head and the timing chain and camshaft sprocket. The radiator should be removed as well (see Chapter 3).

10   There are several ways to extract the lifters from the bores. A special tool designed to grip and remove lifters is manufactured by many tool companies and is widely available, but it may not be required in every case. On newer engines without a lot of varnish buildup, the lifters can often be removed with a small magnet. A pair of locking pliers can also be used to grip the top of each lifter, but do this only if the lifters are to be replaced with new ones. **Caution:** *The top lip of the lifter is easily broken if too much pressure is applied by the locking pliers.*

11   Before removing the lifters, arrange to store them in a clearly-labeled box to ensure that they are reinstalled in their original locations. Remove the lifters and store them where they will not get dirty. Do not attempt to withdraw the camshaft with the lifters in place.

12   Remove the distributor (see Chapter 5).
13   Remove the oil pump (see Section 12).
14   Remove the fuel pump (see Chapter 4).
15   Thread a long bolt into the camshaft sprocket bolt hole to use as a handle and give you more leverage when removing the camshaft from the block.

16   Carefully pull the camshaft straight out. Support the cam near the block so the lobes do not nick or gouge the bearings as it is withdrawn. If you feel resistance, don't pull harder or you'll damage the bearings. Lift and rotate the camshaft until it aligns with the bore.

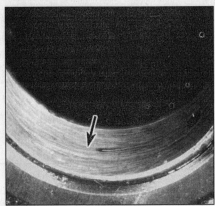

10.17 If the camshaft bearing surfaces are excessively worn or damaged, like the one shown here, the block should be taken to an automotive machine shop to have new bearings installed - this job can be done at home, but it is very involved and requires special tools

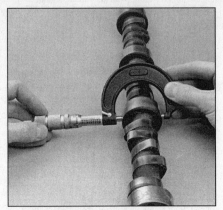

10.18 The camshaft bearing journal diameters are checked to pinpoint excessive wear and out-of-round conditions

10.21 If the bottom of any lifter is worn concave, scratched or galled, REPLACE THE ENTIRE SET AND THE CAMSHAFT WITH NEW PARTS

Inspection

Camshaft and bearings

Refer to illustrations 10.17 and 10.18

17   After the camshaft has been removed from the engine, cleaned with solvent and dried, inspect the bearing journals for uneven wear, pitting and evidence of seizure. If the journals are damaged, the bearing inserts in the block are probably damaged as well **(see illustration)**. Both the camshaft and bearings will have to be replaced with new ones.

18   Measure the bearing journals with a micrometer to determine if they are excessively worn or out-of-round **(see illustration)**.

19   Check the camshaft lobes for heat discoloration, score marks, chipped areas, pitting and uneven wear. If the lobes are in good condition and the valve-lift measurements are as specified, the camshaft can be reused.

Lifters

Refer to illustration 10.21

20   Clean the lifters with solvent and dry them thoroughly without mixing them up.

21   Check each lifter wall, pushrod seat and foot for scuffing, score marks and uneven wear. Each lifter foot (the surface that rides on the cam lobe) must be slightly convex, although this can be difficult to determine by eye (holding the foot of the lifter against the wall (side) of another lifter and rocking the lifter back-and-forth will usually determine if the lifter foot is convex). If the base of the lifter is concave or damaged **(see illustration)**, the lifters and camshaft must be replaced. If the lifter walls are damaged or worn (which is not very likely), inspect the lifter bores in the engine block as well. If the pushrod seats are worn, check the pushrod ends.

22   If new lifters are being installed, a new camshaft must also be installed. If a new camshaft is installed, then use new lifters as well. Never install used lifters unless the original camshaft is used and the lifters can be installed in their original locations.

Bearing replacement

23   Camshaft bearing replacement requires special tools and expertise that place it beyond the scope of the average home mechanic. The tool for bearing removal/installation is available at stores that carry automotive tools, possibly even found at a tool rental business. It is advisable though, if the bearings are bad, that the engine should be removed and the block taken to an automotive machine shop to ensure that the job is done correctly.

Installation

Refer to illustration 10.24

24   Lubricate the camshaft bearing journals and cam lobes with camshaft installation lube **(see illustration)**.

25   Slide the camshaft into the engine. Support the cam near the block and be careful not to scrape or nick the bearings.

26   Lubricate the lifters with clean engine oil and install them in the block. If the original lifters are being reinstalled, be sure to return them to their original locations. If a new camshaft was installed, be sure to install new lifters as well.

27   The remaining installation steps are the reverse of removal. Refer to Chapter 1 for valve adjustment procedures on 1974 to 1980 models with mechanical lifters.

28   Before starting and running the engine, change the oil and install a new oil filter (see Chapter 1).

11   Cylinder head - removal and installation

Removal

1   Drain the cooling system (see Chapter 1) and remove the engine cover (see Section 2).

2   Remove the air cleaner, disconnect the accelerator linkage and unplug the vacuum line from the distributor and the carburetor (see Chapter 4).

3   Disconnect the spark plug wires from the spark plugs.

4   Disconnect the heater hoses from the engine.

5   Disconnect the electrical connector from the coolant temperature sending unit (see Chapter 3).

6   Disconnect the exhaust pipe from the exhaust manifold (see Section 7).

7   Where applicable, disconnect the diverter valve vacuum line and remove the air tube assembly.

8   Remove the manifolds (see Section 7).

9   Remove the rocker arm cover, shaft and rocker arms (see Sections 4 and 5).

10   Remove the pushrods in sequence and label them so they can be installed in their original locations. A numbered box or rack will keep them properly organized.

11   Remove the cylinder head bolts, loosening the bolts in the reverse order of the tightening sequence **(see illustration 11.15)** and lift the cylinder head from the engine block. It may be necessary to break the gasket's seal by prying up the cylinder head. Try to find a casting protrusion to pry against. **Caution:** *Do not wedge any tools between the cylinder head and block gasket mating surfaces.*

Installation

Refer to illustration 11.15

Note: *Clean all the head bolts and run a bottoming tap through the bolt holes in the block and a die down each head bolt (clamp the bolt head in a vise while doing this). Lightly lubricate the bolts before assembly.*

12   Scrape away all traces of gasket material from the mounting surfaces of the cylinder head and the block. Also clean the cylinder head-to-manifold mating surfaces. While scraping, stuff rags into the cylinder bores and pushrod holes in the block to keep gasket material from falling in. If any material does fall in, use a vacuum cleaner to remove it. Gasket removal solvents are available from auto parts stores and may prove helpful. After all gasket material is removed, wipe the surfaces clean with a rag soaked in lacquer thinner or acetone.

13   Position the head gasket over the dowel pins on the block, making sure that it is facing the right direction and that the correct

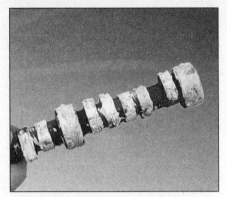

10.24 Be sure to apply camshaft installation lube to the cam lobes and bearing journals before installing the camshaft

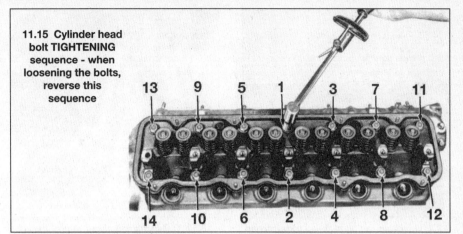

11.15 Cylinder head bolt TIGHTENING sequence - when loosening the bolts, reverse this sequence

surface is exposed. Gaskets are often marked "FRONT" and "THIS SIDE UP" to ensure correct installation.

14   Carefully lower the cylinder head onto the block.

15   Coat the cylinder head retaining bolts with light engine oil and thread the bolts into the block. Tighten the bolts in the proper sequence **(see illustration)**. Work up to the final recommended torque listed in this

12.3a Remove the mounting bolts (arrows) and detach the pump assembly from the block

Chapter's Specifications in three steps.

16   The remainder of the installation procedure is the reverse of removal. After installation, see Chapter 1 for the valve adjustment procedure.

12   Oil pump - removal and installation

Removal

Refer to illustrations 12.3a and 12.3b

1   Disconnect the cable from the negative battery terminal.

2   The oil pump is externally mounted to the block and is located on the right side (passenger's side) of the cylinder block near the engine mount. Drain the oil and remove the oil filter.

3   Remove the mounting bolts and detach the pump assembly from the block **(see illustrations)**. **Note:** *The oil pump can be disassembled **(see illustration)** and rebuilt, but the cost of a new or re-manufactured oil pump is probably less than the individual parts would be. See Chapter 2B for the inspection procedures for checking oil pump clearances but use this Chapter's Specifications. We recommend replacing the oil pump if there's any doubt at all about its condition.*

Installation

4   Clean the gasket mating surfaces on the block and pump. Position the pump on the engine block with a new gasket. Install the mounting bolts and tighten them to the torque listed in this Chapter's Specifications.

5   Run the engine and check for oil pressure and leaks.

13   Oil pan - removal and installation

Note: *On 1978 and earlier models, refer to the special procedure outlined in Chapter 2B, Section 13.*

Removal

Refer to illustrations 13.6 and 13.7

1   Disconnect the battery cable from the negative battery terminal.

2   Raise the vehicle and support it securely on jackstands (see Chapter 1).

3   Drain the engine oil (see Chapter 1).

4   Remove the engine-to-transmission brace.

5   On vehicles equipped with an automatic transmission, remove the torque converter inspection cover.

6   Remove the oil pan mounting bolts **(see illustration)**, separate the oil pan from the engine and remove the oil pan. If the pan is

2A

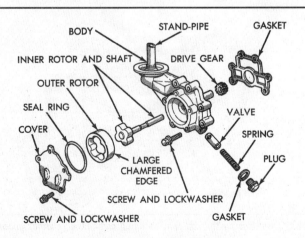

12.3b Oil pump - exploded view

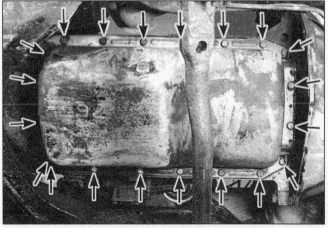

13.6 Remove the oil pan bolts (arrows)

13.7 Unscrew the oil pick-up tube/screen assembly from the block and either clean or replace it at this time

13.12 Checking the alignment of the oil pump pick-up tube (it must be the specified distance from the inside edge of the block)

stuck to the block, strike it with a rubber mallet to break the gasket seal. Do not pry between the pan and block or damage to the gasket surfaces may occur and oil leaks will develop. **Note:** *It may be necessary to rotate the crankshaft when removing the oil pan to get the crankshaft counterweights in a position so the pan can be moved past them.*

7   The oil pick-up tube and screen should be cleaned at this time. Unscrew the tube/screen assembly **(see illustration)** from the block and clean it with solvent. Check the condition of the screen and replace it if necessary.

8   Clean the oil pan with solvent and wipe it dry with a clean cloth.

Installation

Refer to illustrations 13.12 and 13.13

9   Scrape away all traces of gasket material from the mounting surfaces of the oil pan and the block. Gasket removal solvents are available from auto parts stores and may prove helpful. After all gasket material is removed, wipe the surfaces clean with a rag soaked in lacquer thinner or acetone.

10   Check the oil pan mounting flange for distortion and straighten it, if necessary.

11   Thread the oil pick-up tube and screen into the block, applying thread sealant to the threads, and tighten it until it is it tight, then proceed to the next Step to position it correctly. **Caution:** *Be absolutely certain that the pick-up screen is properly tightened so that no air can be sucked into the oiling system at this connection; otherwise, the oil pump could lose its prime and serious engine damage could result.*

12   Check the position of the tube/screen assembly **(see illustration)**. The screen should be about 1-1/8 inch from the inside edge of the block. Turn the screen as necessary so the dimension is correct. Also, the screen must touch the bottom of the oil pan when the oil pan is installed. But also be certain that the pick-up tube is tightly threaded into the pump body so there is no possibility that any air can be sucked into the oiling system at that connection.

13   Install the oil pan, using new gaskets.

Attach the side rail gaskets to the pan with contact cement-type gasket adhesive. Apply a bead of RTV sealant to the four points where the gaskets join **(see illustration)**.

14   Install the bolts and tighten them to the torque listed in this Chapter's Specifications, starting from the center and working out in each direction. Work up to the final torque in three steps.

15   The rest of the installation procedure is the reverse of removal. Fill the crankcase with the proper amount of oil and, after starting the engine, check carefully for leaks at the oil pan gasket sealing surfaces.

14   Rear main oil seal - removal and installation

Note: *If you're installing a new seal during overhaul, ignore the steps in this procedure that concern removal of external parts. Also, since the crankshaft is already removed, it's not necessary to use any special tools to remove the upper seal half; remove and install the upper seal half the same way as the lower seal half.*

Removal

Refer to illustrations 14.3, 14.4a and 14.4b

1   To replace the crankshaft rear oil seal, new replacement split rubber seal halves should be used in place of any rope-type seals installed during original engine assembly.

2   Remove the oil pan (see Section 13), then remove the bolts and detach the rear seal retainer from the rear main bearing cap.

3   Remove a lower rope-type oil seal by prying from the side **(see illustration)** with a small screwdriver. **Note:** *Most replacement seals are of split-rubber type composition. The seals make it possible to replace the upper rear seal without removing the crankshaft. The seal must be used as an upper and lower set and cannot be combined with the original rope-type oil seal.*

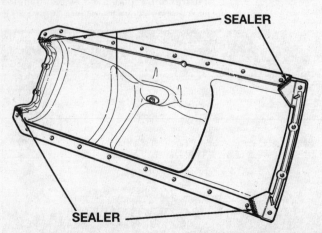

13.13 RTV-type sealant should be applied to the gasket-to-seal junctions before installing the oil pan (attach the side rail gaskets to the oil pan with contact cement-type gasket adhesive)

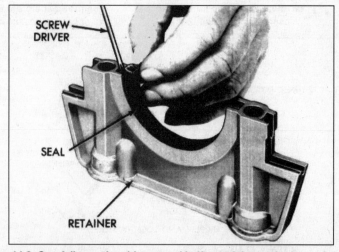

14.3 Carefully pry the old rear seal half out of the retainer groove

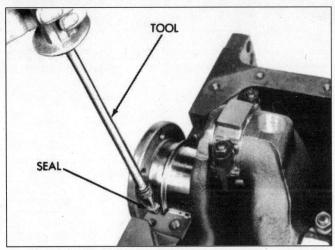

14.4a If a rope-type seal is installed, a special seal puller is recommended for removal of the upper half of the rear main seal (a sheet-metal or wood screw and a pair of pliers will work just as well)

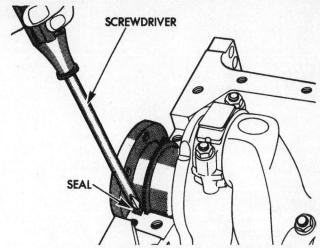

14.4b If a rubber seal is installed, push on one end with a screwdriver until the other end protrudes enough to be grasped and removed with a pair of pliers

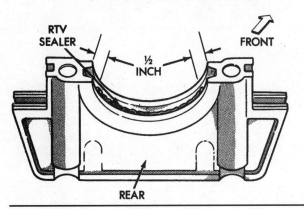

14.8 A 1/8-inch bead of RTV sealant should be applied to the retainer groove, to within 1/2-inch of each end

4   If a rope-type seal is installed, remove the upper rope piece by screwing a puller type tool, available at most auto parts stores, into the end of the seal (see illustration), being careful not to damage the crankshaft. Pull the seal out with the tool while an assistant is rotating the crankshaft. Caution: *Be very careful not to nick or gouge the crankshaft seal contact surface as this is*

done. If a rubber seal is installed, push one end (see illustration) until the other end moves out far enough to be grasped with a pair of pliers.

Installation

Refer to illustrations 14.8, 14.10 and 14.12
5   Wipe the crankshaft surface clean, then oil lightly before installing a new seal.
6   Lubricate the lip of one new seal half with engine oil.
7   Insert the new seal half into the engine block while holding the seal (with paint stripe to the rear - lip facing forward) tightly against

the crankshaft to make sure that the sharp edge of the groove in the block does not shave or nick the back of the seal. Install seal in the block groove. Rotate the crankshaft, if necessary, while sliding the seal into the groove. Care must be exercised not to damage the sealing lip.
8   Apply a 1/8-inch bead of RTV sealant to the bottom of the retainer groove, starting and finishing 1/2-inch from the ends of the groove (see illustration), then install the other half of the seal in the retainer.
9   Install the other half of the seal into the lower seal retainer with the paint stripe to the rear (lip facing forward).
10   Install the two side seals into the grooves in the seal retainer (see illustration) with contact cement-type gasket adhesive.
11   Lightly grease the retainer side seals for ease of assembly into the block.
12   Install the lower seal retainer after applying a small amount of RTV sealant around each bolt hole (see illustration). Caution: *DON'T allow sealant to get on the seal ends or lip.* Tighten the bolts to the torque listed in this Chapter's Specifications.
13   The remainder of installation is the reverse of removal. Add the correct type and quantity of oil to the engine.
14   Start the engine and check for oil leaks.

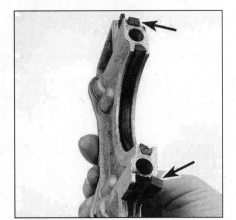

14.10 When installing the two side seals (arrows), be sure to use a contact cement-type gasket sealer to hold them in place during installation

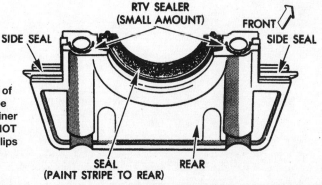

14.12 A small amount of RTV sealant should be applied to the seal retainer mating surfaces - DO NOT get any sealant on the lips or ends of the seal!

Notes

Chapter 2 Part B
V6 and V8 engines

Contents

Specifications

General

Cylinder numbers (front-to-rear)

V6 engine

Left (driver's) side.. 1-3-5

Right side 2-4-6

V8 engines

Left (driver's) side.. 1-3-5-7

Right side 2-4-6-8

Bore and stroke

239 cubic inch V6... 3.91 x 3.31 inches

318 cubic inch V8... 3.91 x 3.31 inches

360 cubic inch V8... 4.00 x 3.58 inches

400 cubic inch V8... 4.342 x 3.375 inches

440 cubic inch V8... 4.320 x 3.750 inches

Firing order

V6 engine.. 1-6-5-4-3-2

V8 engines 1-8-4-3-6-5-7-2

Distributor rotation (viewed from above)

V6 and small-block V8 engines .... Clockwise

Big-block V8 engines .... Counterclockwise

Minimum compression

Engine warm, spark plugs removed and a wide-open-throttle......... 100 psi

Maximum variation between cylinders............................. 25 psi

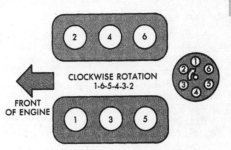

Cylinder location and distributor rotation diagram - V6 engines

The blackened terminal shown on the distributor cap indicates the Number One spark plug wire position

318, 360 engines   400, 440 engines

Cylinder location and distributor rotation diagram - V8 engines

Camshaft

Journal diameter

V6 engine

No. 1 ....

No. 2

1988 to 1991 .... 1.967 to 1.968 inches

1992 on .... 1.982 to 1.983 inches

No. 3 .... 1.951 to 1.952 inches

No. 4 .... 1.5605 to 1.5615 inches

V8 engines

No. 1 .... 1.998 to 1.999 inches

No. 2 .... 1.982 to 1.983 inches

No. 3 .... 1.967 to 1.968 inches

No. 4 .... 1.951 to 1.952 inches

No. 5

Small-block.. 1.5605 to 1.5615 inches

Big-block .... 1.748 to 1.749 inches

End play

Standard... 0.002 to 0.010 inch

Service limit.. 0.010 inch

Note: The No. 1 V6 journal diameter 1.998 to 1.999 inches appears adjacent to No. 1 line.

Oil pump

Minimum pressure at curb idle	8 psi
Operating pressure	30-to-80 psi at 2000 rpm
Oil pump cover surface variation	0.0015 inch maximum
Outer rotor thickness	
V6 and small-block V8	0.825 inch minimum
Big-block V8	0.943 inch minimum
Outer rotor diameter	2.469 inches minimum
Inner rotor thickness	
V6 and small-block	0.825 inch minimum
Big-block V8	0.943 inch minimum
Rotor-to-cover clearance	0.004 inch maximum
Outer rotor-to-body clearance	0.014 inch maximum
Rotor tip clearance	
V6	0.008 inch maximum
V8	0.010 inch maximum

Valve lift

V6 engine	
1988 through 1992	
Intake	0.373 inch
Exhaust	0.400 inch
1993 on	
Intake	0.432 inch
Exhaust	0.432 inch
318 V8 engine	
1974 through 1992	
Intake	0.373 inch
Exhaust	0.400 inch
1993 on	
Intake	0.432 inch
Exhaust	0.432 inch
360 V8 engine	
Intake	0.410 inch
Exhaust	0.410 inch
400 and 440 V8 engines	
Intake	0.434 inch
Exhaust	0.430 inch

Torque specifications*

	Ft-lbs (unless otherwise indicated)
Camshaft sprocket bolt	
V6 and small-block V8	50
Big-block V8	35
Camshaft thrust plate bolts (V6 and small-block V8 only)	18
Cylinder head bolts	
V6 engine	105
Small-block V8	
1974 through 1978	95
1979 through 1980	105
1981 through 1984	95
1985 on	105
Big-block V8	70
Engine mount-to frame nuts	
V6 and 1993 V8 engines	30
1992 and earlier V8 engines	65
Engine mount stud nut	75
Exhaust manifold bolts (V6 and small-block V8)	20
Exhaust manifold nuts	
V6 and small-block V8	15
Big-block V8	30
Exhaust pipe flange nuts	24
Flywheel/driveplate bolt	55
Intake manifold bolts	
Carbureted models	45
Fuel injected models	
Step 1	72 in-lbs
Step 2	72 in-lbs
Step 3	144 in-lbs
Step 4	144 in-lbs

Torque specifications*

	Ft-lbs (unless otherwise indicated)
Intake manifold/valley pan end bolts (big-block V8 only)	120 in-lbs
Intake plenum pan bolts (fuel-injected models only)	
Step 1	24 in-lbs
Step 2	48 in-lbs
Step 3	84 in-lbs
Oil pan bolts	200 in-lbs
Oil pump mounting bolts	30
Rocker arm shaft bracket bolts	
V6 and small-block V8	200 in-lbs
Big-block V8	25
Roller valve lifter retainer bolts	200 in-lbs
Timing chain cover bolts	
V6 and small-block V8	35
Big-block V8	200 in-lbs
Main bearing cap bolts	85
Pulley-to-damper bolts	
V6 and small-block V8	200 in lbs
Big-block V8	108 in-lbs
Rear oil seal retainer bolts (big-block V8 only)	25
Timing cover bolts	
V6 and small-block V8	35
Big-block V8	17
Rocker arm shaft bolts	200 in-lbs
Valve cover	
Nuts	
1974 through 1980	40 in-lbs
1981 through 1993	80 in-lbs
1994 on	95 in-lbs
Studs	115 in-lbs
Vibration damper-to-crankshaft bolt	
V6	135
Small-block V8	
1987 and earlier	100
1988 on	135
Big-block V8	135
Water pump housing bolts (big-block V8 only)	30

*Note: *Refer to Chapter 2, Part C for additional specifications.*

1 General information

This part of Chapter 2 is devoted to in-vehicle repair procedures for V6 and V8 engines. All information concerning engine removal and installation and engine block and cylinder head overhaul can be found in Part C of this Chapter.

Since the repair procedures included in this Part are based on the assumption that the engine is still installed in the vehicle, if they are being used during a complete engine overhaul (with the engine already out of the vehicle and on a stand) many of the steps included here will not apply.

The specifications included in this Part of Chapter 2 apply only to the procedures found here. The specifications necessary for rebuilding the block and cylinder heads are included in Part C.

Three basic engine designs, with many similarities, are covered in this Chapter.

The 318 cubic-inch V8 (5.2L) has been in the car and truck line for two decades, along with its larger brother, the 360 V8. These two engines are referred to as "small-block V8s".

The 239-cubic-inch V6 (3.9L) was introduced in 1988 to replace the venerable 225

straight six, and is a "modular" engine design that is basically a 318 V8 with two cylinders removed.

The 400 and 440 cubic-inch V8s (6.6L and 7.3L) were in the truck line until 1978, and are referred to as "big-block V8s". Though larger, they share many common characteristics with other V8s used in Dodge, Plymouth and Chrysler vehicles.

2 Repair operations possible with the engine in the vehicle

Refer to illustrations 2.3a and 2.3b

Many major repair operations can be accomplished without removing the engine from the vehicle.

Clean the engine compartment and the exterior of the engine with some type of pressure washer before any work is done. A clean engine will make the job easier and will help keep dirt out of the internal areas of the engine.

Almost all work to the engine will require the removal of the engine cover. This is done by adjusting both front seats as far back as possible, unbolting the two brackets at the rear of the cover (some models also have a

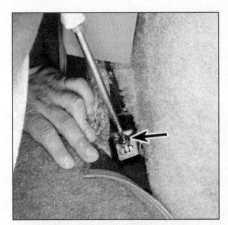

2.3a Remove the screws (arrow) at the rear of the engine cover

third bracket on the driver's side of the cover), and releasing the two forward latches **(see illustrations).**

If oil or coolant leaks develop, indicating a need for gasket or seal replacement, the repairs can generally be made with the engine in the vehicle. The oil pan gasket, the cylinder head gaskets, intake and exhaust manifold gaskets, timing cover gaskets and the crankshaft oil seals are all accessible with

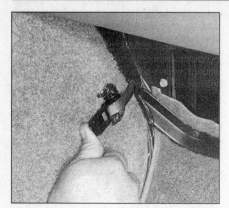

2.3b Release the two latches at the front of the engine cover

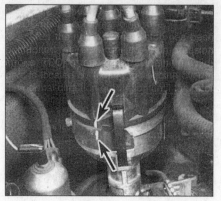

3.4 Mark the base of the distributor below the number one spark plug terminal (arrows)

3.6 On V6 and small-block V8 engines, align the full-width groove on the damper with the "0" or "TDC" mark on the timing cover - on a big-block V8, the timing scale is a sheetmetal plate attached to the front cover on the passenger side - on most van models, the marks must be observed from below

the engine in place.

Exterior engine components, such as the water pump, the starter motor, the alternator, the distributor and the carburetor or fuel injection, as well as the intake and exhaust manifolds, can be removed for repair with the engine in place.

Since the cylinder heads can be removed without pulling the engine, valve component servicing can also be accomplished with the engine in the vehicle.

Replacement of, repairs to or inspection of the timing chain and sprockets and the oil pump are all possible with the engine in place.

In extreme cases caused by a lack of necessary equipment, repair or replacement of piston rings, pistons, connecting rods and rod bearings is possible with the engine in the vehicle. However, this practice is not recommended because of the cleaning and preparation work that must be done to the components involved.

3   Top Dead Center (TDC) for number one piston - locating

Refer to illustrations 3.4, 3.6 and 3.7

1   Top Dead Center (TDC) is the highest point in the cylinder that each piston reaches as it travels up the cylinder bore. Each piston reaches TDC on the compression stroke and again on the exhaust stroke, but TDC generally refers to piston position on the compression stroke. The timing marks are referenced to the number one piston at TDC on the compression stroke.

2   Positioning the pistons at TDC is an essential part of many procedures such as camshaft removal, timing chain replacement and distributor removal.

3   To bring any piston to TDC, the crankshaft must be turned using one of the methods outlined below. When looking at the front of the engine, normal crankshaft rotation is clockwise. **Warning:** *Before beginning this procedure, be sure to place the transmission in Neutral and disable the ignition system by disconnecting the coil wire from the distributor cap and grounding it on the engine block.*

a) *The preferred method is to turn the crankshaft with a large socket and breaker bar attached to the large bolt that is threaded into the front of the crankshaft.*

b) *A remote starter switch, which may save some time, can also be used. Attach the switch leads to the S (switch) and B (battery) terminals on the starter solenoid. Once the piston is close to TDC, use a socket and breaker bar as described in the previous paragraph.*

c) *If an assistant is available to turn the ignition switch to the Start position in short bursts, you can get the piston close to TDC without a remote starter switch. Use a socket and breaker bar as described in Paragraph (a) to complete the procedure.*

4   Scribe or paint a small mark on the distributor body directly below the number one spark plug wire terminal in the distributor cap **(see illustration)**.

5   Remove the engine cover (see Section 2). Remove the distributor cap as described in Chapter 1.

6   Turn the crankshaft (see Step 3) until the line on the vibration damper is aligned with the zero or "TDC" mark on the timing indicator **(see illustration)**. **Note:** *On most models, the timing marks are only visible from* **underneath** *the engine, as shown.*

7   The rotor should now be pointing directly at the mark on the distributor base **(see illustration)**. If it is 180-degrees off, the piston is at TDC on the exhaust stroke.

8   If the rotor is 180-degrees off, turn the crankshaft one complete turn (360-degrees) clockwise. The rotor should now be pointing at the mark. When the rotor is pointing at the number-one spark plug wire terminal in the distributor cap (which is indicated by the mark on the distributor body) and the timing marks are aligned, the number one piston is at TDC on the compression stroke.

9   After the number one piston has been positioned at TDC on the compression stroke, TDC for any of the remaining cylinders is located by turning the engine over in the normal direction of rotation until the rotor

is pointing to the spark plug wire terminal in the distributor cap (you will have to install and remove the distributor cap as you are turning the engine over) for the piston you wish to bring to TDC.

4   Valve covers - removal and installation

Removal

Refer to illustrations 4.4, 4.6 and 4.7

1   Disconnect the battery cable from the negative battery terminal and remove the engine cover (see Section 2).

2   Remove the air cleaner assembly (see Chapter 4).

3   Detach any air injection (AIR) system components that are in the way, if equipped (see Chapter 6), and the long oil-filler tube from the left valve cover.

4   Remove the breather tube or PCV valve and hose **(see illustration)**.

5   Label the spark plug wires and position

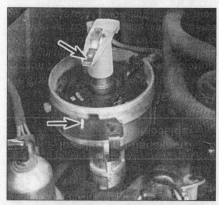

3.7 The tip of the rotor should now align with the mark made on the distributor body below the number one terminal (arrows)

4.4 Remove the PCV valve or breather tube and move the spark plug wires aside

1 *PCV valve*
2 *Spark plug wires*

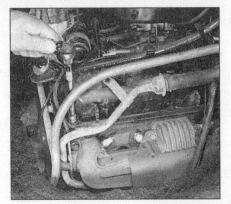

4.6 Remove the valve cover mounting bolts

4.7 Move the wires and hoses aside and guide the valve cover out

the brackets/wires out of the way.

6 Remove the valve cover mounting bolts **(see illustration)**.

7 Remove the valve cover **(see illustration)**. Note: *If the cover is stuck to the head, bump the cover with a block of wood and a hammer to release it. If it still will not come loose, try to slip a flexible putty knife between the head and cover to break the seal. Don't pry at the cover-to-head joint, as damage to the sealing surface and cover flange will result and oil leaks will develop.*

4.8 Remove all traces of old gasket material

Installation

Refer to illustration 4.8

8 The mating surfaces of each cylinder head and valve cover must be perfectly clean when the covers are installed. Use a gasket scraper to remove all traces of sealant or old gasket **(see illustration)**, then wipe the mating surfaces with a cloth saturated with lacquer thinner or acetone. If there is sealant or oil on the mating surfaces when the cover is installed, oil leaks may develop.

9 Make sure any threaded holes are clean. Run a tap into them to remove corrosion and restore damaged threads.

10 Mate the new gaskets to the covers before the covers are installed. Apply a thin coat of RTV sealant to the cover flange, then position the gasket inside the cover lip and allow the sealant to set up so the gasket adheres to the cover (if the sealant is not allowed to set, the gasket may fall out of the cover as it is installed on the engine). Note: *A better-sealing valve cover gasket has been available from Dodge since 1991. The new gasket is silicone and steel with compression limiters. Because it is thicker it requires different fasteners. A kit (with gaskets and fasteners) is available from Dodge for all 3.9L V6 engines, and one for all 318/360 V8 engines.*

11 Carefully position the cover on the head and install the nuts/bolts.

12 Tighten the bolts in three steps to the

torque listed in this Chapter's Specifications. **Caution:** *DON'T over-tighten the valve cover bolts.*

13 The remaining installation steps are the reverse of removal.

14 Start the engine and check carefully for oil leaks as the engine warms up.

5 Rocker arms, shafts and pushrods - removal, inspection and installation

Removal

Refer to illustrations 5.2a, 5.2b, 5.2c, 5.2d and 5.3

Note: *The 1992 and later "Magnum" engines have individual rocker arms instead of rocker shafts. The same cautions about keeping the parts together for each valve still apply. The pushrods, rocker arms and pivots should be kept in order, i.e. "intake 1, exhaust 1" etc.*

1 Remove the valve covers from the cylinder heads (see Section 4).

2 Loosen the rocker arm shaft bolts a little at a time (or rocker-arm bolts on Magnum engines), working back and forth until they're all loose. Lift off the rocker arm shaft with the bolts in place and, if they are removed from the shafts, keep the rocker arms in order (the rocker arms and shafts must be reinstalled in their original positions) **(see illustrations)**.

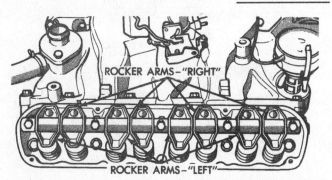

5.2a The correct rocker arm locations (left and right) on the V8 rocker shaft - note the locations of the five bolts along the shaft that secure the shaft to the cylinder head

5.2b V6 engines have only four bolts to each rocker shaft

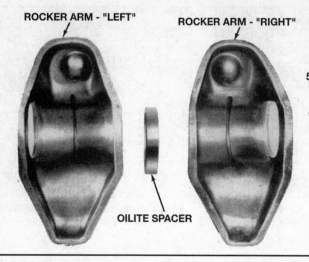

ROCKER ARM - "LEFT"   ROCKER ARM - "RIGHT"

5.2c Big-block engines
use "Oilite" spacers
between rocker arms
for the same cylinders

OILITE SPACER

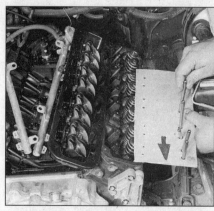

5.3 A perforated sheet of cardboard can
be used to store the pushrods to ensure
that they're reinstalled in their original
locations - note the label indicating
the front of the engine

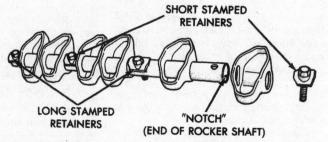

SHORT STAMPED
RETAINERS

5.2d V6 and small-
block V8 engines have
different-size rocker
shaft retainers -
mark them for proper
reassembly

LONG STAMPED
RETAINERS

"NOTCH"
(END OF ROCKER SHAFT)

Note 1: *We recommend removing the rocker arms from the shafts for cleaning and inspection. The best way to keep the rocker arms in order once they're removed is to tie a piece of heavy wire (such as an straightened coathanger wire) around a short dowel or similar object, then slide each rocker arm, starting from the engine-front end, onto the end of the wire in the same orientation as it was on the shaft. When finished, you'll have the rocker arms stored in the correct order in a*

way that will allow them to be cleaned and inspected easily. Just remember that the end of the wire with the dowel represents the front of the engine. Mark each dowel (as well as the rocker shaft) to indicate which side of the engine it is from. Also, mark the rocker shafts to indicate which end faces front.

Note 2: *On big-block (400 and 440) V8 engines, there is an "Oilite" spacer used between rocker arms that are on the same cylinder. On the 318 and 360, there are no*

"Oilite" spacers, but the spacers under the bolts are not the same size and must be reinstalled in their original locations, so note where they go and, if necessary, make a drawing showing the locations of the long and short spacers under the bolts **(see illustrations).**

3   Remove the pushrods and store them separately to make sure they don't get mixed up during installation **(see illustration). Caution:** *All valve train components must go back in their original positions. Organize and store the parts in a way that they won't get mixed up.*

Inspection

Refer to illustration 5.5

4   Check each rocker arm for wear, cracks and other damage, especially where the pushrods and valve stems contact the rocker arm faces. Also check the rocker-arm pivots for wear on Magnum engines.

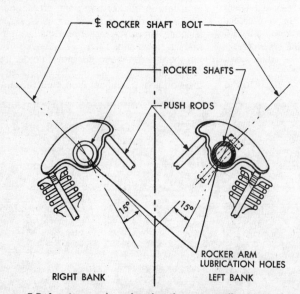

₵ ROCKER SHAFT BOLT

ROCKER SHAFTS

PUSH RODS

15°   15°

ROCKER ARM
LUBRICATION HOLES

RIGHT BANK   LEFT BANK

5.5 A cutaway view showing the proper alignment of
the oiling holes in the rocker arm shaft

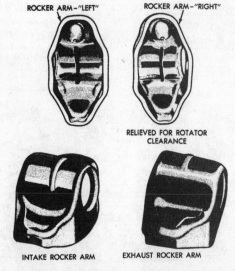

ROCKER ARM-"LEFT"   ROCKER ARM-"RIGHT"

RELIEVED FOR ROTATOR
CLEARANCE

INTAKE ROCKER ARM   EXHAUST ROCKER ARM

5.9 Rocker arm identification details showing the offset of
the pushrod on both right and left side rocker arms

5.11 Lubricate the pushrod ends and the valve stems with a moly-based grease before installing the rocker arms

6.5 This is what the air hose adapter that fits into the spark plug hole looks like - they're commonly available from auto parts stores

6.7 Once the spring is depressed, the keepers can be removed with a small magnet or needle-nose pliers (a magnet is preferred to prevent dropping the keepers)

5   Make sure the oil-feed holes in each rocker arm shaft **(see illustration)** are not plugged so that each rocker arm gets proper lubrication.

6   Check each rocker arm pivot area and wear area on the shaft for excessive wear, scoring, cracks and galling. If the rocker arms or shafts are worn or damaged, replace them with new ones. **Note:** *Keep in mind that there is no valve adjustment on these engines, so excessive wear or damage in the valve train can easily result in excessive valve clearance, which in turn will cause valve "tapping" or "clattering" noises when the engine is running.*

7   Inspect the pushrods for cracks and excessive wear at the ends. Roll each pushrod across a piece of plate glass to see if it is bent (if it wobbles, it is bent). Replace the pushrods if any of these conditions are present.

Installation

Refer to illustrations 5.9 and 5.11

8   Assemble the rocker arms onto the rocker arm shafts in the same orientation as they were originally.

9   If replacing rocker arms, note that, on engines with exhaust valve rotators, exhaust rockers must have a clearance relief **(see illustration)**. **Caution:** *When replacing any of the valvetrain components on 3.9L V6 engines, make sure the replacement parts are for the exact year of your engine. There have been three variations of pushrods/lifters/rocker arms and the parts are not interchangeable. Some pushrods have a hole at each end, some do not, some rocker arms have an oil hole on the pushrod end, and the Magnum engines have roller lifters with longer pushrods. Match the design of the replacement part to your old part when making any replacement, or serious oiling problems could develop.*

10   Lubricate the lower end of each pushrod with clean engine oil or moly-base grease and install them in their original locations. Make sure each pushrod seats completely in the lifter socket.

11   Apply moly-base grease to the ends of

the valve stems and the upper ends of the pushrods before positioning the rocker arm shafts onto the pedestals on the engine **(see illustration)**. Install the shafts so the notch on the end of each shaft is pointing toward the centerline of the engine, and toward the front of the engine on the driver's (left) side or toward the rear on the passenger's (right) side. **Caution:** *Be sure the wear areas on the shaft are facing down, as they originally were. DON'T try to gain some extra service life by turning the shaft over, since the oil holes in the shaft must face down.*

12   Tighten the rocker-arm shaft bolts evenly, in several stages, working back and forth. Apply moly-base grease to the contact surfaces to prevent damage before engine oil pressure builds up. Tighten the bolts to the torque listed in this Chapter's Specifications.

13   Refer to Section 4 and install the valve covers. Start the engine, listen for unusual valve train noise and check for oil leaks at the valve cover joints, then reinstall the engine cover.

---

6   Valve springs, retainers and seals - replacement

Refer to illustrations 6.5, 6.7, 6.14, 6.16 and 6.18

Note: *Broken valve springs and defective valve stem seals can be replaced without removing the cylinder head. Two special tools and a compressed-air source are normally required to perform this operation, so read through this Section carefully and rent or buy the tools before beginning the job.*

1   Remove the valve covers (see Section 4).

2   Remove the spark plugs (see Chapter 1).

3   Turn the crankshaft until the number one piston is at top dead center on the compression stroke (see Section 3).

4   Remove the rocker arm shaft (see Section 5) or individual rockers (Magnum engines). If you intend to use a lever-type spring compressor, remove the rocker arms

from the shaft and re-install the shaft without the rocker arms. For Magnum engines, a different tool is commonly available that bolts to the rocker-arm stud.

5   Thread an adapter into the spark plug hole and connect an air hose from a compressed air source to it **(see illustration)**. Most auto parts stores can supply the air hose adapter. **Note:** *Many cylinder compression gauges utilize a screw-in fitting that may work with your air hose quick-disconnect fitting.*

6   Apply compressed air to the cylinder. The valves should be held in place by the air pressure. If the valve faces or seats are in poor condition, leaks may prevent the air pressure from retaining the valves - valve reconditioning is indicated.

7   Stuff shop rags into the cylinder head holes around the valves to prevent parts and tools from falling into the engine, then use a valve-spring compressor to compress the spring. Remove the keepers with small needle-nose pliers or a magnet **(see illustration)**. **Note:** *Several different types of tools are available for compressing the valve springs with the head in place. One type, shown here, grips the lower spring coils and presses on the retainer as the knob is turned, while the lever-type utilizes the rocker arm shaft for leverage. Both types work very well, although the lever type is usually less expensive.*

8   Remove the valve spring and retainer (intake valve) or rotator (exhaust valve). **Caution:** *Different-length springs are used on the intake and exhaust valves - accidentally installing an intake spring on an exhaust valve can cause major engine damage.*

9   Remove the old valve stem seals. **Caution:** *The exhaust valve oil seal is referred to by the manufacturer as an oil shield. The intake valve seal fits over the valve guide, while the exhaust valve "shield" is what's commonly referred to in the industry as an "umbrella-type" seal. If you install an exhaust valve shield on an intake valve, high oil consumption will result.*

2B

6.14 Be sure to install the seals on the correct valve stems

1   *Exhaust seal (shield)*
2   *Intake seal*

10   Wrap a rubber band or tape around the top of the valve stem so the valve won't fall into the combustion chamber, then release the air pressure.

11   Inspect the valve stem for damage. Rotate the valve in the guide and check the end for eccentric movement, which would indicate that the valve is bent.

12   Move the valve up-and-down in the guide and make sure it doesn't bind. If the valve stem binds, either the valve is bent or the guide is damaged. In either case, the head will have to be removed for repair.

13   Reapply air pressure to the cylinder to retain the valve in the closed position, then remove the tape or rubber band from the valve stem.

14   If you're working on an exhaust valve, install the new shield on the valve stem and push it down to the top of the valve guide **(see illustration)**.

15   If you're working on an intake valve, push a new valve stem seal down over the valve guide, using the valve stem as a guide, but don't force the seal against the top of the guide.

16   Install the spring and retainer or rotator in position over the valve **(see illustration)**. **Caution:** *Different length springs are used on*

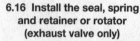

6.16 Install the seal, spring and retainer or rotator (exhaust valve only)

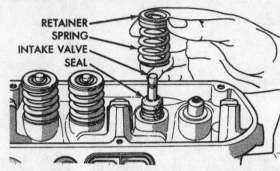

the intake and exhaust valves - accidentally installing an intake spring on an exhaust valve can cause major engine damage!

17   Compress the valve spring assembly only enough to install the keepers in the valve stem.

18   Position the keepers in the valve stem groove. Apply a small dab of grease to the inside of each keeper to hold it in place if necessary **(see illustration)**. Remove the pressure from the spring tool and make sure the keepers are seated.

19   Disconnect the air hose and remove the adapter from the spark plug hole.

20   Repeat the above procedure on the remaining cylinders, following the firing order sequence (see the Specifications). Bring each piston to top dead center on the compression stroke before applying air pressure.

21   Reinstall the rocker arm assemblies and the valve covers.

22   Start the engine, then check for oil leaks and unusual sounds coming from the valve cover area.

---

7   Intake manifold - removal and installation

Removal

Carbureted models

1   Disconnect the battery cable from the negative battery terminal and remove the engine cover (see Section 2).

2   On V6 and small-block V8 engines, drain

the cooling system (see Chapter 1). **Note:** *On big-block engines, no coolant is routed through the intake manifold, so it's not necessary to drain the cooling system.*

3   Remove the air cleaner assembly. **Note:** *If you intend to resurface, clean or replace the manifold, remove the carburetor (see Chapter 4).*

4   Remove the alternator (see Chapter 5). On air-conditioned models, you'll probably also need to unbolt the air-conditioning compressor and set it aside. **Caution:** *DO NOT disconnect the refrigerant lines.*

5   Label and then disconnect fuel lines, wires and vacuum hoses that run from the vehicle to the intake manifold connections and carburetor.

6   On V6 and small-block V8 engines, disconnect the heater hose and water pump bypass hose from the intake manifold (see Chapter 3).

7   Disconnect the accelerator linkage (see Chapter 4) and, if equipped, the cruise control linkage. Then, on models with automatic transmissions, disconnect the throttle rod at the carburetor (see Chapter 7B). Then proceed to Step 21.

Fuel-injected models

Refer to illustrations 7.10, 7.11 and 7.12

8   Relieve the fuel system pressure (see Chapter 4).

9   Remove the serpentine drivebelt (see Chapter 1).

10   One of the alternator/compressor bracket bolts is under the power steering pump, so

6.18 Apply small dab of grease to each keeper as shown here before installation - it'll hold them in place on the valve stem as the spring is released

7.10 Access the power steering pump mounting bolts through the holes in the pulley

7.11 One bracket bolt is hidden under the idler pulley - remove the pulley, the bolts and the bracket

reach through the holes in the pulley with an extension to remove the three power-steering-pump mounting bolts (see illustration). Lay the pump aside, but upright to avoid spilling fluid.

11 Remove the bolts retaining the alternator/compressor bracket to the front of the engine. Note: *Remove the idler pulley, which covers access to one of the bolts, then remove the bracket (see illustration).*

12 Label and disconnect the electrical connectors and vacuum hoses from the intake manifold, plenum and throttle body (see illustration).

13 Disconnect the intake air temperature sender and pull the whole engine wiring harness up and over the intake manifold to the rear of the engine.

14 On models with multi-port fuel injection, unbolt the two fuel rails. The passenger-side fuel rail is attached with Torx fasteners (see Chapter 4) while the driver's side has standard bolts. The two fuel rails can be pulled straight up with the injectors still attached, but it will take some force to dislodge the injectors from the intake manifold.

15 Remove the upper radiator hose from the engine, then disconnect the heater hose and water pump bypass hose from the intake manifold.

16 Disconnect the accelerator linkage (see Chapter 4) and, if equipped, the cruise control linkage. Then, on models with automatic transmissions, disconnect the kickdown cable at the throttle body.

17 Disconnect the spark plug wires.

18 Unbolt the vacuum distribution block from the driver's side of the manifold and move it to the opposite side of the engine.

19 Remove the two bolts holding the EGR tube to the passenger-side exhaust manifold (see Section 8).

20 Unbolt the EGR valve from the back of the intake manifold, which will give better access to get a wrench on the EGR tube's fitting that is threaded into the rear of the intake manifold, just below the EGR valve itself.

All engines

Refer to illustrations 7.21 and 7.22

21 Loosen the intake manifold mounting bolts (see illustrations 7.31a and 7.31b) in 1/4-turn increments until they can be removed by hand. Note: *V6 and small-block V8 engines use six bolts on each side of the manifold (12 total); big-block engines use only four bolts per side (8 total).* The manifold will probably be stuck to the cylinder heads and force may be required to break the gasket seal. A prybar can be positioned to pry up a casting projection at the front of the manifold (see illustration) to break the bond made by the gasket. Caution: *Do not pry between the block and manifold or the heads and manifold or damage to the gasket sealing surfaces may result and vacuum leaks could develop.*

22 Remove the intake manifold (see illustration). As the manifold is lifted from the engine, be sure to check for and disconnect anything still attached or wrapped over the

7.12 Disconnect each fuel injector electrical connector and label it clearly so it can be installed on the proper injector during reassembly

manifold. On big-block engines, remove the end bolts and lift off the engine valley cover, which is integral with the intake manifold gasket.

Installation

Refer to illustrations 7.25, 7.27, 7.28, 7.31a and 7.31b

All engines

Note: *The mating surfaces of the cylinder heads, block and manifold must be perfectly clean when the manifold is installed. Gasket removal solvents in aerosol cans are available at most auto parts stores and may be helpful when removing old gasket material that is stuck to the heads and manifold. Be sure to follow the directions printed on the container.*

23 Remove carbon deposits from the exhaust crossover passages. Use a gasket scraper to remove all traces of sealant and old gasket material, then wipe the mating surfaces with a cloth saturated with lacquer thinner or acetone. If there is old sealant or oil on the mating surfaces when the manifold is installed, oil or vacuum leaks may develop. Cover the lifter valley with shop rags to keep debris out of the engine. Use a vacuum cleaner to remove any gasket material that

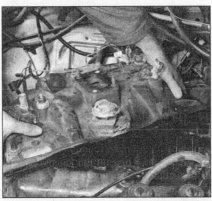

7.22 The manifold is heavy and must be removed from an awkward position; be sure to get a good hold on the manifold before trying to lift it off the engine (typical carbureted engine shown)

7.21 Pry up on the corner of the manifold, if necessary

falls into the intake ports in the heads.

24 Use a tap of the correct size to chase the threads in the bolt holes, then use compressed air (if available) to remove the debris from the holes. Warning: *Wear safety glasses or a face shield to protect your eyes when using compressed air.*

25 On late-model fuel-injected engines, the intake manifold has a stamped, sheetmetal plate on the bottom, called the plenum pan (see illustration). If you have the intake manifold off for any reason, it's a good idea to replace the gasket under this pan. Use RTV sealant on the gasket and tighten the bolts to the torque listed in this Chapter's Specifications, from the center out to the ends, in three steps.

V6 and small-block V8 engines only

26 Apply a thin coat of RTV sealant to the cylinder-head side of the new intake manifold gaskets.

27 Position the side gaskets on the cylinder heads. Look to see if the gaskets are marked LT for left or RT for right (see illustration). If they are, this will ensure proper installation. Make sure they are installed on the correct side and all intake port openings, coolant passage holes and bolt holes are aligned correctly.

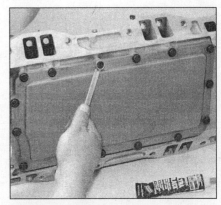

7.25 On fuel-injected engines, remove the plenum pan from under the manifold, clean the gasket surfaces, and replace the gasket, using RTV sealant

2B

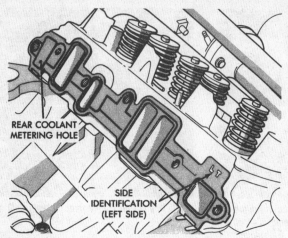

7.27 The gaskets must be installed on the proper side; some may be marked, as shown, but you'll often need to line up the holes with the ports on the head to be sure they're installed correctly

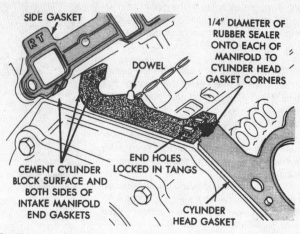

7.28 Fit the end seals over the dowels and engage the end tangs with the holes - apply a bead of RTV sealant to each of the four corners where the manifold gaskets and end seals meet

28   Install the front and rear end seals on the block. Apply a thin, uniform coating of quick-dry cement to the intake manifold end seals and the cylinder block contact surfaces. Engage the dowels (if equipped) and the end tangs **(see illustration)**. Refer to the instructions with the gasket set for further information. Apply RTV sealant at the four corners where the gaskets meet.

Big-block V8 engine only

29   Apply a bead of RTV sealant to each of the four gasket-surface corners where the cylinder head meets the engine block, then apply a thin coat of RTV to the gasket surfaces on the block, between the cylinder heads; coat both sides of the intake manifold sealing surfaces with a brush-on gasket sealer, then install the gasket/valley pan, tightening the end bolts to the torque listed in this Chapter's Specifications.

All engines

30   Carefully set the manifold in place. **Caution:** *Do not disturb the gaskets and DO NOT move the manifold fore-and-aft after it contacts the front and rear seals or the gaskets will be pushed out of place and you may not notice the problem until you see the oil leak.*

31   Install the intake manifold bolts and hand tighten them. Then tighten the bolts, following the tightening sequence **(see illustrations)**, to the torque listed in this Chapter's Specifications. Do not over torque the bolts or the gaskets may leak.

32   The remaining installation steps are the reverse of removal. Change the engine oil and filter. On V6 and small-block V8 engines, add coolant. Start the engine and check carefully for oil, vacuum and coolant leaks at the intake manifold joints. Refer to Chapter 2C for vacuum-leak detection procedures.

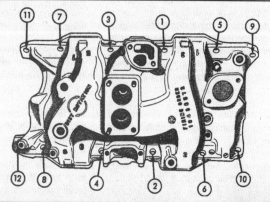

7.31a Intake manifold bolt-tightening sequence - V6 and small-block V8 engine

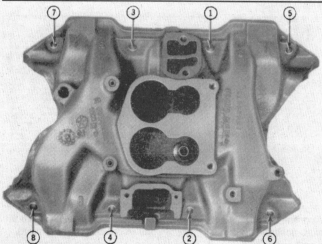

7.31b Intake manifold bolt tightening sequence (big-block engine)

8.4 Remove the nuts (arrows) and detach the air pre-heater stove (if equipped) - also unbolt the air injection tube (if equipped)

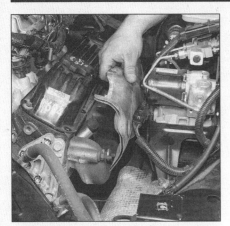

8.7 On models with heat shields, remove the nuts from the studs and remove the heat shields

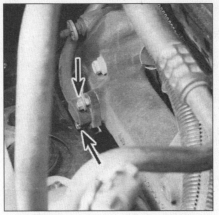

8.8 Remove the two bolts (arrows) retaining the EGR tube to the passenger-side exhaust manifold

8.10 Remove the exhaust pipe-to-manifold nuts

8 Exhaust manifolds - removal and installation

Removal

Refer to illustrations 8.4, 8.7, 8.8 and 8.10
Warning: *Allow the engine to cool completely before performing this procedure.*

1 Disconnect the battery cable from the negative battery terminal and remove the engine cover (see Section 2).

2 Disconnect the spark plug wires and remove the spark plugs (see Chapter 1). If there is any danger of mixing the plug wires up, it's recommended that they be labeled with pieces of numbered tape.

3 Remove the air cleaner assembly and duct for access to the manifold.

4 Detach the air injection hose/tube and remove the air pre-heater stove, if equipped **(see illustration)**.

5 On 1980 and later models equipped with an oxygen sensor, disconnect the wiring harness to the sensor.

6 Unbolt the air injection tube, if equipped, from the exhaust manifold.

7 Remove the exhaust manifold heat shields, if equipped **(see illustration)**.

8 On later models, disconnect the EGR tube from the intake manifold, and the two bolts on the passenger-side exhaust manifold retaining the other end of the EGR tube **(see illustration)**.

9 Set the parking brake and block the rear wheels. Raise the front of the vehicle and support it securely on jackstands.

10 Disconnect the exhaust pipe from the manifold outlets **(see illustration)**. **Note:** *Often a short period of soaking with penetrating oil is necessary to remove frozen exhaust attaching nuts/bolts.*

11 Loosen the outer fasteners first, then the center ones to separate the manifold from the head. **Note:** *The studs that retain the heat shields have large spacer washers (that go between the studs and the shields), save these washers and remember which studs they go on. If the studs come out with the*

nuts, install new studs. Apply sealant to the coarse-thread ends or coolant leaks may develop.

Installation

12 Installation is basically the reverse of the removal procedure. Clean the manifold and head gasket surfaces and check the manifold heat-control valve for ease of operation. If the valve is stuck, free it with penetrating oil.

13 Install the fasteners. Tighten them to the torque listed in this Chapter's Specifications. Work from the center to the ends and approach the final torque in three steps.

14 Apply anti-seize compound to the exhaust manifold-to-exhaust pipe bolts and tighten them securely. Also apply anti-seize to the EGR fittings.

9 Crankshaft front oil seal - replacement

Removal

Refer to illustrations 9.2 9.4, 9.6 and 9.7

1 Remove the drivebelt(s) see Chapter 1.
2 Remove the bolts **(see illustration)** and

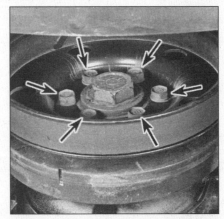

9.2 Remove the crankshaft pulley bolts (arrows) and the large bolt in the center of the vibration damper (viewed from below)

separate the crankshaft pulley(s) from the vibration damper.

3 Remove the large vibration damper bolt **(see illustration 9.2)**. To keep the crankshaft from turning, remove the starter (see Chapter 5) and have an assistant jam a large screwdriver against the ring gear teeth.

4 Use a vibration-damper puller (commonly available from auto parts stores) to detach the vibration damper **(see illustration)**. **Caution:** *Do not use a puller with jaws that grip the outer edge of the damper. The puller must be the type that utilizes bolts to apply force to the damper hub only.*

5 On all big-block engines and 1977 and earlier small-block engines, remove the timing chain cover (see Section 10). On V6 engines and 1978 and later small-block V8 engines, the seal can be replaced with the cover installed, although replacement is usually easier with the cover removed.

6 If the seal is being replaced with the timing chain cover removed (which is the only way on big-block engines and earlier small-block engines), remove the cover (see Section 10) and support the cover on top of two blocks of wood and drive the seal out from the back (later small-block engines) or front (big-block engines and earlier small-block

9.4 Use a bolt-type puller to remove the vibration damper

2B

9.6 Drive the old seal out of the cover (timing cover removed) with a hammer and punch, being careful to hit only the seal, not gouge the housing

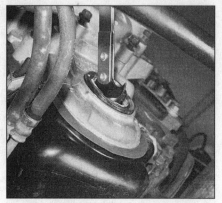

9.7 Use a seal puller with the timing cover installed on V6 and late-model V8 engines

9.9a Use a seal driver or large-diameter pipe to drive the new seal into the cover - from the backside on big-blocks and early small-blocks

engines) with a hammer and punch **(see illustration)**. **Caution:** *Be careful not to scratch, gouge or distort the area that the seal fits into or a leak will develop.*

7   If the seal is being removed while the cover is still attached to the engine block (V6 and later small-block V8 engines only), carefully pry the seal out of the cover with a seal puller or a large screwdriver **(see illustration)**. **Caution:** *Be careful not to scratch, gouge or distort the area that the seal fits into or a leak will develop.*

Installation

Refer to illustrations 9.9a, 9.9b and 9.10

8   Clean the bore to remove any old seal material and corrosion. Position the new seal in the bore with the seal lip (usually the side with the spring) facing IN (toward the inside of the engine). A small amount of oil applied to the outer edge of the new seal will make installation easier - don't overdo it!

9   After either method of seal removal is complete and the new seal is ready to be installed, drive the seal into the bore with a large socket and hammer until it's completely seated **(see illustrations)**. Select a socket that's the same outside diameter as the seal and make sure the new seal is pressed into

place until it bottoms against the cover flange.
10   Check the surface of the damper that the oil seal rides on **(see illustration)**. If the surface has been grooved from long-time contact with the seal, there is a press-on sleeve available to put a new surface on the damper for the seal to ride on. This sleeve is pressed into place with a hammer and a block of wood. These sleeves are commonly available from auto parts stores.
11   Lubricate the seal lips with engine oil and reinstall the vibration damper.
12   The remainder of installation is the reverse of removal.

---

10   Timing cover, chain and sprockets - removal, inspection and installation

Timing chain slack check (cover on engine)

Note: *This procedure will allow you to check the amount of slack in the timing chain without removing the timing chain cover from the engine.*
1   Disconnect the negative battery cable from the battery and remove the engine cover

(see Section 2).
2   Place the engine at TDC (Top Dead Center) (see Section 3).
3   Keeping the piston on the compression stroke, place the number one piston about 30-degrees before TDC (BTDC).
4   Remove the distributor cap (see Chapter 1).
5   You must rotate the crankshaft very slowly by hand for accuracy on this check, so get a 1/2-inch drive breaker bar, extension and correct-size socket to use on the large bolt in the center of the crankshaft pulley.
6   Turn the crankshaft clockwise until the number one piston is at TDC. This will take up the slack on the left side of the timing chain.
7   Mark the position of the distributor rotor on the distributor housing.
8   Slowly turn the crankshaft counterclockwise until the slightest movement is seen at the distributor rotor. Stop and note how far the number one piston has moved away from the TDC mark by looking at the ignition timing marks.
9   If the mark has moved more than 10-degrees, the timing chain is probably worn excessively. Remove the timing chain cover for a more accurate check.

Cover removal

Refer to illustrations 10.15
10   Remove the drivebelt(s) and detach any accessories such as the power steering pump, alternator and air conditioning compressor that block access to the timing chain cover. Leave the hoses connected to the A/C compressor and power steering pump and tie the units aside. Refer to Chapters 3, 5 and 10 for additional information. Unbolt the accessory brackets from the front of the engine.
11   On V6 and small-block V8 engines, remove the water pump (see Chapter 3). On big-block engines, unbolt and remove the water-pump housing from the engine block - the water pump can remain bolted to the water pump housing, although, if the water pump is in questionable condition, now is an excellent time to replace it (see Chapter 3).

9.9b On later engines, the seal can be driven in with the cover still in place on the engine

9.10 If the sealing surface of the damper hub has a wear groove from contact with the seal, repair sleeves are available from auto parts stores

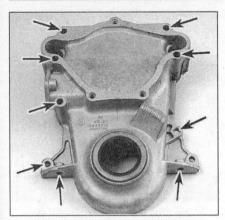

10.15 Timing chain cover bolt hole locations on V6 and small-block V8 engines (removed from engine for clarity)

Note: *On some big-block engines, one of the water-pump housing bolts cannot be removed from its hole because of interference with a heater-hose fitting. Remove all the other bolts first, then loosen the trapped bolt a little at a time, while moving the housing away from the engine block. If the bolt must be removed (for example, to replace the housing), unscrew the heater-hose fitting from the housing.*

12 On carbureted small-block V8 engines, remove the fuel pump and cap the lines (see Chapter 4).

13 Remove the crankshaft pulley and damper (see Section 9).

14 Remove the front two oil pan bolts, that thread into the timing chain cover - they're most easily accessed from below. **Note:** *Even though this procedure can be done without the removal of the oil pan, it is extremely difficult on some models and oil pan removal may actually simplify the job (see Section 13). Use a utility knife or single-edged razor blade to cut the cover free from the front of the oil pan gasket (on small-block engines, only cut along the top edge - do not cut the rounded area). If the pan has been in place for an extended period of time, it is likely the pan gasket will break when the cover is removed. **Note:** 1993 engines use a metal-reinforced, one-piece oil pan gasket. It is recommended on these models not to cut the old pan gasket. Remove the oil pan (see Section 13) and install a new one-piece gasket.*

15 Remove the remaining mounting bolts and separate the timing chain cover from the block and oil pan **(see illustration)**. The cover may be stuck; if so, use a putty knife to break the gasket seal. The cover is easily damaged, so DO NOT attempt to pry it off. **Note:** *Big-block engines have a bolt-on timing tab. Note the location and be sure to reinstall it in the same way.*

16 Remove the oil slinger, if equipped, from the end of the crankshaft. Note how it's installed so you can reinstall it the same way on reassembly. It's a good idea to replace the crankshaft front seal at this time (see Section 9).

10.17 Position a ruler over the chain to measure slack while applying pressure with a torque wrench to the camshaft sprocket bolt

10.20 With the timing marks aligned, remove the camshaft sprocket bolt

1 Timing marks
2 Camshaft sprocket bolt

Timing chain and gear inspection (cover removed from engine)

Refer to illustration 10.17

17 Attach a socket and torque wrench to the camshaft sprocket bolt and apply force in the normal direction of crankshaft rotation (30 ft-lbs if the cylinder head is still in position complete with rocker arms, or 15 ft-lbs if the cylinder head has been removed) **(see illustration)**. Don't allow the crankshaft to rotate. If necessary, jam the crankshaft sprocket so that it can't move. Using a ruler, note the amount of movement of the chain. If it exceeds 1/8-inch a new timing chain will be required. **Note:** *Whenever a new timing chain is required, the entire set (chain, camshaft and crankshaft sprockets) must be replaced as an assembly.*

18 Inspect the camshaft gear for damage or wear. The camshaft gear on some models is steel, but most original-equipment cam gears will be an aluminum gear with a nylon coating on the teeth. This nylon coating may be cracked or be breaking off in small pieces. These pieces tend to end up in the oil pan and may eventually plug the oil pump pickup screen. If the pieces have come off the camshaft gear, the oil pan should be removed to properly clean or replace the oil pump pickup screen.

19 Inspect the crankshaft gear for damage or wear. The crankshaft gear is a steel gear, but the teeth can be grooved or worn enough to cause a poor meshing of the gear and the chain.

10.21a The sprockets on the camshaft and crankshaft can be removed with a two or three-jaw puller . . .

10.21b . . . or two screwdrivers

Chain removal

Refer to illustrations 10.20, 10.21a and 10.21b

20 Be sure the timing marks are aligned **(see illustration)**. Hold a prybar through one of the camshaft sprocket holes to prevent the camshaft from turning. Remove the bolt from the camshaft sprocket, then detach the fuel pump eccentric (carbureted small-block V8 engines only).

21 The sprockets on the camshaft and crankshaft can be removed with a two or three-jaw puller or by using two screwdrivers, but be careful not to damage the threads in the end of the crankshaft **(see illustrations)**. **Note:** *If the timing chain cover oil seal has been leaking, refer to Section 9 and install a new one.*

2B

10.26a Slip the chain and camshaft sprocket in place over the crankshaft sprocket with the timing mark (arrow) at the bottom

10.26b The keyway must align with the key (arrows)

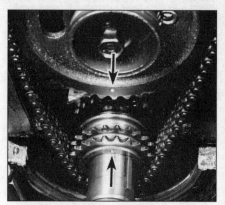

10.26c The timing marks MUST align as shown here

10.31 Be sure to engage the oil seal tang in the oil seal hole in the pan

Installation

Refer to illustrations 10.26a, 10.26b, 10.26c and 10.31

22   Use a gasket scraper to remove all traces of old gasket material and sealant from the cover and engine block (and the water pump housing and engine block on big-block engines). Stuff a shop rag into the opening at the front of the oil pan to keep debris out of the engine. On small-block engines, remove the rubber semi-circular seal from the oil pan. Wipe the cover and block sealing surfaces with a cloth saturated with lacquer thinner or acetone.

23   On big-block engines, check the cover flange for distortion, particularly around the bolt holes. Straighten any distortion by placing the sealing flange on a block of wood and tapping it with the rounded end of a ball-peen hammer.

24   If you're installing a new crankshaft sprocket, be sure to align the keyway in the crankshaft sprocket with the Woodruff key in the end of the crankshaft. **Note:** *Timing chains must be replaced as a set with the camshaft and crankshaft gears. Never put a new chain on old gears.* Align the sprocket with the Woodruff key and press the sprocket onto the crankshaft with the vibration damper bolt, a large socket and some washers or tap it gently into place until it is completely seated. **Caution:** *If resistance is encountered, do not hammer the sprocket onto the crankshaft. It may eventually move onto the shaft, but it may be cracked in the process and fail later, causing extensive engine damage.*

25   On 1998 and later V6 engines, insert a large flat-bladed screwdriver between the crankshaft sprocket and the timing chain tensioner assembly. Compress the tensioner shoe until the hole in the shoe aligns with the bracket hole. Slide a suitable size pin into the hole to lock the tensioner shoe in place. Remove the screwdriver.

26   Loop the new chain over the camshaft sprocket, then turn the sprocket until the timing mark is at the bottom. Mesh the chain with the crankshaft sprocket and position the camshaft sprocket on the end of the cam. If necessary, turn the camshaft so the key fits into the sprocket keyway with the timing

mark in the 6 o'clock position. When the chain is installed, the timing marks MUST align as shown **(see illustrations)**.

27   Apply a thread locking compound to the camshaft sprocket bolt threads, then install the fuel pump eccentric (small-block engines) and tighten the bolt to the torque listed in this Chapter's Specifications.

28   Lubricate the chain with clean engine oil.

29   On 1998 and later V6 engines, insert a large flat-bladed screwdriver between the timing chain and the tensioner assembly. Compress the tensioner shoe and remove the pin from the tensioner shoe. Remove the screwdriver.

30   Check for cracks and deformation of the oil pan gasket before installing the timing cover. If the gasket has deteriorated or is damaged, it must be replaced before reinstalling the timing cover. On small-block engines, the rubber semi-circular seal must be replaced whenever the oil pan is removed. On all models, if the portion of oil pan gasket that extends beyond the block is damaged, you may be able to repair the area without replacing the entire oil pan gasket. First, using a utility knife or single-edged razor blade, trim off the front portion of the gasket at the point where it meets the engine block, then trim off the front portion of the new oil pan rail gasket so it's exactly the same size. Cover the exposed inside area of the oil pan with a rag, then clean all traces of gasket material off the area where the gasket was removed. Attach the new gasket piece to the front of the oil pan with contact-cement-type gasket adhesive and seal the mating edges of the new/old gaskets with RTV sealant. **Note:** *1993 engines use a metal-reinforced, one-piece oil pan gasket. It is recommended on these models not to cut the old pan gasket, remove the oil pan (see Section 13) and install a new one-piece gasket.*

31   On V6 and small-block V8 engines, clean all traces of old gasket material off the semi-circular area at the front of the oil pan, finishing off with lacquer thinner or acetone to de-grease the area. Install a new semi-circular rubber seal to the oil pan; be sure to pull the oil seal "nubs" securely through the holes at the front of the oil pan **(see illustration)**.

32   Apply a thin layer of RTV sealant to both

sides of the new cover gasket and the corners of the pan and block, then position the new cover gasket on the engine. The sealant will hold it in place. **Note:** *On V6 and small-block V8 engines, be sure to apply a dab of RTV to the two points where the semi-circular rubber seal meets the oil pan side-rail gaskets as well as the two points where the oil pan side-rail gaskets meet the block.*

33   Install the timing chain cover on the block. On big-block engines, tighten the bolts, a little at a time, until you reach the torque listed in this Chapter's Specifications. On V6 and small-block V8 engines, install all front cover bolts, except the ones that also secure the water pump, tightening them finger-tight.

34   On V6 and small-block V8 engines, install the water pump and tighten the timing chain cover bolts to the torque listed in this Chapter's Specifications. On big-block engines, install the water pump housing, using new gaskets and tightening the bolts to the torque listed in this Chapter's Specifications.

35   Install the two oil pan bolts, bringing the oil pan up against the timing chain cover.

36   Lubricate the oil seal contact surface of the vibration damper hub with moly-base grease or clean engine oil, then install the damper on the end of the crankshaft. The keyway in the damper must be aligned with the Woodruff key in the crankshaft nose. If the damper cannot be seated by hand, slip the large washer over the bolt, install the bolt and tighten it to pull the damper into place. Tighten the bolt to the torque listed in this Chapter's Specifications.

11.3 When checking the camshaft lobe lift, the dial indicator plunger must be positioned directly above and in-line with the pushrod

37 The remaining installation steps are the reverse of removal.

38 Add coolant and check the oil level. Run the engine and check for oil and coolant leaks, then reinstall the engine cover.

11 Camshaft and lifters - removal, inspection and installation

Camshaft lobe lift check

Refer to illustration 11.3

1 To determine the extent of cam lobe wear, the lobe lift should be checked prior to camshaft removal.

2 Remove the valve covers (see Section 4).

3 Beginning with the number one cylinder, mount a dial indicator on the engine and position the plunger against the top surface of the first rocker arm. Set the number one cylinder at TDC on the compression stroke (see Section 3). The plunger should be directly above and in line with the pushrod **(see illustration)**.

4 Zero the dial indicator, then very slowly turn the crankshaft in the normal direction of rotation (clockwise) until the indicator needle stops and begins to move in the opposite direction. The point at which it stops indicates maximum cam lobe lift.

5 Record this figure for future reference, then reposition the piston at TDC on the compression stroke.

6 Move the dial indicator to the other number one cylinder rocker arm and repeat the check. Be sure to record the results for each valve.

7 Repeat the check for the remaining valves. Since each piston must be at TDC on the compression stroke for this procedure, work from cylinder-to-cylinder following the firing order sequence. For instance, after checking the number one cylinder rocker arms, turn the engine slowly until the distributor's rotor points to the secondary terminal for the cylinder next in the firing order (number six on V6 engines, number eight on V8s).

8 After the check is complete, multiply all the recorded figures by 1.5 to obtain the total valve lift for each valve (the rocker arms have

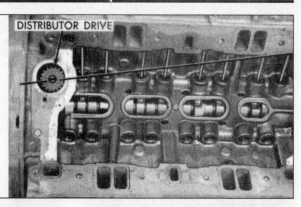

11.10 Pull up on the distributor driveshaft gear to remove it - upon installation, with the number one cylinder at TDC, be sure the slot is aligned as shown (small-block engines) or parallel to the crankshaft (big-block engines)

11.12a The lifters in an engine that has accumulated many miles may have to be removed with a special tool (arrow) - store them in an organized manner to make sure they're reinstalled in their original locations

a 1.5:1 ratio). Compare the results to the specifications in this Chapter. If the valve lift is 0.003 inch less than specified, cam lobe wear has occurred and a new camshaft should be installed.

Removal

Refer to illustrations 11.10, 11.12a, 11.12b, 11.12c, 11.12d, 11.13 and 11.15

9 Refer to the appropriate Sections and remove the intake manifold (V6 and small-block V8 engines only), valve covers, rocker arms, pushrods and the timing chain and camshaft sprockets. **Note:** *On big-block engines, the lifters can be removed through the pushrod openings in the cylinder head with the use of a special removal tool that's available from most auto parts stores; therefore, intake manifold removal may not be necessary.*

10 Remove the distributor drive **(see illustration)**. Also remove the grille, radiator and air conditioning condenser (see Chapter 3). **Note:** *To provide enough clearance for camshaft removal on some models, it will be necessary to unbolt and remove the vertical hood latch support bar.*

11 There are several ways to extract the lifters from the bores. A special tool designed to grip and remove lifters is manufactured by many tool companies and is widely available, but it may not be required in every case. On newer engines without a lot of varnish buildup, the lifters can often be removed with a small

11.12b Engines with roller lifters have a retainer that must be removed before taking out the yokes or lifters - arrows indicate bolts holding the retainer to the block

11.12c Remove the lifter yoke (arrow) and lifters and store them in order

magnet or even with your fingers. A machinist's scribe with a bent end can be used to pull the lifters out by positioning the point under the retainer ring inside the top of each lifter. **Caution:** *Do not use pliers to remove the lifters unless you intend to replace them with new ones (along with the camshaft). The pliers will damage the precision machined and hardened lifters, rendering them useless.*

12 Before removing the lifters, arrange to store them in a clearly labeled box to ensure that they are reinstalled in their original locations **(see illustration)**. On engines equipped with roller lifters, the lifter-aligning yokes and yoke retainer must be removed before the lifters are withdrawn **(see illustrations)**.

2B

11.12d Note the paint marks (arrows) that indicate which side of the roller lifters face the valley - apply marks if none are visible

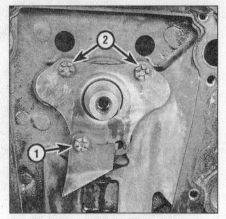

11.13 Remove the camshaft thrust plate and oil tab

1   Oil tab bolt
2   Thrust plate bolts

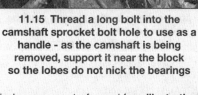

11.15 Thread a long bolt into the camshaft sprocket bolt hole to use as a handle - as the camshaft is being removed, support it near the block so the lobes do not nick the bearings

Remove the lifters and store them where they will not get dirty. The roller-type lifters should have paint dabs showing which side of the lifter faces the lifter "valley" **(see illustration)**. If they aren't marked, apply some paint dabs before removing the lifters. The lifters must be installed the same way to aim the oil feed holes properly. **Caution:** *Do not attempt to withdraw the camshaft with the lifters in place.*

13   On V6 and small-block V8 engines, unbolt and remove the camshaft thrust plate and oil tab **(see illustration)**. Note how the oil tab is installed so you can return it to its original location on reassembly. On big-block engines, remove the oil pump (see Section 14).

14   On all engines, thread a long bolt into the camshaft sprocket bolt hole to use as a handle when removing the camshaft from the block.

15   Carefully withdraw the camshaft from the block. Support the cam near the block so the lobes do not nick or gouge the bearings as it is withdrawn **(see illustration)**.

Inspection

Refer to illustration 11.17

Camshaft and bearings

16   After the camshaft has been removed from the engine, cleaned with solvent and dried, inspect the bearing journals for uneven wear, pitting and evidence of seizure. If the journals are damaged, the bearing inserts in the block are probably damaged as well. Both the camshaft and bearings will have to be replaced. **Note:** *Camshaft bearing replacement requires special tools and expertise that place it beyond the scope of the average home mechanic. The tool for bearing removal/installation is available at stores that carry automotive tools, possibly even found at a tool rental business. It is advisable though, if the bearings are bad, that the engine should be removed and the block taken to an automotive machine shop to ensure that the job is done with precise tools and to specs.*

17   Measure the bearing journals with a micrometer to determine if they are exces-

sively worn or out-of-round **(see illustration)**.

18   Check the camshaft lobes for heat discoloration, score marks, chipped areas, pitting and uneven wear. If the lobes are in good condition and if the lobe lift measurements are as specified in this Chapter, the camshaft can be reused.

Flat tappet lifters

Refer to illustrations 11.20a, 11.20b, 11.20c and 11.20d

Note: *Some engines are fitted with 0.008 inch oversize lifters at the factory. These may be identified by a 3/8-inch diamond-shaped stamp on the top pad at the front of the engine and a flat ground on the outside surface of each oversize lifter bore.*

19   Clean the lifters with solvent and dry them thoroughly without mixing them up.

20   Check each lifter wall, pushrod seat and foot for scuffing, score marks and uneven wear. Each lifter foot (the surface that rides on the cam lobe) must be slightly convex, although this can be difficult to determine by eye. If the base of the lifter is concave, the

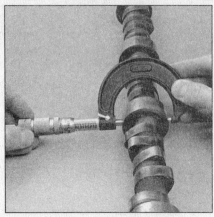

11.17 Check the diameter of each camshaft bearing journal to pinpoint excessive wear and out-of-round conditions

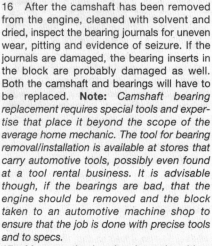

11.20a if the bottom of any lifters is worn concave, scratched or galled, replace the entire set with new lifters

11.20b The foot of each lifter should be slightly convex - the side of another lifter can be used as a straightedge to check it; if it appears flat, it is worn and must not be reused

11.20c If the lifters are pitted or rough, they shouldn't be reused

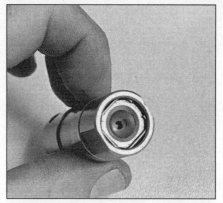

11.20d Check the pushrod seat (arrow) in the top of each lifter for wear

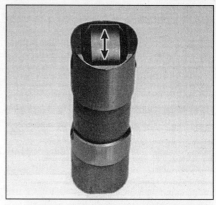

11.22 The roller on the roller lifters must turn freely - check for wear and excessive play as well

lifters and camshaft must be replaced. If the lifter walls are damaged or worn (which is not very likely), inspect the lifter bores in the engine block as well. If the pushrod seats are worn, check the pushrod ends **(see illustrations)**.

21   If new lifters are being installed, a new camshaft must also be installed. If a new camshaft is installed, then use new lifters as well. Never install used lifters unless the original camshaft is used and the lifters can be installed in their original locations.

Roller lifters

Refer to illustration 11.22

22   Check the rollers carefully for wear and damage and make sure they turn freely without excessive play **(see illustration)**. The inspection procedure for conventional lifters also applies to roller lifters.

23   Used roller lifters can be reinstalled with a new camshaft and the original camshaft can be used if new lifters are installed, provided the used parts are in good condition. **Caution:** *When replacing any of the valvetrain components on 3.9L V6 engines, make sure the replacement parts are for the exact year of your engine. There have been three variations of pushrods/lifters/rocker arms and the parts are not interchangeable. Some pushrods have a hole at each end, some do not, some rocker arms have an oil hole on the pushrod end, and the Magnum engines have roller lifters with longer pushrods. Match the design of the replacement part to your old part when making any replacement, or serious oiling problems could develop.*

Installation

Refer to illustrations 11.24 and 11.25

24   Lubricate the camshaft bearing journals and cam lobes with moly-base grease or engine assembly lube **(see illustration)**.

25   Position a camshaft holding tool, available at most auto parts stores, through the distributor hole **(see illustration)**. This tool will prevent the camshaft from being pushed in too far and prevents the welch plug in the back of the block from being knocked out. The tool should remain in place until the timing chain has been installed. **Caution:** *If the tool is unavailable, work carefully to avoid*

knocking the plug out. If the welch plug seal is damaged, a massive oil leak will occur.

26   Slide the camshaft slowly and gently into the engine. Support the cam near the block and be careful not to scrape or nick the bearings. On V6 and small-block V8 engines, install the camshaft thrust plate and oil tab.

27   On all engines, install the timing chain and sprockets (see Section 10). Align the timing marks on the crankshaft and camshaft sprockets.

28   Install the distributor drive gear **(see illustration 11.10)**. On big-block engines, install the oil pump.

29   On all engines, lubricate the lifters with clean engine oil and install them in the block. If the original lifters are being reinstalled, be sure to return them to their original locations. If a new flat-tappet camshaft was installed, be sure to install new lifters as well. Lubricate the lifters with clean engine oil and install them in the block. If roller lifters are being reinstalled, be sure to return them to their original locations, and with the paint marks facing the valley; make sure the oil-feed holes on the side of the lifter body faces UP away from the crankshaft. Install the lifter yokes with their arrows pointing toward the camshaft. Install the lifter yoke retainer and tightened the bolts to the torque listed in this Chapter's Specifications.

30   The remaining installation steps are the reverse of removal. **Note:** *When installing the distributor, the rotor will point to number 4 secondary terminal on a V6 and to number 6 on a V8 with the camshaft and crankshaft sprocket timing marks aligned.*

31   Change the oil, add a break-in compound, available at most auto parts stores, and install a new oil filter (see Chapter 1).

32   Start the engine, check for oil pressure and leaks and set the ignition timing. **Caution:** *Do not run the engine above a fast idle until all the hydraulic lifters have filled with oil and become quiet .*

33   If a new camshaft and lifters have been installed, the engine should be brought to operating temperature and run at a fast idle for 15 to 20 minutes to "break in" the components. Then change the oil and filter once more.

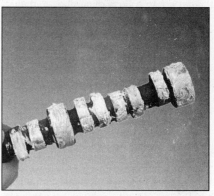

11.24 Be sure to apply moly-based grease or engine assembly lube to the cam lobes and bearing journals before installing the camshaft

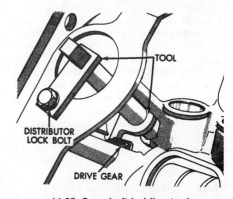

11.25 Camshaft holding tool (installed position)

12   Cylinder heads - removal and installation

Removal

Refer to illustrations 12.6 and 12.7

1   Drain the cooling system (see Chapter 1) and remove the engine cover (see Section 2).

2   Remove the valve covers (see Section 4).

3   Remove the intake manifold (see Section 7).

12.6 To avoid mixing up the head bolts, use a new gasket to transfer the bolt pattern to a piece of cardboard, then punch holes to accept the bolts

12.7 Pry on a casting protrusion to break the head loose

12.10 Remove all traces of old gasket material

12.12 A die should be used to remove sealant and corrosion from the bolt threads prior to installation

12.13 Install the new head gasket over the dowels (arrow); there is a dowel at each end of the cylinder head

4   Detach both exhaust manifolds from the cylinder heads (see Section 8).

5   Remove the rocker arm shaft assembly and pushrods (see Section 5). **Caution:** *Again, as mentioned in Section 5, keep all the parts in order so they are reinstalled in the same location.*

6   Loosen the head bolts in 1/4-turn increments until they can be removed by hand. Reverse the sequence shown in **illustration 12.16a or 12.16b. Note:** *On all engines, there will be different-length head bolts for different locations, so store the bolts in a cardboard holder* **(see illustration)** *or some type of container as they are removed. This will ensure that the bolts are reinstalled in their original holes.*

7   Lift the heads off the engine. If resistance is felt, do not pry between the head and block as damage to the mating surfaces will result. To dislodge the head, place a block of wood against the end of it and strike the wood block with a hammer, or lift on a casting protrusion **(see illustration)**. Store the heads on blocks of wood to prevent damage to the gasket sealing surfaces.

8   Cylinder head disassembly and inspection procedures are covered in detail in Chapter 2, Part C.

Installation

Refer to illustrations 12.10, 12.12, 12.13, 12.16a, 12.16b and 12.16c

9   The mating surfaces of the cylinder heads and block must be perfectly clean when the heads are installed. Gasket removal solvents are available at auto parts stores and may prove helpful.

10   Use a gasket scraper to remove all traces of carbon and old gasket material **(see illustration)**, then wipe the mating surfaces with a cloth saturated with lacquer thinner or acetone. If there is oil on the mating surfaces when the heads are installed, the gaskets may not seal correctly and leaks may develop. When working on the block, cover the lifter valley with shop rags to keep debris out of the engine. Use a vacuum cleaner to remove any debris that falls into the cylinders.

11   Check the block and head mating surfaces for nicks, deep scratches and other damage. If damage is slight, it can be removed with emery cloth. If it is excessive, machining may be the only alternative.

12   Use a tap of the correct size to chase the threads in the head bolt holes in the block. Mount each bolt in a vise and run a die down the threads to remove corrosion and restore the threads **(see illustration)**. Dirt, corrosion, sealant and damaged threads will affect torque readings.

13   Position the new gaskets over the dowels in the block **(see illustration)**.

14   Carefully position the heads on the block without disturbing the gaskets.

15   Before installing the head bolts, coat the threads with a non-hardening sealant such as Permatex No. 2.

16   Install the bolts in their original locations and tighten them finger tight. Following the recommended sequence **(see illustrations)**, tighten the bolts in several steps to the torque listed in this Chapter's Specifications.

17   The remaining installation steps are the reverse of removal.

18   Add coolant and change the oil and filter (see Chapter 1). Start the engine, set the ignition timing and check the engine for proper operation and coolant or oil leaks.

13   Oil pan - removal and installation

Removal

1   Disconnect the battery cable from the negative battery terminal. Raise the vehicle and support it securely on jackstands (see Chapter 1). Remove the engine cover (see Section 2).

2   Drain the engine oil and replace the oil filter (see Chapter 1). Remove the engine oil dipstick.

3   Unbolt and remove the lower engine-to-transmission support struts and the torque converter inspection cover (see Chapter 7, Part B).

4   Remove the starter (see Chapter 5).

5   On single-exhaust systems that use a cross-over pipe under the oil pan, disconnect the exhaust pipe from the exhaust manifolds (see Section 8) and lower the pipe as far as possible. **Note:** *On models with fuel injection, disconnect the oxygen sensor from the exhaust pipe, remove the bolts/nuts and separate the exhaust pipe from both exhaust*

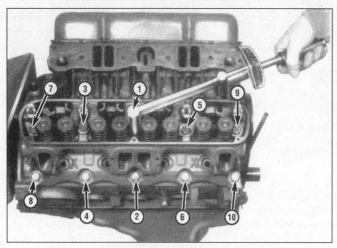

12.16a Cylinder head tightening sequence for the small-block V8 engine

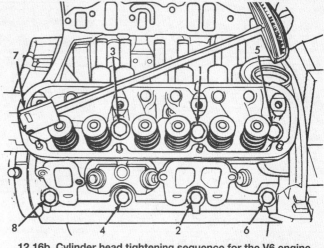

12.16b Cylinder head tightening sequence for the V6 engine

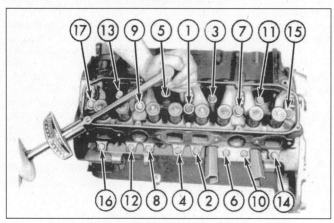

12.16c Cylinder head tightening sequence for the big-block V8 engine

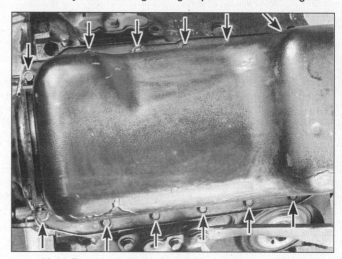

13.14 Remove the bolts (arrows) around the perimeter of the oil pan (typical)

2B

manifolds (see Section 8). *In some cases, exhaust hangers further back may have to be disconnected to allow the exhaust system to hang down low enough to allow oil pan removal.*

6　Depending on the year and model of van, it may be necessary to raise the rear of the engine in order to allow sufficient room for the pan to slide out the rear.

1971 through 1978 models

7　On these models, the transmission and flywheel must be removed to provide room for oil pan removal. The problem is that the engine must be supported to remove the transmission, and not by jacking under the oil pan you are trying to remove. The factory method is to install a support framework inside the van to hold the engine from above. Such a fixture can be built from heavy-gauge threaded pipe and fittings, or with lengths of 4x4 wood. **Note:** *In some cases an engine hoist with a long arm can be used, working through the passenger side door opening.*

8　Remove the driveshaft and transmission (see Chapter 8 and Chapter 7A or 7B).

9　Remove the flywheel or driveplate (see

Section 16).

10　Raise the rear of the engine two inches, but as you raise it, keep watching the clearance between the cooling fan blades and the radiator core to check for interference. If necessary, remove the fan (see Chapter 5).

11　Remove the oil pan mounting bolts, then lower the pan from the engine, twisting it slightly to the driver's side to clear the oil pump pickup. **Note:** *If there is interference at the front of the engine, rotate the crankshaft until the counterweights are are positioned to facilitate pan removal.*

1979 and later models

Refer to illustration 13.14

12　Support the front of the engine with a jack under the vibration damper.

13　Remove the throughbolts from the left and right engine mounts and raise the engine, watching for interference between the cooling fan and the radiator core. If necessary, remove the fan.

14　Remove the oil pan mounting bolts **(see illustration)**, then lower the pan from the engine, twisting it slightly to the driver's side to clear the oil pump pick-up. *Note: If there is*

interference at the front of the engine, rotate the crankshaft until the counterweights are positioned to facilitate pan removal.

Installation

Refer to illustrations 13.17 and 13.21

15　Wash out the oil pan with solvent.

16　Thoroughly clean the mounting surfaces of the oil pan and engine block of old gasket material and sealer. If the oil pan is distorted at the bolt-hole areas, straighten the flange by supporting it from below on a 1x4 wood block and tapping the bolt holes with the rounded end of a ball-peen hammer. Wipe the gasket surfaces clean with a rag soaked in lacquer thinner or acetone.

Note: *Connecting rod bolts from 1990 on are slightly longer than previous. If your engine is apart to replace a rod, rod bolts, or even a new short-block, and you use your pre-1990 oil pan, there can be interference that will cause a tapping noise in the engine. If the newer components are to be used, "pinch in" the top rear corner of the dipstick guide in the pan for clearance. Move it in with a hammer and punch 1/8-inch closer to the pan side, and extend this indentation down 1/4-inch.*

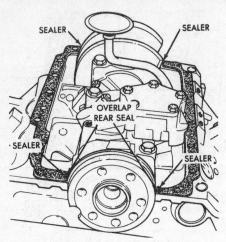

13.17 Apply RTV sealer to the four corners, as shown

13.21 Place the pan and gasket into position over the alignment studs (arrow) and install several pan bolts finger-tight

14.2 Remove the bolts (arrows) and lower the oil pump

17   On V6 and small-block V8 engines, install the side-rail gaskets on the engine. Attach them with contact-cement-type adhesive. Attach the rubber end seals to the oil pan, pulling the "nubs" through the holes in the pan. Apply a bead of RTV sealant in the four corners where the side-rail gaskets meet the front and rear rubber gaskets **(see illustration)**. **Note:** *1993 engines have one-piece, metal-reinforced oil pan gaskets.*

18   On big-block engines, apply a small dab of RTV sealant to the points where the front cover meets the engine block and where the rear seal retainer meets the engine block. Then attach the oil pan gasket to the engine block with contact-cement-type adhesive.

19   Prepare four pan alignment dowels from 5/16-inch bolts 1.5-inches long. Cut off the bolt heads and slot the ends with a hacksaw.

20   Install the four alignment dowels into the foremost and rearmost pairs of pan bolt holes in the block. Apply a dab of RTV on both sides where the pan meets the rear main cap, and up front at the juncture of the front cover and block.

21   Lift the pan into position with its gasket in place, slipping it over the alignment dowels and being careful not to disturb the gasket. Install several bolts finger tight **(see illustration)**.

22   Check that the gasket isn't sticking out anywhere around the block's perimeter.

When all the bolts are in place, replace the alignment dowels with four pan bolts.

23   Starting at the ends and alternating from side-to-side towards the center, tighten the bolts to the torque listed in this Chapter's Specifications.

24   The remainder of the installation procedure is the reverse of removal.

25   Add the proper type and quantity of oil (see Chapter 1), start the engine and check for leaks before placing the vehicle back in service.

14   Oil pump - removal, inspection and installation

V6 and small-block V8 engines

Removal

Refer to illustration 14.2

1   Remove the oil pan (see Section 13).

2   While supporting the oil pump, remove the pump-to-rear main bearing cap bolts **(see illustration)**.

3   Lower the pump and pickup screen assembly from the vehicle and clean the gasket surfaces.

Inspection

Refer to illustrations 14.6, 14.7a, 14.7b and 14.7c

4   Remove the oil pump cover.

5   Put a straightedge across the inner sur-

face of the oil pump cover and try to insert a 0.0015-inch feeler gauge under it. If the gauge fits, the oil pump assembly should be replaced.

6   Remove the outer rotor **(see illustration)** and measure its thickness with a micrometer. If it is less than the minimum thickness listed in this Chapter's Specifications, the pump assembly should be replaced. The inner rotor should meet this same minimum thickness.

7   With feeler gauges, measure the clearance between the outer rotor and the body, between the inner and outer rotors, and clearance between the rotors and cover **(see illustrations)**. Compare these measurements to this Chapter's Specifications.

Installation

Refer to illustrations 14.10a and 14.10b

8   Reassemble the oil pump and thread the oil pickup tube and screen into the oil pump and tighten it. **Caution:** *Be absolutely certain that the pickup screen is properly tightened so that no air can be sucked into the oiling system at this connection.*

9   Prime the pump by pouring clean motor oil into the pickup tube, while turning the pump by hand.

10   Position the pump on the engine with a new gasket and make sure the pump driveshaft is aligned with the oil pump **(see illustrations)**.

11   Install the mounting bolts and tighten them to the torque listed in this Chapter's Specifications.

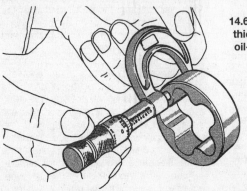

14.6 Measure the thickness of the oil-pump rotors

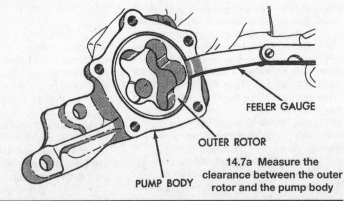

FEELER GAUGE

OUTER ROTOR

PUMP BODY

14.7a Measure the clearance between the outer rotor and the pump body

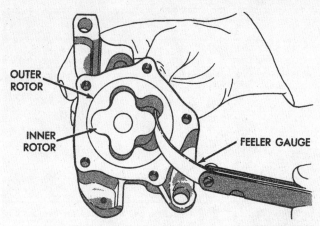

14.7b Measure the clearance between the inner and outer rotors at the rotor tips

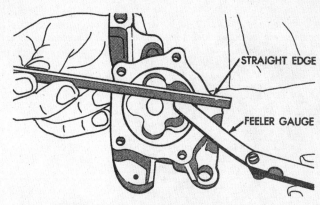

14.7c Using a straightedge, measure the rotor-to-cover clearance

12   Install the oil pan and add oil.

13   Run the engine and check for oil pressure and leaks.

Big-block V8 engine

Refer to illustration 14.15

14   Remove the oil filter (see Chapter 1).

15   Remove the bolts and the oil pump assembly from the engine **(see illustration)**. The drive-gear extension may bind a bit on the way out, so rock the assembly back and forth to permit easier removal.

16   Using a scraper, remove all traces of old gasket material from the gasket surfaces of the pump and the engine block. Gasket removal solvents are available from auto parts stores and may prove helpful. After the gasket is removed, wipe the gasket surfaces clean with a rag soaked in lacquer thinner or acetone.

17   Prime the pump by pouring clean motor oil into the pickup tube, while turning the pump by hand.

18   Coat the new oil pump gasket with RTV sealant, then place it on the oil pump. Install the bolts through the pump to hold the gasket in place, then install the oil pump on the block, tightening the bolts to the torque listed in this Chapter's Specifications.

19   Replace the oil filter and add the proper type and quantity of oil (see Chapter 1).

15   Rear main oil seal - replacement

All engines

Refer to illustrations 15.4, 15.5, 15.6, 15.7a and 15.7b

Note: *If you're installing a new seal during a complete engine overhaul, ignore the steps in this procedure that concern removal of external parts. Also, since the crankshaft is already removed, it's not necessary to use any special tools to remove the upper seal half; remove and install the upper seal half the same way as the lower seal half.*

1   Remove the oil pan (see Section 13).

2   Remove the oil pump on V6 and small-

14.10a Inspect the end of the driveshaft for excessive wear and replace it, if necessary

block V8 engines (see Section 14).

3   The rear main seal can be replaced with the engine in the vehicle. There are two different types of material used by the manufacturer. The original material used on most older models is the rope-type, two-piece seal. The newer version of the rear main seal is also a two-piece seal, but made from Viton rubber. Most aftermarket part suppliers have substituted the rubber for the older rope-type, because it works better than the original

14.15 The oil pump on big-block engines is located on the lower right side of the block - remove the bolts (arrows) to remove it

14.10b The oil pump driveshaft (arrow) must align with the pump

design, but the seal must be used as a complete upper and lower set. The rope and rubber styles of seals CANNOT be combined.

4   On V6 and small-block V8 engines, remove the bolts and detach the rear main bearing cap from the engine **(see illustration)**. On big-block engines, unbolt and remove the rear oil seal retainer that's located behind the rear main bearing cap. **Note:** *A twelve-point socket is necessary to remove the two seal retainer bolts on big-block engines.*

15.4 Remove the bolts (arrows) and detach the rear main bearing cap from the engine

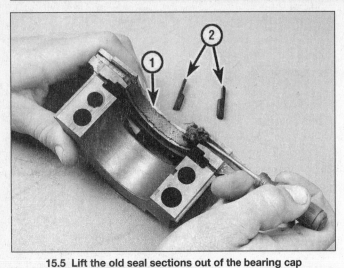

15.5 Lift the old seal sections out of the bearing cap
with a small screwdriver

1   Rear main seal (rope or rubber)
2   Side seals

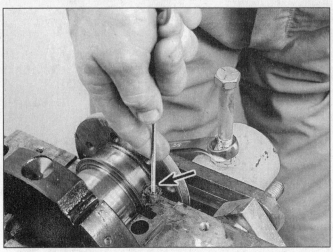

15.6 Screw a removal tool into one end of the old upper seal
section and pull it out (engine removed for clarity)

5   Remove the lower half of the oil seal from the bearing cap or retainer. On small-block engines equipped with side seals, also remove the side seals from the cap by prying them from the side with a small screwdriver and grasping the seals with needle-nose pliers, if necessary **(see illustration)**.

6   Remove upper half of the rope oil seal (in the engine block) by screwing a tool (KD-492, or equivalent) into the end of the seal **(see illustration)**, being careful not to damage the crankshaft. Pull the seal out with the tool while rotating the crankshaft. If a neoprene rubber-type seal is installed on your engine, use a small brass punch to carefully press on one end of the upper half of the seal until the other side protrudes, then pull the seal out from the other side with needle-nose pliers. Clean the bearing cap and engine block surfaces carefully to degrease them and remove any sealant. **Note:** On V6 and small block V8 engines, it may be easier to remove the upper rear seal when the two

main bearing caps ahead of the rear cap are loosened slightly. These caps should be retorqued to Specifications after the new seal and rear cap are installed.

7   On V6 and small-block V8 engines equipped with cap side seals, insert the seals into the slots in bearing cap and press them carefully into place. **Note :** On many aftermarket seals, the side seals are integral with the lower seal half. Generally, the side seal with the yellow paint goes into right side of the cap (when installed on the engine) and the narrow sealing edge should face up. Make sure the seals are fully seated, that they match the contours of the cap and they protrude slightly from the cap; then, using a new, single-edged razor blade, cut the ends flush with side edges of the cap **(see illustration)**. On 1992 and later Magnum V6 and small-block V8 engines, there are no side seals. Apply a dab of RTV sealant to each side of the cap, adjacent to the seal. **(see illustration)**. **Caution:** Do not apply RTV to the seal itself.

8   On all engines, lightly oil the lips of the new (neoprene rubber) crankshaft seals. **Caution:** Always wipe the crankshaft surface clean. then oil it lightly before installing a new seal.

9   Rotate a new seal half into cylinder block with the paint stripe toward the rear (the seal lip must face toward the front of the engine). **Caution:** Hold your thumb firmly against the outside diameter of the seal as you're rotating it into place. This will prevent the seal outside diameter from being shaved from contact with the sharp edge of the engine block. If the seal gets shaved, you may wind up with an oil leak.

10   Place the other seal half in the bearing cap or seal retainer with the paint stripe toward the rear.

V6 and small-block V8 engines

11   Install the rear main bearing cap and tighten the bolts to the torque listed in this Chapter's Specifications. **Note:** 1991and later V6 engines have a dowel pin to help

15.7a On V6 and small-block V8 engines, after installing the rear
main bearing cap, trim the side seals flush with the cap

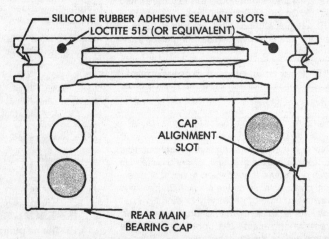

15.7b Apply a small drop of Loctite 515, or equivalent, to either
side of the cap on 1992 and later engines

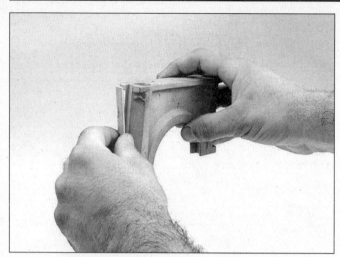

15.16 On big-block V8 models, after immersing the side seals in mineral spirits or diesel fuel, install them on the sides of the retainer, apply RTV to the base of the retainer and install the retainer quickly

16.3a Before removing the flywheel, mark its relationship to the crankshaft

align the rear main cap, and 1992 and later models have a groove in the main cap for the oil pan seal.

12   Install the oil pump and oil pan.

13   The remainder of installation is the reverse of removal. Fill the pan with oil, run the engine and check for leaks.

Big-block engines

Refer to illustration 15.16

14   Clean the seal retainer side grooves in preparation for installing the new side seals. **Caution:** *Perform the following operations as rapidly as possible. These side seals are made from a material that expands quickly when oiled.*

15   Apply mineral spirits or diesel fuel to the side seals. Do not apply any sealer to side seals, as this will prevent the oil from swelling the seals. **Caution:** *Failure to pre-oil the seals will result in an oil leak.*

16   Install the side seals immediately in the seal retainer grooves **(see illustration)**. Apply RTV sealant to the bottom of the retainer on both sides.

17   Install seal retainer and tighten the screws to the torque listed in this Chapter's Specifications.

18   The remainder of installation is the reverse of removal. Fill the pan with oil, run the engine and check for leaks.

16   Flywheel/driveplate - removal and installation

Removal

Refer to illustrations 16.3a and 16.3b

1   Raise the vehicle and support it securely on jackstands, then refer to Chapter 7 and remove the transmission.

2   If you're working on a model with a manual transmission, remove the pressure plate and clutch disc **(see Chapter 8)**. Now is

a good time to check/replace the clutch components and pilot bearing.

3   Use paint or a center punch to make alignment marks on the flywheel/driveplate and crankshaft to ensure correct alignment during reinstallation **(see illustrations)**.

4   Remove the bolts that secure the flywheel/driveplate to the crankshaft. If the crankshaft turns, hold the flywheel/driveplate with a prybar or wedge a screwdriver into the ring gear teeth.

5   Remove the flywheel/driveplate from the crankshaft. If you're removing a flywheel, be careful because it's fairly heavy. Be sure to support it while removing the last bolt. **Caution:** *The teeth may be sharp, wear gloves or use rags to protect your hands.*

Inspection

6   Clean the flywheel to remove grease and oil. Inspect the surface for cracks, rivet grooves, burned areas and score marks. Light scoring can be removed with emery cloth. Check for cracked and broken ring gear teeth. **Note:** *On automatic transmission-equipped vehicles, the ring gear teeth are part of the torque converter. Check the condition of the teeth before reassembly.* Lay the flywheel on a flat surface and use a straightedge to check for warpage. If any defects are evident, take the flywheel to an automotive machine shop to have it resurfaced. Clean and inspect the mating surfaces of the flywheel/driveplate and the crankshaft.

Installation

7   Position the flywheel/driveplate against the crankshaft. Align the marks made during removal. Note that some models have an alignment dowel or staggered bolt holes to ensure correct installation. Before installing the bolts, apply thread-locking compound to the threads.

8   Wedge a screwdriver in the ring gear teeth to keep the flywheel/driveplate from

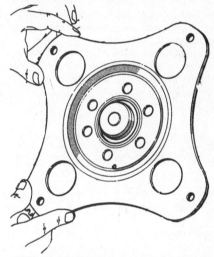

16.3b The bolt holes on most driveplates are offset to ensure correct installation

turning as you tighten the bolts to the torque listed in this Chapter's Specifications. Follow a criss-cross pattern and work up to the final torque in three or four steps.

9   The remainder of installation is the reverse of the removal procedure.

17   Engine mounts - check and replacement

Refer to illustrations 17.7a and 17.7b

1   Engine mounts seldom require attention, but broken or deteriorated mounts should be replaced immediately or the added strain placed on the driveline components may cause damage.

Check

2   During the check, the engine must be raised slightly to remove the weight from the mounts.

3   Raise the vehicle and support it securely on jackstands, then position the jack under the engine oil pan. Place a large block of wood between the jack head and the oil pan, then carefully raise the engine just enough to take the weight off the mounts.

4   Check the mounts to see if the rubber is cracked, hardened or separated from the metal plates. Sometimes the rubber will split right down the center. Rubber preservative may be applied to the mounts to slow deterioration.

5   Check for relative movement between the mount brackets and the engine or (use a large screwdriver or pry bar to attempt to move the mounts). If movement is noted, lower the engine and tighten the mount fasteners.

Replacement

6   Disconnect the cable from the negative battery terminal. Raise the vehicle and support it securely on jackstands.

7   Raise the engine slightly, then remove the mount-to-bracket nuts **(see illustrations)**.

8   Remove the mount from the engine compartment.

9   Installation is the reverse of removal. Tighten the engine mount nuts to the torque listed in this Chapter's Specifications.

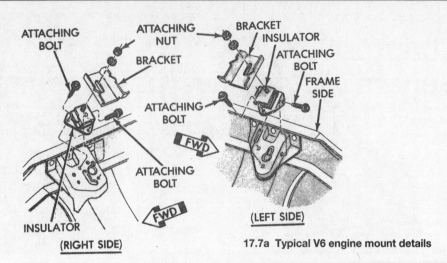

17.7a Typical V6 engine mount details

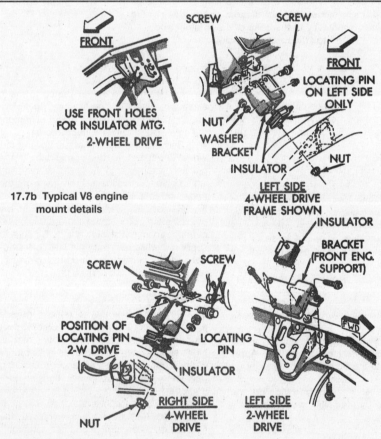

17.7b Typical V8 engine mount details

Chapter 2 Part C
General engine overhaul procedures

Contents

2C

Specifications

Inline six-cylinder engine

General
Oil pressure	30 to 70 psi above 2000 rpm
Cylinder compression	100 psi minimum; no more than 25 psi variation among cylinders

Cylinder block
Cylinder bore diameter (standard)	3.400 to 3.402 inches
Taper limit	0.010 inch
Out-of-round limit	0.005 inch
Deck (head gasket surface) warpage limit	0.00075 inch times the span length of the deck (maximum)

Pistons and rings
Piston-to-cylinder bore clearance	0.0005 to 0.0015 inch
Piston ring side clearance	
Compression rings	0.0015 to 0.0030 inch
Oil ring (steel rails)	0.0002 to 0.005 inch
Piston ring end gap	
Compression rings	0.010 to 0.020 inch
Oil ring (steel rails)	0.015 to 0.055 inch

Crankshaft and connecting rods
Crankshaft main journal	
Diameter (standard)	2.7495 to 2.7505 inches
Out-of-round/taper limit	0.001 inch
Main bearing oil clearance	
Standard	
1971 through 1976	0.0005 to 0.0015 inch
1977 through 1980	0.0002 to 0.0022 Inch
1981 through 1986	0.0010 to 0.0025 inch
Service limit	0.0025 inch
Crankshaft connecting rod journal	
Diameter	2.1865 to 2.1875 inches
Taper limit	0.001 inch
Out-of-round limit	0.001 inch
Standard bearing oil clearance	
1971 through 1976	0.0005 to 0.002 inch
1977	0.0005 to 0.0025 inch
1978 and later	0.0010 to 0.0022 inch
Oil clearance service limit	0.0025 inch

Crankshaft and connecting rods (continued)

Connecting rod side clearance
    1971 through 1976 .... 0.006 to 0.012 inch
    1977 and 1978 .... 0.006 to 0.025 inch
    1979 .... 0.002 to 0.022 inch
    1980 .... 0.002 to 0.009 inch
    1981 and later .... 0.007 to 0.013 inch
Crankshaft endplay
    1971 through 1976 .... 0.002 to 0.007 inch
    1977 through 1980 .... 0.002 to 0.009 inch
    1981 and later .... 0.0035 to 0.0095 inch

Cylinder head and valve train

Head gasket surface warpage limit .... 0.00075 inch times the span length of the head (maximum)
Valve stem diameter (standard)
    Intake .... 0.372 to 0.373 inch
    Exhaust .... 0.371 to 0.372 inch
Valve stem-to-guide clearance
    Intake
        Standard .... 0.001 to 0.003 inch
        Service limit .... 0.008 inch
    Exhaust
        Standard .... 0.002 to 0.004 inch
        Service limit .... 0.008 inch
Valve spring free length .... 1-59/64 inches
Valve spring installed height .... 1-5/8 to 1-11/16 inches
Valve seat width
    Intake .... 0.070 to 0.090 inch
    Exhaust .... 0.040 to 0.060 inch

3.9L V6 and 318, 360 cubic inch small-block V8 engines

General

Oil pressure .... 30 to 80 psi at 3000 rpm
Cylinder compression .... 100 psi minimum; no more than 25 psi variation among cylinders

Cylinder block

Cylinder bore diameter
    V6 and 318 V8 .... 3.910 to 3.912 inches
    360 V8 .... 4.000 to 4.002 inches
Taper limit .... 0.010 inch
Out-of-round limit .... 0.005 inch
Deck (head gasket surface) warpage limit .... 0.00075 inch times the span length of the deck (maximum)

Pistons and rings

Piston-to-cylinder bore clearance .... 0.0005 to 0.0015 inch
Piston ring side clearance
    Compression rings .... 0.0015 to 0.0030 inch
    Oil ring (steel rails) .... 0.002 to 0.005 inch
Piston ring end gap
    Compression rings .... 0.010 to 0.020 inch
    Oil ring (steel rails)
        V6 .... 0.010 to 0.050 inch
        318 V8 .... 0.010 to 0.062 inch
        360 V8 .... 0.015 to 0.062 inch

Crankshaft and connecting rods

Main journal
    Diameter
        3.9L V6 and 318 V8 .... 2.4995 to 2.5005 inches
        360 V8 .... 2.8095 to 2.8105 inches
    Taper/out-of-round limits .... 0.001 inch
Main bearing oil clearance
    No. 1 journal (1981 and later only)
        Standard .... 0.0005 to 0.0015 inch
        Service limit .... 0.0015 inch
    All other journals (and No. 1 journal on 1980 and earlier models)
        Standard .... 0.0005 to 0.0020 inch
        Service limit .... 0.0025 inch
Connecting rod journal
    Diameter .... 2.124 to 2.125 inches
    Taper/out-of-round limits .... 0.001 inch

Connecting rod bearing oil clearance
  3.9L V6
    Standard .... 0.0005 to 0.0022 inch
    Service limit... 0.0022 inch
  318 V8
    1971 through 1978
      Standard .... 0.0002 to 0.0022 inch
      Service limit .... 0.003 inch
    1979 and later
      Standard .... 0.0005 to 0.0025 inch
      Service limit .... 0.0025 inch
  360 V8
    Standard .... 0.0005 to 0.0025 inch
    Service limit... 0.0025 inch
Connecting rod side clearance... 0.006 to 0.014 inch
Crankshaft endplay
  3.9L V6 and 318 V8 .... 0.002 to 0.007 inch
  360 V8 .... 0.002 to 0.009 inch

Cylinder head and valve train

Head gasket surface warpage limit .... 0.00075 inch times the span length of the head (maximum)
Valve stem diameter (standard)
  1991 and earlier V6, 318 V8; 360 V8 through 1993
    Intake .... 0.372 to 0.373 inch
    Exhaust .... 0.371 to 0.372 inch
  1992 and later V6, 318 V8; 1994 and later 360 V8 (intake
    and exhaust) .... 0.311 to 0.312 inch
Valve stem-to-guide clearance
  Intake
    Standard .... 0.001 to 0.003 inch
    Service limit... 0.008 inch
  Exhaust
    Standard .... 0.002 to 0.004 inch
    Service limit... 0.008 inch
Valve spring free length
  Intake.. 1-13/16 inches
  Exhaust
    Without rotators .... 2.00 inches
    With rotators .... 1-13/16 inches
Valve spring installed height
  Intake.. 1-5/8 to 1-11/16 inches
  Exhaust
    Without rotators .... 1-5/8 to 1-11/16 inches
    With rotators .... 1-29/64 to 1-33/64 inches
Valve seat width
  V6 and 318 V8 to 1991, 360 V8 to 1992
    Intake .... 0.065 to 0.085 inch
    Exhaust .... 0.080 to 0.100 inch
  1992 and later V6 and 318, 1993 and later 360 V8
    Intake .... 0.040 to 0.060 inch
    Exhaust .... 0.060 to 0.080 inch

Big-block (400, 440) V8 engines

General

Oil pressure... 30 to 80 psi at 2000 rpm
Cylinder compression pressure... 100 psi minimum; no more than 25 psi variation among cylinders

Cylinder block

Cylinder bore diameter
  400 V8 .... 4.342 to 4.344 inches
  440 V8 .... 4.320 to 4.322 inches
Taper limit.. 0.010 inch
Out-of-round.. 0.005 inch
Deck (cylinder head gasket surface) warpage limit .... 0.00075 inch times the span length of the deck (maximum)

Pistons and rings

Piston-to-cylinder clearance.. 0.0003 to 0.0013 inch
Piston ring side clearance
  Compression rings .... 0.0015 to 0.003 inch
  Oil ring (steel rails)... 0.0002 to 0.005 inch

Pistons and rings (continued)

Piston ring end gap	
Compression rings	0.013 to 0.023 inch
Oil ring (steel rails)	0.015 to 0.055 inch

Crankshaft and connecting rods

Main journal diameter	
400 V8	2.4625 to 2.6255 inches
440 V8	2.7495 to 2.7505 inches
Main journal taper/out-of-round limits	0.001 inch
Main bearing oil clearance	
Standard	0.0005 to 0.002 inch
Service limit	0.0025 inch
Connecting rod journal	
Diameter	2.375 to 2.376 inches
Taper/out-of-round limits	0.001 inch
Connecting rod bearing oil clearance	
Standard	0.0005 to 0.0030 inch
Service limit	0.0030 inch
Connecting rod side clearance	0.009 to 0.017 inch
Crankshaft endplay	0.002 to 0.009 inch

Cylinder head and valve train

Head gasket surface warpage limit	0.00075 inch times the span length of the head (maximum)
Valve stem diameter	
400 V8	
Intake	0.3723 to 0.3730 inch
Exhaust	0.3713 to 0.3730 inch
440 V8, intake and exhaust	0.3718 to 0.3725 inch
Valve stem-to-guide clearance	
Intake	
Standard	0.0011 to 0.0028 inch
Service limit	0.008 inch
Exhaust	
Standard	0.0011 to 0.0038 inch
Service limit (all models)	0.008 inch
Valve spring free length	2-37/64 inches
Valve spring installed height	1-53/64 to 1-57/64 inches
Valve seat width (intake and exhaust)	0.040 to 0.060 inch

Torque specifications*

	Ft-lbs
Main bearing cap bolts (all engines)	85
Connecting rod cap nuts (all engines)	45
Crankshaft rear main seal retainer bolts	
Inline six-cylinder engines	30
400/440 (big block) V8 engines	25

*Note: *Refer to either Chapter 2, Part A or Part B for additional specifications.*

1   General information

Included in this portion of Chapter 2 are the general overhaul procedures for cylinder heads and internal engine components. The information ranges from advice concerning preparation for an overhaul and the purchase of replacement parts to detailed, step-by-step procedures covering removal and installation of internal engine components and the inspection of parts.

The following Sections have been written based on the assumption that the engine has been removed from the vehicle. For information concerning in-vehicle engine repair, as well as removal and installation of the external components necessary for the overhaul, see Parts A and B of this Chapter and Section 9 of this Part.

The Specifications included here in Part C are only those necessary for the inspection and overhaul procedures which follow. Refer to Parts A and B for additional specifications.

2   Engine removal - methods and precautions

If you have decided that an engine must be removed for overhaul or major repair work, several preliminary steps should be taken.

Locating a suitable work area is extremely important. A shop is, of course, the most desirable place to work. Adequate work space, along with storage space for the vehicle, will be needed. If a shop or garage is not available, at the very least a flat, level, clean work surface made of concrete or asphalt is required.

Cleaning the engine compartment and engine before beginning the removal procedure will help keep tools clean and organized.

An engine hoist will be needed. Make sure that the equipment is rated in excess of the combined weight of the engine and its accessories. Safety is of primary importance, considering the potential hazards involved in lifting the engine out of the vehicle.

If the engine is being removed by a novice, a helper should be available. Advice and aid from someone more experienced would also be helpful. There are many instances when one person cannot simultaneously perform all of the operations required when lifting the engine out of the vehicle.

Plan the operation ahead of time. Arrange for or obtain all of the tools and equipment you will need prior to beginning the job. Some of the equipment necessary to perform engine removal and installation safely and with relative ease are (in addition to an engine hoist) a heavy duty floor jack, complete sets of

3.4a The oil pressure sending unit on V6 and V8 engines is located at the rear of the engine, just behind the intake manifold (adjacent to the distributor on small-block engines)

3.4b On inline six-cylinder engines, the sending unit is screwed into the oil pump housing, just below the rear of the oil filter. The type of sending unit shown here is for a system that uses an indicator light on the dash, while the canister-type shown for the V8 engines in the previous illustration is for a system that uses a gauge

3.4c Unscrew the oil pressure sending unit and install a pressure gauge in its place. On gauge-type sending units, there are flats below the canister to unscrew it with a wrench. For sending units on systems with an indicator light, it's best to purchase a special socket designed for removing sending units - this will prevent damaging the sending unit during removal

wrenches and sockets as described in the front of this manual, wooden blocks and plenty of rags and cleaning solvent for mopping up spilled oil, coolant and gasoline. If the hoist is to be rented, make sure that you arrange for it in advance and perform beforehand all of the operations possible without it. This will save you money and time.

Plan for the vehicle to be out of use for a considerable amount of time. A machine shop will be required to perform some of the work which the do-it-yourselfer cannot accomplish due to a lack of special equipment. These shops often have a busy schedule, so it would be wise to consult them before removing the engine in order to accurately estimate the amount of time required to rebuild or repair components that may need work.

Always use extreme caution when removing and installing the engine. Serious injury can result from careless actions. Plan ahead, take your time and a job of this nature, although major, can be accomplished successfully.

3 Engine overhaul - general information

Refer to illustrations 3.4a, 3.4b and 3.4c

It is not always easy to determine when, or if, an engine should be completely overhauled, as a number of factors must be considered.

High mileage is not necessarily an indication that an overhaul is needed, while low mileage does not preclude the need for an overhaul. Frequency of servicing is probably the most important consideration. An engine that has had regular and frequent oil and filter changes, as well as other required maintenance, will most likely give many thousands of miles of reliable service. Conversely, a neglected engine may require an overhaul very early in its life.

Excessive oil consumption is an indication that piston rings and/or valve guides are in need of attention. Make sure that oil leaks are not responsible before deciding that the rings and/or guides are bad. Test the cylinder compression (see Section 5) or have a leakdown test performed by an experienced tune-up mechanic to determine the extent of the work required.

If the engine is making obvious knocking or rumbling noises, the connecting rod and/or main bearings are probably at fault. To accurately test oil pressure, temporarily connect a mechanical oil pressure gauge in place of the oil pressure sending unit **(see illustrations)**. Compare the reading to the pressure listed in this Chapter's Specifications. If the pressure is extremely low, the bearings and/or oil pump are probably worn out.

Loss of power, rough running, excessive valve train noise and high fuel consumption rates may also point to the need for an overhaul, especially if they are all present at the same time. If a complete tune-up does not remedy the situation, major mechanical work is the only solution.

An engine overhaul involves restoring the internal parts to the specifications of a new engine. During an overhaul, the piston rings are replaced and the cylinder walls are reconditioned (rebored and/or honed). If a re-bore is done, new pistons are required. The main bearings, connecting rod bearings and camshaft bearings are generally replaced with new ones and, if necessary, the crankshaft may be reground to restore the journals. Generally, the valves are serviced as well, since they are usually in less-than-perfect condition at this point. While the engine is being overhauled, other components, such as the distributor, starter and alternator, can be rebuilt as well. The end result should be a like-new engine that will give many thousands of trouble-free miles. **Note:** *Critical cooling system components such as the hoses, the drivebelts, the thermostat and the water pump MUST be replaced with new parts when an engine is overhauled. The radiator should be professionally cleaned. If in doubt, replace it with a new one. Also, we do not recommend overhauling the oil pump - always install a new one when an engine is rebuilt.*

Before beginning the engine overhaul, read through the entire procedure to familiarize yourself with the scope and requirements of the job. Overhauling an engine is not difficult if you take your time and follow all procedures carefully. But it is time consuming. Plan on the vehicle being tied up for a minimum of two weeks, especially if parts must be taken to an automotive machine shop for repair or reconditioning. Check on availability of parts and make sure that any necessary special tools and equipment are obtained in advance. Most work can be done with typical hand tools, although a number of precision measuring tools are required for inspecting parts to determine if they must be replaced. Often an automotive machine shop will handle the inspection of parts and offer advice concerning reconditioning and replacement. **Note:** *Always wait until the engine has been completely disassembled and all components, especially the engine block, have been inspected before deciding what service and repair operations must be performed by an automotive machine shop. Since the block's condition will be the major factor to consider when determining whether to overhaul the original engine or buy a rebuilt one, never purchase parts or have machine work done on other components until the block has been thoroughly inspected. As a general rule, time is the primary cost of an overhaul, so it does not pay to install worn or sub-standard parts.*

As a final note, to ensure maximum life and minimum trouble from a rebuilt engine, everything must be assembled with care in a spotlessly clean environment.

2C

5.4 A compression gauge with a threaded fitting for the spark plug hole is preferred over the type that requires hand pressure to maintain the seal - be sure to open the throttle valve as far as possible during the compression check

4   Top Dead Center (TDC) for number one piston - locating

For information on finding Top Dead Center for the number one piston, refer to Chapter 2A for the inline six-cylinder engine, and Chapter 2B for the V6 and V8 engines.

5   Compression check

Refer to illustration 5.4

1   A compression check will tell you what mechanical condition the pistons, rings, valves and head gaskets of your engine are in. Specifically, it can tell you if the compression is down due to leakage caused by worn piston rings, defective valves and seats or a blown head gasket. **Note:** *The engine must be at normal operating temperature for this check and the battery must be fully charged.*

2   Begin by cleaning the area around the spark plugs before you remove them (compressed air works best for this). This will prevent dirt from getting into the cylinders as the compression check is being done. Remove all of the spark plugs from the engine.

3   Block the throttle wide open and disconnect the primary wires from the coil.

4   With the compression gauge in the number one spark plug hole, crank the engine over at least four compression strokes and watch the gauge **(see illustration)**. The compression should build up quickly in a healthy engine. Low compression on the first stroke, followed by gradually increasing pressure on successive strokes, indicates worn piston rings. A low compression reading on the first stroke, which does not build up during successive strokes, indicates leaking valves or a blown head gasket (a cracked head could also be the cause). Record the highest gauge reading obtained.

5   Repeat the procedure for the remaining cylinders and compare the results to the Specifications.

6   Add some engine oil (about three squirts from a plunger-type oil can) to each cylinder, through the spark plug hole, and repeat the test.

7   If the compression increases after the oil is added, the piston rings are definitely worn. If the compression does not increase significantly, the leakage is occurring at the valves or head gasket. Leakage past the valves may be caused by burned valve seats and/or faces or warped, cracked or bent valves.

8   If two adjacent cylinders have equally low compression, there is a strong possibility that the head gasket between them is blown. The appearance of coolant in the combustion chambers or the oil would verify this condition. Generally, coolant will cause the oil on the dipstick to look light-colored (milky).

9   If the compression is unusually high, the combustion chambers are probably coated with carbon deposits. If that is the case, the cylinder heads should be removed and decarbonized.

10   If compression is way down or varies greatly between cylinders, it would be a good idea to have a leak-down test performed by an automotive repair shop. This test will pinpoint exactly where the leakage is occurring and how severe it is.

6   Vacuum gauge diagnostic checks

A vacuum gauge provides valuable information about what is going on in the engine at a low cost. You can check for worn rings or cylinder walls, leaking head or intake manifold gaskets, incorrect carburetor adjustments, restricted exhaust, stuck or burned valves, weak valve springs, improper ignition or valve timing and ignition problems.

Unfortunately, vacuum gauge readings are easy to misinterpret, so they should be used in conjunction with other tests to confirm the diagnosis.

Both the gauge readings and the rate of needle movement are important for accurate interpretation. Most gauges measure vacuum in inches of mercury (in-Hg). As vacuum increases (or atmospheric pressure decreases), the reading will increase. Also, for every 1,000-foot increase in elevation above sea level, the gauge readings will decrease about one inch of mercury.

Connect the vacuum gauge directly to intake manifold vacuum, not to ported (carburetor) vacuum. Be sure no hoses are left disconnected during the test or false readings will result.

Before you begin the test, allow the engine to warm up completely. Block the wheels and set the parking brake. With the transmission in Park, start the engine and allow it to run at normal idle speed. **Warning:** *Carefully inspect the fan blades for cracks or damage before starting the engine. Keep your hands and the vacuum tester clear of the fan and do not stand in front of the vehicle or in line with the fan when the engine is running.*

Read the vacuum gauge; an average, healthy engine should normally produce about 17 to 22 inches of vacuum with a fairly steady needle. Refer to the following vacuum gauge readings and what they indicate about the engines condition:

1   A low, steady reading usually indicates a leaking gasket between the intake manifold and carburetor or throttle body, a leaky vacuum hose, late ignition timing or incorrect camshaft timing. Check ignition timing with a timing light and eliminate all other possible causes, utilizing the tests provided in this Chapter before you remove the timing chain cover to check the timing marks.

2   If the reading is three to eight inches below normal and it fluctuates at that low reading, suspect an intake manifold gasket leak at an intake port.

3   If the needle has regular drops of about two to four inches at a steady rate, the valves are probably leaking. Perform a compression or leak-down test to confirm this.

4   An irregular drop or down-flick of the needle can be caused by a sticking valve or an ignition misfire. Perform a compression or leak-down test and read the spark plugs.

5   A rapid vibration of about four inches-Hg vibration at idle combined with exhaust smoke indicates worn valve guides. Perform a leak-down test to confirm this. If the rapid vibration occurs with an increase in engine speed, check for a leaking intake manifold gasket or head gasket, weak valve springs, burned valves or ignition misfire.

6   A slight fluctuation, say one inch up and down, may mean ignition problems. Check all the usual tune-up items and, if necessary, run the engine on an ignition analyzer.

7   If there is a large fluctuation, perform a compression or leak-down test to look for a weak or dead cylinder or a blown head gasket.

8   If the needle moves slowly through a wide range, check for a clogged PCV system, incorrect idle fuel mixture, carburetor/throttle body or intake manifold gasket leaks.

9   Check for a slow return after revving the engine by quickly snapping the throttle open until the engine reaches about 2,500 rpm and let it shut. Normally the reading should drop to near zero, rise above normal idle reading (about 5 in-Hg over) and then return to the previous idle reading. If the vacuum returns slowly and doesn't peak when the throttle is snapped shut, the rings may be worn. If there is a long delay, look for a restricted exhaust system (often the muffler or catalytic converter). An easy way to check this is to temporarily disconnect the exhaust ahead of the suspected part and re-test.

Vacuum leak diagnosis procedure

Refer to illustrations 6.15, 6.16a and 6.16b

10   If you suspect a vacuum leak, or have just replaced the intake manifold gaskets and want to check for proper sealing, a simple check can be made with a small household propane torch.

11   Place the transmission in Park and apply

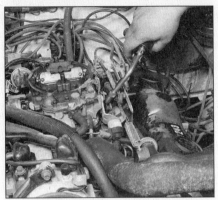

6.15 Use an UNLIT torch to introduce propane around the base of the carburetor; if idle speed increases, a vacuum leak is indicated

6.16a Also check for vacuum leaks around the intake manifold gasket . . .

6.16b . . . and the vacuum hoses and fittings

2C

the emergency brake.

12   Remove the air cleaner assembly for better access to the carburetor or throttle body and intake manifold.

13   Hook up a vacuum gauge and if available, a tachometer.

14   With the engine warmed up and idling, slightly open the valve on the propane bottle. **Warning:** *Do NOT light the torch under any circumstances.*

15   Bring the tip of the torch nozzle around the throttle shaft and the base of the carburetor or throttle body **(see illustration)**. Allow the torch tip to move slowly around these areas. If the idle speed or vacuum increases, it means that there is a vacuum leak wherever the torch is pointing. The extra combustible gas from the torch makes a temporarily-richer mixture if a vacuum leak sucks it into the engine.

16   Move the torch tip slowly around the edges of the intake manifold, especially around the ports at the cylinder head **(see illustration)**. If there is no increase in vacuum or idle speed, there are no vacuum leaks at the intake. However, vacuum leaks can also occur at a variety of vacuum hoses, which can also be checked with an unlit propane torch **(see illustration)**.

7   Engine rebuilding alternatives

The do-it-yourselfer is faced with a number of options when performing an engine overhaul. The decision to replace the engine block, piston/connecting rod assemblies and crankshaft depends on a number of factors, with the number one consideration being the condition of the block. Other considerations are cost, access to machine shop facilities, parts availability, time required to complete the project and the extent of prior mechanical experience on the part of the do-it-yourselfer.

Some of the rebuilding alternatives include:

Individual parts - If the inspection procedures reveal that the engine block and most engine components are in reusable condition, purchasing individual parts may be

the most economical alternative. The block, crankshaft and piston/connecting rod assemblies should all be inspected carefully. Even if the block shows little wear, the cylinder bores should be surface honed.

Crankshaft kit - This rebuild package consists of a reground crankshaft and a matched set of pistons and connecting rods. The pistons will already be installed on the connecting rods. Piston rings and the necessary bearings will be included in the kit. These kits are commonly available for standard cylinder bores, as well as for engine blocks which have been bored to a standard oversize.

Short block - A short block consists of an engine block with a camshaft, timing chain and gears, crankshaft and piston/connecting rod assemblies already installed. All new bearings are incorporated and all clearances will be correct. The existing cylinder head(s), valve train components and external parts can be bolted to the short block with little or no machine shop work necessary.

Long block - A long block consists of a short block plus an oil pump, cylinder head(s) and valve train components. All components are installed with new bearings, seals and gaskets incorporated throughout. The installation of manifolds and external parts is all that is necessary.

Give careful thought to which alternative is best for you and discuss the situation with local automotive machine shops, auto parts dealers or parts store counterpeople before ordering or purchasing replacement parts.

8   Engine - removal and installation

Refer to illustrations 8.1, 8.10, 8.16, 8.17, 8.19, 8.20, 8.23 and 8.26

Warning: *Gasoline is extremely flammable, so take extra precautions when disconnecting any part of the fuel system. Don't smoke or allow open flames or bare light bulbs in or near the work area and don't work in a garage where a natural gas appliance (such as a clothes dryer or water heater) is installed. If you spill gasoline on your skin, rinse it off immediately. Have a fire extinguisher rated for*

8.1 Unbolt the air conditioning compressor and set it aside without disconnecting the hoses, if possible

gasoline fires handy and know how to use it! Also, the air conditioning system is under high pressure - have a dealer service department or service station discharge the system and recapture the refrigerant before disconnecting any of the hoses or fittings.

Note: *Read through the following steps carefully and familiarize yourself with the procedure before beginning work.*

Removal

1   On air-conditioned models, inspect the compressor mounting to determine if the compressor can be unbolted and moved aside without disconnecting the refrigerant lines **(see illustration)**. If this is not possible, have the system discharged by a dealer or service station (see **Warning** above).

2   Remove the hood (see Chapter 11) and engine cover (see chapter 2B, section 2).

3   Remove the air cleaner assembly and all hoses connected to it (see Chapter 4).

4   Drain the cooling system and remove the drivebelts (see Chapter 1). **Note:** *Mark all the accessory drivebelts with markers to indicate which components they drive, i.e. alternator, air conditioning compressor, power steering pump, and also number the belts to indicate which order they go onto the pulleys. This will eliminate confusion on reassembly, which might be days later.*

5   Remove the radiator, shroud and fan,

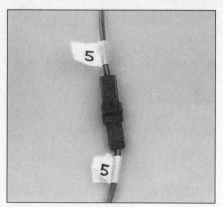

8.10 Label each wire before unplugging the connector

8.16 Support the transmission with a length of chain (as shown) or a floor jack to keep it in place when the engine is separated for removal

8.17 On inline six-cylinder and small-block V8 engines, unbolt and remove the engine-to-transmission brace (arrow) from each side of the engine

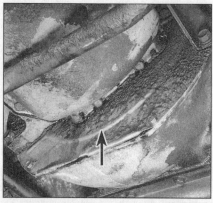

8.19 Remove the bolts and take off the torque converter inspection cover (arrows)

8.20 On automatic-transmission models, remove the four bolts that attach the driveplate to the torque converter - you'll have to rotate the engine for access to each one

and air-conditioning condenser (see Chapter 3).

6   Detach the radiator and heater hoses from the engine.

7   Disconnect the accelerator cable (see Chapter 4) and throttle rod (automatic transmission only - see Chapter 7B) from the carburetor.

8   Remove the power steering pump and brackets (if equipped) without disconnecting the hoses and tie it out of the way (see Chapter 10).

9   Remove the alternator (see Chapter 5).

10   Label and disconnect all wires from the engine **(see illustration)**. Masking tape and/or a touch-up paint applicator work well for marking items. **Note:** *Take instant photos or sketch the locations of components and brackets to help with reassembly.*

11   Disconnect the fuel lines at the engine (see Chapter 4) and plug the lines to prevent fuel loss. On fuel-injected models, refer to Chap-ter 4 for the proper procedure to relieve the fuel system pressure.

12   Label and remove all vacuum lines from the intake manifold.

13   Raise the vehicle and support it securely on jackstands.

14   Drain the engine oil (see Chapter 1).

15   Disconnect the exhaust pipe(s) from the exhaust manifold(s).

16   On 1979 and later models, support the transmission with a floor jack or use a length of chain to keep it in place when the engine is separated for removal **(see illustration)**. **Note:** *On 1971 through 1978 models, the transmission should be removed first (while the engine is supported with a hoist), then the oil pan and pump (refer to Chapter 2B, Section 13).*

17   Remove the engine-to-transmission brace(s), if so equipped **(see illustration)**.

18   Disconnect the wires from the starter solenoid and remove the starter (see Chapter 5).

19   Remove the flywheel or torque converter inspection cover **(see illustration)**.

20   If equipped with an automatic transmission, remove the torque converter-to-driveplate bolts **(see illustration)**. In order to turn the engine to reach all the bolts, reinstall the bolt in the front end of the crankshaft and use

it to turn the driveplate to the needed positions. **Note:** *There is no bolt in the front of the crankshaft on inline six-cylinder engines; you will have to get a 3/4-inch x 16 threads-per-inch bolt that is about 1-1/2 or 2-inches long.*

21   Remove the transmission-to-engine bolts.

22   Attach an engine hoist to the engine and take the weight off the engine mounts. **Note:** *The more components removed from the engine while still in the van, the easier it is to remove the engine through the front. It is best to remove the oil pan, exhaust manifolds, valve covers and intake manifold (see Chapter 2A or 2B for in-vehicle procedures), but removal is even easier if the cylinder heads are also removed first.*

23   The end of the hoist must be very close to the engine **(see illustration)**, and as parallel to the ground as possible, since the engine has to come straight forward. If the hoist is at an angle, raise the vehicle higher to lessen the angle (but no higher than can be safely supported on jackstands).

24   Remove the engine mount through-bolts or stud nuts (see Chapter 2A or 2B).

25   Lift the engine slightly and separate the engine from the transmission. Carefully work it forward to separate it from the transmission. If you're working on a vehicle with an automatic transmission, be sure the torque

converter stays in the transmission (clamp a pair of locking pliers to the housing to keep the converter from sliding out). If you're working on a vehicle with a manual transmission,

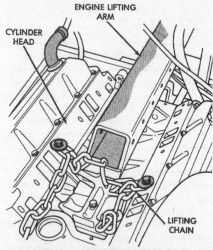

8.23 Keep the hoist arm chained closely to the engine - do not lift the engine by the intake manifold

8.26 The crane arm must be fairly level for the engine to be removed straight out the front of the van - it is easier if the cylinder heads are removed also

9.3a Before removing the engine accessories and their attaching brackets, note how they're attached (front view of a small-block V8 engine shown)

9.3b Small-block V8 engine - left side view

9.3c Small-block V8 engine - right side view

2C

the input shaft must be completely disengaged from the clutch.

26   Check to make sure everything is disconnected, then pull the engine straight out of the vehicle (see illustration). Warning: *Do not place any part of your body under the engine when it is supported only by a hoist or other lifting device.*

27   Remove the flywheel/driveplate and mount the engine on an engine stand or set the engine on the floor and support it so it doesn't tip over. Then disconnect the engine hoist.

Installation

28   Check the engine mounts. If they're worn or damaged, replace them.

29   On manual transmission models, inspect the clutch components (see Chapter 8). On automatic models, inspect the converter seal and bushing.

30   Apply a dab of grease to the pilot bushing on manual transmission models.

31   Attach the hoist to the engine and carefully lower the engine into the engine compartment.

32   Carefully guide the engine into place. Follow the procedure outlined in Chapter 7 for transmission attachment. Caution: *Do not*

use the bolts to force the engine and transmission into alignment. It may crack or damage major components.*

33   Align the engine to the engine mounts. Install the through bolts and tighten them securely.

34   Reinstall the remaining components in the reverse order of removal.

35   Add coolant, oil, power steering and transmission fluid as needed (see Chapter 1).

36   Run the engine and check for proper operation and leaks. Shut off the engine and recheck fluid levels.

9   Engine overhaul - disassembly sequence

Refer to illustrations 9.3a, 9.3b and 9.3c

1   It is much easier to disassemble and work on the engine if it is mounted on a portable engine stand. These stands can often be rented quite cheaply from an equipment rental yard. Before the engine is mounted on a stand, the flywheel/driveplate should be removed from the engine (refer to Chapter 2A or 2B).

2   If a stand is not available, it is possible

to disassemble the engine with it blocked up on a sturdy workbench or on the floor. Be extra careful not to tip or drop the engine when working without a stand.

3   If you are going to obtain a rebuilt engine, all external components must come off first, to be transferred to the replacement engine (see illustrations), just as they will if you are doing a complete engine overhaul yourself. These include:

Alternator and brackets
Emissions control components
Distributor, spark plug wires and spark plugs
Thermostat and housing cover
Water pump
Carburetor
Intake/exhaust manifolds
Oil pump assembly
Engine mounts
Clutch and flywheel/driveplate

Note: *When removing the external components from the engine, pay close attention to details that may be helpful or important during installation. Note the installed position of gaskets, seals, spacers, pins, washers, bolts and other small items.*

4   If you are obtaining a short block, which

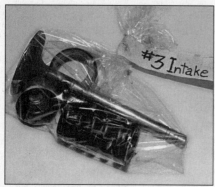

10.2 A small plastic bag, with an appropriate label, can be used to store the valve train components so they can be kept together and reinstalled in the original position

10.3a Use a valve spring compressor to compress the spring, then remove the keepers from the valve stem with a magnet or needle-nose pliers

10.3b If the valve won't pull through the guide, deburr the edge of the stem end and the area around the top of the keeper groove with a file or whetstone

consists of the engine block, crankshaft, pistons and connecting rods all assembled, then the cylinder heads, oil pan and oil pump will have to be removed as well. See *Engine rebuilding alternatives* for additional information regarding the different possibilities to be considered.

5   If you are planning a complete overhaul, the engine must be disassembled and the internal components removed in the general following order:

> *Valve cover(s)*
> *Intake and exhaust manifolds*
> *Rocker arms and pushrods*
> *Valve lifters*
> *Cylinder head(s)*
> *Oil pan*
> *Timing chain cover*
> *Timing chain and sprockets*
> *Oil pump*
> *Camshaft*
> *Piston/connecting rod assemblies*
> *Crankshaft and main bearings*

6   Critical cooling system components such as the hoses, the drivebelts, the thermostat and the water pump MUST be replaced with new parts when an engine is overhauled. If you are buying a rebuilt engine or short-block, some rebuilders will not guarantee their engines unless you have proof that the radiator has been professionally cleaned. Also, we do not recommend overhauling the oil pump - always install a new one when an engine is rebuilt.

7   Before beginning the disassembly and overhaul procedures, make sure the following items are available:

> *Common hand tools*
> *Small cardboard boxes or plastic bags for storing parts*
> *Gasket scraper*
> *Ridge reamer*
> *Vibration damper puller*
> *Micrometers*
> *Telescoping gauges*
> *Dial indicator set*
> *Valve spring compressor*
> *Cylinder surfacing hone*
> *Piston ring groove cleaning tool*

> *Electric drill motor*
> *Tap and die set*
> *Wire brushes*
> *Oil gallery brushes*
> *Cleaning solvent*

10   Cylinder head - disassembly

Refer to illustrations 10.2, 10.3a and 10.3b
Note: *New and rebuilt cylinder heads are commonly available for most engines at dealerships and auto parts stores. Due to the fact that some specialized tools are necessary for the disassembly and inspection procedures, and replacement parts may not be readily available, it may be more practical and economical for the home mechanic to purchase replacement heads rather than taking the time to disassemble, inspect and recondition the originals.*

1   Cylinder head disassembly involves removal of the intake and exhaust valves and related components. If they are still in place, remove the rocker arms. Label the parts or store them separately so they can be reinstalled in their original locations.

2   Before the valves are removed, arrange to label and store them, along with their related components, so they can be kept separate and reinstalled in the same valve guides they are removed from **(see illustration)**.

3   Compress the springs on the first valve with a spring compressor and remove the keepers **(see illustration)**. Carefully release the valve spring compressor and remove the retainer and (if used) rotators, the shield, the springs and the spring seat or shims (if used). Remove the oil seal(s) from the valve stem and the umbrella-type seal from over the guide boss (if used), then pull the valve from the head. If the valve binds in the guide (won't pull through), push it back into the head and deburr the area around the keeper groove with a fine file or whetstone **(see illustration)**.

4   Repeat the procedure for the remaining valves. Remember to keep all the parts for each valve together so they can be reinstalled in the same locations.

5   Once the valves and related compo-

nents have been removed and stored in an organized manner, the head should be thoroughly cleaned and inspected. If a complete engine overhaul is being done, finish the engine disassembly procedures before beginning the cylinder head cleaning and inspection process.

11   Cylinder head - cleaning and inspection

1   Thorough cleaning of the cylinder heads and related valve train components, followed by a detailed inspection, will enable you to decide how much valve service work must be done during the engine overhaul.

Cleaning

2   Scrape away all traces of old gasket material and sealing compound from the head gasket, intake manifold and exhaust manifold sealing surfaces. Be very careful not to gouge the cylinder head. Special gasket removal solvents, which soften gaskets and make removal much easier, are available at auto parts stores.

3   Remove any built up scale from the coolant passages.

4   Run a stiff wire brush through the various holes to remove any deposits that may have formed in them.

5   Run an appropriate size tap into each of the threaded holes to remove any corrosion and thread sealant that may be present. If compressed air is available, use it to clear the holes of debris produced by this operation.

6   Check the condition of the spark plug threads.

7   Clean the cylinder head with solvent and dry it thoroughly. Compressed air will speed the drying process and ensure that all holes and recessed areas are clean. **Note:** *Decarbonizing chemicals are available and may prove very useful when cleaning cylinder heads and valve train components. They are very caustic and should be used with caution. Be sure to follow the instructions on the container.*

8   Clean the rocker arms, shafts and

11.12 Check the cylinder head gasket surface for warpage by trying to slip a feeler gauge under the straightedge (see this Chapter's Specifications for the maximum warpage allowed and use a feeler gauge of that thickness)

pushrods with solvent and dry them thoroughly (don't mix them up during the cleaning process). Compressed air will speed the drying process and can be used to clean out the oil passages.

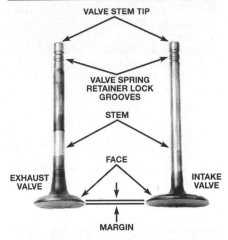

11.15 Check for valve wear at the points shown here

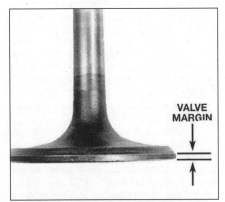

11.16 The margin width on each valve must be as specified (if no margin exists, the valve cannot be reused)

11.14 A dial indicator can be used to determine the valve stem-to-guide clearance - move the valve stem as indicated by the arrows

9 Clean all the valve springs, shields, keepers and retainers (or rotators) with solvent and dry them thoroughly. Do the components from one valve at a time to avoid mixing up the parts.

10 Scrape off any heavy deposits that may have formed on the valves, then use a wire brush mounted in a drill motor to remove deposits from the valve heads and stems. Again, make sure the valves do not get mixed up.

Inspection
Cylinder head
Refer to illustrations 11.12 and 11.14

11 Inspect the head very carefully for cracks, evidence of coolant leakage and other damage. If cracks are found, a new cylinder head should be obtained.

12 Using a straightedge and feeler gauge, check the head gasket mating surface for warpage **(see illustration)**. If the warpage exceeds the limit specified in this Chapter, it can be resurfaced at an automotive machine shop.

13 Examine the valve seats in each of the combustion chambers. If they are pitted, cracked or burned, the head will require valve

service with special equipment that is beyond the scope of most home mechanics, especially from an equipment stand point.

14 Check the valve stem-to-guide clearance by measuring the lateral movement of the valve stem with a dial indicator attached securely to the head **(see illustration)**. The valve must be in the guide and approximately 1/16-inch off the seat. The total valve stem movement indicated by the gauge needle must be divided by two to obtain the actual clearance. After this is done, if there is still some doubt regarding the condition of the valve guides they should be checked by an automotive machine shop (the cost should be minimal).

Valves
Refer to illustrations 11.15 and 11.16

15 Carefully inspect each valve face for uneven wear **(see illustration)**, deformation, cracks, pits and burned spots. Check the valve stem for scuffing and galling and the neck for cracks. Rotate the valve and check for any obvious indication that it is bent. Look for pits and excessive wear on the end of the stem. The presence of any of these conditions indicates the need for valve service by an automotive machine shop.

16 Measure the margin width on each valve. Any valve with a margin that is narrower than 0.047 (3/64) inch should be discarded and replaced with a new one **(see illustration)**.

Valve components
Refer to illustrations 11.17, 11.18 and 11.19

17 Check each valve spring for wear (on the ends) and pits. Measure the free length and compare it to the Specifications in this Chapter **(see illustration)**. Any springs that are shorter than specified have sagged and should not be reused. The tension of all springs should be checked with a special fixture before deciding that they are suitable for use in a rebuilt engine (take the springs to an automotive machine shop for this check).

18 Stand each spring on a flat surface and check it for squareness **(see illustration)**. If any of the springs are distorted or sagged, replace all of them with new parts.

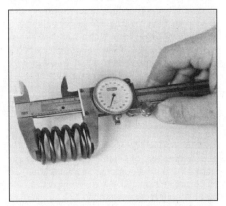

11.17 Measure the free length of each valve spring with a dial or vernier caliper

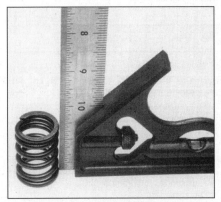

11.18 Check each valve spring for squareness - if it is bent, it should be replaced

2C

11.19 The exhaust valve rotators can be checked by turning the inner and outer sections in opposite directions to feel for smooth movement and excessive play

19   Check the spring retainers (or rotators) and keepers for obvious wear **(see illustration)** and cracks. Any questionable parts should be replaced with new ones, as extensive damage will occur if they fail during engine operation.

Rocker arm components

20   Check the rocker arm faces (the areas that contact the pushrod ends and valve stems) for pits, wear, galling, score marks and rough spots. Check the rocker arm pivot contact areas and shafts. Look for cracks in each rocker arm.
21   Inspect the pushrod ends for scuffing and excessive wear. Roll each pushrod on a flat surface, such as plate glass, to determine if it is bent.
22   Check the rocker arm bolt holes in the cylinder heads for damaged threads.
23   Any damaged or excessively worn parts must be replaced with new ones.
24   If the inspection process indicates that the valve components are in generally poor condition and worn beyond the limits specified in this Chapter, which is usually the case in an engine that is being overhauled, reassemble the valves in the cylinder head and refer to Section 12 for valve servicing recommendations.

12   Valves - servicing

1   Because of the complex nature or the job and the special tools and equipment needed, servicing of the valves, the valve seats and the valve guides (commonly known as a "valve job") is best left to a professional.
2   The home mechanic can remove and disassemble the heads, do the initial cleaning and inspection, then reassemble and deliver the heads to a dealer service department or an automotive machine shop for the actual valve servicing.
3   The dealer service department, or automotive machine shop, will remove the valves and springs, recondition or replace the valves

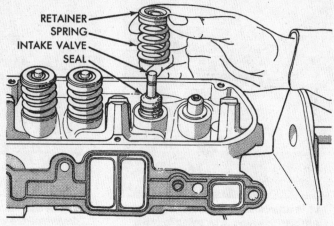

13.5 The intake valves have retainers; the exhaust valves on some earlier models may use rotators

RETAINER
SPRING
INTAKE VALVE
SEAL

13.6 Apply a small dab of grease to each keeper as shown here before installation - it'll hold them in place on the valve stem as the spring is released

and valve seats, recondition the valve guides, check and replace the valve springs, spring retainers or rotators and keepers (as necessary), replace the valve seals with new ones, reassemble the valve components and make sure the installed spring height is correct. The cylinder head gasket surface will also be resurfaced if it is warped.
4   After the valve job has been performed by a professional, the head will be in like-new condition. When the head is returned, be sure to clean it again before installation on the engine to remove any metal particles and abrasive grit that may still be present from the valve service or head resurfacing operations. Use compressed air, if available, to blow out all the oil holes and passages.

13   Cylinder head - reassembly

Refer to illustrations 13.5, 13.6 and 13.8
1   Regardless of whether or not the heads were sent to an automotive machine shop for valve servicing, make sure they are clean before beginning reassembly.
2   If the heads were sent out for valve servicing, the valves and related components will already be in place. Begin the reassembly

13.8 Valve spring installed height is the distance from the spring seat on the head to the top of the spring

procedure with Step 8.
3   Beginning at one end of the head, lubricate and install the first valve. Apply moly-base grease or clean engine oil to the valve stem.
4   Different types of valve stem oil seals are used on the intake and exhaust. On some applications, an umbrella type seal which extends down over the valve guide boss is used over the valve stem.
5   Drop the spring seat or shim(s), if used, over the valve guide and set the valve spring and retainer (or rotator) in place **(see illustration)**.
6   Compress the springs with a valve spring compressor. Position the keepers in the upper groove, then slowly release the compressor and make sure the keepers seat properly. Apply a small dab of grease to each keeper to hold it in place if necessary **(see illustration)**.
7   Repeat the procedure for the remaining valves. Be sure to return the components to their original locations - do not mix them up!
8   Check the installed valve spring height with a ruler graduated in 1/32-inch increments **(see illustration)** or a dial caliper. If the heads were sent out for service work, the installed height should be correct (but don't automatically assume that it is). The mea-

14.1 A ridge reamer is required to remove the ridge from the top of each cylinder - do this before removing the pistons!

14.3 Check the connecting rod side clearance (endplay) with a feeler gauge as shown here

14.4 The connecting rods should be marked with numbers at the parting line of each rod and cap - the number corresponds to the engine cylinder number - on reassembly, be sure the numbers are on the same side, facing each other, as shown, and the rod/piston goes back into the correct cylinder

surement is taken from the top of each spring seat or shim(s) to the bottom of the retainer/rotator. If the height is greater than listed in this Chapter's specifications, shims can be added under the springs to correct it. **Caution:** *Do not, under any circumstances, shim the springs to the point where the installed height is less than specified. A condition called "coil bind" will be caused when the spring is compressed.*

14 Piston/connecting rod assembly - removal

Refer to illustrations 14.1, 14.3, 14.4 and 14.5
Note: *Prior to removing the piston/connecting rod assemblies, remove the cylinder head(s), the oil pan and the oil pump, if it is mounted in the crankcase, and the oil pump pickup screen assembly by referring to the appropriate Sections in Chapter 2, Part A or Part B.*

1   Completely remove the ridge at the top of each cylinder with a ridge reaming tool **(see illustration)**. Follow the manufacturer's instructions provided with the tool. Failure to remove the ridge before attempting to remove the piston/connecting rod assemblies will result in piston breakage.
2   After the cylinder ridges have been removed, turn the engine upside-down so the crankshaft is facing up.
3   Before the connecting rods are removed, check the endplay (side clearance) with feeler gauges. Slide them between the first connecting rod and the crankshaft throw until the play is removed **(see illustration)**. The endplay is equal to the thickness of the feeler gauge(s). If the endplay exceeds the service limit listed in this Chapter's Specifications, new connecting rods will be required. If new rods (or a new crankshaft) are installed, the endplay may fall under the specified minimum. If it does, the rods will have to be machined to restore it - consult an automotive machine shop for advice if necessary. Repeat the procedure for the remaining connecting rods.
4   Check the connecting rods and caps for

identification marks **(see illustration)**. If they are not plainly marked, use a small center-punch to make the appropriate number of indentations on each rod and cap.
5   Loosen each of the connecting rod cap nuts 1/2-turn at a time until they can be removed by hand. Remove the number one connecting rod cap and bearing insert. Do not drop the bearing insert out of the cap. Slip a short length of plastic or rubber hose over each connecting rod cap bolt to protect the crankshaft journal and cylinder wall when the piston is removed **(see illustration)**. Push the connecting rod/piston assembly out through the top of the engine. Use a wooden hammer handle to push on the upper bearing insert in the connecting rod. If resistance is felt, double-check to make sure that all of the ridge was removed from the cylinder.
6   Repeat the procedure for the remaining cylinders. After removal, reassemble the connecting rod caps and bearing inserts in their respective connecting rods and install the cap nuts finger tight. Leaving the old bearing inserts in place until reassembly will help prevent the connecting rod bearing surfaces from being accidentally nicked or gouged.

15 Crankshaft - removal

Refer to illustrations 15.2, 15.3, 15.4a and 15.4b
Note: *The crankshaft can be removed only after the engine has been removed from the vehicle. It is assumed that the flywheel or driveplate, vibration damper or crankshaft pulley, timing chain, oil pan, oil pump and piston/connecting rod assemblies have already been removed.*

1   Before the crankshaft is removed, check the endplay. Mount a dial indicator with the stem in line with the crankshaft and touching one of the crank throws.
2   Push the crankshaft all the way to the rear and zero the dial indicator. Next, pry the crankshaft to the front as far as possible and check the reading on the dial indicator **(see illustration)**. The distance that it moves is the

14.5 To prevent damage to the crankshaft journals and cylinder walls, slip sections of hose over the rod bolts before removing the piston/rod assemblies

endplay. If it is greater than the maximum listed in this Chapter's Specifications, check the crankshaft thrust surfaces for wear. If no wear is evident, new main bearings should correct the endplay.

15.2 Checking crankshaft endplay with a dial indicator

2C

15.3 Checking crankshaft endplay with a feeler gauge

15.4a The main bearing caps are marked to indicate their locations (arrows) - they should be numbered consecutively from the front of the engine to the rear

15.4b If the main cap marks aren't clear, mark the caps with number stamping dies or a center-punch

3   If a dial indicator is not available, feeler gauges can be used. Gently pry or push the crankshaft all the way to the front of the engine. Slip feeler gauges between the crankshaft and the front face of the thrust main bearing to determine the clearance **(see illustration)**.

4   Check the main bearing caps to see if they are marked to indicate their locations. They should be numbered consecutively from the front of the engine to the rear **(see illustration)**. If they aren't, mark them with number stamping dies or a center-punch **(see illustration)**.

5   Loosen each of the main bearing cap bolts 1/4-turn at a time each, until they can be removed by hand.

6   Gently tap the caps with a soft-face hammer, then separate them from the engine block. If necessary, use the bolts as levers to remove the caps. Try not to drop the bearing inserts if they come out with the caps.

7   Carefully lift the crankshaft out of the engine. It is a good idea to have an assistant available, since the crankshaft is quite heavy. With the bearing inserts in place in the engine block and main bearing caps, return the caps to their respective locations on the engine block and tighten the bolts finger tight.

16 Engine block - cleaning

Refer to illustrations 16.1a, 16.1b, 16.4, 16.8 and 16.10

1   Using the wide end of a punch **(see illustration)** tap in on the outer edge of the core plug to turn the plug sideways in the bore. Then, using a pair of pliers, pull the core plug from the engine block **(see illustration)**. Don't worry about the condition of the old core plugs as they are being removed because they will be replaced on reassembly with new plugs.

2   Using a gasket scraper, remove all traces of gasket material from the engine block.

3   Remove the main bearing caps or cap assembly and separate the bearing inserts from the caps and the engine block. Tag the bearings, indicating which cylinder they were removed from and whether they were in the cap or the block, then set them aside.

4   Remove all of the threaded oil gallery plugs from the block **(see illustration)**. The plugs are usually very tight - they may have

to be drilled out and the holes retapped. If drilled out or damaged during removal, use new plugs when the engine is reassembled.

5   If the engine is extremely dirty it should be taken to an automotive machine shop to be steam cleaned or hot tanked.

6   After the block is returned, clean all oil holes and oil galleries one more time. Brushes specifically designed for this purpose are available at most auto parts stores. Flush the passages with warm water until the water runs clear, dry the block thoroughly and wipe all machined surfaces with a light, rust preventive oil. If you have access to compressed air, use it to speed the drying process and to blow out all the oil holes and galleries. **Warning:** *Wear eye protection when using compressed air!*

7   If the block isn't extremely dirty or sludged up, you can do an adequate cleaning job with hot soapy water and a stiff brush. Take plenty of time and do a thorough job. Regardless of the cleaning method used, be sure to clean all oil holes and galleries very thoroughly, dry the block completely and coat all machined surfaces with light oil.

16.1a A hammer and a large punch can be used to knock the core plugs sideways in their bores

16.1b Pull the core plugs from the block with pliers

16.4 Remove all the oil galley plugs/bolts (arrows) to more thoroughly clean all debris from the oiling system (typical galley plugs shown)

16.8 All bolt holes in the block - particularly the main bearing cap and head bolt holes - should be cleaned and restored with a tap (be sure to remove debris from the holes after this is done)

16.10 A large socket on an extension can be used to drive the new core plugs into the bores

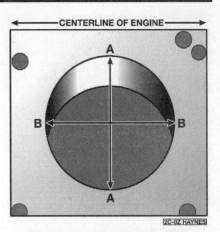

17.4a Measure the diameter of each cylinder just under the wear ridge (A), at the center (B) and at the bottom

8   The threaded holes in the block must be clean to ensure accurate torque readings during reassembly. Run the proper size tap into each of the holes to remove rust, corrosion, thread sealant or sludge and restore damaged threads **(see illustration)**. If possible, use compressed air to clear the holes of debris produced by this operation. Now is a good time to clean the threads on the head bolts and the main bearing cap bolts as well.
9   Reinstall the main bearing caps and tighten the bolts finger tight.
10   After coating the sealing surfaces of the new core plugs with Permatex no. 2 sealant, install them in the engine block **(see illustration)**. Make sure they're driven in straight or leakage could result. Special tools are available for this purpose, but a large socket, with an outside diameter that will just slip into the core plug, a 1/2-inch drive extension and a hammer will work just as well.
11   Apply non-hardening sealant (such as Permatex no. 2 or Teflon tape) to the new oil gallery plugs and thread them into the holes in the block. Make sure they're tightened securely.
12   If the engine isn't going to be reassembled right away, cover it with a large plastic trash bag to keep it clean.

17   Engine block - inspection

Refer to illustrations 17.4a, 17.4b and 17.4c

1   Before the block is inspected, it should be cleaned (see Section 14).
2   Visually check the block for cracks, rust and corrosion. Look for stripped threads in the threaded holes. It's also a good idea to have the block checked for hidden cracks by an automotive machine shop that has the special equipment to do this type of work. If defects are found, have the block repaired, if possible, or replaced.
3   Check the cylinder bores for scuffing and scoring.
4   Check the cylinders for taper and out-

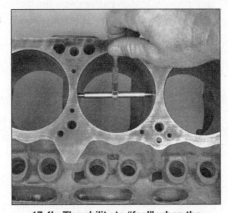

17.4b The ability to "feel" when the telescoping gauge is at the correct point will be developed over time, so work slowly and repeat the check until you're satisfied the bore measurement is accurate

of-round conditions as follows **(see illustrations)**:
5   Measure the diameter of each cylinder at the top (just under the ridge area), center and bottom of the cylinder bore, parallel to the crankshaft axis.
6   Next measure each cylinder's diameter at the same three locations perpendicular to the crankshaft axis.
7   The taper of the cylinder is the difference between the bore diameter at the top of the cylinder and the diameter at the bottom. The out-of-round specification of the cylinder bore is the difference between the parallel and perpendicular readings. Compare your results to those listed in this Chapter's Specifications.
8   If the cylinder walls are badly scuffed or scored, or if they're out-of-round or tapered beyond the specified limits, have the engine block rebored and honed at an automotive machine shop.
9   If a rebore is done, new pistons and rings will be required.
10   Using a precision straightedge and feeler gauge, check the block deck (the surface that mates with the cylinder head(s) for distortion as you did with the cylinder head(s)

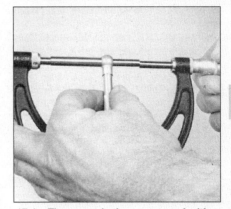

17.4c The gauge is then measured with a micrometer to determine the bore size

(see illustration 11.12). If it's distorted beyond the specified limit, it can be resurfaced by an automotive machine shop.
11   If the cylinders are in reasonably good condition and not worn to the outside of the limits, and if the piston-to-cylinder clearances can be maintained properly (see Section 19), then they don't have to be rebored. Honing is all that's necessary (see Section 18).

18   Cylinder honing

Refer to illustrations 18.3a and 18.3b

1   Prior to engine reassembly, the cylinder bores must be honed so the new piston rings will seat correctly and provide the best possible combustion chamber seal. **Note:** *If you do not have the tools or do not want to tackle the honing operation, most automotive machine shops will do it for a reasonable fee.*
2   Before honing the cylinders, install the main bearing caps and tighten the bolts to the specified torque.
3   Two types of cylinder hones are commonly available - the flex hone or "bottle brush" type and the more traditional surfacing hone with spring-loaded stones. Both will do the job, but for the less experienced mechanic the "bottle brush" hone will proba-

2C

18.3a A "bottle brush" hone will produce better results if you've never honed cylinders before

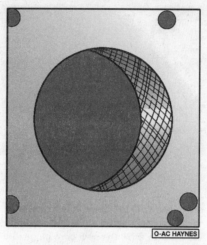

18.3b The cylinder hone should leave a smooth, crosshatch pattern with the lines intersecting at approximately a 60-degree angle

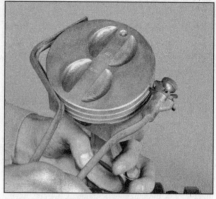

19.4a The piston ring grooves can be cleaned with a special tool, as shown here . . .

19.4b . . . or a section of a broken ring

bly be easier to use. You will also need plenty of light oil or honing oil, some rags and an electric drill motor. Proceed as follows:

a) *Mount the hone in the drill motor, compress the stones and slip it into the first cylinder* **(see illustration)**.

b) *Lubricate the cylinder with plenty of oil, turn on the drill and move the hone up-and-down in the cylinder at a pace which will produce a fine crosshatch pattern on the cylinder walls. Ideally, the crosshatch lines should intersect at approximately a 60-degree angle* **(see illustrations)**. *Be sure to use plenty of lubricant and do not take off any more material than is absolutely necessary to produce the desired finish.* **Note:** *Piston ring manufacturers may specify a smaller crosshatch angle than the traditional 60-degrees, typically 45-degrees, so read and follow any instructions printed on the piston ring packages.*

c) *Do not withdraw the hone from the cylinder while it is running. Instead, shut off the drill and continue moving the hone up-and-down in the cylinder until it comes to a complete stop, then compress the stones and withdraw the hone. If you are using a - "bottle brush" type hone, stop the drill motor, then turn the chuck in the normal direction of rotation while withdrawing the hone from the cylinder.*

d) *Wipe the oil out of the cylinder and repeat the procedure for the remaining cylinders.*

4   After the honing job is complete, chamfer the top edges of the cylinder bores with a small file so the rings will not catch when the pistons are installed. Be very careful not to nick the cylinder walls with the end of the file.

5   The entire engine block must be washed again very thoroughly with warm, soapy water to remove all traces of the abrasive grit produced during the honing operation. **Note:** *The bores can be considered clean when a white cloth - dampened with clean engine oil - used to wipe down the bores does not pick up any more honing residue, which will show up as gray areas on the cloth. Be sure to run a*

brush through all oil holes and galleries and flush them with running water.

6   After rinsing, dry the block and apply a coat of light rust preventive oil to all machined surfaces. Wrap the block in a plastic trash bag to keep it clean and set it aside until reassembly.

19   Piston/connecting rod assembly - inspection

Refer to illustrations 19.4a, 19.4b and 19.10

1   Before the inspection process can be carried out, the piston/connecting rod assemblies must be cleaned and the original piston rings removed from the pistons. **Note:** *Always use new piston rings when the engine is reassembled.*

2   Using a piston ring installation tool, carefully remove the rings from the pistons. Be careful not to nick or gouge the pistons in the process.

3   Scrape all traces of carbon from the top of the piston. A hand-held wire brush or a piece of fine emery cloth can be used once the majority of the deposits have been scraped away. Do not, under any circumstances, use a wire brush mounted in a drill motor to remove deposits from the pistons. The piston material is soft and may be eroded away by the wire brush.

4   Use a piston ring groove cleaning tool to remove carbon deposits from the ring grooves. If a tool isn't available, a piece broken off the old ring will do the job. Be very careful to remove only the carbon deposits - don't remove any metal and do not nick or scratch the sides of the ring grooves **(see illustrations)**.

5   Once the deposits have been removed, clean the piston/rod assemblies with solvent and dry them with compressed air (if available). **Warning:** *Wear eye protection when using compressed air!* Make sure the oil

return holes in the back sides of the ring grooves and the oil hole in the lower end of each rod are clear.

6   If the pistons and cylinder walls aren't damaged or worn excessively, and if the engine block is not rebored, new pistons won't be necessary. Normal piston wear appears as even vertical wear on the piston thrust surfaces and slight looseness of the top ring in its groove. New piston rings, however, should always be used when an engine is rebuilt.

7   Carefully inspect each piston for cracks around the skirt, at the pin bosses and at the ring lands.

8   Look for scoring and scuffing on the thrust faces of the skirt, holes in the piston crown and burned areas at the edge of the crown. If the skirt is scored or scuffed, the engine may have been suffering from overheating and/or abnormal combustion, which caused excessively high operating temperatures. The cooling and lubrication systems should be checked thoroughly. A hole in the piston crown is an indication that abnormal combustion (preignition) was occurring. Burned areas at the edge of the piston crown are usually evidence of spark knock (detonation). If any of the above problems exist, the causes must be corrected or the damage will occur again. The causes may include intake air leaks, incorrect fuel/air mixture, incorrect ignition timing and EGR system malfunctions.

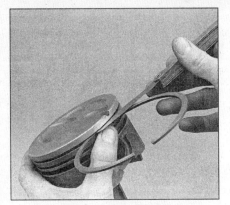

19.10 Check the ring side clearance with a feeler gauge at several points around the groove

20.1 The oil holes should be chamfered so sharp edges don't gouge or scratch the new bearings

20.2 Use a wire or stiff plastic bristle brush to clean the oil passages in the crankshaft - flush the passages out with solvent

9   Corrosion of the piston, in the form of small pits, indicates that coolant is leaking into the combustion chamber and/or the crankcase. Again, the cause must be corrected or the problem may persist in the rebuilt engine.

10   Measure the piston ring side clearance by laying a new piston ring in each ring groove and slipping a feeler gauge in beside it **(see illustration)**. Check the clearance at three or four locations around each groove. Be sure to use the correct ring for each groove - they are different. If the side clearance is greater than specified, new pistons will have to be used.

11   Check the piston-to-bore clearance by measuring the bore (see Section 17) and the piston diameter. Make sure the pistons and bores are correctly matched. Measure the piston across the skirt, at a 90-degree angle to the piston pin, the specified distance down from the top of the piston or the lower edge of the oil ring groove. Subtract the piston diameter from the bore diameter to obtain the clearance. If it's greater than specified, the block will have to be rebored and new pistons and rings installed.

12   Check the piston-to-rod clearance by twisting the piston and rod in opposite directions. Any noticeable play indicates excessive wear, which must be corrected. The piston/connecting rod assemblies should be taken to an automotive machine shop to have the pistons and rods re-sized and new pins installed.

13   If the pistons must be removed from the connecting rods for any reason, they should be taken to an automotive machine shop. While they are there have the connecting rods checked for bend and twist, since automotive machine shops have special equipment for this purpose. **Note:** *Unless new pistons and/or connecting rods must be installed, do not disassemble the pistons and connecting rods.*

14   Check the connecting rods for cracks and other damage. Temporarily remove the rod caps, lift out the old bearing inserts, wipe the rod and cap bearing surfaces clean and inspect them for nicks, gouges and

scratches. After checking the rods, replace the old bearings, slip the caps into place and tighten the nuts finger tight. **Note:** *If the engine is being rebuilt because of a connecting rod knock, be sure to check with an automotive machine shop on the possibility of re-sizing either the small or large end of the rod. If this isn't possible install new rods.*

20   Crankshaft - inspection

Refer to illustrations 20.1, 20.2, 20.4, 20.6a, 20.6b and 20.8

1   Remove all burrs from the crankshaft oil holes with a stone, file or scraper **(see illustration)**.

2   Clean the crankshaft with solvent and dry it with compressed air (if available). Be sure to clean the oil holes with a stiff brush and flush them with solvent **(see illustration)**.

3   Check the main and connecting rod bearing journals for uneven wear, scoring, pits and cracks.

4   Rub a penny across each journal several times **(see illustration)**. If a journal picks up copper from the penny, it's too rough and must be reground.

5   Check the rest of the crankshaft for cracks and other damage. It should be magnafluxed to reveal hidden cracks - an automotive machine shop will handle the procedure.

6   Using a micrometer, measure the diameter of the main and connecting rod journals and compare the results to the Specifications **(see illustrations)**. By measuring the diameter at a number of points around each journal's circumference, you'll be able to determine whether or not the journal is out-of-

20.4 Rubbing a penny lengthwise on each journal will reveal its condition - if copper rubs off and is embedded in the crankshaft, the journals should be reground

20.6a Measure the diameter of each crankshaft journal at several points to detect taper and out-of-round conditions

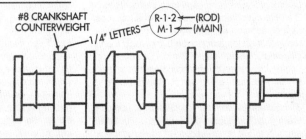

20.6b The V8 engine crankshaft may have a letter/number stamped into the counterweight of the number eight journal - this marking will reveal the location (journal number) of the undersize rod (R) or (M) main journal bearing(s)

#8 CRANKSHAFT COUNTERWEIGHT

1/4" LETTERS

R-1-2 (ROD)
M-1 (MAIN)

20.8 If the seals have worn grooves in the crankshaft journals, or if the seal contact surfaces are nicked or scratched, the new seals will leak

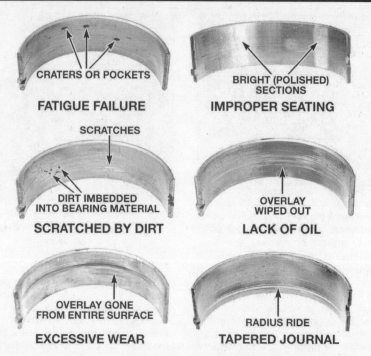

21.1 Before discarding the used bearings, examine them for indications of any possible problems with the crankshaft, noting the bearing location it came from. Here are some typical bearing failures

round. Take the measurement at each end of the journal, near the crank throws, to determine if the journal is tapered. Crankshaft runout should be checked also, but large V-blocks and a dial indicator are needed to do it correctly. If you don't have the equipment, have a machine shop check the runout.

7   If the crankshaft journals are damaged, tapered, out-of-round or worn beyond the limits given in the Specifications, have the crankshaft reground by an automotive machine shop. Be sure to use the correct size bearing inserts if the crankshaft is reconditioned.

8   Check the oil seal journals at each end of the crankshaft for wear and damage **(see illustration)**. If the seal has worn a groove in the journal, or if it's nicked or scratched, the new seal may leak when the engine is reassembled. In some cases, an automotive machine shop may be able to repair the journal by pressing on a thin sleeve. If repair isn't feasible, a new or different crankshaft should be installed.

9   Examine the main and rod bearing inserts (see Section 21).

21 Main and connecting rod bearings - inspection and selection

Inspection

Refer to illustration 21.1

1   Even though the main and connecting rod bearings should be replaced with new ones during the engine overhaul, the old bearings should be retained for close examination, as they may reveal valuable information about the condition of the engine **(see illustrations)**.

2   Bearing failure occurs because of lack of lubrication, the presence of dirt or other foreign particles, overloading the engine and corrosion. Regardless of the cause of bearing failure, it must be corrected before the engine is reassembled to prevent it from happening again.

3   When examining the bearings, remove them from the engine block, the main bearing caps, the connecting rods and the rod caps

and lay them out on a clean surface in the same general position as their location in the engine. This will enable you to match any bearing problems with the corresponding crankshaft journal.

4   Dirt and other foreign particles get into the engine in a variety of ways. It may be left in the engine during assembly, or it may pass through filters or the PCV system. It may get into the oil, and from there into the bearings. Metal chips from machining operations and normal engine wear are often present. Abrasives are sometimes left in engine components after reconditioning, especially when parts are not thoroughly cleaned using the proper cleaning methods. Whatever the source, these foreign objects often end up embedded in the soft bearing material and are easily recognized. Large particles will not embed in the bearing and will score or gouge the bearing and journal. The best prevention for this cause of bearing failure is to clean all parts thoroughly and keep everything spotlessly clean during engine assembly. Frequent and regular engine oil and filter changes are also recommended.

5   Lack of lubrication (or lubrication breakdown) has a number of interrelated causes. Excessive heat (which thins the oil), overloading (which squeezes the oil from the bearing face) and oil leakage or throw-off (from excessive bearing clearances, worn oil pump or high engine speeds) all contribute to lubrication breakdown. Blocked oil passages, which usually are the result of misaligned oil holes in a bearing shell, will also oil starve a bearing and destroy it. When lack of lubrication is the cause of bearing failure, the bearing material is wiped or extruded from the

steel backing of the bearing. Temperatures may increase to the point where the steel backing turns blue from overheating.

6   Driving habits can have a definite effect on bearing life. Low speed operation in too high a gear (lugging the engine) puts very high loads on bearings, which tends to squeeze out the oil film. These loads cause the bearings to flex, which produces fine cracks in the bearing face (fatigue failure). Eventually the bearing material will loosen in pieces and tear away from the steel backing. Short trip driving leads to corrosion of bearings because insufficient engine heat is produced to drive off the condensed water and corrosive gases. These products collect in the engine oil, forming acid and sludge. As the oil is carried to the engine bearings, the acid attacks and corrodes the bearing material.

7   Incorrect bearing installation during engine assembly will lead to bearing failure as well. Tight fitting bearings leave insufficient bearing oil clearance and will result in oil starvation. Dirt or foreign particles trapped behind a bearing insert result in high spots on the bearing which lead to failure.

8   If you need to use a STANDARD size main bearing, install one that has the same number as the original bearing.

9   If you need to use a STANDARD size rod bearing, install one that has the same number as the number stamped into the connecting rod cap.

10   Remember, the oil clearance is the final judge when selecting new bearing sizes for either main or rod bearings. If you have any questions or are unsure which bearings to use, get help from a dealer parts or service department or other parts supplier.

22 Engine overhaul - reassembly sequence

1   Before beginning engine reassembly, make sure you have all the necessary new parts, gaskets and seals as well as the following items on hand:

Common hand tools
A torque wrench
Piston ring installation tool
Piston ring compressor
Short lengths of rubber or plastic hose to fit over connecting rod bolts,
Plastigage
Feeler gauges
A fine-tooth file
New engine oil
Engine assembly lube or moly-base grease
Gasket sealant
Thread locking compound

2   In order to save time and avoid problems, engine reassembly must be done in the following general order:

Piston rings
Rear main oil seal (V6 and small-block V8 engines)
Crankshaft and main bearings
Piston/connecting rod assemblies
Rear main oil seal (inline six-cylinder and big-block V8 engines)
Cylinder head(s) and rocker arms or lifters
Camshaft
Timing chain and gears and cover

The following parts should be assembled after engine installation:

Oil pump
Oil pick-up tube
Oil pan
Valve cover(s)
Intake and exhaust manifolds
Flywheel/driveplate

23 Piston rings - installation

Refer to illustrations 23.3, 23.4, 23.5, 23.10a, 23.10b and 23.13

1   Before installing the new piston rings, the ring end gaps must be checked. It's assumed that the piston ring side clearance has been

23.10a Installing the spacer/expander in the oil control ring groove . .

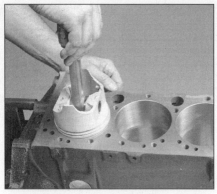

23.3 When checking piston ring end gap, the ring must be square in the cylinder bore (this is done by pushing the ring down with the top of a piston as shown)

checked and verified correct (see Section 19).

2   Lay out the piston/connecting rod assemblies and the new ring sets so the ring sets will be matched with the same piston and cylinder during the end gap measurement and engine assembly.

3   Insert the top (number one) ring into the first cylinder and square it up with the cylinder walls by pushing it in with the top of the piston **(see illustration)**. The ring should be near the bottom of the cylinder, at the lower limit of ring travel.

4   To measure the end gap, slip feeler gauges between the ends of the ring until a gauge equal to the gap width is found **(see illustration)**. The feeler gauge should slide between the ring ends with a slight amount of drag. Compare the measurement to the Specifications. If the gap is larger or smaller than specified, double-check to make sure you have the correct rings before proceeding.

5   If the gap is too small, it must be enlarged or the ring ends may come in contact with each other during engine operation, which can cause serious damage to the engine. The end gap can be increased by filing the ring ends very carefully with a fine file. Mount the file in a vise equipped with soft jaws, slip the ring over the file with the ends contacting the file face and slowly move the ring to remove material from the ends **(see illustration)**. **Caution:** *When performing this operation, file only from the outside in.*

6   Excess end gap is not critical unless it is greater than 0.040-inch. Again, double-check to make sure you have the correct rings for your engine.

7   Repeat the procedure for the rest of the rings. Remember to keep rings, pistons and cylinders matched up.

8   Repeat the procedure for each ring that will be installed in the first cylinder and for each ring in the remaining cylinders. Remember to keep rings, pistons and cylinders matched up.

9   Once the ring end gaps have been checked/corrected, the rings can be installed on the pistons.

10   The oil control ring (lowest one on the piston) is usually installed first. It's composed

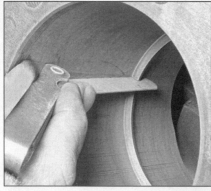

23.4 With the ring square in the cylinder, measure the end gap with a feeler gauge

23.5 If the end gap is too small, clamp a file in a vise and file the ring ends by pushing the ring over the file toward the vise - after filing to correct size, deburr the ring ends with fine emery cloth or a whetstone

of three separate components. Slip the spacer/expander into the groove **(see illustration)**. If an anti-rotation tang is used, make sure it's inserted into the drilled hole in the ring groove. Next, install the lower side rail. Don't use a piston ring installation tool on the oil ring side rails, as they may be damaged. Instead, place one end of the side rail into the groove between the spa-cer/expander and the ring land, hold it firmly in place and slide a finger around the piston while pushing the rail into the groove **(see illustration)**. Next, install the upper side rail in the same manner.

23.10b . . . followed by the side rails - DO NOT use a piston ring installation tool when installing the oil ring side rails

2C

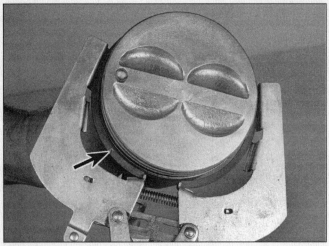

23.13 Installing the compression rings with a ring expander - the mark (arrow) must face up

24.10 Lay the Plastigage strips (arrow) on the main bearing journals, parallel to the crankshaft centerline

11   After the three oil ring components have been installed, check to make sure that both the upper and lower side rails can be turned smoothly in the ring groove.

12   The number two (middle) ring is installed next. It's usually stamped with a mark which must face up, toward the top of the piston. **Note:** *Always follow the instructions printed on the ring package or box - different manufacturers may require different approaches. Do not mix up the top and middle rings, as they have different cross sections.*

13   Use a piston ring installation tool and make sure the identification mark is facing the top of the piston, then slip the ring into the middle groove on the piston. Don't expand the ring any more than necessary to slide it over the piston **(see illustration)**.

14   Install the number one (top) ring in the same manner. Make sure the mark is facing up. Be careful not to confuse the number one and number two rings.

15   Repeat the procedure for the remaining pistons and rings.

24   Crankshaft - installation and main bearing oil clearance check

1   Crankshaft installation is the first major step in engine reassembly. It's assumed at this point that the engine block and crankshaft have been cleaned, inspected and repaired or reconditioned.

2   Position the engine with the bottom facing up.

3   Remove the main bearing cap bolts and lift out the caps. Lay the caps out in the proper order to ensure correct installation.

4   If they're still in place, remove the old bearing inserts from the block and the main bearing caps. Wipe the main bearing surfaces of the block and caps with a clean, lint-free cloth. They must be kept spotlessly clean!

Main bearing oil clearance check

Refer to illustration 24.10 and 24.14

5   Clean the back sides of the new main bearing inserts and lay the bearing half with the oil groove and hole in each main bearing saddle in the block (on all engines covered by this manual, the bearings with oil holes go in the block and those without oil holes go in the caps). Lay the other bearing half from each bearing set in the corresponding main bearing cap. Make sure the tab on each bearing insert fits into the recess in the block or cap. Also, the oil holes in the block must line up with the oil holes in the bearing insert. **Caution:** *Do not hammer the bearings into place and don't nick or gouge the bearing faces. No lubrication should be used at this time.*

6   The thrust bearings must be installed in the number three main bearing saddle.

7   Clean the faces of the bearings in the block and the crankshaft main bearing journals with a clean, lint-free cloth. Check or clean the oil holes in the crankshaft, as any dirt here can go only one way - straight through the new bearings.

8   Once you're certain the crankshaft is clean, carefully lay it in position in the main bearings. No lubricant should be used at this time.

9   Before the crankshaft can be permanently installed, the main bearing oil clearance must be checked.

10   Trim several pieces of the appropriate size Plastigage (they must be slightly shorter than the width of the main bearings) and place one piece on each crankshaft main bearing journal, parallel with the journal axis **(see illustration)**.

11   Clean the faces of the bearings in the caps and install the caps in their respective positions (don't mix them up) with the arrows pointing toward the front of the engine. Don't disturb the Plastigage. Apply a light coat of

oil to the bolt threads and the under sides of the bolt heads, then install them.

12   Working from the center out, tighten the main bearing cap bolts, in three steps, to the torque listed in this Chapter's Specifications. Don't rotate the crankshaft at any time during this operation!

13   Remove the bolts and carefully lift off the main bearing caps. Keep them in order. Don't disturb the Plastigage or rotate the crankshaft. If any of the main bearing caps are difficult to remove, tap them gently from side-to-side with a soft-face hammer to loosen them.

14   Compare the width of the crushed Plastigage on each journal to the scale printed on the Plastigage envelope to obtain the main bearing oil clearance **(see illustration)**. Check the Specifications to make sure it's correct.

15   If the clearance is not as specified, the bearing inserts may be the wrong size which means different ones will be required (see Section 21). Before deciding that different inserts are needed, make sure that no dirt or oil was between the bearing inserts and the caps or block when the clearance was measured. If the Plastigage is noticeably wider at one end than the other, the journal may be tapered (see Section 20).

16   Carefully scrape all traces of the Plastigage material off the main bearing journals and/or the bearing faces. Don't nick or scratch the bearing faces.

Final crankshaft installation

17   Carefully lift the crankshaft out of the engine. Clean the bearing faces in the block, then apply a thin, uniform layer of clean moly-base grease or engine assembly lube to each of the bearing surfaces. Coat the thrust washers as well.

18   Install the rear main oil seal halves into the engine block and rear main bearing cap (V6 and small-block V8 engines) or seal

24.14 Compare the width of the crushed Plastigage to the scale on the envelope to determine the main bearing oil clearance (always take the measurement at the widest point of the Plastigage); be sure to use the correct scale - standard and metric ones are included

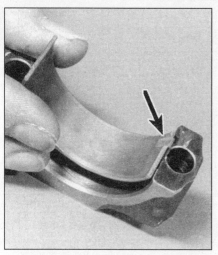

25.3 Make sure the bearing tang fits securely into the notch in the rod cap

2C

retainer (inline six-cylinder and big-block V8 engines) (refer to Chapter 2A or 2B for details). It's not necessary to "shoehorn" the upper seal half into the engine - simply press it into place while the crankshaft is removed. Lubricate the crankshaft surfaces that contact the oil seals with engine assembly lube or clean engine oil.

19   Make sure the crankshaft journals are clean, then lay the crankshaft back in place in the block. Clean the faces of the bearings in the caps or cap assembly, then apply the same lubricant to them. Install the caps in their respective positions with the arrows pointing toward the front of the engine. **Note:** *If you haven't already done so, be sure to install the thrust washers in the number three main bearing saddle.*

20   Apply a light coat of oil to the bolt threads and the under sides of the bolt heads, then install them. Tighten all except the number three cap bolts (the one with the thrust washers on the block side) to the torque listed in this Chapter's Specifications (work from the center out and approach the final torque in three steps). Tighten the number three cap bolts to 10-to-12 ft-lbs. Tap the

ends of the crankshaft forward and backward with a lead or brass hammer to line up the thrust washer and crankshaft surfaces. Retighten all main bearing cap bolts to the specified torque, following the recommended sequence.

21   Rotate the crankshaft a number of times by hand to check for any obvious binding.

22   Check the crankshaft endplay with a feeler gauge or a dial indicator (see Section 15). The endplay should be correct if the crankshaft thrust faces aren't worn or damaged and new thrust washers have been installed.

23   Install a new rear main oil seal, then bolt the retainer to the block (inline six-cylinder and big-block V8 engines) (see Chapter 2A or 2B).

25   Pistons/connecting rods - installation and rod bearing oil clearance check

Refer to illustrations 25.3, 25.5, 25.9, 25.11, 25.13 and 25.17

1   Before installing the piston/connecting

rod assemblies, the cylinder walls must be perfectly clean, the top edge of each cylinder must be chamfered, and the crankshaft must be in place.

2   Remove the cap from the end of the number one connecting rod (refer to the marks made during removal). Remove the original bearing inserts and wipe the bearing surfaces of the connecting rod and cap with a clean, lint-free cloth. They must be kept spotlessly clean.

Connecting rod bearing oil clearance check

3   Clean the back side of the new upper bearing insert, then lay it in place in the connecting rod. Make sure the tab on the bearing fits into the recess in the rod **(see illustration)** so the oil holes line up. Don't hammer the bearing insert into place and be very careful not to nick or gouge the bearing face. Don't lubricate the bearing at this time.

4   Clean the back side of the other bearing insert and install it in the rod cap. Again, make sure the tab on the bearing fits into the recess in the cap, and don't apply any lubricant. It's critically important that the mating surfaces of the bearing and connecting rod are perfectly clean and oil free when they're assembled.

5   Position the piston ring gaps at staggered intervals around the piston **(see illustration)**.

6   Slip a section of plastic or rubber hose over each connecting rod cap bolt **(see illustration 14.5)**.

7   Lubricate the piston and rings with clean engine oil and attach a piston ring compressor to the piston. Leave the skirt protruding about 1/4-inch to guide the piston into the cylinder. The rings must be compressed until they're flush with the piston.

8   Rotate the crankshaft until the number one connecting rod journal is at BDC (bottom dead center) and apply a coat of engine oil to the cylinder walls.

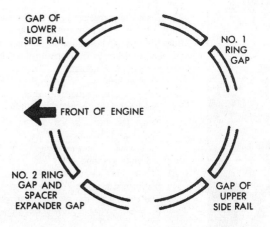

GAP OF LOWER SIDE RAIL

NO. 1 RING GAP

FRONT OF ENGINE

NO. 2 RING GAP AND SPACER EXPANDER GAP

GAP OF UPPER SIDE RAIL

25.5 Position the ring end gaps as shown here before installing the pistons in the block

25.9 When installed, the mark/notch (arrow) in the piston must face the front of the engine

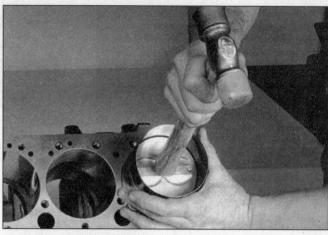

25.11 The piston can be driven gently into the cylinder bore with the end of a wooden or plastic hammer handle

9   With the mark on top of the piston **(see illustration)** facing the front (timing chain end) of the engine, gently insert the piston/connecting rod assembly into the number one cylinder bore and rest the bottom edge of the ring compressor on the engine block.

10   Tap the top edge of the ring compressor to make sure it's contacting the block around its entire circumference.

11   Gently tap on the top of the piston with the end of a wooden or plastic hammer handle **(see illustration)** while guiding the end of the connecting rod into place on the crankshaft journal. The piston rings may try to pop out of the ring compressor just before entering the cylinder bore, so keep some pressure on the ring compressor. Work slowly, and if any resistance is felt as the piston enters the cylinder, stop immediately. Find out what's hanging up and fix it before proceeding. Do not, for any reason, force the piston into the cylinder - you might break a ring and/or the piston.

12   Once the piston/connecting rod assembly is installed, the connecting rod bearing oil clearance must be checked before the rod cap is permanently bolted in place.

13   Cut a piece of the appropriate size Plastigage slightly shorter than the width of the connecting rod bearing and lay it in place on the number one connecting rod journal, parallel with the journal axis **(see illustration)**.

14   Clean the connecting rod cap bearing face, remove the protective hoses from the connecting rod bolts and install the rod cap. Make sure the mating mark on the cap is on the same side as the mark on the connecting rod. Check the cap to make sure the front mark is facing the timing chain end of the engine.

15   Apply a light coat of oil to the undersides of the nuts, then install and tighten them to the torque listed in this Chapter's Specifications. Use a thin-wall socket to avoid erroneous torque readings that can result if the socket is wedged between the rod cap and nut. If the socket tends to wedge itself between the nut and the cap, lift up on it slightly until it no longer contacts the cap. Do not rotate the crankshaft at any time during this operation.

16   Remove the nuts and detach the rod cap, being very careful not to disturb the Plastigage.

17   Compare the width of the crushed Plas-

tigage to the scale printed on the Plastigage envelope to obtain the oil clearance **(see illustration)**. Compare it to the Specifications to make sure the clearance is correct.

18   If the clearance is not as specified, the bearing inserts may be the wrong size (which means different ones will be required). Before deciding that different inserts are needed, make sure that no dirt or oil was between the bearing inserts and the connecting rod or cap when the clearance was measured. Also, recheck the journal diameter. If the Plastigage was wider at one end than the other, the journal may be tapered (see Section 20).

Final connecting rod installation

19   Carefully scrape all traces of the Plastigage material off the rod journal and/or bearing face. Be very careful not to scratch the bearing - use your fingernail or the edge of a credit card.

20   Make sure the bearing faces are perfectly clean, then apply a uniform layer of clean moly-base grease or engine assembly lube to both of them. You'll have to push the piston into the cylinder to expose the face of the bearing insert in the connecting rod - be sure to slip the protective hoses over the rod bolts first.

21   Slide the connecting rod back into place on the journal, remove the protective hoses from the rod cap bolts, install the rod cap and tighten the nuts to the torque listed in this Chapter's Specifications.

22   Repeat the entire procedure for the remaining pistons/connecting rods.

23   The important points to remember are:

 a) *Keep the back sides of the bearing inserts and the insides of the connecting rods and caps perfectly clean when assembling them.*

 b) *Make sure you have the correct piston/rod assembly for each cylinder.*

 c) *The dimple on the piston must face the front (timing chain end) of the engine.*

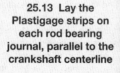

25.13 Lay the Plastigage strips on each rod bearing journal, parallel to the crankshaft centerline

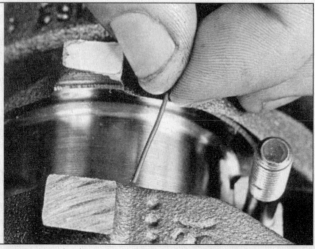

d) *Lubricate the cylinder walls with clean oil.*

e) *Lubricate the bearing faces when installing the rod caps after the oil clearance has been checked.*

24   After all the piston/connecting rod assemblies have been properly installed, rotate the crankshaft a number of times by hand to check for any obvious binding.

25   As a final step, the connecting rod side clearance (endplay) must be checked (see Section 14).

26   Compare the measured side clearance to the Specifications to make sure it's correct. If it was correct before disassembly and the original crankshaft and rods were reinstalled, it should still be right. If new rods or a new crankshaft were installed, the side clearance may be inadequate. If so, the rods will have to be removed and taken to an automotive machine shop for re-sizing.

27   Adjust the oil pickup tube and screen the the correct distance above the bottom of the oil pan (see Chapter 2A or 2B) and reinstall the oil pan.

28   The rest of the assembly procedure is the reverse of the removal procedure.

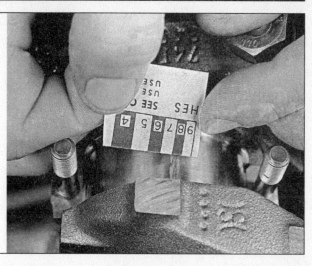

25.17 Measuring the width of the crushed Plastigage to determine the rod bearing oil clearance (be sure to use the correct scale - standard and metric ones are included)

26   Initial start-up and break-in after overhaul

Warning: *Have a fire extinguisher handy when starting the engine for the first time.*

1   Once the engine has been installed in the vehicle, double-check the engine oil and coolant levels. Add one pint of Chrysler Crankcase Conditioner (part no. 4318002) or equivalent.

2   Start the engine. It may take a few moments for the gasoline to reach the carburetor but the engine should start without a great deal of effort.

3   After the engine starts, it should be allowed to warm up to normal operating temperature. While the engine is warming up, make a thorough check for oil and coolant leaks. **Caution:** *Do not run the engine above a fast idle until all the hydraulic lifters have filled with oil and become quiet again.*

4   Shut the engine off and recheck the engine oil and coolant levels.

5   Drive the vehicle to an area with minimum traffic, accelerate from 30 to 50 mph, then allow the vehicle to slow to 30 mph with the throttle closed. Repeat the procedure 10 or 12 times. This will load the piston rings and cause them to seat properly against the cylinder walls. Check again for oil and coolant leaks.

6   Drive the vehicle gently for the first 500 miles (no sustained high speeds) and keep a constant check on the oil level. It is not unusual for an engine to use oil during the break-in period.

7   At approximately 500 to 600 miles, change the oil and filter.

8   For the next few hundred miles, drive the vehicle normally. Do not pamper it or abuse it.

9   After 2000 miles, change the oil and filter again and consider the engine fully broken in.

2C

Notes

Chapter 3
Cooling, heating and air conditioning systems

Contents

3

Specifications

General

Coolant capacity	See Chapter 1
Drivebelt tension	See Chapter 1
Radiator pressure cap rating	14 to 18 psi
Thermostat opening temperature	185 to 195-degrees F

Torque specifications

	Ft-lbs (unless otherwise indicated)
Fan-to-water pump attaching bolts	200 in-lbs
Viscous fan drive-to-fan blade assembly attaching bolts	20
Thermostat housing bolts	
1984 and earlier	30
1985 on	200 in-lbs
Water pump attaching bolts	30

1 General information

The cooling system consists of a radiator and coolant reserve system, a radiator pressure cap, a thermostat (of varying temperatures depending on year and model), a four, five or seven blade fan, and a crankshaft pulley-driven water pump.

The radiator cooling fan is mounted on the front of the water pump. There are two types of fans used; one incorporates a fluid drive fan clutch (viscous fan clutch) which saves horsepower and reduces noise, the other a basic fan with a solid hub attaching to the water pump. On either type a fan shroud is mounted on the rear of the radiator to direct air flow through the radiator.

The system is pressurized by a spring-loaded radiator cap, which, by maintaining pressure, increases the boiling point of the coolant. If the coolant temperature goes above this increased boiling point, the extra pressure in the system forces the radiator cap valve off its seat and exposes the overflow pipe or hose. The overflow pipe/hose leads to a coolant recovery system. This consists of a plastic reservoir into which the coolant that normally escapes due to expansion is retained. When the engine cools, the excess coolant is drawn back into the radiator by the vacuum created as the system cools, maintaining the system at full capacity. This is a continuous process and provided the level in the reservoir is correctly maintained, it is not necessary to add coolant to the radiator.

Coolant in the right side of the radiator circulates up the lower radiator hose to the water pump, where it is forced through the water passages in the cylinder block. The coolant then travels up into the cylinder head, circulates around the combustion chambers and valve seats, travels out of the cylinder head past the open thermostat into the upper radiator hose and back into the radiator.

When the engine is cold, the thermostat restricts the circulation of coolant to the

2.4 An inexpensive hydrometer can be used to test the condition of your coolant

3.7 Loosen the clamp (arrow) and disconnect the upper radiator hose from the thermostat cover (V8 shown)

3.8 Remove the bolts and lift the thermostat cover off (inline six-cylinder engine shown)

engine. When the minimum operating temperature is reached, the thermostat begins to open, allowing coolant to return to the radiator.

Automatic transmission-equipped models have a cooler element incorporated into the radiator to cool the transmission fluid.

The heating system works by directing air through the heater core mounted in the dash and then to the interior of the vehicle by a system of ducts. Temperature is controlled by mixing heated air with fresh air, using a system of flapper doors in the ducts, and a blower motor.

Air conditioning is an optional accessory, consisting of an evaporator core located under the dash, a condenser in front of the radiator, a receiver/drier in the engine compartment and a belt-driven compressor mounted at the front of the engine.

2   Antifreeze - general information

Refer to illustration 2.4
Warning: *Do not allow antifreeze to come in contact with your skin or painted surfaces of the vehicle. Rinse off spills immediately with plenty of water. Antifreeze is highly toxic if ingested. Never leave antifreeze lying around in an open container or in puddles on the floor; children and pets are attracted by it's sweet smell and may drink it. Check with local authorities about disposing of used antifreeze. Many communities have collection centers which will see that antifreeze is disposed of safely. Never dump used anti-freeze on the ground or pour it into drains.*
Note: *Non-toxic antifreeze is now available at most auto parts stores, but even these types should be disposed of properly.*

The cooling system should be filled with a water/ethylene glycol based antifreeze solution which will prevent freezing down to at least -20-degrees F (even lower in cold climates). It also provides protection against corrosion and increases the coolant boiling point.

The cooling system should be drained,

flushed and refilled at least every other year (see Chapter 1). The use of antifreeze solutions for periods of longer than two years is likely to cause damage and encourage the formation of rust and scale in the system.

Before adding antifreeze to the system, check all hose connections. Antifreeze can leak through very minute openings.

The exact mixture of antifreeze to water which you should use depends on the relative weather conditions. The mixture should contain at least 50-percent antifreeze, but should never contain more than 70-percent antifreeze. Consult the mixture ratio chart on the antifreeze container before adding coolant. Hydrometers are available at most auto parts stores to test the coolant **(see illustration)**. Use antifreeze which meets the vehicle manufacturer's specifications.

3   Thermostat - check and replacement

Refer to illustrations 3.7, 3.8, 3.10 and 3.11
Warning: *The engine must be completely cool when this procedure is performed.*
Note: *Don't drive the vehicle without a thermostat! The computer (if so equipped) may stay in open loop and emissions and fuel economy will suffer.*

Check

1   Before condemning the thermostat, check the coolant level, drivebelt tension and temperature gauge (or light) operation.
2   If the engine takes a long time to warm up, the thermostat is probably stuck open. Replace the thermostat.
3   If the engine runs hot, check the temperature of the upper radiator hose. If the hose isn't hot, the thermostat is probably stuck shut. Replace the thermostat.
4   If the upper radiator hose is hot, it means the coolant is circulating and the thermostat is open. Refer to the *Troubleshooting* section for the cause of overheating.
5   If an engine has been overheated, you

may find damage such as leaking head gaskets, scuffed pistons and warped or cracked cylinder heads.

Replacement

6   Drain coolant (about 1 gallon) from the radiator, until the coolant level is below the thermostat housing.
7   Disconnect the upper radiator hose from the thermostat cover **(see illustration)**.
8   Remove the bolts and lift the cover off **(see illustration)**. It may be necessary to tap the cover with a soft-face hammer to break the gasket seal.
9   On 1992 and later V6 and 5.2L V8 engines, the thermostat housing is hidden behind a large cast-aluminum bracket holding the air conditioning compressor and alternator. Loosen and remove the alternator drivebelt (see Chapter 1), and unbolt the alternator support bracket and the alternator (see Chapter 5) for access to the thermostat.
10   Note how it's installed, then remove the thermostat **(see illustration)**. Be sure to use a replacement thermostat with the correct opening temperature (see this Chapter's Specifications).
11   Use a scraper or putty knife to remove all traces of old gasket material and sealant

3.10 Note how it's installed, then remove the thermostat

3.11 Remove all traces of old gasket material and sealant, but avoid gouging the mating surfaces

4.7 Remove the fan-to-water pump hub bolts (arrows) - two additional bolts cannot be seen on the other side of the pulley

4.8 If the engine has a non-viscous fan, a spacer will be used

from the mating surfaces **(see illustration)**. **Caution:** *Be careful not to gouge or damage the gasket surfaces, because a leak could develop after assembly. Make sure no gasket material falls into the coolant passages; it is a good idea to stuff a rag in the passage. Wipe the mating surfaces with a rag saturated with lacquer thinner or acetone.*

12   Install the thermostat and make sure the correct end faces out - the spring is directed toward the engine **(see illustration 3.10).**

13   Apply a thin coat of RTV sealant to both sides of the new gasket and position it on the engine side, over the thermostat, and make sure the gasket holes line up with the bolt holes in the housing.

14   Carefully position the cover and install the bolts. Tighten them to the torque listed in this Chapter's Specifications - do not over-tighten the bolts or the cover may crack or become distorted. **Note:** *On some later models, the thermostat cover is stamped "Front" - this must face the front of the engine during installation for there to be adequate clearance for hose installation.*

15   Reattach the radiator hose to the cover and tighten the clamp - now may be a good time to check and replace the hoses and clamps (see Chapter 1).

16   Refer to Chapter 1 and refill the system, then run the engine and check carefully for leaks.

17   Repeat steps 1 through 5 to be sure the repairs corrected the previous problem(s).

4   Engine cooling fan and clutch - check, removal and installation

Warning: *Keep hands, tools and clothing away from the fan. To avoid injury or damage DO NOT operate the engine with a damaged fan. Do not attempt to repair fan blades - replace a damaged fan with a new one.*

Check

Viscous clutch models only

1   Symptoms of failure of the fan clutch are; continuous noisy operation, looseness leading to vibration and evidence of silicone fluid leaks.

2   Rock the fan back and forth by hand to check for excessive bearing play.

3   With the engine cold, turn the blades by hand. The fan should turn freely.

4   Visually inspect for substantial fluid leakage from the clutch assembly, a deformed bi-metal spring or grease leakage from the cooling fan bearing. If any of these conditions exist, replace the fan clutch.

5   When the engine is warmed up, turn off the ignition switch and disconnect the cable from the negative battery terminal. Turn the fan by hand. Some resistance should be felt. If the fan turns easily, replace the fan clutch.

Removal and installation

All 1991 and earlier models and 1992 and later 5.9L V8 models

Refer to illustrations 4.7 and 4.8

6   Loosen the drivebelt tension (see Chapter 1).

7   Remove the fan-to-water pump hub bolts **(see illustration)**.

8   Remove the fan assembly from the engine. **Note:** *If the engine has a non-viscous fan there will be a spacer used. Be careful not to lose it once removed* **(see illustration)**.

1992 and later V6 and 5.2L V8 models

Refer to illustrations 4.9 and 4.11

9   A special tool, obtainable at most auto parts stores, is required to remove the fan on these models. The fan attaches to the water pump with a large threaded nut that is part of the fan drive **(see Illustration)**. The tool holds the four bolts of the water pump pulley while a large wrench is used to loosen the large drive nut. Sometimes it is possible to hold the water pump pulley by applying considerable hand pressure to the serpentine belt while the large nut is turned (right-hand threads), but it may require the tool if the fan has been installed for years.

10   Drain the radiator, disconnect the upper radiator hose from the radiator, remove the coolant recovery tank and set it aside.

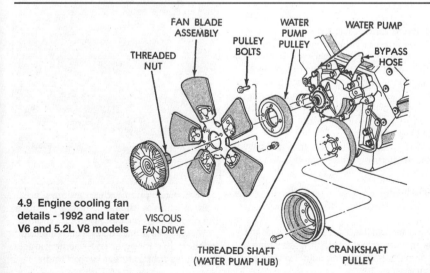

4.9 Engine cooling fan details - 1992 and later V6 and 5.2L V8 models

FAN BLADE ASSEMBLY

THREADED NUT

PULLEY BOLTS

WATER PUMP PULLEY

WATER PUMP

BYPASS HOSE

VISCOUS FAN DRIVE

THREADED SHAFT (WATER PUMP HUB)

CRANKSHAFT PULLEY

4.11 Remove the mounting bolts (arrows) to the upper fan shroud and lift out the upper half - the lower half does not interfere with fan removal

11   Remove the upper fan shroud mounting bolts; two that mount to the radiator support and two that mount the the lower shroud, and remove the upper shroud half **(see illustration)**.

12   Unthread the cooling fan from the water pump hub (see Step 9) and remove the fan, being careful not to damage the radiator fins.

All models

Refer to illustration 4.13

13   The fan clutch (if equipped) can be unbolted from the fan blade assembly for replacement **(see illustration). Caution:** *To prevent silicone fluid from draining from the clutch assembly into the fan drive bearing and ruining the lubricant, DON'T place the drive unit in a position with the rear of the shaft pointing down.*

14   Installation is the reverse of removal.

5   Radiator and coolant reservoir - removal and installation

Warning: *The engine must be completely cool when this procedure is performed.*

Radiator

Refer to illustrations 5.2a, 5.2b, 5.2c, 5.3, 5.4, 5.5, 5.6 and 5.8

Removal

1   Disconnect the battery cable at the negative battery terminal.

2   Drain the cooling system as described in Chapter 1, then disconnect the overflow hose and the upper and lower radiator hoses from the radiator **(see illustrations)**. Refer to the coolant Warning in Section 2.

3   If equipped with an automatic transmission or engine oil cooler, remove the cooler lines from the radiator **(see illustration)** - be careful not to damage the lines or fittings. Plug the ends of the disconnected lines to prevent leakage and stop dirt from entering the system. Have a drip pan ready to catch any spills. **Note:** *Some models may have an*

4.13 Remove the fan-to-fan clutch mounting bolts (arrows)

5.2b Loosen the clamp (arrow) on the upper radiator hose . . .

optional external transmission cooler, which is mounted in front of the radiator.

4   Disconnect the vacuum lines at the valve located at the top of the radiator **(see illustration)**, if equipped.

5   Remove the bolt on the bracket holding the oil filler tube **(see illustration)**, and tie the filler tube out of the way.

6   Remove the radiator mounting bolts **(see illustration). Note:** *On some models, one of the left-side radiator mounting bolts may be obscured by the coolant reservoir. Remove the reservoir for access to the bolt.*

7   Unbolt the upper and lower halves of the fan shroud (see Section 4) and remove them.

8   Lift the radiator from the engine compartment. Take care not to contact the fan blades **(see illustration)**.

9   Prior to installation of the radiator, replace any damaged hose clamps and radiator hoses.

Installation

10   Radiator installation is the reverse of removal.

11   After installation, fill the system with the proper mixture of antifreeze, and also check the automatic transmission fluid level, where applicable (see Chapter 1).

5.2a Disconnect the overflow hose from the radiator filler neck

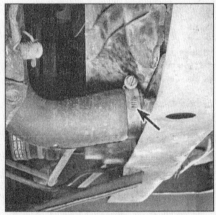

5.2c . . . and the clamp (arrow) on the lower radiator hose and separate the hoses from the radiator

5.3 If equipped with an automatic transmission or engine oil cooler, remove the cooler lines (arrows) from the radiator - always use a backup wrench to avoid twisting off a fitting or line - some models have rubber hoses with clamps instead of fittings

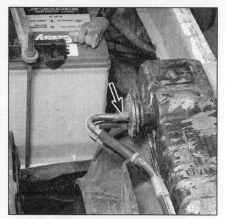

5.4 Disconnect the vacuum lines at the valve located at the top of the radiator (arrow)

5.5 Unbolt the oil filler tube bracket and tie the tube out of the way

5.6 Remove the radiator mounting bolts (arrows) - there are two more in the same location on the other side of the radiator

Coolant reservoir

Refer to illustration 5.12

Removal and installation

12   Remove the coolant overflow hose from the reservoir, and remove the two screws **(see illustration)**. Pull the reservoir up and out.

13   Prior to installation make sure the reservoir is clean and free of debris which could be drawn into the radiator (wash it with soapy water and a brush if necessary, then rinse thoroughly).

14   Installation is the reverse of removal.

---

6   Water pump - check

Refer to illustrations 6.2 and 6.4

1   Water pump failure can cause overheating and serious damage to the engine. There are three ways to check the operation of the water pump while it is installed on the engine. If any one of the following quick-checks indicates water pump problems, it should be replaced immediately.

2   A seal protects the water pump impeller shaft bearing from contamination by engine coolant. If this seal fails, a weep hole in the water pump snout will leak coolant **(see illustration)** (an inspection mirror can be used to look at the underside of the pump if the hole isn't on top). If the weep hole is leaking, shaft bearing failure will follow. Replace the water pump immediately.

3   Besides contamination by coolant after a seal failure, the water pump impeller shaft bearing can also be prematurely worn out by an improperly-tensioned drivebelt. When the bearing wears out, it emits a high pitched squealing sound. If such a noise is coming from the water pump during engine operation, the shaft bearing has failed - replace the water pump immediately. Don't mistake drivebelt slippage, which causes a squealing sound, for water pump bearing failure.

4   To identify excessive bearing wear before the bearing actually fails, grasp the water pump pulley and try to force it up-and-

5.8 Carefully angle the radiator out to avoid damaging any of the cooling fins

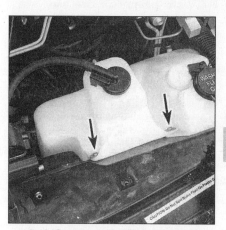

5.12 On some models the coolant reservoir is secured to the right side of the radiator support by two screws (arrows) - on other models the reservoir simply unclips and slides off the mount (power distribution center removed for clarity)

down or from side-to-side **(see illustration)**. If the pulley can be moved either horizontally or vertically, the bearing is nearing the end of its service life. Replace the water pump.

5   It is possible for a water pump to be bad, even if it doesn't howl or leak water. Sometimes the fins on the back of the impeller can corrode away until the pump is no longer effective. The only way to check for this is to remove the pump for examination.

6.2 Check the weep hole (arrow) for leakage (pump removed for clarity)

6.4 Check the pump for loose or rough bearings (this can be done with the fan and pulley in place)

3

7.5 Detach the coolant hoses from the water pump (arrows) - V8 engine shown

7 Water pump - removal and installation

Refer to illustrations 7.5, 7.7a, 7.7b and 7.8
Warning: *Wait until the engine is completely cool before beginning this procedure.*

Removal

1   Disconnect the battery cable from the negative battery terminal.
2   Drain the coolant (see Chapter 1).
3   Loosen the water pump pulley bolts or fan mounting bolts, then remove the drivebelts (see Chapter 1). **Note:** *On some late-model engines with serpentine drivebelts, take note of the belt routing. Improper reinstallation could cause the water pump to turn the wrong way and cause overheating.*
4   Remove the fan and shroud assembly (see Section 5) and water pump pulley.
5   Detach the coolant hoses from the water pump **(see illustration)**.
6   On many models the compressor, alternator, power steering and air pump brackets will be attached to the water pump. Where necessary, loosen or remove the components so the brackets can be moved aside and the water pump bolts removed. **Note:** *Do not disconnect the power steering or air con-*

7.8 Remove all traces of old gasket material - use care to avoid gouging the soft aluminum

7.7a Remove the water pump mounting bolts and detach the pump (if it is difficult to remove, use a soft-face hammer to break the seal made by the gasket)

ditioning compressor hoses; unbolt the units and tie them aside.
7   Remove the water pump mounting bolts and detach it. It may be necessary to tap the pump with a soft-face hammer to break the gasket seal **(see illustrations)**. Inspect the pump's impeller blades on the backside for corrosion. If any fins are missing or badly corroded, replace the pump with a new one.

Installation

8   Clean the sealing surfaces of all gasket material on both the water pump and block **(see illustration)**. Wipe the mating surfaces with a rag saturated with lacquer thinner or acetone.
9   Apply a thin layer of RTV sealant to both sides of the new gasket.
10   Install the gasket on the water pump.
11   Place the water pump in position and install the bolts finger-tight. Use caution to ensure that the gasket doesn't slip out of position. Remember to replace any mounting brackets secured by the water pump mounting bolts. Tighten the bolts to the torque listed in this Chapter's Specifications.
12   Install the water pump pulley, fan assembly and pulley bolts.

8.1a The coolant-temperature sending unit (arrow) is located near the front of the intake manifold (inline six-cylinder shown)

7.7b Water pump mounting bolt locations (arrows) - big-block V8 engine shown

13   Install the coolant hoses and hose clamps. Tighten the hose clamps securely.
14   Install the drivebelts (see Chapter 1) and tighten the fan/pulley mounting bolts securely.
15   Add coolant to the specified level (see Chapter 1).
16   Reconnect the battery cable to the negative battery terminal.
17   Start the engine and check for the proper coolant level and the water pump and hoses for leaks.

8 Coolant temperature sending unit - check and replacement

Refer to illustrations 8.1a and 8.1b
Warning: *Wait until the engine is completely cool before beginning this procedure.*

Check

1   The coolant temperature indicator system is composed of a light or temperature gauge mounted in the dash and a coolant temperature sending unit mounted on the engine **(see illustrations)**. Some vehicles have more than one sending unit, but only one is used for the indicator system and the other is used to send temperature information to the computer.

8.1b Coolant temperature sending unit location (arrow) - late model V8 shown (the other temperature sensor at right is for the computer)

9.3 Connect a voltmeter to the heater blower motor connector (arrow) by backprobing, and check the running voltage at each blower switch position

2   If an overheating indication occurs, check the coolant level in the system and then make sure the wiring between the light or gauge and the sending unit is secure and all fuses are intact.
3   When the ignition switch is turned on and the starter motor is turning, the indicator light (if equipped) should be on (overheated engine indication).
4   If the light is not on, the bulb may be burned out, the ignition switch may be faulty or the circuit may be open. Test the circuit by grounding the wire to the sending unit while the ignition is on (engine NOT running). If the gauge deflects full scale or the light comes on, replace the sending unit.
5   As soon as the engine starts, the light should go out and remain out unless the engine overheats. Failure of the light to go out may be due to a grounded wire between the light and the sending unit, a defective sending unit or a faulty ignition switch. Check the coolant to make sure it's the proper type. **Note:** *Plain water may have too low a boiling point to activate the sending unit.*

Replacement

6   Make sure the engine is cool before removing the defective sending unit. There

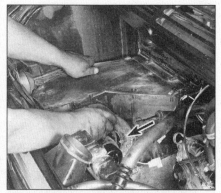

10.1 Remove the three screws (one indicated here by an arrow) retaining the blower motor to the housing (heater core/evaporator housing removed here for clarity)

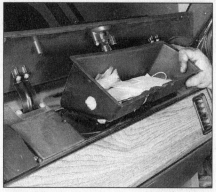

9.6 Bend the clips and pull the glove box insert from the dash

will be some coolant loss as the unit is removed, so be prepared to catch it. Refer to the coolant Warning in Section 2.
7   Prepare the new sending unit by coating the threads with sealant or wrapping them with Teflon tape. Disconnect the electrical connector and simply unscrew the sensor from the engine and install the replacement.
8   Check the coolant level after the replacement unit has been installed and top up the system, if necessary (see Chapter 1). Check now for proper operation of the gauge and sending unit.

9   Blower motor and circuit - check

Refer to illustrations 9.3, 9.6 and 9.7
1   Check the fuse and all connections in the circuit for looseness and corrosion. Make sure the battery is fully charged.
2   With the transmission in Park, and the parking brake securely set, turn the ignition switch to the Run position. It isn't necessary to start the engine.
3   Backprobe the blower motor electrical connector with two small paper clips (straightened out) and connect a voltmeter to the blower motor connector and ground **(see illustration)**.
4   Move the blower switch through each of its positions and note the voltage readings. Changes in voltage indicate that the motor speeds will also vary as the switch is moved to the different positions.
5   If there is voltage present, but the blower motor does not operate, the blower motor is probably faulty. Disconnect the blower motor connector and hook one side to a chassis ground and the other to a fused source of battery voltage. If the blower doesn't operate, it is faulty.
6   If there was no voltage present at the blower motor at one or more speeds, and the motor itself tested OK, check the blower motor resistor, located inside the passenger compartment. Bend the clips on the glove box plastic liner and pull it out of the dash **(see illustration)** for access to the resistor.
7   Backprobe the connector to the resistor while it is connected and look for varying

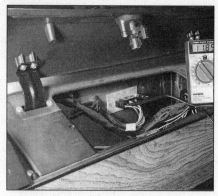

9.7 Check for varying voltage by backprobing the resistor with a voltmeter

voltages at each position **(see illustration)**.
8   Disconnect the electrical connector from the blower motor resistor. With the ignition on, check for voltage at each of the terminals in the connector as the switch is moved to the different positions. If the voltmeter responds correctly to the switch then the resistor is probably faulty. If there is no voltage present from the switch then, the switch, control panel or related wiring is probably faulty.

10   Blower motor - removal and installation

Refer to illustrations 10.1, 10.2 and 10.3
1   Unplug the electrical connector to the blower motor and remove the three nuts from the engine side of the blower housing **(see illustration)**. **Note:** *On some models, it may be easier to remove the upper half of the fan shroud before removing the blower motor (see Section 5).*
2   Pull the blower motor and fan straight out, being careful not to bend the fan assembly. If the nuts are frozen on the three motor-mounting studs, causing the studs to turn, the blower can be removed by taking out the screws holding the blower mounting plate to the blower housing **(see illustration)**. If this is done, the studs can be held on the inside

10.2 If the nuts are frozen on the blower mounting studs, remove the motor with the mounting plate (arrow)

3

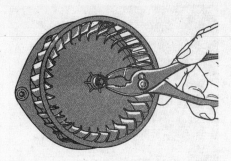

10.3 Remove the fan by squeezing the spring clip together and slipping the fan off the shaft

with locking pliers, the nuts removed, and the studs cleaned and the flat heads epoxied in place in the cover plate for future use.

3   To remove the fan from the blower motor, squeeze the spring clip together and slip the fan off the shaft (see illustration).

4   Install the fan onto the motor and install the blower motor into the heater housing.

11 Heater core - removal and installation

Refer to illustrations 11.3, 11.4a, 11.4b, 11.4c, 11.5, 11.6, 11.7, 11.8a, 11.8b, 11.9, 11.10 and 11.11

Warning 1: *1995 and later models covered by this manual are equipped with a Supplemental Restraint System (SRS), more commonly known as an airbag(s). All 1995 and later models are equipped with a driver side airbag and all 1998 and later models are equipped with a passenger side airbag, located in the instrument panel. Always disconnect the negative battery cable and wait at least two minutes before working in the vicinity of any airbag system component to avoid the possibility of accidental deployment of the airbag, which could cause personal injury (see Chapter 12, Section 28). Do not use electrical test equipment on the airbag system wiring or tamper with it in any way.*

11.3 Loosen the hose clamps and disconnect the heater hoses from the heater core tubes at the firewall - arrow indicates one of the two clamps

Warning 2: *The air conditioning system is under high pressure. DO NOT loosen any fittings or remove any components until after the system has been discharged. Air conditioning refrigerant should be properly discharged into an EPA-approved container at a dealership service department or an automotive air conditioning facility. Always wear eye protection when disconnecting air conditioning system fittings.*

Removal

1   Disconnect the battery cable at the negative battery terminal.

2   Drain the cooling system (see Chapter 1).

3   Disconnect the heater hoses at the heater core inlet and outlet on the engine side of the firewall, and plug the open fittings (see illustration). On later models, the hoses may connect to plastic Y-tubes, not directly to the engine.

4   From the inside of the van, remove the glove box (see illustration 9.6) for access to the upper mounting nuts (see illustration). Below the ducting, remove two nuts on the lower firewall and one long screw (see illustrations).

5   If equipped with air conditioning, disconnect the refrigerant lines at the expansion

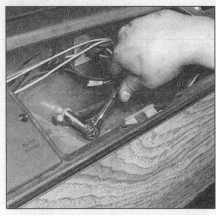

11.4a Access to the upper interior mounting nuts for the heater core box is with the glove box liner removed

valve on the front of the heater core/evaporator housing, and plug the open lines to prevent entry of contaminants (see illustration).

6   Remove the two bolts in the receiver/drier bracket and move the wires, hoses and receiver/drier assembly out of the way (see illustration).

7   On the engine compartment side, remove the nuts from the studs projecting through the firewall (see illustration).

8   On 1971 to 1978 models, the heater/evaporator housing is held to the blower housing with a sheetmetal plate in front and two bolts to a crossbar on the blower housing. Remove these bolts (see illustration). On later models, there are two screws holding the heater core/evaporator housing to the blower housing below it. Remove these screws (see illustration).

9   With all fasteners removed, the heater core/evaporator housing can be angled out of the engine compartment (see illustration).

10   After removing the heater housing, remove the control door arm and the screws retaining the housing cover (see illustration).

11   With a small socket and a long extension, remove the two retaining screws at the bottom of the heater core and slide the core out of the housing for repairs or replacement (see illustration).

11.4b Remove these two nuts (arrows) . . .

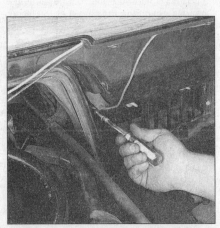

11.4c . . . and one long interior screw

11.5 Disconnect the air conditioning lines at the front of the heater core housing

11.6 Remove these bolts (arrows) from the receiver/drier bracket and pull the hoses, wires and receiver/drier towards the driver's side

11.7 Remove the nuts on the engine compartment side of the firewall

Installation

12 Installation is the reverse of removal.
13 Refill the cooling system (see Chapter 1).
14 Start the engine and check for proper operation. If equipped with air conditioning, have the air conditioning system evacuated, recharged and leak-checked at a dealership service department or an automotive air conditioning repair facility.

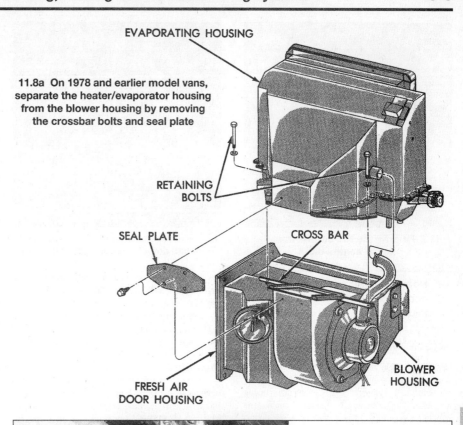

EVAPORATING HOUSING

11.8a On 1978 and earlier model vans, separate the heater/evaporator housing from the blower housing by removing the crossbar bolts and seal plate

RETAINING BOLTS

SEAL PLATE

CROSS BAR

FRESH AIR DOOR HOUSING

BLOWER HOUSING

11.8b On 1979 and later model vans, remove these screws between the heater core housing and the blower housing below it - one is shown with arrow, the other is on the fender side of the housings

11.9 The heater core/evaporator housing can be angled out of the engine compartment

11.10 Remove the nut and control door arm (arrow), then take out the screws retaining the housing top

11.11 Remove two small retaining screws at the bottom of the housing and the core can be pulled out

3

12.2 Remove the two screws from the heater control bezel

12.3 Pull the bezel forward - for complete removal, detach the wiring to the lighter

12.4 Remove the screws and pull the control assembly out to detach the cable and electrical connector

12 Heater and air conditioning control assembly - removal and installation

Refer to illustrations 12.2, 12.3, 12.4, 12.6a and 12.6b
Warning: *1995 and later models covered by this manual are equipped with a Supplemental Restraint System (SRS), more commonly known as an airbag(s). All 1995 and later models are equipped with a driver side airbag and all 1998 and later models are equipped with a passenger side airbag, located in the instrument panel. Always disconnect the negative battery cable and wait at least two minutes before working in the vicinity of any airbag system component to avoid the possibility of accidental deployment of the airbag, which could cause personal injury (see Chapter 12, Section 28). Do not use electrical test equipment on the airbag system wiring or tamper with it in any way.*

1971 through 1997 models
Removal
1   Disconnect the cable from the negative terminal of the battery.

2   Remove the two screws on the trim bezel that surrounds the heater/air conditioning control assembly **(see illustration)**.
3   Pull the trim bezel forward. It can be pulled far enough away to access the control assembly, but for removal, the electrical connector to the lighter must be removed **(see illustration)**.
4   Remove the control assembly retaining screws and pull the unit from the dash **(see illustration)**. It can be pulled out just far enough to allow disconnecting the control cable end and wiring for the dash illumination and the blower speed switch plug.
5   Disconnect the control cable, electrical connections and vacuum lines (on air conditioned models) from the control head. **Note:** *When disconnecting the vacuum lines, they may be in one large cluster or attached individually. Be careful when removing the vacuum lines to avoid cracking the plastic connectors and causing a vacuum leak (possibly within the control head).*
6   On early-model vans, the cables attach with simple clamps retained by screws, but later models have cable-ends that slip over control door arms and plastic "flags" that secure the cable housing **(see illustrations)**.

Installation
7   To install the control assembly, reverse the removal procedure. **Caution:** *When reconnecting vacuum lines to the control assembly, do not use any lubricant to make them slip on easier. If necessary, use a drop of plain water to make reconnection easier.*

1998 and later models
Removal
8   Disconnect the cable from the negative terminal of the battery.
9   Remove the engine cover (see Chapter 2B, Section 2).
10   Reach up between the upper instrument panel engine housing extension and the bottom of the instrument panel. Unplug the vacuum harness connection from the heater/air conditioning control assembly.
11   Remove the instrument cluster bezel (see Chapter 12).
12   Remove the three screws on the heater/air conditioning control assembly.
13   Pull the heater/air conditioning control assembly away from the instrument panel and unplug the electrical connectors from the heater/air conditioning control assembly. Remove the control assembly.

Installation
14   To install the control assembly, reverse the removal procedure. **Caution:** *When connecting the vacuum lines to the control assembly, do not use any lubricant to make them slip on easier. If necessary, use a drop of plain water to make reconnection easier.*

13 Air conditioning and heating system - check and maintenance

Refer to illustration 13.1
Warning: *The air conditioning system is under high pressure. DO NOT loosen any fittings or remove any components until after the system has been discharged. Air conditioning refrigerant should be properly discharged into an EPA-approved recovery con-*

12.6a Control cables are typically retained by a flag retainer (left arrow) and a clip (right arrow) over the shaft - depress the tab on the flag to pull out

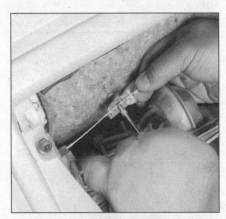

12.6b The self-adjusting clip can be removed or installed by twisting it off/on the wire by using a 1/4-inch ID tube over the bottom tube or a 3/16-inch rod inside the tube as shown

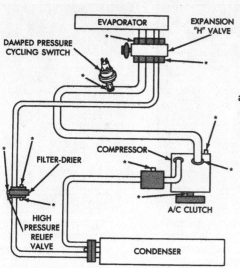

13.1 Diagram of basic air conditioning system, showing possible leak locations (*) (front air conditioning unit only)

13.8 Inspect the sight glass (left arrow) - if the refrigerant looks foamy, it's low (but if the refrigerant appears clear, the system is properly charged) - the right arrow points to the low-pressure cycling switch

tainer at a dealership service department or an automotive air conditioning repair facility. Always wear eye protection when disconnecting air conditioning system fittings.

1   The following maintenance steps should be performed on a regular basis to ensure that the air conditioner continues to operate at peak efficiency.

a) Check the tension of the drivebelt and adjust if necessary (see Chapter 1).
b) Check the condition of the hoses. Look for cracks, hardening and deterioration. Look at potential leak areas for signs of refrigerant oil leaking out (see illustration). **Warning:** Do not replace air conditioning hoses until the system has been discharged by a dealership or air conditioning repair facility.
c) Check the fins of the condenser for leaves, bugs and other foreign material. A soft brush and compressed air can be used to remove them.
d) Check the wire harness for correct routing, broken wires, damaged insulation, etc. Make sure the electrical connectors are clean and tight.
e) Maintain the correct refrigerant charge.

2   The system should be run for about 10 minutes at least once a month. This is particularly important during the winter months because long-term non-use can cause hardening of the internal seals.

3   Because of the complexity of the air conditioning system and the special equipment required to effectively work on it, accurate troubleshooting of the system should be left to a certified air conditioning technician.

4   If the air conditioning system doesn't operate at all, check the fuse panel.

5   The most common cause of poor cooling is simply a low system refrigerant charge. If a noticeable drop in cool air output occurs, the following quick check will help you determine if the refrigerant level is low. For more complete information on the air conditioning system, refer to the *Haynes Automotive Heating and Air Conditioning Manual*.

Check

Refer to illustrations 13.8, 13.9 and 13.10

6   Warm the engine up to normal operating temperature.

7   Place the air conditioning temperature selector at the coldest setting and the blower at the highest setting. Open the doors (to make sure the air conditioning system does not cycle off as soon as it cools the passenger compartment).

8   With the compressor engaged - the clutch will make an audible click and the center of the clutch will rotate - inspect the sight glass (see illustration). If the refrigerant looks foamy, it's low. Charge the system as described later in this Section. If the refrigerant appears clear, the compressor discharge line is warm and the compressor inlet pipe is cool, the system is properly charged.

9   Place a thermometer in the dashboard vent nearest the evaporator and add refrigerant to the system until the indicated temperature is around 40 to 45-degrees F (see illustration). If the ambient (outside) air temperature is very high, say 110-degrees F, the duct air temperature may be as high as 60-degrees F, but generally the air conditioning is 30 to 50-degrees F cooler than the ambient air.

10   Dodge vans have used refrigerant R-12 in air conditioning systems until 1994. Starting in 1995, the new, environmentally-friendly R-134A has been used. Because of recent Federal regulations enacted by the Environmental Protection Agency, 12-ounce cans of R-12 refrigerant may not be available in your area for home charging of air conditioning systems. If this is the case, it will be necessary to take the vehicle to a licensed air conditioning technician for charging. Home charging kits and 12-ounce cans *are* available for R-134A systems. **Caution:** *Although the components of R-12 and R-134A systems may seem similar, the two refrigerants and their refrigerant oils are NOT compatible. Seals, O-rings and other parts can be damaged by using the wrong refrigerant. Never introduce R-12 to an R-134A system, or vice-versa.*

11   Warm the engine up to normal operating temperature.

12   Place the air conditioning temperature selector at the coldest setting and put the blower at the highest setting. Open the doors (to make sure the air conditioning system doesn't cycle off as soon as it cools the passenger compartment).

13   With the compressor engaged - the clutch will make an audible click and the center of the clutch will rotate - inspect the sight glass, if equipped. If the refrigerant looks foamy, it's low. Charge the system. If the refrigerant appears clear, the system is properly charged.

14   The earliest warning that a system is low on refrigerant is the air coming out of the ducts inside the vehicle. If the air isn't as cold as it used to be, the system probably needs a charge.

15   Further inspection or testing of the system is beyond the scope of the home mechanic and should be left to a professional.

13.9 Insert a thermometer in the center duct while operating the air conditioning system - the output should be 35 to 40-degrees F below the ambient air temperature

3

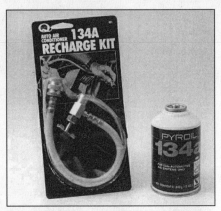

13.16 A basic charging kit for 134A systems is available at most auto parts stores - it must say 134A (not R-12) and so should the 12-ounce can of refrigerant

13.19 Add R-134A refrigerant to the low-side port only - the procedure is easier if you wrap the can with a warm, wet towel to prevent icing

14.3a Disconnect the electrical connector (arrow) from the air conditioning compressor . . .

Adding refrigerant (R-134A systems)

Refer to illustrations 13.16 and 13.19

16   Buy an R-134A charging kit at an auto parts store. A charging kit includes a 14-ounce can of refrigerant, a tap valve and a short section of hose that can be attached between the tap valve and the system low side service valve **(see illustration)**. Because one can of refrigerant may not be sufficient to bring the system charge up to the proper level, it's a good idea to buy an additional can. Make sure that one of the cans contains red refrigerant dye. If the system is leaking, the red dye will leak out with the refrigerant and help you pinpoint the location of the leak. **Warning:** *Never add more than two cans of refrigerant to the system.*

17   Hook up the charging kit by following the manufacturer's instructions. **Warning:** *DO NOT hook the charging kit hose to the system high side!* The fittings on the charging kit are designed to fit **only** on the low side of the system.

18   Back off the valve handle on the charging kit and screw the kit onto the refrigerant can, making sure first that the O-ring or rubber seal inside the threaded portion of the kit is in place. **Warning:** *Wear protective eyewear when dealing with pressurized refrigerant cans.*

19   Remove the dust cap from the low-side charging connection and attach the quick-connect fitting on the kit hose **(see illustration)**.

20   Warm up the engine and turn on the air conditioner. Keep the charging kit hose away from the fan and other moving parts. **Note:** *The charging process requires the compressor to be running. Your compressor may cycle off if the pressure is low due to a low charge. If the clutch cycles off, you can unplug the connector from the low-pressure cycling switch* **(see illustration 13.8)** *and attach a jumper wire between the terminals. This will keep the compressor ON.*

21   Turn the valve handle on the kit until the

stem pierces the can, then back the handle out to release the refrigerant. You should be able to hear the rush of gas. Add refrigerant to the low side of the system until the sight glass is free of bubbles. Allow stabilization time between each addition.

22   If you have an accurate thermometer, you can place it in the center air conditioning duct inside the vehicle and keep track of the air temperature **(see illustration 13.9)**. A charged system that is working properly should put out air that is 40-degrees F. If the ambient (outside) air temperature is very high, say 110-degrees F, the duct air temperature may be as high as 60-degrees F, but generally the air conditioning is 30 to 50-degrees F cooler than the ambient air.

23   When the can is empty, turn the valve handle to the closed position and release the connection from the low-side port. Replace the dust cap.

24   Remove the charging kit from the can and store the kit for future use with the piercing valve in the UP position, to prevent inadvertently piercing the can on the next use.

Heating systems

25   If the carpet under the heater core is damp, or if antifreeze vapor or steam is coming through the vents, the heater core is leaking. Remove it (see Section 11) and install a new unit (most radiator shops will not repair a leaking heater core).

26   If the air coming out of the heater vents isn't hot, the problem could stem from any of the following causes:

a) *The thermostat is stuck open, preventing the engine coolant from warming up enough to carry heat to the heater core. Replace the thermostat (see Section 3).*

b) *A heater hose is blocked, preventing the flow of coolant through the heater core. Feel both heater hoses at the firewall. They should be hot. If one of them is cold, there is an obstruction in one of the hoses or in the heater core, or the heater control valve is shut (located in the hot water hose to the heater core,*

along the passenger side of the engine on 1980 and earlier models). Detach the hoses and back flush the heater core with a water hose. If the heater core is clear but circulation is impeded, remove the two hoses and flush them out with a water hose.

c) *If flushing fails to remove the blockage from the heater core, the core must be replaced (see Section 11).*

14   Air conditioning compressor - removal and installation

Refer to illustrations 14.3a, 14.3b and 14.5

Removal

Warning: *The air conditioning system is under high pressure. DO NOT loosen any fittings or remove any components until after the system has been discharged. Air conditioning refrigerant should be properly discharged into an EPA-approved container at a dealership service department or an automotive air conditioning repair facility. Always wear eye protection when disconnecting air conditioning system fittings.*

14.3b . . . and remove the refrigerant line fitting bolts (arrows) from the top of the compressor

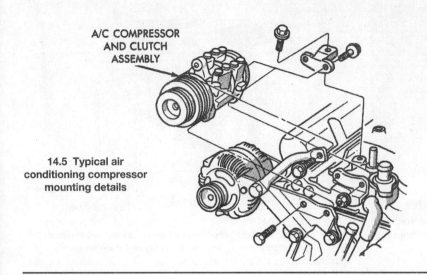

14.5 Typical air conditioning compressor mounting details

A/C COMPRESSOR AND CLUTCH ASSEMBLY

16.3 Near the right side of the radiator top, remove the one bolt securing both lines to the top of the condenser

Note: *The receiver/drier (see Section 15) should be replaced whenever the compressor is replaced.*

1   Have the air conditioning system discharged (see Warning above). Disconnect the cable at the negative battery terminal.

2   Clean the top of the compressor thoroughly around the refrigerant line fittings.

3   Unplug the electrical connector from the air conditioning compressor and disconnect the suction and discharge lines from the top of the compressor **(see illustrations)**. Plug the open fittings to prevent the entry of dirt and moisture.

4   Remove the drivebelt(s) (refer to Chapter 1).

5   Remove the compressor-to-bracket bolts and nuts and lift the compressor from the engine compartment **(see illustration)**.

Installation

6   If a new compressor is being installed, pour the oil from the old compressor into a graduated container and add that exact amount of new refrigerant oil to the new compressor. Also follow any directions included with the new compressor. **Note:** *Some replacement compressors come with new refrigerant oil in them.*

7   Place the compressor in position on the bracket and install the nuts and bolts finger-tight. Once all the compressor mounting nuts and bolts are installed, tighten them securely.

8   Install the drivebelt(s) (see Chapter 1).

9   Reconnect the electrical connector to the compressor. Install the line fitting bolt(s) to the compressor, use new O-rings lubricated with clean refrigerant oil (O-rings and oil must be compatible with the specific type of refrigerant in your system), and tighten the bolts securely.

10   Reconnect the battery cable to the negative battery terminal.

11   Have the system evacuated, recharged and leak tested by a dealership service department or an automotive air conditioning repair facility.

15   Air conditioning receiver/drier - removal and installation

Removal

Warning: *The air conditioning system is under high pressure. DO NOT loosen any fittings or remove any components until after the system has been discharged. Air conditioning refrigerant should be properly discharged into an EPA-approved container at a dealership service department or an automotive air conditioning repair facility. Always wear eye protection when disconnecting air conditioning system fittings.*

1   Have the air conditioning system discharged (see **Warning** above). Disconnect the battery cable at the negative battery terminal.

2   Disconnect the refrigerant inlet and outlet lines **(see illustrations 11.5 and 11.6)**. Cap or plug the open lines immediately to prevent the entry of dirt or moisture.

3   Remove the mounting bolts and slide the receiver/drier assembly up and out of the compartment, then remove the mounting bracket from the receiver/drier.

16.4 Remove the center brace ahead of the condenser

Installation

4   If you are replacing the receiver/drier with a new one, add one ounce of fresh refrigerant oil to the new unit (oil must be compatible with the specific type of refrigerant in your system).

5   Place the new receiver/drier into position, install the mounting bracket, bolts and tighten it securely.

6   Install the inlet and outlet lines, using clean refrigerant oil on the new O-rings. Tighten the through-bolt securely.

7   Connect the cable to the negative terminal of the battery.

8   Have the system evacuated, recharged and leak tested by a dealership service department or an automotive air conditioning repair facility.

16   Air conditioning condenser - removal and installation

Refer to illustrations 16.3, 16.4 and 16.5
Warning: *The air conditioning system is under high pressure. DO NOT loosen any fittings or remove any components until after the system has been discharged. Air conditioning refrigerant should be properly discharged into an EPA-approved container at a dealership service department or an automotive air conditioning repair facility. Always wear eye protection when disconnecting air conditioning system fittings.*

Removal

1   Have the air conditioning system discharged (see **Warning** above). Disconnect the battery cable at the negative battery terminal.

2   Remove the grille (see Chapter 11).

3   Disconnect the refrigerant line fittings and cap the open fittings to prevent the entry of dirt and moisture. One bolt secures both lines to the top of the condenser **(see illustration)**.

4   Remove the four bolts holding the center brace of the radiator support **(see illustration)**.

3

16.5 Four bolts (one shown here) secure the condenser from the grille side to the radiator support

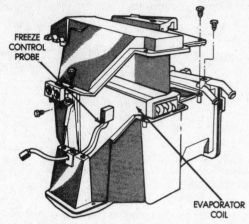

17.5 On later models, remove the freeze control probe from the evaporator core after the housing cover is removed

5   Unbolt the condenser from the radiator support **(see illustration)**.

6   Tilt the condenser forward and lift the condenser from the vehicle. **Caution:** *The condenser is made of aluminum - be careful not to damage it during removal.*

Installation

7   Installation is the reverse of removal. Be sure to use new O-rings on the refrigerant line fittings (lubricate the O-rings with clean refrigerant oil). If a new condenser is installed, add one ounce of new refrigerant oil to the system.

8   Have the system evacuated, recharged and leak tested by a dealership service department or an automotive air conditioning repair facility.

---

17 Air conditioning evaporator/expansion valve (front unit) - removal and installation

---

Warning: *The air conditioning system is under high pressure. DO NOT loosen any fit-*

tings or remove any components until after the system has been discharged. Air conditioning refrigerant should be properly discharged into an EPA-approved container at a dealership service department or an automotive air conditioning repair facility. Always wear eye protection when disconnecting air conditioning system fittings.

Removal

Refer to illustrations 17.5, 17.6 and 17.7

1   All models require complete removal of the heater/evaporator case for either heater core or evaporator core removal.

2   Have the air conditioning system discharged (see **Warning** above). Disconnect the cable from the negative battery terminal.

3   Drain the cooling system (see Chapter 1).

4   Follow the procedures in Section 11 for removing the heater/evaporator case from the vehicle. **Note:** *Plug all refrigerant lines to keep moisture out of the system.*

5   Disassemble the heater/evaporator case by removing the screws and taking off the cover. On later models, there may be a

freeze-control probe stuck in the forward fins of the evaporator core **(see illustration)**.

6   Remove the freeze control probe if present and slide the evaporator core out of the case **(see illustration)**.

7   Check the core over carefully for signs of leaks. The efficiency can be improved slightly if any bent fins are straightened with a "fin comb" **(see illustration)**.

8   Be sure to replace all of the aluminum gaskets with new ones upon reassembly.

Installation

Refer to illustration 17.10

9   Installation is the reverse of the removal procedure. Be sure to re-fill the cooling system (see Chapter 1). Use new O-rings on all connections, lubricated by refrigerant oil.

10   When reinstalling the expansion valve, use new aluminum gaskets, being careful not to scratch the sealing surfaces or dent the gaskets during assembly. Bolt the expansion valve to the evaporator coil first, then the lines to the expansion valve **(see illustration)**.

11   If a new evaporator has been installed, add one ounce of refrigerant oil. Have the

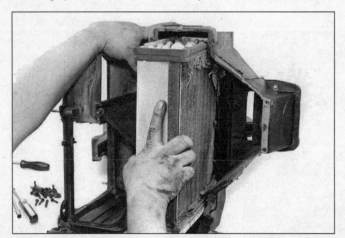

17.6 Slide the evaporator core out - the core can be cleaned with soapy water and a hose, blowing from the interior side through the fins

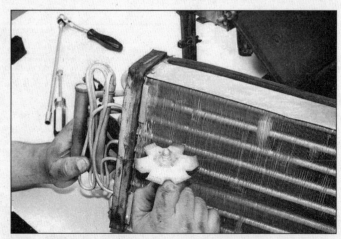

17.7 A plastic "fin comb" can be used to straighten bent fins - the tool has a head for various fins-per-inch spacings

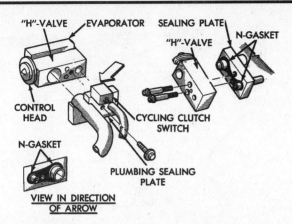

17.10 Remove the two Torx screws to separate the expansion valve - always replace the aluminum gaskets with new ones upon reassembly

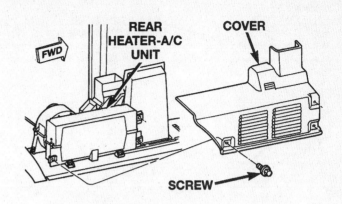

18.6 Remove the screws on the air conditioning evaporator/heater unit and remove the cover

system evacuated, recharged and leak tested by a dealership service department or an automotive air conditioning repair facility.

18 Air conditioning evaporator/heater unit (rear) — removal and installation

Warning 1: *The air conditioning system is under high pressure. DO NOT loosen any fittings or remove any components until after the system has been discharged. Air conditioning refrigerant should be properly discharged into an EPA-approved container at a dealership service department or automotive air conditioning repair facility. Always wear eye protection when disconnecting air conditioning system fittings.*
Warning 2: *Wait until the engine is completely cool before beginning this procedure.*

Removal

Refer to illustrations 18.6, 18.8, 18.9a and 18.9b

1   Have the air conditioning system discharged (see the **WARNING** above).

18.8 Release the hose clamps and disconnect the two heater hoses, loosen the flange nut and release the connector on the two refrigerant lines and disconnect them from the air conditioning evaporator/heater unit fittings

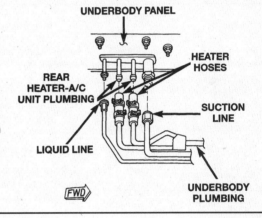

2   Drain the cooling system (see Chapter 1).
3   Disconnect the cable from the negative terminal of the battery.
4   Remove the rear bench seat on models so equipped.
5   Remove the vertical and horizontal ducts (see Section 19).
6   Remove the air conditioning evaporator/heater unit cover **(see illustration)**.
7   Raise the rear of the vehicle and support it securely on jackstands.

8   Working under the vehicle, clamp off the heater hoses, then disconnect the two heater hoses and the two refrigerant lines from the air conditioning evaporator/heater unit fittings **(see illustration)**. **Note:** *Plug the refrigerant lines to keep moisture out of the system.*
9   Working under the vehicle, remove the three nuts on the plumbing plate studs to the underbody panel **(see illustration)**. Remove the one screw securing the unit to the underbody **(see illustration)**.

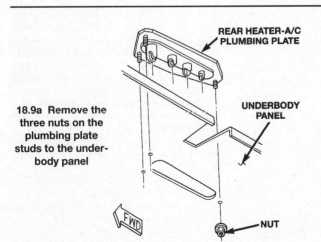

18.9a Remove the three nuts on the plumbing plate studs to the underbody panel

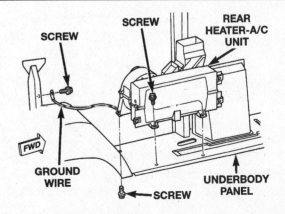

18.9b Remove the one screw securing the unit to the underbody and the one screw on the ground wire

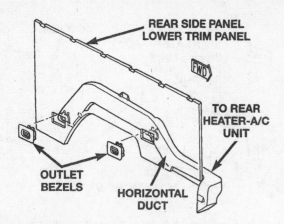

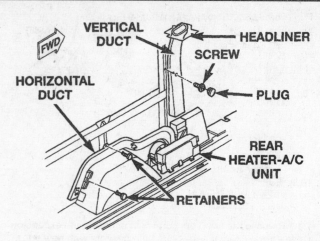

19.2 Remove the mounting screws and remove the rear lower side trim panel

19.3 Remove the two retainers securing the horizontal duct to the body panel, then gently pry out the mounting screw plug

10   Lower the vehicle.

11   Remove the vertical and horizontal ducts from the air conditioning evaporator/heater unit.

12   Remove the screw securing the ground wire to the body panel (**see illustration 18.9b**).

13   Remove the two screws securing the unit to the underbody panel (**see illustration 18.9b**).

14   Lift the unit straight up far enough for the plumbing to clear the opening in the underbody panel.

15   Disconnect the electrical connector from the unit and remove the unit from the vehicle.

Installation

16   Installation is the reverse of removal.

17   Run the engine with the heater units set to the maximum heat position, then let the engine cool down and check the coolant level, adding as necessary to bring it to the appropriate level (see Chapter 1).

18   Have the system evacuated, recharged and leak tested by a dealership service department or an automotive air conditioning repair facility.

19   Air conditioning unit air ducts (rear) - removal and installation

Horizontal duct

Refer to illustrations 19.2 and 19.3

1   Remove the outlet bezels from the rear lower side trim panel.

2   Remove the mounting screws and remove the rear lower side trim panel (**see illustration**).

3   Remove the two retainers securing the horizontal duct to the body panel (**see illustration**).

4   Carefully remove the horizontal duct off the rear unit duct outlet and remove the duct from the vehicle.

5   Installation is the reverse of removal.

Vertical duct

6   Remove the outlet bezels from the rear lower side trim panel.

7   Remove the mounting screws and remove the rear lower side trim panel (**see illustration 19.2**).

8   Carefully pry out the mounting screw plug (**see illustration 19.3**).

9   Loosen, but don't completely remove the screw securing the vertical duct to the rear side panel. Gently pull the vertical duct away from the panel making sure the screw is backed-out sufficiently (it can stay with the duct).

10   Grasp the top of the vertical duct with one hand and press upward on the inboard side of the duct opening in the headliner with the other hand. Carefully pull the top of the vertical duct downward and inboard until it unsnaps from the headliner duct opening.

11   Lift the vertical duct off the rear unit duct outlet and remove it from the vehicle.

12   Installation is the reverse of removal.

20   Combination coil (rear unit) — removal and installation

Warning 1: *The air conditioning system is under high pressure. DO NOT loosen any fittings or remove any components until after the system has been discharged. Air conditioning refrigerant should be properly discharged into an EPA-approved container at a dealership service department or automotive air conditioning repair facility. Always wear eye protection when disconnecting air conditioning system fittings.*

Warning 2: *Wait until the engine is completely cool before beginning this procedure.*

Removal

Refer to illustrations 20.10 and 20.11

1   Have the air conditioning system discharged (see the **WARNING** above).

2   Drain the cooling system (see Chapter 1).

3   Disconnect the cable from the negative terminal of the battery.

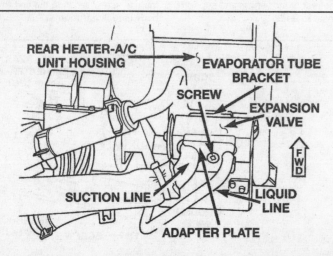

20.10 Remove the screw from the adapter plate securing the liquid and suction lines to the expansion valve

4   Remove the rear bench seat on models so equipped.

5   Remove the vertical and horizontal ducts (see Section 19).

6   Remove the cover from the air conditioning evaporator/heater unit **(see illustration 18.6)**.

7   Raise the rear of the vehicle and support it securely on jackstands.

8   Working under the vehicle, disconnect the two heater hoses and the two refrigerant lines from the air conditioning evaporator/heater unit fittings **(see illustration 18.8)**. **Note:** *Plug the refrigerant lines to keep moisture out of the system.*

9   Remove the screws and the upper housing from the air conditioning evaporator/heater unit.

10   Remove the screw from the adapter plate securing the liquid and suction lines to the expansion valve **(see illustration)**. Remove the two screws securing the expansion valve to the unit evaporator tube bracket and remove the expansion valve. **Note:** *Plug the refrigerant lines to keep moisture out of the system*. Remove the expansion valve from the unit.

11   Carefully lift the combination coil out of the lower housing **(see illustration)**.

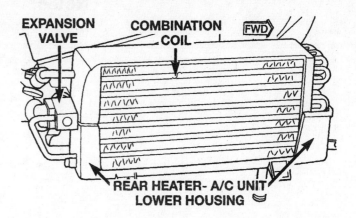

20.11 Carefully lift the combination coil out of the lower housing

Installation

12   Installation is the reverse of removal.

13   Run the engine with the heater units set to the maximum heat position, then let the engine cool down and check the coolant level, adding as necessary to bring it to the appropriate level (see Chapter 1).

14   Have the system evacuated, recharged and leak tested by a dealership service department or an automotive air conditioning repair facility.

3

Notes

Chapter 4 Part A
Fuel and exhaust systems - carbureted engines

Contents

4A

1 General information

The fuel system on all models consists of a fuel tank (or tanks) mounted in various locations on the chassis, a mechanically operated fuel pump and a carburetor. A combination of metal and rubber fuel hoses is used to connect these components.

The carburetor is either a one-, two- or four-barrel type, depending on engine displacement and year of production. Later models are equipped with electronic feedback carburetors. These carburetors are linked with a variety of sensors and output actuators to help control the amount of emissions.

Here is a list of the various carburetors used on these models:

One-barrel carburetors

Holley Model 1945 installed on early six-cylinder engines
Holley Model 6145 (this is a feedback version of the Model 1945, used on later six-cylinder engines)
Holley Model 1920 (installed on early inline six-cylinder engines

Two-barrel carburetors

Carter BBD (installed on 318 V8 engines)
Holley Model 2245 (installed on 360 and 400 V8 engines)
Holley Model 2210 (installed on 360 V8 engines)
Holley Model 2280 (installed on 318 V8 engines)
Holley Model 6280 (feedback version of the Model 2280, installed on later 318 V8 engines)

Four-barrel carburetors

Carter Thermo-quad (installed on 360, 400 and 440 V8 engines - this is the most common four barrel)
Rochester Quadrajet (installed on some 1985 and later V8 engines)

The fuel system (especially the carburetor) is interrelated with the emissions control systems on all vehicles produced for sale in the United States. Components of the emissions control systems are described in Chapter 6.

2 Fuel pump/fuel pressure check

Refer to illustration 2.1
Warning: *Gasoline is extremely flammable, so take extra precautions when you work on any part of the fuel system. Don't smoke or allow open flames or bare light bulbs near the work area, and don't work in a garage where a natural gas-type appliance (such as a water heater or a clothes dryer) is present. Since gasoline is carcinogenic, wear latex gloves when there's a possibility of being exposed to fuel, and, if you spill any fuel on your skin, rinse it off immediately with soap and water. Mop up any spills immediately and do not store fuel-soaked rags where they could ignite. When you perform any kind of work on the fuel system, wear safety glasses and have a Class B type fire extinguisher on hand.*
Note: *It is a good idea to check the fuel pump and lines for any obvious damage or fuel leakage. Also, check all hoses from the tank to the fuel pump, particularly the suction hoses at the fuel tank and pump which, if they have cracked, may not allow fuel to the fuel pump. It is also possible for the fuel pump*

2.1 Install a fuel pressure gauge onto a T-fitting and connect the assembly between the fuel inlet and the carburetor bowl inlet valve at the carburetor

4.5 If the vehicle is equipped with a fuel tank shield, unscrew the bolts (arrows)

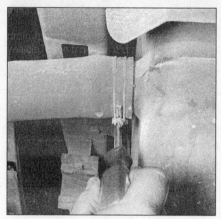

4.6 Loosen the clamp and detach the fuel filler hose from the tank

diaphragm to rupture internally and leak fuel into the crankcase. If you have excess fuel consumption or fuel smell, and you can't find any external fuel leaks, check the condition of the engine oil for any signs of fuel mixing with the engine oil; the engine oil level will usually be abnormally high and the oil will be thinned out and have a fuel smell.

1   Remove the engine cover (see Chapter 2B, Section 2). Disconnect the fuel line from the carburetor and install a T-fitting. Connect a fuel pressure gauge to the T-fitting with a section of fuel hose that is no longer than six inches **(see illustration)**.

2   Start the engine and allow it to idle. The pressure on the gauge should be 2-1/2 to 8 psi. It should remain constant and return to zero slowly when the engine is shut off. **Note:** *If the engine will not start, crank the engine until you get a reading on the gauge.*

3   An instant pressure drop indicates a faulty outlet valve and the fuel pump must be replaced.

4   If the pressure is too high, check the air vent to see if it is plugged before replacing the pump.

5   If the pressure is too low, be sure the fuel hoses and lines are in good shape and not plugged. Detach the fuel inlet line from the fuel pump and blow through it with low pressure compressed air to see if the line or in-tank filter is plugged. If you can hear bubbles in the tank (an assistant may be required), replace the pump.

---

3   Fuel lines and fittings - repair and replacement

Warning: *Gasoline is extremely flammable, so take extra precautions when you work on any part of the fuel system. Don't smoke or allow open flames or bare light bulbs near the work area, and don't work in a garage where a natural gas-type appliance (such as a water heater or a clothes dryer) is present. Since gasoline is carcinogenic, wear latex gloves*

when there's a possibility of being exposed to fuel, and, if you spill any fuel on your skin, rinse it off immediately with soap and water. Mop up any spills immediately and do not store fuel-soaked rags where they could ignite. When you perform any kind of work on the fuel system, wear safety glasses and have a Class B type fire extinguisher on hand.

Inspection

1   Once in a while, you will have to raise the vehicle to service or replace some component (an exhaust pipe hanger, for example). Whenever you work under the vehicle, always inspect the fuel lines and fittings for possible damage or deterioration.

2   Check all hoses and pipes for cracks, kinks, deformation or obstructions.

3   Make sure all hose and pipe clips attach their associated hoses or pipes securely to the underside of the vehicle.

4   Verify all hose clamps attaching rubber hoses to metal fuel lines or pipes are snug enough to assure a tight fit between the hoses and pipes.

Replacement

5   If you must replace any damaged sections, use hoses approved for use in fuel systems or pipes made from steel only (it's best to use an original-type pipe from a dealer that's already flared and pre-bent). Do not install substitutes constructed from inferior or inappropriate material, as this could result in a fuel leak and a fire.

6   Always, before detaching or disassembling any part of the fuel line system, note the routing of all hoses and pipes and the orientation of all clamps and clips to assure that replacement sections are installed in exactly the same manner.

7   Before detaching any part of the fuel system, be sure to relieve the fuel tank pressure by removing the fuel filler cap.

8   Always use new hose clamps after loosening or removing them.

9   While you're under the vehicle, it's a good idea to check the following related components:

a) *Check the condition of the fuel filter - make sure that it's not clogged or damaged (see Chapter 1).*

b) *Inspect the evaporative emission control (EVAP) system. Verify that all hoses are attached and in good condition (see Chapter 6).*

---

4   Fuel tank - removal and installation

1971 through 1997 models

Refer to illustrations 4.5, 4.6, 4.7, 4.10a and 4.10b

Warning: *Gasoline is extremely flammable, so take extra precautions when you work on any part of the fuel system. Don't smoke or allow open flames or bare light bulbs near the work area, and don't work in a garage where a natural gas-type appliance (such as a water heater or a clothes dryer) is present. Since gasoline is carcinogenic, wear latex gloves when there's a possibility of being exposed to fuel, and, if you spill any fuel on your skin, rinse it off immediately with soap and water. Mop up any spills immediately and do not store fuel-soaked rags where they could ignite. When you perform any kind of work on the fuel system, wear safety glasses and have a Class B type fire extinguisher on hand.*

Note: *The following procedure is much easier to perform if the fuel tank is empty. Some tanks have a drain plug for this purpose. If the tank does not have a drain plug, drain the fuel into an approved fuel container using a commercially available siphoning kit (NEVER start the siphoning action by mouth) or wait until the fuel tank is nearly empty, if possible.*

1   Remove the fuel tank filler cap to relieve fuel tank pressure.

2   Detach the cable from the negative terminal of the battery.

3   If the tank still has fuel in it, you can drain it at the fuel feed line after raising the vehicle. If the tank has a drain plug, remove it and allow the fuel to drain into an approved gasoline container.

4   Raise the vehicle and place it securely

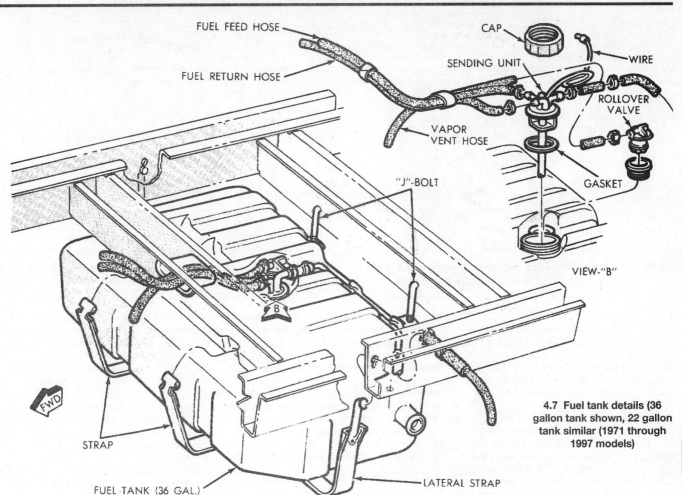

4.7 Fuel tank details (36 gallon tank shown, 22 gallon tank similar (1971 through 1997 models)

on jackstands.

5   If the fuel tank is equipped with a protective shield, remove the bolts and detach it from the underside of the chassis (see illustration).

6   Disconnect the fuel filler hose from the tank (see illustration).

7   Disconnect the fuel lines and the vapor return line. **Note:** *The fuel feed and return lines and the vapor return line are different diameters, so reattachment is simplified. If*

you have any doubts, however, clearly label the three lines and the fittings. Be sure to plug the hoses to prevent leakage and contamination of the fuel system (see illustration).

8   If there is still fuel in the tank, siphon it out from the fuel feed port. Remember - NEVER start the siphoning action by mouth! Use a siphoning kit, which can be purchased at most auto parts stores.

9   Support the fuel tank with a floor jack (unless it's an in-cab mounted tank). Position

a piece of wood between the jack head and the fuel tank to protect the tank.

10   Disconnect the fuel tank retaining straps and pivot them down until they are hanging out of the way (see illustrations).

11   Lower the tank enough to disconnect the wires and ground strap from the fuel pump/fuel gauge sending unit, if you have not already done so.

12   Remove the tank from the vehicle.

13   Installation is the reverse of removal.

1998 and later models

Refer to illustrations 4.19, 4.22, 4.23 and 4.24

14   Refer to the previous **WARNING** and **NOTE** in the 1971 through 1997 procedure.

15   Remove the fuel tank fill cap to relieve fuel tank pressure. On some models it is necessary to depress the spring loaded flap located in the fuel fill tube below the fuel fill cap. Use a non-metallic (plastic) tool and depress the flap to relieve all residual fuel tank pressure.

16   Detach the cable from the negative terminal of the battery.

17   If the fuel tank has fuel in it, you can drain it at the fuel vent line after raising the vehicle.

18   Raise the rear of the vehicle and support

4.10a Unscrew the fuel tank strap bolts (arrows)

4.10b Location of the fuel tank strap side mounts

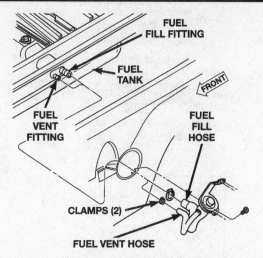

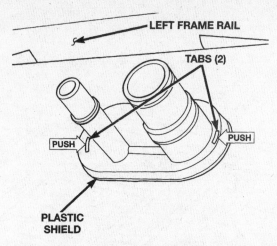

4.19 Loosen the clamps and disconnect the fuel vent line from the fuel tank fitting at the outer part of the frame

4.22 Push the two tabs inward the release the fill/vent fittings from the plastic shield

it securely on jackstands.

19   Disconnect the fuel vent line from the fuel tank fitting at the outer part of the frame **(see illustration)**. Install a hose into the fitting and siphon the fuel into an approved gasoline container. **Warning:** *Don't start the siphoning action be mouth. Use a siphoning kit (available at most auto parts stores).*

20   Disconnect the fuel fill hose from the fitting on the fuel tank **(see illustration 4.19)**.

21   **Caution:** *Do not damage the fuel vent and fuel fill fittings on the fuel tank when removing and lowering the fuel tank. These fittings are attached to the tank and are not serviceable.* Support the fuel tank with a floor jack. Position a piece of wood between the jack head and the fuel tank to protect the tank.

22   Push the two tabs inward the release the fill/vent fittings from the plastic shield before lowering the tank **(see illustration)**.

23   Disconnect the fuel tank retaining straps and pivot them down until they are hanging out of the way **(see illustration)**. Guide the fuel vent and fuel fill fitting assembly through the plastic shield and the frame access hole.

24   Lower the tank enough to disconnect the wires from the fuel pump/fuel sending unit, if you have not already done so. Disconnect the fuel vapor lines from both rollover valve fittings **(see illustration)**.

25   Remove the tank from the vehicle.

26   Installation is the reverse of removal.

5   Fuel tank cleaning and repair - general information

1   All repairs to the fuel tank or filler neck should be carried out by a professional who has experience in this critical and potentially dangerous work. Even after cleaning and

flushing of the fuel system, explosive fumes can remain and ignite during repair of the tank. **Note:** *The fuel tank on some models is made of plastic and cannot be repaired.*

2   If the fuel tank is removed from the vehicle, it should not be placed in an area where sparks or open flames could ignite the fumes coming out of the tank. Be especially careful inside garages where a natural gas-type appliance is located, because the pilot light could cause an explosion.

6   Fuel pump - removal and installation

Refer to illustrations 6.3 and 6.4

Warning: *Gasoline is extremely flammable, so take extra precautions when you work on any part of the fuel system. Don't smoke or*

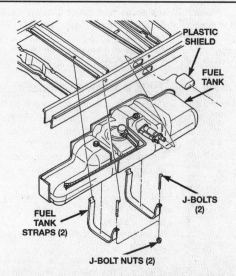

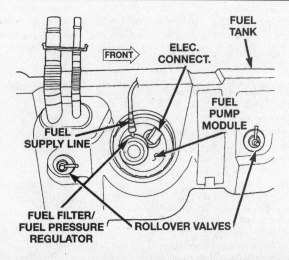

4.23 Disconnect the fuel tank retaining straps and pivot them down until they are hanging out of the way

4.24 Lower the tank enough to disconnect the wires from the fuel pump/fuel sending unit, then disconnect the fuel vapor lines from both rollover valve fittings

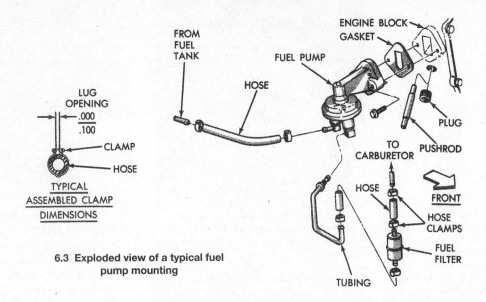

LUG
OPENING
.000
.100
CLAMP
HOSE
TYPICAL
ASSEMBLED CLAMP
DIMENSIONS

FROM
FUEL
TANK

HOSE

ENGINE BLOCK
GASKET

FUEL PUMP

PLUG

PUSHROD

TO
CARBURETOR

HOSE

FRONT

HOSE
CLAMPS

FUEL
FILTER

TUBING

6.3 Exploded view of a typical fuel pump mounting

allow open flames or bare light bulbs near the work area, and don't work in a garage where a natural gas-type appliance (such as a water heater or a clothes dryer) is present. Since gasoline is carcinogenic, wear latex gloves when there's a possibility of being exposed to fuel, and, if you spill any fuel on your skin, rinse it off immediately with soap and water. Mop up any spills immediately and do not store fuel-soaked rags where they could ignite. When you perform any kind of work on the fuel system, wear safety glasses and have a Class B type fire extinguisher on hand.

1   On V8 engines, the fuel pump is located on the right (passenger's) side of the engine, towards the front of the engine, either on or near the timing chain cover. On six-cylinder engines, the pump is also located on the right side of the engine, but it's closer to the center of the engine.

2   Loosen the right front wheel lug nuts. Raise the front of the vehicle and support it securely on jackstands. Remove the wheel. Place rags under the fuel pump to catch any gasoline which is spilled during removal.

3   Loosen the fuel line fittings and detach the lines from the pump. A flare-nut wrench along with a backup wrench should be used on the pressure-side fitting to prevent damage to the line and fittings **(see illustration)**.

4   Unbolt and remove the fuel pump **(see illustration)**. Remove all traces of gasket material from the mounting surface on the engine block.

5   Before installation, coat both sides of the gasket with RTV sealant, position the gasket and fuel pump against the block and install the bolts, tightening them securely.

6   Attach the lines to the pump and tighten the pressure fitting securely (use a flare-nut wrench, if one is available, to prevent damage to the fittings). Use a new hose clamp on the inlet hose.

6.4 Remove the two bolts (arrows) and detach the pump from the engine block

7   Run the engine and check for leaks. Install the wheel and lug nuts, lower the vehicle and tighten the lug nuts to the torque listed in the Chapter 1 Specifications.

7   Fuel level sending unit - check and replacement

Warning: Gasoline is extremely flammable, so take extra precautions when you work on any part of the fuel system. Don't smoke or allow open flames or bare light bulbs near the work area, and don't work in a garage where a natural gas-type appliance (such as a water heater or a clothes dryer) is present. Since gasoline is carcinogenic, wear latex gloves when there's a possibility of being exposed to fuel, and, if you spill any fuel on your skin, rinse it off immediately with soap and water. Mop up any spills immediately and do not store fuel-soaked rags where they could ignite. When you perform any kind of work on the fuel system, wear safety glasses and have a Class B type fire extinguisher on hand.

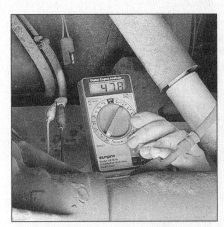

7.4 Connect the leads of an ohmmeter to the sending unit terminals and check the resistance of the sending unit (this particular fuel tank is almost empty, as indicated by the high resistance value)

Check

Refer to illustrations 7.4 and 7.7

1   Before performing any tests on the fuel level sending unit, completely fill the tank with fuel.

2   Raise the vehicle and secure it with jackstands.

3   Disconnect the fuel level sending unit electrical connector located on the fuel tank, usually at the rear.

4   Connect the probes of an ohmmeter to the electrical connector of the sending unit and check for resistance **(see illustration)**. Use the 200-ohm scale on the ohmmeter.

5   With the fuel tank completely full, the resistance should be about 5.0 to 8.0 ohms.

6   Reconnect the electrical connector, lower the vehicle and drive it until the tank is nearly empty.

7   Check the resistance. The resistance of the sending unit should be about 60 to 75

4A

7.7 Connect the probes of the ohmmeter to the terminals of the sending unit, then move the float through the complete range of travel and observe the resistance. The sending unit should indicate low resistance when it reaches the top of its travel distance (full tank) and then high resistance when it reaches the bottom of its travel (empty tank)

7.11 Use a brass punch to tap the sending unit lock ring until the tabs align with the indentations on the fuel tank ring

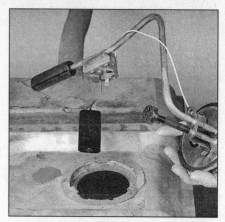

7.12 Carefully lift the sending unit from the fuel tank without damaging the float

ohms. **Note:** *If there is any doubt concerning the resistance of the sending unit, remove the sending unit* (see Steps 9 through 14) *and check it by moving the float through its entire range of motion* **(see illustration)**.
8   If the readings are incorrect, replace the sending unit.

Replacement

Refer to illustrations 7.11 and 7.12
9   Remove the fuel tank from the vehicle (see Section 4).
10   If necessary, carefully peel back the insulation material from the top of the fuel tank without damaging it.
11   Using a brass punch and hammer, tap the sending unit lock ring **(see illustration)** until the tabs align with the indents on the fuel tank retaining ring.

12   Lift the sending unit from the tank. Carefully angle the sending unit out of the opening without damaging the fuel level float located at the bottom of the assembly **(see illustration)**.
13   Disconnect the electrical connector from the sending unit.
14   Installation is the reverse of removal.

8   Air cleaner housing - removal and installation

Refer to illustration 8.2
1   Remove the engine cover (see Chapter 2B, Section 2). Remove the air filter from the air cleaner housing (see Chapter 1).
2   Disconnect any vacuum hose or electrical connectors that are attached to the air cleaner **(see illustration)** and mark them with paint or marked pieces of tape for reassembly purposes.
3   Lift the air cleaner housing from the engine.
4   Installation is the reverse of removal.

9   Accelerator cable - removal, installation and adjustment

Refer to illustrations 9.2, 9.3 and 9.5

Removal

1   Remove the engine cover (see Chapter 2B, Section 2).
2   Remove the accelerator cable from the throttle linkage by prying the retainer clip off the shaft using a small screwdriver **(see illustration)**.
3   Loosen the accelerator cable locknut and remove the cable from the bracket assembly **(see illustration)**.
4   Detach the screws and the clips retaining the lower instrument trim panel on the driver's side and remove the trim piece (see Chapter 11).
5   Pull the cable end out and then up from the accelerator pedal recess **(see illustration)**.
6   Push the cable through the firewall into the engine compartment. To disconnect the cable at the firewall, remove the cable retainer clip, if equipped, then use pliers to squeeze

8.2 Using a small screwdriver, pry the tabs on the electrical connectors and separate them from the computer (models with a Spark Control Computer)

9.2 Remove the retainer clip from the throttle shaft and slide the accelerator cable off the shaft

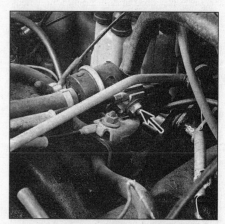

9.3 Loosen the clamp nut (arrow) and remove the cable from the clamp

the tabs on the cable retainer from inside the vehicle (see illustration 9.5, View Z).

Installation

7 Installation is the reverse of removal. **Note:** *To prevent possible interference, flexible components (hoses, wires, etc.) must not be routed within two inches of moving parts, unless routing is controlled.*

8 Operate the accelerator pedal and check for any binding condition by completely opening and closing the throttle.

9 If necessary, at the engine compartment side of the firewall, apply sealant around the accelerator cable to prevent water from entering the passenger compartment.

Adjustment

10 Warm the engine until normal operating temperature is attained and allow the engine to idle. Set the specified idle speed, if necessary.

11 Loosen the cable clamp (see illustration 9.3) and slide the accelerator cable away from the carburetor until all the cable slack has been removed, but don't move the throttle lever on the carburetor.

12 Slide the cable back toward the carburetor to allow 1/4-inch play in the cable tension. Tighten the clamp assembly.

13 Raise the engine rpms and then allow the engine to settle back down to idle. Check for proper adjustment and carburetor throttle actuation and re-adjust the cable if necessary.

14 Shut off the engine and remove the air cleaner assembly. Press the accelerator all the way to the floor while you observe the movement of the throttle plate(s). The throttle plate(s) must be fully open (completely vertical). If not, readjustment is necessary.

---

10 Carburetor - diagnosis and overhaul

Warning: *Gasoline is extremely flammable, so take extra precautions when you work on any part of the fuel system. Don't smoke or allow open flames or bare light bulbs near the work area, and don't work in a garage where a natural gas-type appliance (such as a water heater or a clothes dryer) is present. Since gasoline is carcinogenic, wear latex gloves when there's a possibility of being exposed to fuel, and, if you spill any fuel on your skin, rinse it off immediately with soap and water. Mop up any spills immediately and do not store fuel-soaked rags where they could ignite. When you perform any kind of work on the fuel system, wear safety glasses and have a Class B type fire extinguisher on hand.*

Diagnosis

1 A thorough road test and check of carburetor adjustments should be done before any major carburetor service. Specifications for some adjustments are listed on the *Vehicle Emissions Control Information* (VECI) label found in the engine compartment or on the air cleaner housing.

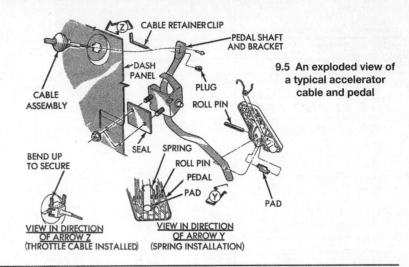

9.5 An exploded view of a typical accelerator cable and pedal

2 Carburetor problems usually show up as flooding, hard starting, stalling, severe backfiring and poor acceleration. A carburetor that's leaking fuel and/or covered with wet looking deposits definitely needs attention.

3 Some performance complaints directed at the carburetor are actually a result of loose, out-of-adjustment or malfunctioning engine or electrical components. Others develop when vacuum hoses leak, are disconnected or are incorrectly routed. The proper approach to analyzing carburetor problems should include the following items:

a) *Inspect all vacuum hoses and actuators for leaks and correct installation (see Chapters 1 and 6).*
b) *Tighten the intake manifold and carburetor mounting nuts/bolts evenly and securely.*
c) *Perform a compression test and vacuum test (see Chapter 2C).*
d) *Clean or replace the spark plugs as necessary (see Chapter 1).*
e) *Check the spark plug wires (see Chapter 1).*
f) *Inspect the ignition primary wires.*
g) *Check the ignition timing (follow the instructions printed on the Emissions Control Information label).*
h) *Check the fuel pump pressure/volume (see Section 2).*
i) *Check the heat control valve in the air cleaner for proper operation (see Chapter 1).*
j) *Check/replace the air filter element (see Chapter 1).*
k) *Check the PCV system (see Chapter 6).*
l) *Check/replace the fuel filter (see Chapter 1). Also, the strainer in the tank could be restricted.*
m) *Check for a plugged exhaust system.*
n) *Check EGR valve operation (see Chapter 6).*
o) *Check the choke - it should be completely open at normal engine operating temperature (see Chapter 1).*
p) *Check for fuel leaks and kinked or dented fuel lines (see Chapters 1 and 4)*

q) *Check accelerator pump operation with the engine off (remove the air cleaner cover and operate the throttle as you look into the carburetor throat - you should see a stream of gasoline enter the carburetor).*
r) *Check for incorrect fuel or bad gasoline.*
s) *Check the valve clearances (if applicable) and camshaft lobe lift (see Chapters 1 and 2)*
t) *Have a dealer service department or repair shop check the electronic engine and carburetor controls.*

4 Diagnosing carburetor problems may require that the engine be started and run with the air cleaner off. While running the engine without the air cleaner, backfires are possible. This situation is likely to occur if the carburetor is malfunctioning, but just the removal of the air cleaner can lean the fuel/air mixture enough to produce an engine backfire. **Warning:** *Do not position any part of your body, especially your face, directly over the carburetor during inspection and servicing procedures. Wear eye protection!*

Overhaul

Refer to illustrations 10.7a through 10.7f

5 Once it's determined that the carburetor needs an overhaul, several options are available. If you're going to attempt to overhaul the carburetor yourself, first obtain a good-quality carburetor rebuild kit (which will include all necessary gaskets, internal parts, instructions and a parts list). You'll also need some special solvent and a means of blowing out the internal passages of the carburetor with air.

6 An alternative is to obtain a new or rebuilt carburetor. They are readily available from dealers and auto parts stores. Make absolutely sure the exchange carburetor is identical to the original. A tag is usually attached to the top of the carburetor or a number is stamped on the float bowl. It will help determine the exact type of carburetor you have. When obtaining a rebuilt carburetor or a rebuild kit, make sure the kit or carburetor matches your application exactly. Seemingly insignificant differences can make a

Index Number	Part Name
1	Choke Plate
2	Choke Shaft & Lever Assembly
3	Fast Idle Adjusting Screw
4	Throttle Body Screw & L.W.
5	Dashpot Bracket Screw
6	Choke Diaphragm Bracket Screw
7	Pump Rod Clamp Screw
8	Power Valve Diaphragm Body Screw & L.W. (Duty Cycle Solenoid)
9	Throttle Plate Screw
10	Curb Idle Adjusting Screw
11	ECS Hinge Retaining Screw
12	Retaining Plate Screw
13	Choke Plate Screw
14	Wire Hold-Down Clamp Screw
15	ECS Vent Cover Screw
16	Air Horn Screw & L.W.
17	Idle Solenoid Bracket Screw & L.W
18	Solenoid Adjusting Screw
19	Power Valve Lever Adjusting Screw
20	ECS Vent Valve Adjusting Screw
21	Fuel Inlet Fitting (Needle & Seat Assy.) Gasket
22	Throttle Body Gasket
23	Main Body Gasket
24	Switch Vent Solenoid Gasket
25	Power Valve Spacer Gasket
26	Throttle Plate
27	Throttle Body & Shaft Assy.
28	Idle Adjusting Needle
29	Float & Hinge Assy.
30	Float Hinge Shaft
31	Fuel Inlet Needle & Seat Assy.
32	Choke Diaphragm Hose Connector
33	Spark Tube Hose Connector
34	Main Jet
35	ECS Vent Valve
36	Switch Vent Seal
37	Power Valve Diaphragm & Stem Assy.
38	Power Valve Piston
39	Power Valve Assembly (Vacuum)
40	Power Valve Assembly (Mech.)
41	Idle Adj. Needle "0" Ring
42	Pump-Rod Seal
43	Pump Discharge Valve Checkball
44	ECS Lever Hinge Pin
45	Choke Diaphragm Link Pin
46	Pump Stem & Retainer Plate Assy.
47	Pump Operating Link
48	Choke Diaphragm Link
49	Idle Needle Limiter Cap or Plug
50	ECS Vent Cover
51	Power Valve Diaphragm Body
52	Accelerator Pump Cup
53	Choke Diaphragm Assy.
54	Fast Idle Cam Retainer
55	Float Shaft Retainer
56	Power Valve Piston Retainer
57	Wire Tie-Down Strap*
58	Idle Adj. Needle Spring
59	Power Valve Lever Adj. Screw Spring
60	Curb Idle Adj. Screw Spring
61	Fast Idle Adj. Screw Spring
62	ECS Vent Spring
63	Power Valve Diaphragm Spring
64	Power Valve Piston Spring
65	Power Valve Lever Return Spring
66	Power Valve Rod Drive Spring
67	Switch Vent Solenoid Spring
68	Idle Solenoid Adj. Screw Spring
69	Pump Operating Lever Nut
70	Dashpot Lock Nut
71	Pump Discharge Valve Weight
72	Power Valve Diaphragm Spacer*
73	Fast Idle Cam
74	Power Valve Rod
75	Fast Idle Rod
76	Pump Operating Rod
77	Pump Operating Lever Nut L.W.
78	Insulator Washer*
79	Dashpot Bracket

Index Number	Part Name
80	Pump Rod Clamp
81	Wire Hold-Down Clamp
82	Choke Diaphragm Vacuum Hose
83	Idle Screw Insulator*
84	ECS Vent Lever
85	Pump Operating Lever
86	Power Valve Lever
87	Idle Stop Switch*
88	Idle Stop Solenoid
89	Duty Cycle Solenoid
90	Switch Vent Solenoid & Armature Assy.
91	Dashpot Assy.
92	A/C Speed-Up Solenoid*
93	T.P.T. Assy.*
94	T.P.T. Bracket*
95	Power Valve Vacuum Hose Clamp
96	Power Valve Vacuum Hose
97	Spark Tube Vacuum Hose

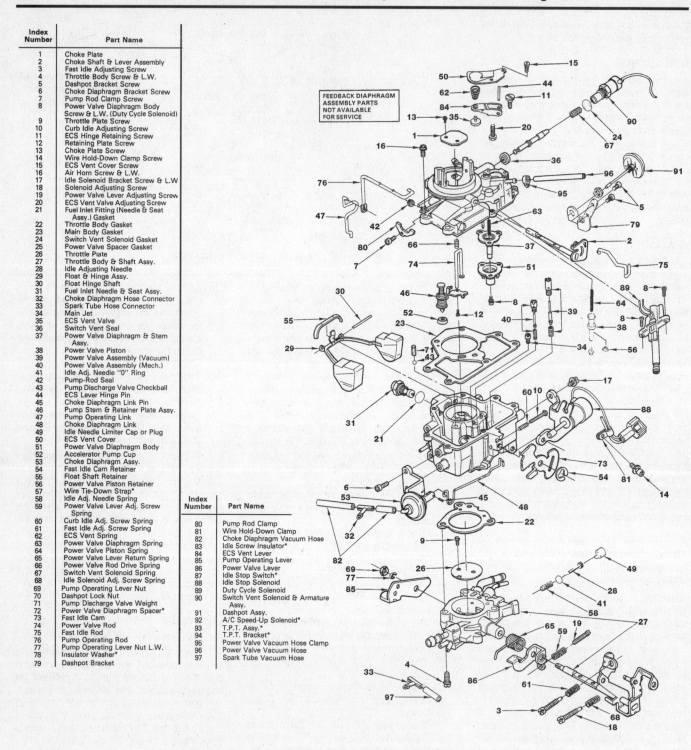

FEEDBACK DIAPHRAGM ASSEMBLY PARTS NOT AVAILABLE FOR SERVICE

10.7a An exploded view of the Holley Models 1945 and 6145 carburetors

---

large difference in engine performance.

7   If you choose to overhaul your own carburetor, allow enough time to disassemble it carefully, soak the necessary parts in the cleaning solvent (usually for at least one-half day or according to the instructions listed on the carburetor cleaner) and reassemble it, which will usually take much longer than dis-assembly **(see illustrations)**. When disassembling the carburetor, match each part with the illustration in the carburetor kit and lay the parts out in order on a clean work surface. Overhauls by inexperienced mechanics can result in an engine which runs poorly or not at all. To avoid this, use care and patience when disassembling the carburetor so you can reassemble it correctly.

8   Because carburetor designs are constantly modified by the manufacturer in order to meet increasingly more stringent emissions regulations, it isn't feasible to include a step-by-step overhaul of each type. You'll receive a detailed, well illustrated set of instructions with the carburetor overhaul kit.

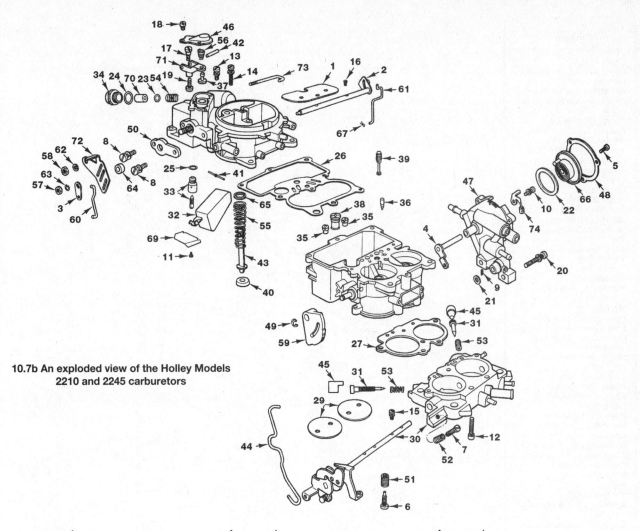

10.7b An exploded view of the Holley Models 2210 and 2245 carburetors

Index Number	Part Name	Index Number	Part Name	Index Number	Part Name
1	Choke Plate	27	Throttle Body Gasket	52	Curb Idle Adjusting Screw Spring
2	Choke Shaft & Lever Assembly	28	Flange Gasket	53	Idle Needle Adjusting Screw Spring
3	Fast Idle Lever	29	Throttle Plate		
4	Choke Housing Lever & Shaft Assembly	30	Throttle Body & Shaft Assembly	54	Fuel Inlet Filter Spring
5	Thermostat Cap Clamp Screw	31	Idle Adjusting Needle	55	Pump Drive Spring
6	Fast Idle Adjusting Screw	32	Float & Hinge Assembly	56	E.C.S. Vent Spring
7	Curb Idle Adjusting Screw	33	Fuel Inlet Needle & Seat Assembly	57	Fast Idle Lever Nut
8	Pump Shaft Lever Screw			58	Pump Lever Nut
9	Choke Piston Adjusting Screw	34	Fuel Inlet Fitting	59	Fast Idle Cam
10	Choke Lever screw & L.W.	35	Main Jet	60	Fast Idle Rod
11	Fuel Bowl Baffle Screw & L.W.	36	Pump Discharge Needle Valve	61	Choke Rod
12	Throttle Body Screw & L.W.	37	Air Vent Valve	62	Pump Lever L.W.
13	Air Horn Screw & L.W. (Short)	38	Power Valve Assembly	63	Fast Idle Lever L.W.
14	Air Horn Screw & L.W. (Long)	39	Power Valve Piston Assembly	64	Washer
15	Throttle Plate Screw & L.W.	40	Pump Piston Cup	65	Pump Drive Spring Washer
16	Choke Plate Screw	41	Float Hinge Pin	66	Thermostat & Cap Assembly
17	E.C.S. Hinge Retainer Screw	42	E.C.S. Lever Hinge Pin	67	Rod Retainer
18	E.C.S. Vent Cover Screw	43	Pump Stem Assembly	68	Choke Vacuum Hose
19	Vent Valve Adjusting Screw	44	Pump Link	69	Fuel Bowl Baffle
20	Choke Housing Screw & L.W.	45	Idle Needle Limiter Cap	70	Fuel Inlet Filter
21	Choke Housing Gasket	46	E.C.S. Vent Cover	71	Vent Lever
22	Thermostat Housing Gasket	47	Choke Housing Assembly	72	Pump Lever
23	Fuel Inlet Filter Gasket	48	Thermostat Cap Retainer Ring	73	Pump Lever Shaft
24	Fuel Inlet Fitting Gasket	49	"E" Ring Retainer	74	Choke Thermostat Lever & Piston Assembly
25	Fuel Inlet Seat Gasket	50	Pump Shaft Retainer		
26	Main Body Gasket	51	Fast Idle Adjusting Screw Spring		

Index Number	Part Name
1	Choke Plate
2	Choke Shaft & Lever Assembly
3	Dechoke Lever
4	Air Cleaner Bracket Screw
5	Air Horn to Main Body Screw & L.W.
6	Throttle Stop Screw
7	Fast Idle Adjusting Screw
8	Bracket Retaining
9	Throttle Body to Main Body Screw & L.W.
10	Choke Plate Screw
11	Throttle Plate Screw
12	Cover Screw
13	Nozzle Bar Screw
14	Fuel Inlet Fitting Gasket
15	Main Body Gasket
16	Nozzle Bar Gasket
17	Throttle Body Gasket
18	E.C.S. Cover Gasket
19	Throttle Plate
20	Throttle Body & Shaft Assembly
21	Throttle Return Spring Bushings
22	Idle Adjusting Needle
23	Float & Hinge Assembly
24	Float Hinge Shaft & Retainer
25	Fuel Inlet Needle & Seat Assembly
26	Nozzle Bar Assembly
27	Main Jet
28	Vent Valve
29	Power Valve Piston
30	Power Valve Assembly
31	Pump Cup
32	Pump Discharge Ball
33	Vent Valve Lever Pin
34	Vent Valve Operating Lever Pin
35	Air Cleaner Ring Retaining Pin
36	Pump Steam & Head Assembly
37	Choke Diaphragm Link
38	Pump Stem Link
39	Pump Link
40	Idle Needle Limiter
41	Accelerator Pump Housing Cover
42	Choke Diaphragm Assembly Complete
43	Fast Idle Cam Retainer
44	Pump Operating Lever Retainer
45	Power Valve Piston Retainer
46	Throttle Stop Screw Spring
47	Fast Idle Adjusting Screw Spring
48	Idle Adjusting Needle Spring
49	Vent Valve Operating Lever Spring
50	Vent Valve Spring
51	Throttle Return Spring
52	Pump Drive Spring
53	Power Valve Piston Spring
54	Choke Shaft Spring
55	Dechoke Lever Nut
56	Air Cleaner Ring
57	Fast Idle Cam
58	Fast Idle Rod
59	Pump Discharge Weight
60	Dechoke Lever Nut L.W.
63	Pump Link Retainer
64	Air Cleaner Bracket
65	Vacuum Hose
66	Float Baffle
67	Pump Lever
68	Vent Valve Lever
69	Vent Valve Operating Lever
70	Pump Operating Lever
71	Pump Operating Lever Shaft
72	Vent Valve Retainer
73	T.P.T. Bracket Screw
74	Power Valve Adjusting Screw
75	Power Valve Assembly (Staged)

Index Number	Part Name
76	Choke Diaphragm Link Pin
77	Roll-Pin
78	Power Valve Stem Retainer
79	Power Valve Spring
80	T.P.T. Lever Nut
81	T.P.T. Assembly Locknut
82	Mechanical Power Valve Stem
83	Mechanical Power Valve Stem Cap
84	T.P.T. Lever Lockwasher
85	T.P.T. Bracket
86	T.P.T. Lever
87	T.P.T. Assembly
88	T.P.T. Wire Assembly
89	Duty-Cycle Solenoid

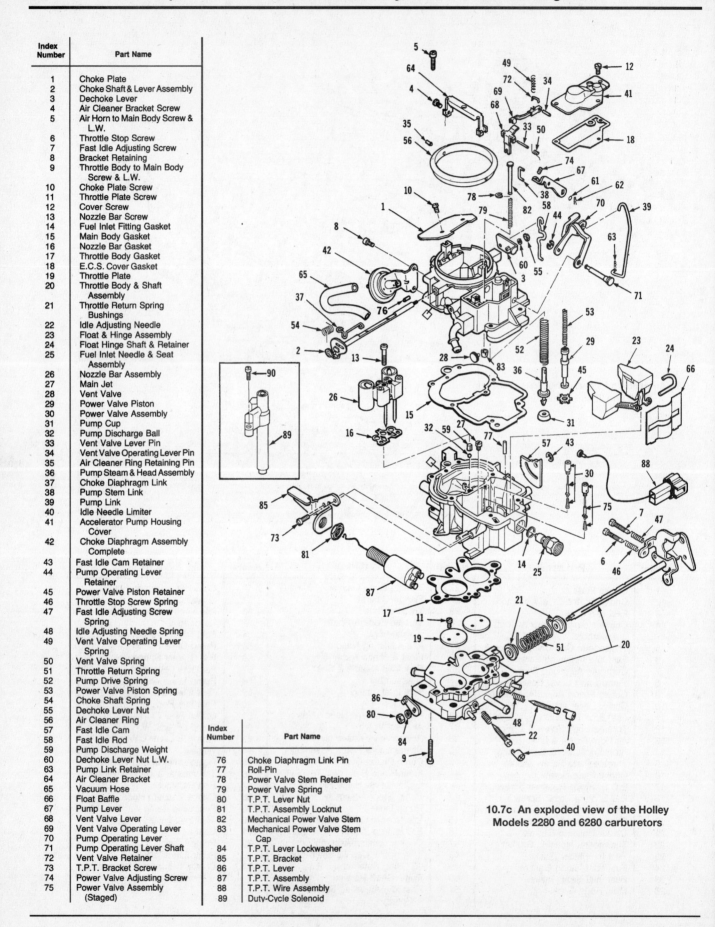

10.7c An exploded view of the Holley Models 2280 and 6280 carburetors

1. Screw & Lockwasher (2)—Altitude Compensator
2. Altitude Compensator Assy. (Some Models)
3. Gasket—Altitude Compensator
4. Screw (2)—Diaphragm Cover
5. Screw & Seal Washer (1)—Cover
6. Cover—Diaphragm
7. Spring—Diaphragm
8. Diaphragm Assy.—Enrichment Valve
9. Housing—Idle Enrichment Valve
10. Valve Seat—Enrichment
11. Valve—Enrichment Seat
12. Spring—Enrichment Valve
13. Gasket—Enrichment Valve Housing
14. Screw—Transducer Bracket
15. Transducer, Idle Stop Switch & Bracket Assy.
16. Hose—Bowl Vent Solenoid
17. Screw & Lockwasher (3)—Bowl Vent Solenoid
18. Bowl Vent Solenoid Assy.
19. Valve—Bowl Vent
20. Gasket—Bowl Vent Solenoid Assy.
21. Fitting—Fuel Inlet
22. Gasket—Inlet Fitting
23. Screw—Rod Retainer
24. Retainer—Choke Connector Rod
25. Rod—Choke Connector
26. Screw—Metering Rod Cover Plate
27. Plate—Cover
28. Screw—Metering Rod Cover Plate
29. Plate—Cover
30. Screw—Plate
31. Plate—Step Up Piston Cover
32. Step Up Piston & Rod Assy.
33. Rod (2)—Step Up
34. Spring—Step Up Piston
35. Retainer—Choke Pull Off Rod
36. Washer—Choke Pull Off Rod
37. Retainer—Choke Pull Off Rod
38. Rod—Choke Pull Off
39. Retainer—Pump Rod
40. Rod—Pump Arm Connector
41. Screw—Pump Arm
42. Arm—Pump
43. Screw—Pump Jet Housing
44. Housing—Pump Jet
45. Gasket—Housing
46. Needle—Pump Disc. Check
47. Screw (10)—Bowl Cover
48. Bowl Cover Assy.
49. Pin (2)—Float Lever
50. Float Assy. (2)
51. Rotary Disc—Fuel Inlet Valve & Gasket (2)
52. Tube—Pump Passage
53. Gasket—Bowl Cover
54. Valve—Bowl Vent - (Mechanical Model)
55. S Link—Pump
56. Check Valve Assy.—Pump Intake
57. Pump Assy.
58. Spring—Pump
59. O Ring (2)—Main Well Seal
60. Bowl Assy.—Fuel
61. Gasket—Throttle Body
62. Lever—Step Up Piston
63. Pin—Lever
64. Screw—Choke Pull Off Bracket
65. Choke Pull Off Assy.
66. Cap (2)—Idle Limiter
67. Needle (2)—Idle Adjusting
68. Spring (2)—Idle Adjusting Needle
69. Throttle Body Assy.

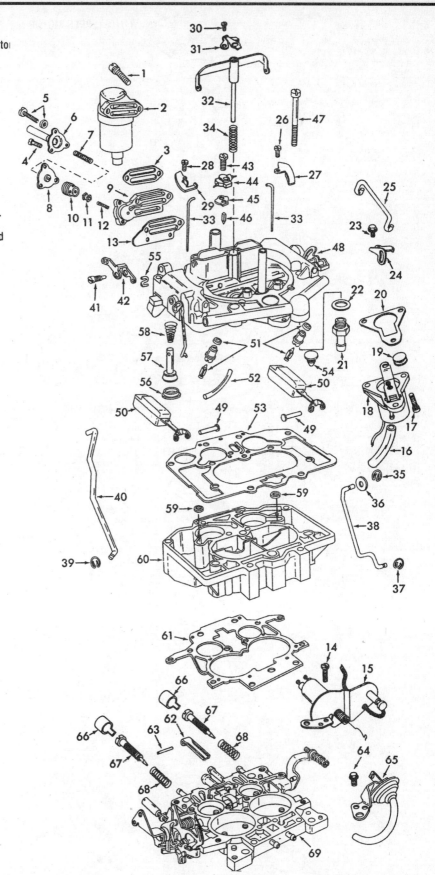

10.7d An exploded view of the Carter ThermoQuad carburetor

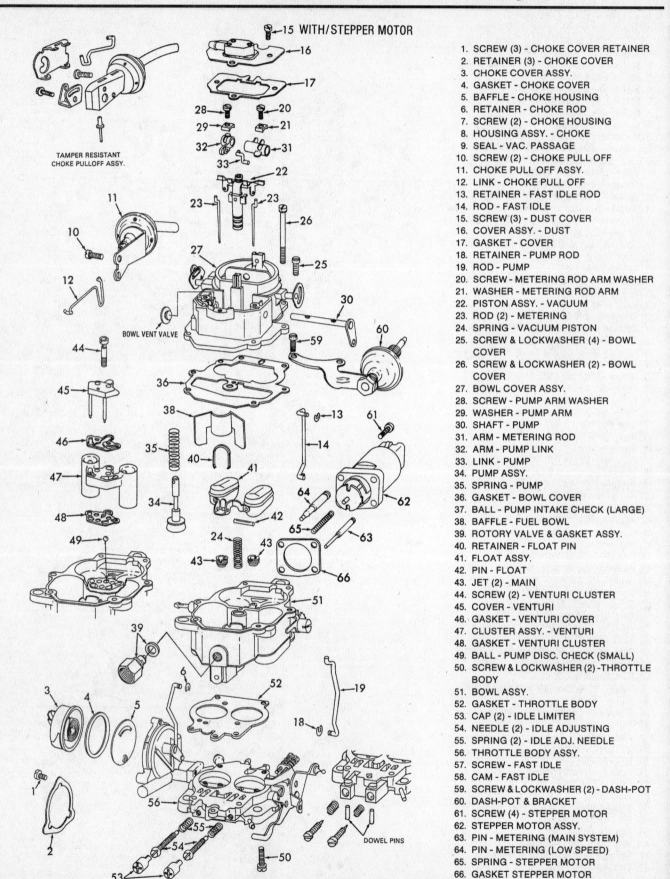

15 WITH/STEPPER MOTOR

TAMPER RESISTANT
CHOKE PULLOFF ASSY.

BOWL VENT VALVE

DOWEL PINS

1. SCREW (3) - CHOKE COVER RETAINER
2. RETAINER (3) - CHOKE COVER
3. CHOKE COVER ASSY.
4. GASKET - CHOKE COVER
5. BAFFLE - CHOKE HOUSING
6. RETAINER - CHOKE ROD
7. SCREW (2) - CHOKE HOUSING
8. HOUSING ASSY. - CHOKE
9. SEAL - VAC. PASSAGE
10. SCREW (2) - CHOKE PULL OFF
11. CHOKE PULL OFF ASSY.
12. LINK - CHOKE PULL OFF
13. RETAINER - FAST IDLE ROD
14. ROD - FAST IDLE
15. SCREW (3) - DUST COVER
16. COVER ASSY. - DUST
17. GASKET - COVER
18. RETAINER - PUMP ROD
19. ROD - PUMP
20. SCREW - METERING ROD ARM WASHER
21. WASHER - METERING ROD ARM
22. PISTON ASSY. - VACUUM
23. ROD (2) - METERING
24. SPRING - VACUUM PISTON
25. SCREW & LOCKWASHER (4) - BOWL
    COVER
26. SCREW & LOCKWASHER (2) - BOWL
    COVER
27. BOWL COVER ASSY.
28. SCREW - PUMP ARM WASHER
29. WASHER - PUMP ARM
30. SHAFT - PUMP
31. ARM - METERING ROD
32. ARM - PUMP LINK
33. LINK - PUMP
34. PUMP ASSY.
35. SPRING - PUMP
36. GASKET - BOWL COVER
37. BALL - PUMP INTAKE CHECK (LARGE)
38. BAFFLE - FUEL BOWL
39. ROTARY VALVE & GASKET ASSY.
40. RETAINER - FLOAT PIN
41. FLOAT ASSY.
42. PIN - FLOAT
43. JET (2) - MAIN
44. SCREW (2) - VENTURI CLUSTER
45. COVER - VENTURI
46. GASKET - VENTURI COVER
47. CLUSTER ASSY. - VENTURI
48. GASKET - VENTURI CLUSTER
49. BALL - PUMP DISC. CHECK (SMALL)
50. SCREW & LOCKWASHER (2) - THROTTLE
    BODY
51. BOWL ASSY.
52. GASKET - THROTTLE BODY
53. CAP (2) - IDLE LIMITER
54. NEEDLE (2) - IDLE ADJUSTING
55. SPRING (2) - IDLE ADJ. NEEDLE
56. THROTTLE BODY ASSY.
57. SCREW - FAST IDLE
58. CAM - FAST IDLE
59. SCREW & LOCKWASHER (2) - DASH-POT
60. DASH-POT & BRACKET
61. SCREW (4) - STEPPER MOTOR
62. STEPPER MOTOR ASSY.
63. PIN - METERING (MAIN SYSTEM)
64. PIN - METERING (LOW SPEED)
65. SPRING - STEPPER MOTOR
66. GASKET STEPPER MOTOR

10.7e An exploded view of the Carter BBD carburetor

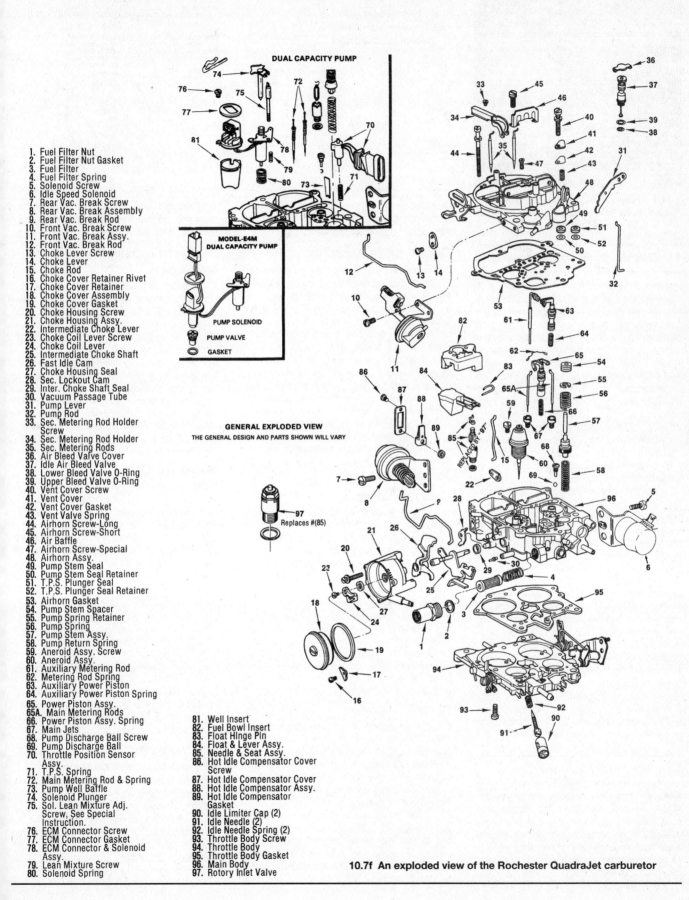

DUAL CAPACITY PUMP

MODEL-E4M
DUAL CAPACITY PUMP

PUMP SOLENOID

PUMP VALVE

GASKET

GENERAL EXPLODED VIEW
THE GENERAL DESIGN AND PARTS SHOWN WILL VARY

97
Replaces #(85)

1. Fuel Filter Nut
2. Fuel Filter Nut Gasket
3. Fuel Filter
4. Fuel Filter Spring
5. Solenoid Screw
6. Idle Speed Solenoid
7. Rear Vac. Break Screw
8. Rear Vac. Break Assembly
9. Rear Vac. Break Rod
10. Front Vac. Break Screw
11. Front Vac. Break Assy.
12. Front Vac. Break Rod
13. Choke Lever Screw
14. Choke Lever
15. Choke Rod
16. Choke Cover Retainer Rivet
17. Choke Cover Retainer
18. Choke Cover Assembly
19. Choke Cover Gasket
20. Choke Housing Screw
21. Choke Housing Assy.
22. Intermediate Choke Lever
23. Choke Coil Lever Screw
24. Choke Coil Lever
25. Intermediate Choke Shaft
26. Fast Idle Cam
27. Choke Housing Seal
28. Sec. Lockout Cam
29. Inter. Choke Shaft Seal
30. Vacuum Passage Tube
31. Pump Lever
32. Pump Rod
33. Sec. Metering Rod Holder
    Screw
34. Sec. Metering Rod Holder
35. Sec. Metering Rods
36. Air Bleed Valve Cover
37. Idle Air Bleed Valve
38. Lower Bleed Valve O-Ring
39. Upper Bleed Valve O-Ring
40. Vent Cover Screw
41. Vent Cover
42. Vent Cover Gasket
43. Vent Valve Spring
44. Airhorn Screw-Long
45. Airhorn Screw-Short
46. Air Baffle
47. Airhorn Screw-Special
48. Airhorn Assy.
49. Pump Stem Seal
50. Pump Stem Seal Retainer
51. T.P.S. Plunger Seal
52. T.P.S. Plunger Seal Retainer
53. Airhorn Gasket
54. Pump Stem Spacer
55. Pump Spring Retainer
56. Pump Spring
57. Pump Stem Assy.
58. Pump Return Spring
59. Aneroid Assy. Screw
60. Aneroid Assy.
61. Auxiliary Metering Rod
62. Metering Rod Spring
63. Auxiliary Power Piston
64. Auxiliary Power Piston Spring
65. Power Piston Assy.
65A. Main Metering Rods
66. Power Piston Assy. Spring
67. Main Jets
68. Pump Discharge Ball Screw
69. Pump Discharge Ball
70. Throttle Position Sensor
    Assy.
71. T.P.S. Spring
72. Main Metering Rod & Spring
73. Pump Well Baffle
74. Solenoid Plunger
75. Sol. Lean Mixture Adj.
    Screw, See Special
    Instruction.
76. ECM Connector Screw
77. ECM Connector Gasket
78. ECM Connector & Solenoid
    Assy.
79. Lean Mixture Screw
80. Solenoid Spring

81. Well Insert
82. Fuel Bowl Insert
83. Float Hinge Pin
84. Float & Lever Assy.
85. Needle & Seat Assy.
86. Hot Idle Compensator Cover
    Screw
87. Hot Idle Compensator Cover
88. Hot Idle Compensator Assy.
89. Hot Idle Compensator
    Gasket
90. Idle Limiter Cap (2)
91. Idle Needle (2)
92. Idle Needle Spring (2)
93. Throttle Body Screw
94. Throttle Body
95. Throttle Body Gasket
96. Main Body
97. Rotary Inlet Valve

4A

10.7f An exploded view of the Rochester QuadraJet carburetor

11.1 Apply 12 volts to the choke heater connector - the choke should open in about five minutes

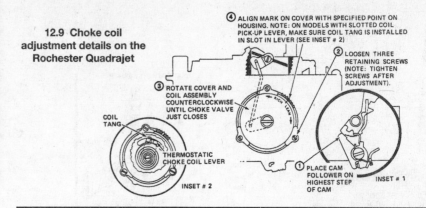

12.9 Choke coil adjustment details on the Rochester Quadrajet

11   Electric choke heater - testing

Refer to illustration 11.1

Caution: *If there is any loss of electrical current to the choke heater, operation of any type, including idling, should be avoided. Loss of power to the choke will cause the choke to remain partly closed during engine operation. A very rich air-to-fuel mixture will be created and result in abnormally high exhaust system temperatures, which may cause damage to the catalytic converter or other underbody parts of the vehicle.* **Note:** *This system is used on most 1974 and later models, except some mid-1980's and later four-barrel carburetors (Rochester Quadrajet). For more information and detailed test procedures for the entire system, refer to Chapter 6.*

1   Remove the engine cover (see Chapter 2B, Section 2). With the engine cold, connect a jumper wire from the choke heater wire to the positive battery terminal. The choke heater housing should begin becoming hot and the choke plate in the carburetor should slowly open after some time. Be sure it opens completely in about five minutes or replacement is necessary **(see illustration)**. **Note:** *As the heater is heating the choke coil, occasionally tap the accelerator to allow the choke to open.*

2   If the choke heater tests okay, but the choke valve does not open in normal operation, check for voltage to the choke heater when the engine is cold and the ignition switch is in the ON position.

3   If no voltage is present in Step 2, repair the open in the power circuit to the choke heater (be sure to check the fuse first!).

12   Carburetor adjustments

Note 1: *These carburetor adjustments are strictly on-vehicle adjustments. During overhaul, refer to the instructions included in the overhaul kit for complete procedures and any additional adjustments that are required.*
Note 2: *For additional information concerning overhaul and adjustments on the Holley and*

Rochester carburetors, refer to the Haynes Holley Carburetor Manual *and the* Haynes Rochester Carburetor Manual.

Choke plate opening

Note: *For more information on the choke system, refer to Chapter 1 and the previous Section on the electric choke heater system.*

1   The choke plate opening is a critical carburetor adjustment directly involved with cold-running conditions and choke enrichment during cranking. The choke unloader is a mechanical device that partially opens the choke valve at wide open throttle to eliminate choke enrichment during hard acceleration.

2   Engines which have been stalled or flooded by excessive choke enrichment can be cleared by the use of the choke unloader. With the throttle valve at wide open throttle, the choke plate should be slightly open to allow a sufficient amount of intake air into the carburetor venturi.

Holley carburetors

3   These models are equipped with a choke thermostat coil mounted in the intake manifold, which actuates the choke via a rod. With the carburetor installed and the engine cold, the coil should hold the choke lightly closed (possibly open slightly, but no more than about 1/16-inch).

4   If the choke is closed tightly or open too far, remove the coil from the intake manifold; the center mount for the thermostat coil should be slotted so you can adjust the coil's tension with a screwdriver. If there's no adjustment at the choke coil's center mount, bend the rod to adjust the choke. If the coil is broken, corroded or deformed, replace it.

Carter carburetors

5   These models are equipped with a choke thermostat coil mounted in the intake manifold, which actuates the choke via a rod. With the carburetor installed and the engine cold, the coil should hold the choke lightly closed (possibly open slightly, but no more than about 1/16-inch).

6   If the choke is closed tightly or open too far, remove the coil from the intake manifold; the center mount for the thermostat coil should be slotted so you can adjust the coil's tension with a screwdriver. If there's no

adjustment at the choke coil's center mount, bend the rod to adjust the choke. If the coil is broken, corroded or deformed, replace it.

Rochester carburetors

2GV models

7   These models are equipped with a choke thermostat coil mounted in the intake manifold, which actuates the choke via a rod. With the carburetor installed and the engine cold, the coil should hold the choke almost, but not quite, completely closed.

8   If the choke is closed tightly or open too far, remove the coil from the intake manifold; the center mount for the thermostat coil should be slotted so you can adjust the coil's tension with a screwdriver. If there's no adjustment at the choke coil's center mount, bend the rod to adjust the choke. If the coil is broken, corroded or deformed, replace it.

Quadrajet models

Refer to illustration 12.9

9   All the Quadrajet carburetors installed on to the models covered in this manual are equipped with an integral choke system. Since choke coils lose their tension over time, you may find, on older coils, that lining up the marks does not provide enough choke. Readjust the coil by rotating the cover counterclockwise until the choke plate closes completely, then rotate it clockwise until the plate just moves away from the fully closed position (about 1/16-inch, measured at the rear of the air horn) **(see illustration)**.

10   Re-tighten the three cover screws. **Note:** *The metal clips beneath the screws must be positioned so they exert their maximum spring force against the cover. If they're installed backwards, the choke cover will have a tendency to rotate, changing the choke setting.*

Idle mixture

Refer to illustrations 12.12a through 12.12e
Note: *Be sure to check the ignition timing, the condition of the spark plugs and ignition wires and, if necessary, perform a complete tune-up before attempting to make the following carburetor adjustment. Often, problems with the mixture are due to a vacuum leak (see Troubleshooting at the front of this manual and the vacuum gauge checks in Chapter 2C).*

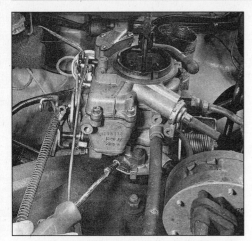

12.12a Adjusting the idle mixture screw on the Holley Model 1945

12.12c Adjusting the idle mixture screw on the Holley Model 6280

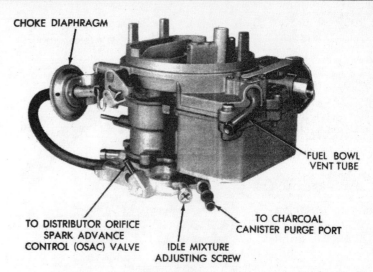

12.12b Component locations on the Holley Model 2245 (note that this idle mixture screw has a limiter cap)

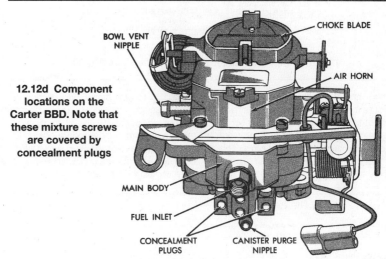

12.12d Component locations on the Carter BBD. Note that these mixture screws are covered by concealment plugs

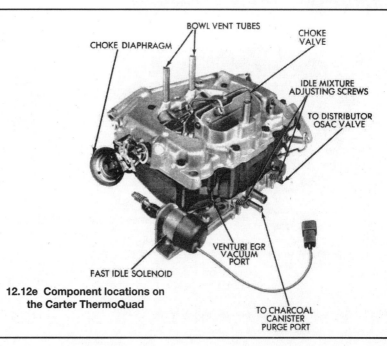

12.12e Component locations on the Carter ThermoQuad

Note: *On carburetors that have the factory concealment plugs installed over the idle mixture screws, do not adjust the idle mixture, as it has been factory pre-set. Also, on models with limiter caps on the mixture screws, do not adjust the mixture any richer than the caps will allow (do not break off the caps). If the engine runs too rich at the factory settings, there's something wrong with the choke or the carburetor is faulty. If the engine is running too lean at the factory settings, there's a vacuum leak or the carburetor is faulty. If necessary, have the system diagnosed by a qualified shop.*

11 After the engine has reached normal operating temperature, set the parking brake and block the wheels, then remove the air cleaner and turn the idle speed adjusting screw out as far as possible without the engine running rough, since the throttle plate(s) must be as nearly closed as possible when adjusting the idle mixture.

12 Working with one idle mixture screw at a time (one-barrel models have only one screw, but all others have two screws), turn the screw clockwise until the idle speed drops a noticeable amount **(see illustrations)**. Now slowly turn the screw out until the maximum idle speed is reached, but no further.

4A

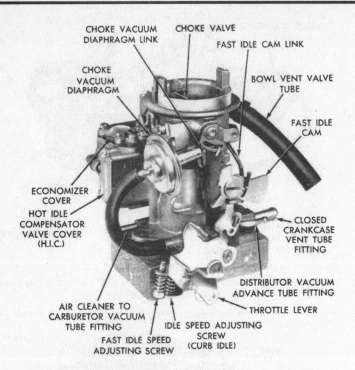

CHOKE VACUUM
DIAPHRAGM LINK

CHOKE VALVE

FAST IDLE CAM LINK

CHOKE
VACUUM
DIAPHRAGM

BOWL VENT VALVE
TUBE

FAST IDLE
CAM

ECONOMIZER
COVER

HOT IDLE
COMPENSATOR
VALVE COVER
(H.I.C.)

CLOSED
CRANKCASE
VENT TUBE
FITTING

DISTRIBUTOR VACUUM
ADVANCE TUBE FITTING

THROTTLE LEVER

AIR CLEANER TO
CARBURETOR VACUUM
TUBE FITTING

IDLE SPEED ADJUSTING
SCREW
(CURB IDLE)

FAST IDLE SPEED
ADJUSTING SCREW

12.17a Component locations on the Holley Model 1920

**12.17b Adjusting the idle speed screw
on the Holley Model 1945**

Idle speed

Refer to illustrations 12.17a through 12.17f
Note: *In order to adjust the idle speed (curb idle speed) on engines equipped with feedback carburetor systems (Models 6280, 6145, BBD, ThermoQuad, Quadrajet etc.), it will be necessary to prevent certain systems (EVAP, EGR, Oxygen sensor feedback, idle stop solenoid, etc.) from operating. Refer to the VECI label in the engine compartment of your vehicle for a list of the steps necessary to make this adjustment.*

14   Engine idle speed is the speed at which the engine operates when no accelerator pedal pressure is applied, as when stopped at a traffic light. This speed is critical to the

13   On two- and four-barrel models, repeat the procedure for the other idle mixture screw. Keep track of the number of turns each screw is off the seat to be sure they are each out approximately the same number of turns. This will verify that the idle circuit is operating properly and balance is maintained. **Note:** *If it was*

necessary to turn the idle speed screw in (to keep the engine running) before setting the idle mixture, adjust the idle speed, then perform the mixture adjustment procedure again. This will ensure that the idle mixture is set with the throttle plates fully closed, so the engine is drawing the fuel mixture only from the idle circuit.

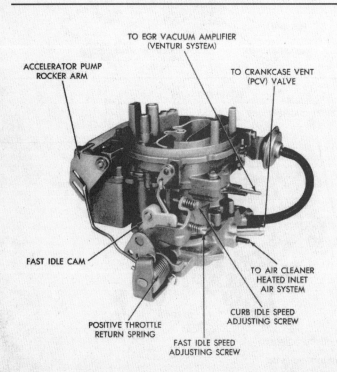

ACCELERATOR PUMP
ROCKER ARM

TO EGR VACUUM AMPLIFIER
(VENTURI SYSTEM)

TO CRANKCASE VENT
(PCV) VALVE

FAST IDLE CAM

TO AIR CLEANER
HEATED INLET
AIR SYSTEM

POSITIVE THROTTLE
RETURN SPRING

FAST IDLE SPEED
ADJUSTING SCREW

CURB IDLE SPEED
ADJUSTING SCREW

12.17c Component locations on the Holley Model 2245

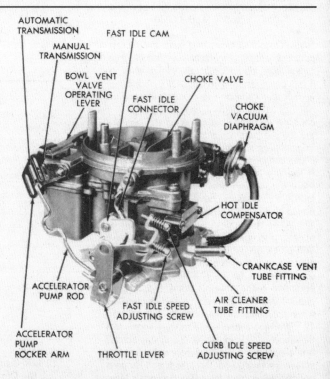

AUTOMATIC
TRANSMISSION

MANUAL
TRANSMISSION

FAST IDLE CAM

BOWL VENT
VALVE
OPERATING
LEVER

FAST IDLE
CONNECTOR

CHOKE VALVE

CHOKE
VACUUM
DIAPHRAGM

HOT IDLE
COMPENSATOR

ACCELERATOR
PUMP ROD

ACCELERATOR
PUMP
ROCKER ARM

THROTTLE LEVER

FAST IDLE SPEED
ADJUSTING SCREW

CRANKCASE VENT
TUBE FITTING

AIR CLEANER
TUBE FITTING

CURB IDLE SPEED
ADJUSTING SCREW

12.17d Component locations on the Holley Model 2210

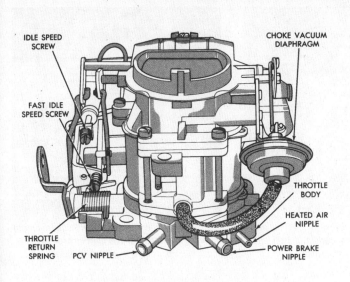

12.17e Component locations on the Carter BBD

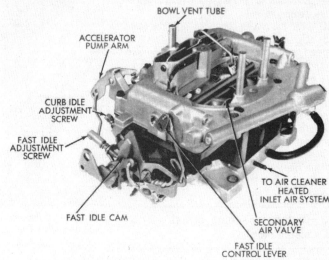

12.17f Component locations on the Carter ThermoQuad

performance of the engine itself, as well as many engine subsystems.

15   A hand-held tachometer must be used when adjusting idle speed to get an accurate reading. The exact hook-up for these meters varies with the manufacturer, so follow the particular directions included.

16   All vehicles covered in this manual should have a tune-up decal or *Emissions Control Information* label located on the inside of the hood. The printed instructions for setting idle speed can be found on this decal, and should be followed since they are for your particular engine.

17   After adjusting the idle mixture, adjust the idle speed. Turn the idle speed adjusting screw **(see Illustrations)** clockwise to increase the idle speed and counterclockwise to decrease it. Set the idle speed to the specifications on the VECI label located in the engine compartment. If the label is missing, set the idle speed to approximately 750 rpm with the transmission in Park or Neutral.

Fast-idle speed

Refer to illustrations 12.18a and 12.18b

Note 1: *To adjust the fast idle speed on engines equipped with feedback carburetor systems (Models 6280, 6145, BBD, Thermoquad, Quadrajet etc.), it may be necessary to prevent certain systems (EVAP, EGR, Oxygen sensor feedback, idle stop solenoid, etc.) from operating. Refer to the VECI label for your particular model for a list of the steps necessary to make this adjustment.*

Note 2: *Many of the locations of the fast idle adjusting screw for the various carburetors are pictured in the preceding illustrations.*

18   Set the parking brake and block the wheels. Remove the air cleaner and place the fast-idle speed adjusting screw on the second highest step of the fast-idle cam **(see illustrations)**.

19   Connect a tachometer according to the tool manufacturer's instructions.

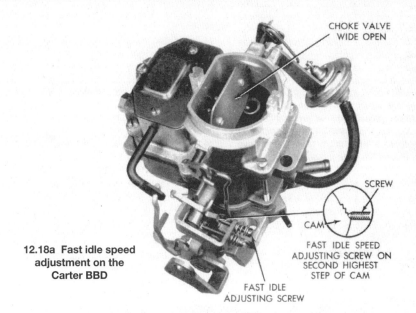

12.18a Fast idle speed adjustment on the Carter BBD

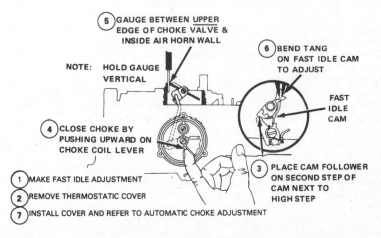

12.18b Fast idle speed adjustment on the Rochester Quadrajet

20   Start and run the engine to normal operating temperature.

21   With the choke fully open, turn the fast-idle adjusting screw until the correct fast-idle speed is obtained. Double-check to make sure the fast idle adjusting screw remains on the second highest step of the cam. Refer to the VECI label in the engine compartment. If the label is not available, set the fast-idle speed to approximately 1,700 rpm.

22   To verify the fast-idle speed is correct, return the engine to idle, then reposition the adjusting screw on the second highest step of the fast-idle cam. Readjust the fast-idle speed, if necessary.

13   Carburetor - removal and installation

Warning: *Gasoline is extremely flammable, so take extra precautions when you work on any part of the fuel system. Don't smoke or allow open flames or bare light bulbs near the work area, and don't work in a garage where a natural gas-type appliance (such as a water heater or a clothes dryer) is present. Since gasoline is carcinogenic, wear latex gloves when there's a possibility of being exposed to fuel, and, if you spill any fuel on your skin, rinse it off immediately with soap and water. Mop up any spills immediately and do not store fuel-soaked rags where they could ignite. When you perform any kind of work on the fuel system, wear safety glasses and have a Class B type fire extinguisher on hand.*

Removal

Refer to illustrations 13.6 and 13.8

1   Remove the fuel filler cap to relieve fuel tank pressure. Remove the engine cover (see Chapter 2B, Section 2).

2   Remove the air cleaner from the carburetor. Be sure to label all vacuum hoses attached to the air cleaner housing.

3   Disconnect the accelerator cable from the throttle lever (see Section 9).

4   If the vehicle is equipped with an automatic transmission, disconnect the kickdown cable from the throttle lever.

5   Clearly label all vacuum hoses and fittings, then disconnect the hoses.

6   Disconnect the fuel line from the carburetor **(see illustration)**. The fuel line fitting is usually very tight. Be sure to use a flare nut wrench on the fuel line nut and hold the inlet fitting with a wrench to prevent it from turning (which would cause the fuel line to twist).

7   Label the wires and terminals, then unplug all the electrical connectors.

8   Remove the mounting fasteners **(see illustration)** and detach the carburetor from the intake manifold. Remove the carburetor mounting gasket. Stuff a rag into the intake manifold openings.

Installation

9   Use a gasket scraper to remove all traces of gasket material and sealant from

13.6 Use a flare-nut wrench and a back-up wrench when detaching the fuel line from the carburetor

the intake manifold (and the carburetor, if it's being reinstalled), then remove the shop rag from the manifold openings. Clean the mating surfaces with lacquer thinner or acetone.

10   Place a new gasket on the intake manifold.

11   Position the carburetor on the gasket and install the mounting fasteners.

12   To prevent carburetor distortion or damage, tighten the fasteners to approximately 16 ft-lbs in a criss-cross pattern, 1/4-turn at a time.

13   The remaining installation steps are the reverse of removal.

14   Check and, if necessary, adjust the idle speed (see Section 12).

15   If the vehicle is equipped with an automatic transmission, refer to Chapter 7B for the kickdown cable adjustment procedure.

16   Start the engine and check carefully for fuel leaks.

14   Electronic Feedback Carburetor (EFC) system - general information

Refer to illustrations 14.1a and 14.1b

Note 1: *Refer to Chapter 5 for additional checks and replacement procedures for components and output actuators incorporated into the SCC system that are not covered in this Section.*

Note 2: *The Electronic Feedback Carburetor systems were installed on some models starting in 1981. The computer, which is mounted on the air cleaner, will usually read* **Electronic Fuel Control** *system.*

1   The Electronic Feedback Carburetor (EFC) emission system relies on an electronic signal, which is generated by an exhaust gas oxygen sensor, to control a variety of devices and keep emissions within limits **(see illustrations)**. The system works in conjunction with a three-way catalyst to control the levels of carbon monoxide (CO), hydrocarbons (HC) and oxides of nitrogen (NO_x). The EFC system also works in conjunction with the Spark Control Computer (SCC) system. The

13.8 Remove the carburetor nuts (arrows) from the base of the carburetor body (other two nuts not visible)

two systems share certain sensors and output actuators; therefore, diagnosing the EFC system will require a thorough check of all the SCC components. Refer to Chapter 5, Section 6 and Section 11 for complete diagnostic procedures for the SCC system.

2   The system operates in two modes: open loop and closed loop. When the engine is cold, the air/fuel mixture is controlled by the computer in accordance with a program designed in at the time of production. The air/fuel mixture during this time will be richer to allow for proper engine warm-up. When the engine is at operating temperature, the system operates in closed loop and the air/fuel mixture is varied depending on the information supplied by the exhaust gas oxygen sensor.

3   The various types of feedback carburetor systems are separated by the different models of feedback carburetors installed on the engine. Here is a list of the feedback carburetors:

> *Carter BBD*
> *Carter Thermoquad*
> *Holley Model 6145*
> *Holley Model 6280*
> *Rochester Quadrajet*

4   Here is a list of the various sensors and output actuators involved with these feedback carburetor systems:

> *Coolant temperature sensor*
> *Charge air temperature sensor*
> *Detonation (knock) sensor*
> *Oxygen sensor*
> *Magnetic pick-up assembly*
> *Vacuum transducer*
> *Mixture control solenoid (also called duty-cycle solenoid, oxygen feedback solenoid)*
> *SCC computer*
> *Throttle position transducer (1976 through 1979 models)*
> *Throttle control system (1980 through 1989 models)*
> *Carburetor switch*

5   Refer to Section 15 for the diagnostic checks for the feedback carburetor system components.

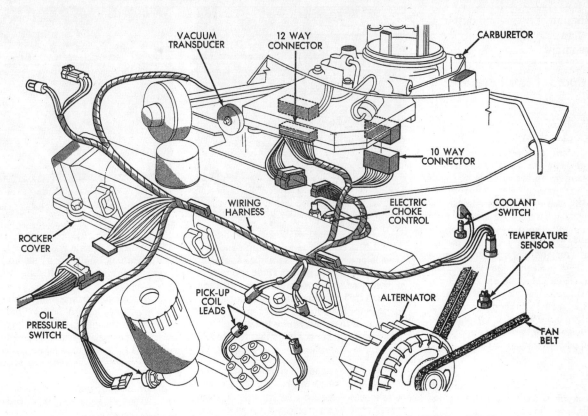

14.1a Typical feedback carburetor system for the six-cylinder engine (1981 model shown)

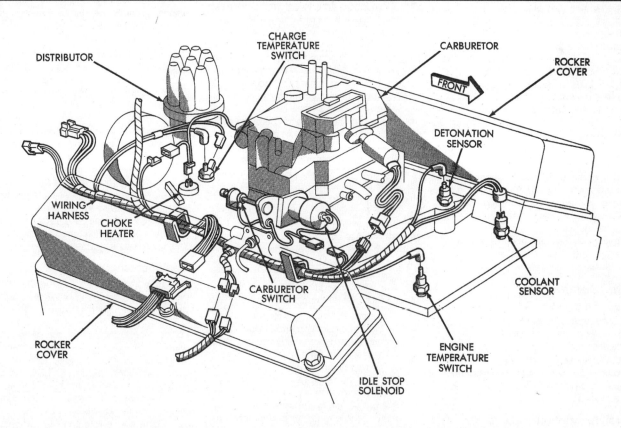

14.1b Typical feedback carburetor system for the V8 engine (1981 model shown)

15 Information sensors and output actuators - check and replacement

Engine coolant temperature sensor

Refer to illustration 15.2

General description

1   The coolant temperature sensor on 1981 through 1987 models is a thermistor (a resistor which varies the value of its resistance in accordance with temperature changes). The change in the resistance values will directly affect the voltage signal from the computer. As the sensor temperature DECREASES, the resistance values will DECREASE. As the sensor temperature INCREASES, the resistance values will INCREASE. **Note:** *On some early feedback carburetor versions the coolant temperature sensor acts as a switch. When the engine is cold, the ohmmeter indicates continuity and when the engine is warmed up, the ohmmeter indicates no continuity (open circuit).*

Check

2   To check the sensor on 1981 through 1987 models, install the probes of the ohmmeter onto the terminals of the coolant temperature sensor while it is completely cold (50 to 80-degrees F = 500 to 1,000 ohms). Next, start the engine and warm it up until it reaches operating temperature **(see illustration)**. The resistance should be higher (180 to 200-degrees F = 1,200 to 1,400 ohms). **Note:** *Access to the coolant temperature sensor makes it difficult to position electrical probes on the terminals. If necessary, remove the sensor and perform the tests in a pan of heated water to simulate the conditions.*

Replacement

3   To remove the sensor, release the locking tab, unplug the electrical connector, then carefully unscrew the sensor. **Caution:** *Handle the coolant sensor with care. Damage to this sensor will affect the operation of the entire feedback carburetor system.*

4   Before installing the new sensor, wrap the threads with Teflon sealing tape to prevent leakage and thread corrosion.

5   Installation is the reverse of removal.

Oxygen sensor

Refer to illustration 15.8

Note 1: *The oxygen sensor is used only on 1981 and later models that are equipped with a feedback carburetor.*

Note 2: *A faulty oxygen sensor is one of the most common causes of high fuel consumption and poor driveability on a vehicle equipped with a feedback carburetor.*

General description and check

6   The oxygen sensor, which is located in the exhaust manifold, monitors the oxygen

15.2 The coolant temperature sensor resistance will increase as the engine temperature increases (1986 models shown). Federal models equipped with the 318 CID (5.2L) and the two-barrel carburetor have the coolant temperature sensor installed where the charge temperature sensor is usually located.

content of the exhaust gas stream. The oxygen content in the exhaust reacts with the oxygen sensor to produce a voltage output which varies from 0.1-volt (high oxygen, lean mixture) to 0.9-volts (low oxygen, rich mixture). The electronic control unit constantly monitors this variable voltage output to determine the ratio of oxygen to fuel in the mixture. The electronic control unit alters the air/fuel mixture ratio by controlling the pulse width (open time) of the mixture control solenoid. A mixture ratio of 14.7 parts air to 1 part fuel is the ideal mixture ratio for minimizing exhaust emissions, thus allowing the catalytic converter to operate at maximum efficiency. It is this ratio of 14.7 to 1 which the electronic control unit and the oxygen sensor attempt to maintain at all times.

7   The oxygen sensor produces no voltage when it is below its normal operating temperature of about 600-degrees F. During this initial period before warm-up, the system operates in open loop mode.

8   If the engine reaches normal operating temperature and/or has been running for two or more minutes, and if the oxygen sensor is producing a steady signal voltage between 0.1 and 0.9-volts **(see illustration)**, the oxygen sensor is working properly.

9   The proper operation of the oxygen sensor depends on four conditions:

a) **Electrical** - *The low voltages generated by the sensor depend upon good, clean connections which should be checked whenever a malfunction of the sensor is suspected or indicated.*

b) **Outside air supply** - *The sensor is designed to allow air circulation to the internal portion of the sensor. Whenever the sensor is removed and installed or replaced, make sure the air passages are not restricted.*

15.8 With the engine warmed to operating temperature, check for a milli-volt signal from the oxygen sensor

c) **Proper operating temperature** - *The electronic control unit will not react to the sensor signal until the sensor reaches approximately 600-degrees F. This factor must be taken into consideration when evaluating the performance of the sensor.*

d) **Unleaded fuel** - *The use of unleaded fuel is essential for proper operation of the sensor. Make sure the fuel you are using is of this type.*

10   In addition to observing the above conditions, special care must be taken whenever the sensor is serviced.

a) *The oxygen sensor has a permanently attached pigtail and electrical connector which should not be removed from the sensor. Damage or removal of the pigtail or electrical connector can adversely affect operation of the sensor.*

b) *Grease, dirt and other contaminants should be kept away from the electrical connector and the louvered end of the sensor.*

c) *Do not use cleaning solvents of any kind on the oxygen sensor.*

d) *Do not drop or roughly handle the sensor.*

e) *The silicone boot must be installed in the correct position to prevent the boot from being melted and to allow the sensor to operate properly.*

Replacement

Note: *Because it is installed in the exhaust manifold or pipe, which contracts when cool, the oxygen sensor may be very difficult to loosen when the engine is cold. Rather than risk damage to the sensor (assuming you are planning to reuse it in another manifold or pipe), start and run the engine for a minute or two, then shut it off. Be careful not to burn yourself during the following procedure.*

11   Disconnect the cable from the negative terminal of the battery.

12   Raise the vehicle and place it securely on jackstands.

13   Carefully disconnect the electrical con-

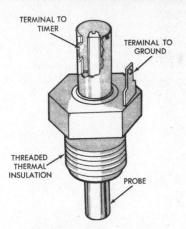

15.20 Diagram of a typical charge temperature sensor

15.24 The mixture-control solenoid connector location on the Rochester Quadrajet - the connector on all models is located on the carburetor air horn

15.29 Tap on the intake manifold near the detonation sensor and confirm that the timing retards

nector from the sensor.

14   Carefully unscrew the sensor from the exhaust manifold. **Caution:** *Excessive force may damage the threads.*

15   Anti-seize compound must be used on the threads of the sensor to facilitate future removal. The threads of new sensors will already be coated with this compound, but if an old sensor is removed and reinstalled, recoat the threads.

16   Install the sensor and tighten it securely.

17   Reconnect the electrical connector of the pigtail lead to the main engine wiring harness.

18   Lower the vehicle and reconnect the cable to the negative terminal of the battery.

Charge temperature switch (sensor)

Check

Refer to illustration 15.20

19   The charge temperature switch or sensor, used on 1981 through 1987 models with a feedback carburetor, monitors the air temperature of the air intake charge. It is located in the intake manifold. To test the charge temperature switch, turn the ignition switch off and disconnect the lead wire from the charge temperature switch.

20   On models equipped with a charge temperature switch, connect one ohmmeter lead to a good ground or the terminal ground **(see illustration)** and the other lead to the center terminal of the switch.

21   On a cold engine, continuity should be indicated. If the charge temperature switch is cooler than 60-degrees F, resistance should be less than 100 ohms. If there is no continuity, the switch is defective and should be replaced. On a hot engine (normal operating temperature), the ohmmeter should indicate no continuity. If continuity exists, the charge temperature switch is defective and should be replaced.

Component replacement

22   The charge temperature switch (sensor) can be replaced by simply unthreading the component and installing the replacement.

The charge temperature switch/sensor is located on the intake manifold, near the carburetor. To prevent a possible vacuum leak, put thread sealing compound on the threads of the new sensor before screwing it in.

Mixture control solenoid

Check

Refer to illustration 15.24

23   The function of the mixture control solenoid is to provide limited regulation of the fuel-air ratio of a feedback carburetor in response to the electronic signals sent by the computer. This is accomplished by metering the main fuel jets in the carburetor with the use of a plunger that extends into the fuel passage and shuts the fuel off (lean) and on (rich), allowing the fuel/air mixture ratio to change. By controlling the duration of this voltage signal, the ratio of power ON-time versus the power OFF-time is called the duty cycle. The mixture control solenoid is of a slightly different design and mounted in different locations on the carburetor, depending upon the type (Quadrajet, Carter BBD, ThermoQuad, etc.).

24   Maintain an engine speed of 1500 rpm. Disconnect the mixture control solenoid connector from the solenoid **(see illustration)**. Average engine speed should increase a minimum of 50 rpm.

25   Reconnect the feedback solenoid connector. The engine speed should slowly return to 1,500 rpm. If rpm does not change as specified, there's a problem in the feedback carburetor system, possibly the mixture-control solenoid, although the oxygen sensor is usually the cause. First check all electrical connections carefully, then check the oxygen sensor. If the oxygen sensor is OK and all connections are OK, the mixture-control solenoid is probably bad or out of adjustment. If you have a Rochester Quadrajet, the Lean and Rich stops may be out of adjustment; see the *Haynes Rochester Carburetor Manual* for the adjustment procedure.

Component replacement

26   The location and the removal procedure differs according to the type of carburetor on the particular feedback carburetor system. Removal and installation will require partial dismantling of the carburetor: If you have a Holley 6280 or 6145, refer to the *Haynes Holley Carburetor Manual* for the procedure. If you have a Rochester Quadrajet, refer to the *Haynes Rochester Carburetor Manual* for the procedure.

Detonation sensor

Check

Refer to illustration 15.29

27   Connect a timing light to the engine, start the engine and run it on the second highest step of the fast-idle cam (at least 1,200 rpm).

28   Connect a hand-held vacuum pump to the vacuum transducer (attached to the computer on the air cleaner) and apply 16 inches of vacuum.

29   Tap lightly on the manifold near the detonation sensor with the end of a wrench or screwdriver and confirm that the ignition retards not more than 11-degrees **(see illustration)**.

30   If there is not any noticeable change in the timing, replace the sensor.

Replacement

31   Disconnect the electrical connector from the sensor and unscrew the sensor from the intake manifold.

32   Installation is the reverse of removal.

16   Exhaust system - servicing and general information

Refer to illustration 16.1

Warning: *Inspection and repair of exhaust system components should be done only after enough time has elapsed after driving the vehicle to allow the system components*

4A

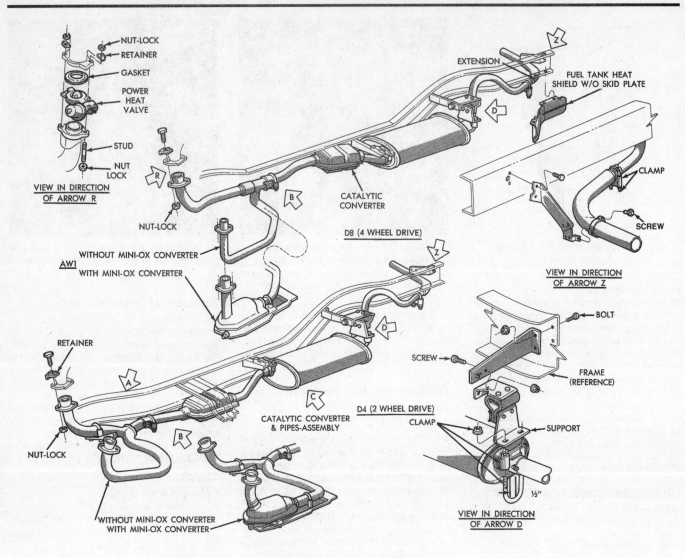

16.1 Typical exhaust system

to cool completely. Also, when working under the vehicle, make sure it is securely supported on jackstands

1   The exhaust system consists of the exhaust manifold(s), the catalytic converter(s), the muffler, the tailpipe and all connecting pipes, brackets, hangers and clamps **(see illustration)**. The exhaust system is attached to the body with mounting brackets and rubber hangers. If any of the parts are improperly installed, excessive noise and vibration will be transmitted to the body.

2   Conduct regular inspections of the exhaust system to keep it safe and quiet. Look for any damaged or bent parts, open seams, holes, loose connections, excessive corrosion or other defects which could allow exhaust fumes to enter the vehicle. Deteriorated exhaust system components should

not be repaired; they should be replaced with new parts.

3   If the exhaust system components are extremely corroded or rusted together, welding equipment will probably be required to remove them. The convenient way to accomplish this is to have a muffler repair shop remove the corroded sections with a cutting torch. If, however, you want to save money by doing it yourself (and you don't have a welding outfit with a cutting torch), simply cut off the old components with a hacksaw. If you have compressed air, special pneumatic cutting chisels can also be used. If you do decide to tackle the job at home, be sure to wear safety goggles to protect your eyes from metal chips and work gloves to protect your hands.

4   Here are some simple guidelines to fol-

low when repairing the exhaust system:

a)   Work from the back to the front when removing exhaust system components.

b)   Apply penetrating oil to the exhaust system component fasteners to make them easier to remove.

c)   Use new gaskets, hangers and clamps when installing exhaust system components.

d)   Apply anti-seize compound to the threads of all exhaust system fasteners during reassembly.

e)   Be sure to allow sufficient clearance between newly installed parts and all points on the underbody to avoid overheating the floor pan and possibly damaging the interior carpet and insulation. Pay particularly close attention to the catalytic converter and heat shield.

Chapter 4 Part B
Fuel and exhaust systems - fuel-injected engines

Contents

Specifications

General

Fuel pressure
TBI systems
With regulator vacuum hose attached.................... 14 to 15 psi
With regulator vacuum hose disconnected.............. 18 to 20 psi
MPFI systems
All except 1993 models with the 5.9L engine
With regulator vacuum hose attached.................... 31 to 33 psi
With regulator vacuum hose disconnected.............. 39 to 41 psi
1993 models with the 5.9L engine......................... 35 to 45 psi
Fuel injector resistance
TBI systems.. 2.0 ohms
MPFI systems... 12.5 ohms

Torque specifications

Throttle body mounting nuts............................... 16 ft-lbs
Fuel rail mounting nuts.................................... 108 in-lbs

1 General information

1988 through 1992 models are equipped with the Throttle Body Injection (TBI) system. This system incorporates two injectors mounted centrally in the throttle body that direct fuel into the intake manifold.

1992 and later models are equipped with a Sequential Multi Port Fuel Injection (MPFI) system. This system uses timed impulses to sequentially inject the fuel directly into the intake ports of each cylinder. The injectors are controlled by the Powertrain Control Module (PCM). The PCM monitors various engine parameters and delivers the exact amount of fuel, in the correct sequence, into the intake ports.

All models are equipped with an electric fuel pump, mounted in the fuel tank. It is necessary to remove the fuel tank for access to the fuel pump. The fuel level sending unit is mounted with the fuel pump and it must be removed from the fuel tank in the same manner.

The exhaust system consists of exhaust manifold(s), a catalytic converter, an exhaust pipe and a muffler. Each of these components is replaceable. For further information regarding the catalytic converter, refer to Chapter 6.

2 Fuel pressure relief procedure

Warning: *Gasoline is extremely flammable, so take extra precautions when you work on any part of the fuel system. Don't smoke or allow open flames or bare light bulbs near the work area, and don't work in a garage where a natural gas-type appliance (such as a water heater or a clothes dryer) is present. Since gasoline is carcinogenic, wear latex gloves when there's a possibility of being exposed to fuel, and, if you spill any fuel on your skin, rinse it off immediately with soap and water. Mop up any spills immediately and do not store fuel-soaked rags where they could* *ignite. The fuel system is under constant pressure, so, if any fuel lines are to be disconnected, the fuel pressure in the system must be relieved first. When you perform any kind of work on the fuel system, wear safety glasses and have a Class B type fire extinguisher on hand.*

1   Detach the cable from the negative battery terminal. Unscrew the fuel filler cap to relieve pressure built up in the fuel tank. Remove the engine cover (see Chapter 2B, Section 2).

TBI systems

Refer to illustration 2.3

2   Unplug the electrical connector from the fuel injector throttle body.

3   Ground the number one injector terminal (-) with a jumper wire securely attached to the battery negative terminal **(see illustration)**.

4   Connect a jumper wire from the battery positive (+) terminal to the other terminal (number 2) on the electrical connector. This will open the injector and depressurize the fuel system. **Caution:** *Do not energize the injectors for more than five seconds. It is recommended that the injectors be bled in several spurts of one to two seconds to make sure the injector is not damaged. The fuel pressure can be heard or observed escaping into the throttle body or combustion chamber. When the sound is no longer heard, the system is depressurized.*

MPFI systems

Refer to illustrations 2.7

5   Remove the cap from the fuel pressure test port located on the fuel rail.

6   Place shop towels under and around the test port to absorb fuel when the fuel pressure is relieved.

7   Using a small screwdriver or pin punch and wearing eye protection, cover the screwdriver and test port with a rag and push the test port valve in to relieve the fuel pressure **(see illustration)**. Absorb the spilled fuel with the shop towels.

8   Install the cap onto the fuel pressure test port.

---

3   Fuel pump/fuel pressure - check

Warning: *Gasoline is extremely flammable, so take extra precautions when you work on any part of the fuel system. Don't smoke or allow open flames or bare light bulbs near the work area, and don't work in a garage where a natural gas-type appliance (such as a water heater or a clothes dryer) is present. Since gasoline is carcinogenic, wear latex gloves when there's a possibility of being exposed to fuel, and, if you spill any fuel on your skin, rinse it off immediately with soap and water. Mop up any spills immediately and do not store fuel-soaked rags where they could ignite. The fuel system is under constant pressure, so, if any fuel lines are to be disconnected, the fuel pressure in the system must be relieved first (see Section 2*

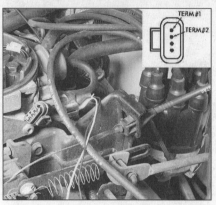

2.3 To relieve the fuel pressure on TBI systems, ground the injector terminal #1 and apply battery voltage to terminal #2 (be careful not to cause a direct short or energize the injector for more than five seconds)

for more information). When you perform any kind of work on the fuel system, wear safety glasses and have a Class B type fire extinguisher on hand.

Note 1: *1990 models with the 5.2L engine are equipped with a throttle body that may cause hard starting when hot and then idle rough. This could be caused by a defective throttle body. If the throttle body date code is 2409 (240th day of 1989) or earlier, remove the throttle body and install a late style unit that is built with a raised injector port that does not damage the injector seal. Consult a dealer parts department for additional information.*

Note 2: *1993 and later 5.9L (360 cu. in.) engines and all 1994 and later engines are equipped with a returnless fuel system. The fuel filter and fuel pressure regulator are combined into a single unit mounted on top of the electric fuel pump. If fuel pressure exceeds 45 psi, an internal diaphragm closes and excess fuel is routed back into the pump module and returned to the fuel tank.*

Preliminary check

1   If you suspect insufficient fuel delivery, first inspect all fuel lines to ensure that the problem is not simply a leak in a line.

3.4a Use a T-fitting and install a fuel pressure gauge into the inlet line (TBI systems)

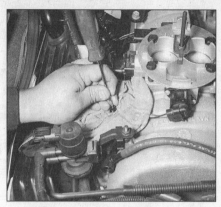

2.7 To relieve fuel pressure on MPFI systems, use a small screwdriver or punch to depress the Schrader valve within the fuel pump test port. Be sure to catch any residual fuel using a rag (cover the screwdriver tip and valve with a rag to prevent fuel spray)

Fuel pump operational check

Note: *On all models, the fuel pump is located inside the fuel tank (see Section 5).*

2   Set the parking brake and have an assistant turn the ignition switch to the ON position while you listen to the fuel pump (inside the fuel tank). You should hear a whirring sound, lasting for a couple of seconds. Start the engine. The whirring sound should now be continuous (although harder to hear with the engine running). If there is no whirring sound, either the fuel pump, the fuel pump relay or the Automatic Shutdown (ASD) relay or circuit is defective (proceed to Step 10).

Pressure check

Refer to illustrations 3.4a, 3.4b, 3.5a, 3.5b, 3.5c, 3.5d, 3.6 and 3.7

3   Relieve the fuel pressure (see Section 2).

4   Attach a fuel pressure gauge as follows:

a) *On TBI systems, disconnect the fuel line from the throttle body and, using a T-fitting, connect a fuel pressure gauge between the fuel inlet hose and the throttle body* **(see illustration)**.

b) *On MPFI systems, remove the cap from the fuel pressure test port located on the*

3.4b Connect the fuel pressure gauge to the test port on the fuel rail (MPFI systems)

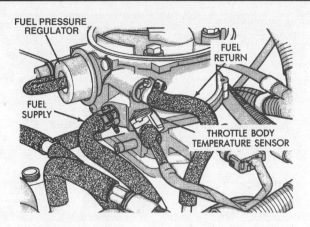

3.5a Location of the fuel pressure regulator on the TBI system

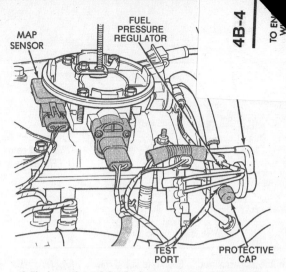

3.5b Location of the fuel pressure regulator on the
3.9L V6 engine (MPFI system)

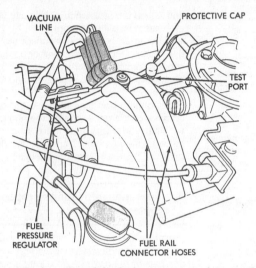

3.5c Location of the fuel pressure regulator on the
5.2L V8 engine (MPFI system)

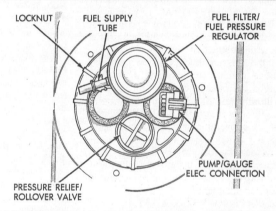

3.5d Location of the fuel pressure regulator on 1993 and later
5.9L engines and all 1994 and later engines

4B

fuel rail **(see illustration)** *and attach a fuel pressure gauge.*

5   Start the engine and check the pressure on the gauge, comparing your reading with the pressure listed in this Chapter's Specifications. Now, detach the vacuum hose from the fuel pressure regulator:

3.6 Pinch the return line and observe the
fuel pressure increase as the pressure
builds up when the fuel is not allowed
to return to the fuel tank

a) *On TBI systems, the fuel pressure regulator is attached to the side of the throttle body* **(see illustration)**.
b) *On MPFI systems, the fuel pressure regulator is attached to the fuel rail* **(see illustrations)**.

With the engine idling, measure the fuel pressure. It should be as listed in this Chapter's Specifications. Reconnect the vacuum hose. **Note:** *The fuel pressure regulator on 1993 and later 5.9L engines and 1994 and later V6 and 5.2L engines is located on the fuel pump/sending unit module along with the fuel filter* **(see illustration)**. *Fuel pressure on these models is not controlled by engine vacuum. Therefore, it will only be possible to check the regulated fuel pressure on these models.*
6   If the fuel pressure is LOW, pinch the fuel return line shut and watch the gauge **(see illustration)**. If the pressure doesn't rise, the fuel pump is defective or there is a leak or restriction in the fuel feed line (replace the fuel filter and repeat the test). If the pressure rises sharply, replace the fuel pressure regulator (see Sections 11 and 12). **Note:** *On MPFI systems it will be necessary to fabricate special rubber fuel lines to be installed*

between the fuel rail and the fuel return line. The braided metal lines cannot be collapsed for testing purposes.
7   If the pressure is within specifications when the vacuum hose is connected to the pressure regulator but does not increase when the vacuum hose is disconnected, check for vacuum at the hose **(see illustration)**. If there is vacuum present, pinch the

3.7 Make sure vacuum is reaching the
fuel pressure regulator

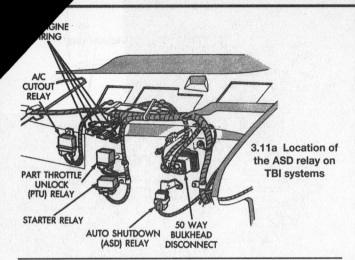

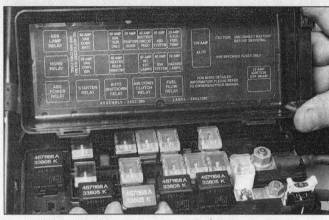

3.11a Location of the ASD relay on TBI systems

3.11b Location of the ASD relay on MPFI systems

return line as described in Step 6 to see if the fuel pump is capable of delivering the required pressure. If the pressure rises to the "vacuum hose disconnected" pressure, replace the pressure regulator (see Sections 11 [TBI] and 12 [MPFI]). If there is no vacuum present at the hose, check the hose for a break or an obstruction.

8   If the pressure at idle was too high (with the hose connected), connect a vacuum gauge to the hose and check for vacuum **(see illustration 3.7)**. If vacuum is present, relieve the fuel pressure (see Section 2), disconnect the fuel return line and blow through it to check for a blockage. If there is no blockage, replace the fuel pressure regulator (see Sections 11 and 12). If there is no vacuum present at the hose, check the hose for a break or an obstruction.

9   If there are no problems with any of the above-listed components, check the fuel pump (see below).

Fuel pump check

10   Working on the power side of the fuel pump electrical connector, check for battery voltage with the ignition key ON (engine not running). Battery voltage should be present. If voltage is available, replace the fuel pump (see Section 5). If no voltage is available, check the main relays (see below).

Main relays check

Refer to illustration 3.11a, 3.11b, 3.12, 3.14 and 3.15

Note 1: *The Automatic Shutdown (ASD) relay and the fuel pump relay must both be tested to insure proper fuel pump operation. Testing procedures for the ASD relay and the fuel pump relay are identical.*

Note 2: *Early style fuel pump relays and ASD relays are equipped with only four terminals. Have the relay checked at a dealer service department or qualified automotive parts store.*

11   To test a relay, first remove it from its location in the engine compartment **(see illustrations)**.

12   Turn the ignition switch to the ON position and verify that there is battery voltage present at the connector **(see illustration)**.

13   If there is no voltage, check the fuel pump fuse. If voltage is not present at the fuse, check the fuel pump circuit.

14   Using an ohmmeter, connect the positive probe (+) to terminal number 86 of the relay and the negative probe (-) to terminal number 85 and check the resistance. There should be approximately 75 ohms resistance **(see illustration)**.

15   Using an ohmmeter, connect the positive probe (+) to terminal number 30 and the negative probe (-) to terminal number 87A

and check the resistance. Continuity should be present **(see illustration)**.

16   Using an ohmmeter, connect the positive probe (+) to terminal number 30 and the negative probe (-) to terminal number 87 and check the resistance. There should be no continuity.

17   Connect battery voltage to the no. 86 terminal of the relay, ground the number 85 terminal and verify that there's continuity between the number 87 and number 30 terminals. If there isn't, replace the relay.

4   Fuel lines and fittings - repair and replacement

Warning: *Gasoline is extremely flammable, so take extra precautions when you work on any part of the fuel system. Don't smoke or allow open flames or bare light bulbs near the work area, and don't work in a garage where a natural gas-type appliance (such as a water heater or a clothes dryer) is present. Since gasoline is carcinogenic, wear latex gloves when there's a possibility of being exposed to fuel, and, if you spill any fuel on your skin, rinse it off immediately with soap and water. Mop up any spills immediately and do not store fuel-soaked rags where they could ignite. The fuel system is under constant*

3.12 Check for battery voltage on the ASD relay electrical connector

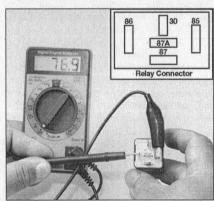

3.14 Check the resistance between terminals number 86 and number 85. It should be approximately 75 ohms

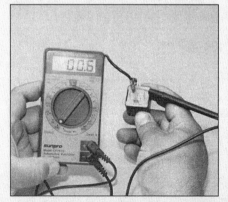

3.15 Also check the resistance between terminals number 87A and 30. Continuity should exist

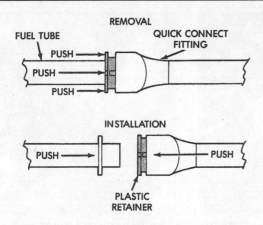

4.11a On plastic ring type fittings, push the tube IN while depressing the plastic retainer ring into the fitting

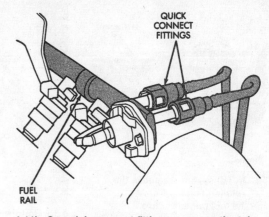

4.11b On quick-connect fittings, squeeze the tabs against the fuel line and pull the line off

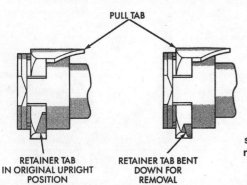

4.11c On TBI systems, bend the retaining tab down with your fingers

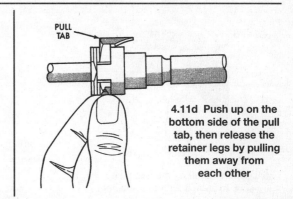

4.11d Push up on the bottom side of the pull tab, then release the retainer legs by pulling them away from each other

pressure, so, if any fuel lines are to be disconnected, the fuel pressure in the system must be relieved first (see Section 2). When you perform any kind of work on the fuel system, wear safety glasses and have a Class B type fire extinguisher on hand.

1   Always relieve the fuel pressure before servicing fuel lines or fittings (see Section 2).

2   The fuel feed, return and vapor lines extend from the fuel tank to the engine compartment. The lines are secured to the underbody with clip and screw assemblies. These lines must be occasionally inspected for leaks, kinks and dents.

3   If evidence of dirt is found in the system or fuel filter during disassembly, the line should be disconnected and blown out. Check the fuel strainer on the fuel gauge sending unit (see Section 5) for damage and deterioration.

4   If replacement of a fuel line or emission line is called for, use tubes/hoses meeting Chrysler specification or equivalent.

Steel tubing

5   Don't use copper or aluminum tubing to replace steel tubing. These materials cannot withstand normal vehicle vibration.

6   Because fuel lines used on fuel-injected vehicles are under high pressure, they require special consideration.

7   Some fuel lines have threaded fittings with O-rings. Any time the fittings are loos-

ened to service or replace components:

a) Use a backup wrench while loosening and tightening the fittings.

b) Check all O-rings for cuts, cracks and deterioration. Replace any that appear hardened, worn or damaged.

c) If the lines are replaced, always use original equipment parts, or parts that meet the original equipment standards specified in this Section.

Flexible hose

Warning: *Use only original equipment replacement hoses or their equivalent. Others may fail from the high pressures of this system.*

8   Don't route fuel hose within four inches of any part of the exhaust system or within ten inches of the catalytic converter. Metal lines and rubber hoses must never be allowed to chafe against the frame. A minimum of 1/4-inch clearance must be maintained around a line or hose to prevent contact with the frame.

Removal and installation

Refer to illustrations 4.11a, 4.11b, 4.11c, 4.11d and 4.11e

9   Relieve the fuel pressure.

10   Remove all fasteners attaching the lines to the vehicle body.

11   There are various methods depending upon the type of quick-disconnect fitting on the fuel line (**see illustrations**). Care-

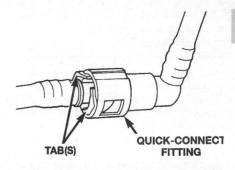

4.11e Plastic two-tab quick-disconnect fitting - squeeze the tabs to release them, then pull the fitting apart

fully remove the fuel lines from the chassis.

Warning: *The plastic ring type fittings are not serviced separately. Do not attempt to service these types of fuel lines in the event the clip or line becomes damaged. Replace the entire fuel line as an assembly.*

12   Installation is the reverse of removal. Be sure to use new O-rings at the threaded fittings (if equipped).

Repair

13   In the event of any fuel line damage (metal or flexible lines) it is necessary to replace the damaged lines with factory replacement parts. Others may fail from the high pressures of this system.

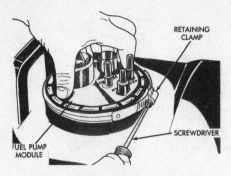

5.5a On TBI systems, remove the retaining clamp from the fuel pump module

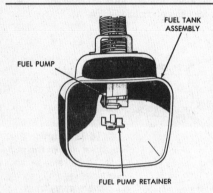

5.8 When installing the fuel pump, be sure to align it with the retainer at the bottom of the tank

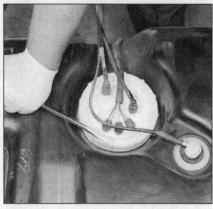

5.5b On threaded type access covers, use a screwdriver or blunt punch and carefully tap the cover counterclockwise

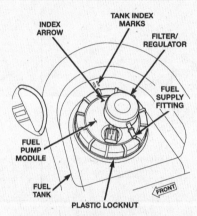

5.12 Turn the plastic locknut counterclockwise to remove it by hand; if the locknut is difficult to turn, use a large pair of adjustable pliers to loosen it. Note the index marks and arrow for correct alignment

5.6 Be careful not to damage the float assembly for the fuel level sending unit while lifting the assembly out of the fuel tank

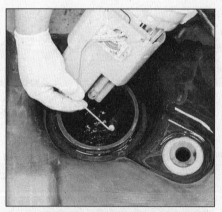

5.13 Carefully remove the fuel pump module from the fuel tank

5   Fuel pump - removal and installation

Refer to illustrations 5.5a, 5.5b, 5.6 and 5.8
Warning: *Gasoline is extremely flammable, so take extra precautions when you work on any part of the fuel system. Don't smoke or allow open flames or bare light bulbs near the work area, and don't work in a garage where a natural gas-type appliance (such as a water heater or a clothes dryer) is present. Since gasoline is carcinogenic, wear latex gloves when there's a possibility of being exposed to fuel, and, if you spill any fuel on your skin, rinse it off immediately with soap and water. Mop up any spills immediately and do not store fuel-soaked rags where they could ignite. The fuel system is under constant pressure, so, if any fuel lines are to be disconnected, the fuel pressure in the system must be relieved first (see Section 2) for more information). When you perform any kind of work on the fuel system, wear safety glasses and have a Class B type fire extinguisher on hand.*

1971 through 1997 models

1   Detach the cable from the negative battery terminal.
2   Relieve the fuel system pressure (see Section 2).
3   Remove the fuel tank (see Chapter 4A).
4   Remove any vent lines or fuel hoses

from the fuel pump access cover.
5   If you're working on a TBI system, loosen the retaining clamp securing the fuel pump module to the tank **(see illustration)**. If you're working on an MPFI system, carefully tap the access cover in a counterclockwise direction **(see illustration)** using a screwdriver. Remove the cover from the assembly.
6   Remove the fuel pump module from the tank **(see illustration)**.
7   In the event of fuel pump failure, the fuel pump/fuel level sending unit must be replaced as a single part. This unit is non-serviceable.
8   Installation is the reverse of removal. Be sure to use a new gasket on the cover plate flange. Be sure to align the module with the retainer at the bottom of the fuel tank on early models with TBI **(see illustration)**.

1998 and later models
Refer to illustrations 5.12 and 5.13
9   Relieve the fuel system pressure (see Section 2).
10   Disconnect the cable from the negative battery terminal.
11   Remove the fuel tank from the vehicle

(see Section 4, Chapter 4A).
12   Turn the plastic locknut counterclockwise to remove it **(see illustration)**. If the locknut is difficult to turn, use a large pair of adjustable pliers to loosen it.
13   Remove the fuel pump module from the fuel tank **(see illustration)**. Angle the assembly slightly to avoid damaging the fuel level sending unit float.
14   In the event of fuel pump failure, the fuel pump/fuel level sending unit must replaced as a single unit. This unit is non-serviceable.
15   Installation is the reverse of removal. Be sure to install a new gasket and make sure it remains in place. Carefully rotate the assembly until the index arrow on the pump module is aligned between the index marks on the fuel tank **(see illustration 5.12)**. Tighten the locknut securely and then carefully rotate the fuel filter/pressure regulator until it is pointed toward the drivers side of the vehicle.

6   Fuel level sending unit - check and replacement

Refer to illustrations 6.2, 6.5 and 6.8
Warning: *Gasoline is extremely flammable, so take extra precautions when you work on any part of the fuel system. Don't smoke or*

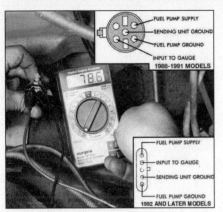

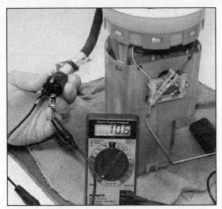

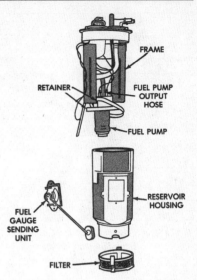

6.2 Working under the vehicle, disconnect the fuel pump/fuel level sending unit electrical connector and check the resistance of the fuel level sending unit by probing the correct wire terminals

6.5 Measure the resistance of the fuel level sending unit with the float raised (full tank) and then with the float near the bottom (empty)

6.8 Exploded view of the fuel pump module components

allow open flames or bare light bulbs near the work area, and don't work in a garage where a natural gas-type appliance (such as a water heater or a clothes dryer) is present. Since gasoline is carcinogenic, wear latex gloves when there's a possibility of being exposed to fuel, and, if you spill any fuel on your skin, rinse it off immediately with soap and water. Mop up any spills immediately and do not store fuel-soaked rags where they could ignite. The fuel system is under constant pressure, so, if any fuel lines are to be disconnected, the fuel pressure in the system must be relieved first (see Section 2). When you perform any kind of work on the fuel system, wear safety glasses and have a Class B type fire extinguisher on hand.

1971 through 1997 models
Check

1   Block the front wheels to prevent the vehicle from rolling, then raise the rear of the vehicle and support it securely on jackstands.
2   Position the probes of an ohmmeter onto the terminals of the fuel level sending unit electrical connector **(see illustration)** and check for resistance.
3   First, check the resistance of the send-

ing unit with the fuel tank completely full. The resistance of the sending unit should be about 120 ohms.
4   Wait until the tank is nearly empty and check the resistance of the unit again. The resistance should be 10 ohms.
5   If the readings are incorrect or there is very little change in resistance as the float travels from full to empty, replace the fuel pump/fuel level sending unit assembly. **Note:** *Another check for the fuel level sending unit is to remove the unit (see Steps 6 through 10) and check the resistance while moving the float from full (arm at highest point of travel) to empty (arm at lowest point of travel* **(see illustration)**.

Replacement

6   Remove the fuel tank (see Chapter 4A).
7   Remove the fuel level sending unit/fuel pump access cover **(see illustrations 5.5a and 5.5b)** and lift the access cover from the fuel tank.
8   Lift the fuel pump/fuel level sending unit from the tank **(see illustration)**. Be careful not to damage the float arm.
9   Installation is the reverse of removal. Be sure to use a new gasket under the sealing flange.

1998 and later models
Check
Refer to illustration 6.12

10   Remove the fuel tank and fuel pump module (see Section 5).
11   Connect the probes of an ohmmeter to the terminals of the fuel pump electrical connector.
12   Position the float in the down (empty) position (see illustration). The resistance of the sending unit should be about 20 ohms.
13   Move the float up to the full position. The resistance should be about 220 ohms.
14   If the readings are incorrect or there is very little change in resistance as the float travels from full to empty, replace the fuel pump/fuel level sending unit assembly.

Replacement
Refer to illustration 6.16, 6.17a, 6.17b and 6.18

15   Remove the fuel tank and fuel pump module (see Section 5).
16   Disconnect the 4-way electrical connector inside the module **(see illustration)**.

4B

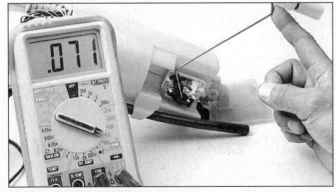

6.12 Measure the resistance of the fuel level sending unit with the float in the down (empty tank) position and then with the float raised (full tank)

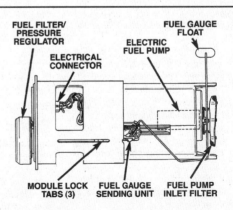

6.16 Fuel pump module components

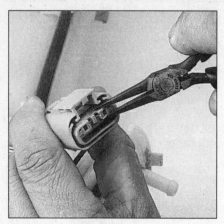

6.17a Carefully remove the locking collar from the connector

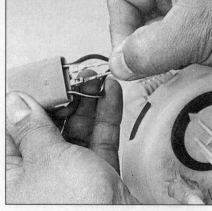

6.17b Pull the two fuel level sending unit wires and terminals out of the electrical connector

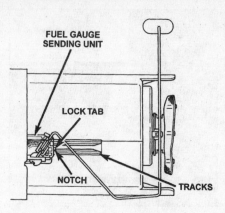

6.18 Pry the lock tab out of the notch and slide the sending unit down the tracks

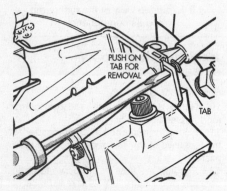

8.3 Push on the tab to disconnect the cable from the bracket

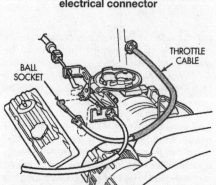

8.4 Typical accelerator cable details on the 5.2L engine

17   Separate the 2 sending unit wires from the 4-way connector as follows **(see illustrations)**:

a) *Remove the locking collar from inside the connector.*

b) *Using a small pick, depress the locking finger on the terminal and push the terminal in.*

c) *Pull the wire and terminal out of the connector from the backside.*

18   Using a small screwdriver or pick, pry the lock tab out of the notch and slide the fuel level sending unit down the tracks **(see illustration)**. Note the routing of the wiring for installation.

19   Slide the level sending unit onto the tracks and engage the lock tab with the notch. Route the wiring as originally installed.

20   Insert the wire terminals into the connector and install the locking collar. Make sure the wire terminals are locked in place by attempting to pull them out from the backside. If necessary carefully bend the locking fingers out before inserting them into the connector.

21   Connect the electrical connector to the fuel pump module and install the module in the fuel tank (see Section 5).

22   The remainder of installation is the reverse of removal.

7   Air cleaner assembly - removal and installation

1   Remove the engine cover (see Chapter 2B, Section 2).

2   Remove the air cleaner cover and filter element (see Chapter 1).

3   Remove the clamps that hold the air intake duct to the air cleaner housing.

4   Remove the bolts that hold the air cleaner housing to the engine compartment.

5   Lift the assembly up and detach it from the fresh air intake duct, then remove it from the engine compartment.

6   Installation is the reverse of removal.

8   Accelerator cable - replacement

Refer to illustrations 8.3, 8.4 and 8.5

Replacement

1   Detach the cable from the negative battery terminal.

2   Using a screwdriver, pry the cable end away from the throttle lever.

3   Push on the tab and release the accelerator cable retainer from the bracket **(see illustration)**.

4   Separate the accelerator cable from the

throttle body assembly **(see illustration)**.

5   Working underneath the dash, detach the cable from the accelerator pedal **(see illustration)**.

6   Pull the grommet from the firewall and pull the cable through the firewall from the engine compartment side.

7   Installation is the reverse of removal.

9   Fuel injection system - general information

Refer to illustrations 9.1a and 9.1b

1988 through 1992 models are equipped with the Throttle Body Injection (TBI) system. This system incorporates two injectors mounted centrally in the throttle body that direct fuel into the intake manifold. Some 1992 and all 1993 and later models are equipped with a Sequential Multi Port Fuel Injection (MPFI) system. This system uses timed impulses to sequentially inject the fuel directly into the intake ports of each cylinder. The Throttle Body Injection (TBI) system and the Multi Port Fuel Injection (MPFI) system **(see illustrations)** consists of three sub-sys-

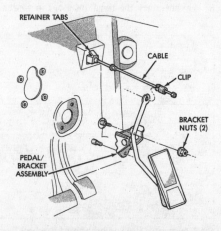

8.5 Exploded view of the accelerator pedal and cable assemblies (typical)

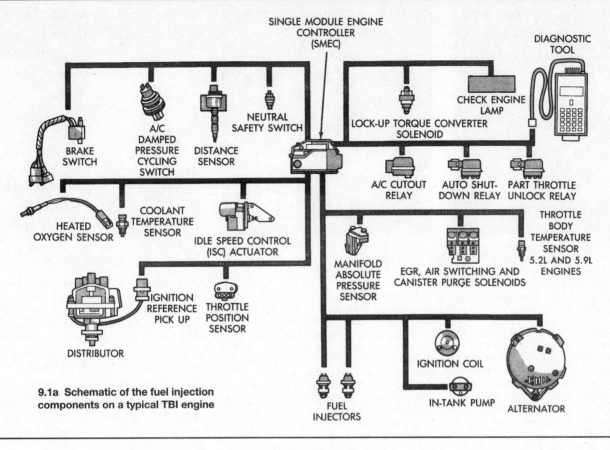

9.1a Schematic of the fuel injection components on a typical TBI engine

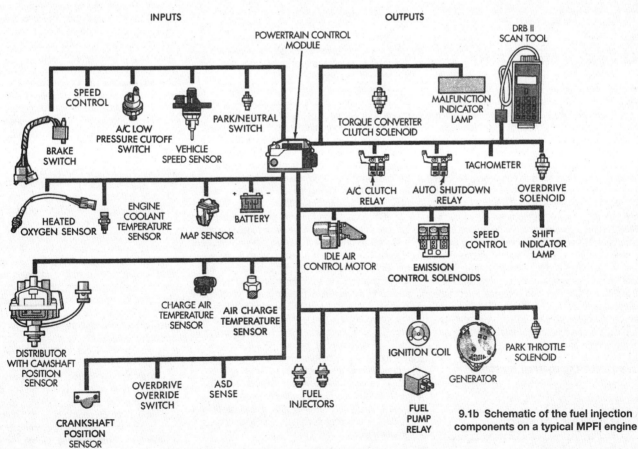

9.1b Schematic of the fuel injection components on a typical MPFI engine

10.7a Use a stethoscope or screwdriver to determine if the injectors are working properly - they should make a steady clicking sound that rises and falls with engine speed changes (TBI system shown)

10.7b Checking the injectors using a stethoscope on an MPFI system

10.8a Measuring the resistance of a throttle body injector

tems: air intake, electronic control and fuel delivery. These systems use a Single Module Engine Controller (SMEC) or a Powertrain Control Module (PCM) along with the sensors (coolant temperature sensor, Throttle Position Sensor (TPS), Manifold Absolute Pressure (MAP) sensor, oxygen sensor, etc.) to determine the proper air/fuel ratio under all operating conditions.

The fuel injection system and the emissions and engine control system are closely interrelated. For additional information, refer to Chapter 6.

Air intake system

The air intake system consists of the air cleaner, the air intake ducts, the throttle body, the idle control system and the intake manifold.

A throttle position sensor is attached to the throttle shaft to monitor changes in the throttle opening.

When the engine is idling, the air/fuel ratio is controlled by the Idle Air Control (IAC) system, which consists of the Single Module Engine Controller (SMEC) or a Powertrain Control Module (PCM) and the Idle Air Control (IAC) motor. The IAC motor is activated by the SMEC/PCM depending upon the running conditions of the engine (air conditioning system, power steering, cold and warm running, etc.). This valve regulates the amount of airflow past the throttle plate and into the intake manifold. The SMEC/PCM receives information from the sensors (vehicle speed, coolant temperature, air conditioning, power steering mode etc.) and adjusts the idle according to the demands of the engine and driver.

Electronic control system

The electronic control system is explained in detail in Chapter 6.

Fuel delivery system

The fuel delivery system consists of

these components: the fuel pump, the pressure regulator, the fuel injectors, the injector resistor and the main relays.

The fuel pump is an in-line, direct drive type. Fuel is drawn through a filter into the pump, flows past the armature through the one-way valve, passes through another filter and is delivered to the injectors. A relief valve prevents excessive pressure build-up by opening in the event of a blockage in the discharge side and allowing fuel to flow from the high to the low pressure side.

The pressure regulator maintains a constant fuel pressure to the injectors. Excess fuel is routed back to the fuel tank through the return line (except on models with a returnless system, in which excess fuel is returned through the fuel pump module [see the Note in Section 3]).

The injectors are solenoid-actuated, constant stroke, pintle types consisting of a solenoid, plunger, needle valve and housing. When current is applied to the solenoid coil, the needle valve raises and pressurized fuel fills the injector housing and squirts out the nozzle. The injection quantity is determined by the length of time the valve is open (the length of time during which current is supplied to the solenoid coils).

Because it determines opening and closing intervals, which in turn determines the air-fuel mixture ratio, injector timing must be quite accurate. To attain the best possible injector response, the current rise time, when voltage is being applied to each injector coil, must be as short as possible. The number of windings in the coil has therefore been reduced to lower the inductance in the coil. However, this creates low coil resistance, which could compromise the durability of the coil. The flow of current in the coil is therefore restricted by a resistor installed in the injector wire harness.

The main relay, which is installed adjacent to the fuse box, is a direct coupler type which contains the relays for the electronic control unit power supply and the fuel pump power supply.

10 Fuel injection system - check

Refer to illustrations 10.7a, 10.7b, 10.8a, 10.8b, 10.9a and 10.9b

Note: *The following procedure is based on the assumption that the fuel pressure is adequate (see Section 3).*

1   Check the ground wire connections on the intake manifold for tightness. Check all wiring harness connectors that are related to the system. Loose connectors and poor grounds can cause many problems that resemble more serious malfunctions.

2   Check to see that the battery is fully charged, as the control unit and sensors depend on an accurate supply voltage in order to properly meter the fuel.

3   Check the air filter element - a dirty or partially blocked filter will severely impede performance and economy (see Chapter 1).

4   If a blown fuse is found, replace it and see if it blows again. If it does, search for a grounded wire in the harness to the fuel pump.

5   Check the air intake duct to the intake manifold for leaks, which will result in an excessively lean mixture. Also check the condition of all vacuum hoses connected to the intake manifold.

6   Remove the air intake duct from the throttle body and check for dirt, carbon or other residue build-up. If it's dirty, clean it with carburetor cleaner and a toothbrush.

7   With the engine running, place an automotive stethoscope against each injector, one at a time, and listen for a clicking sound, indicating operation **(see illustrations)**. If you don't have a stethoscope, place the tip of a screwdriver against the injector and listen through the handle.

8   Unplug the injector electrical connectors and, using an ohmmeter, check the resistance of each injector **(see illustrations)**. Compare the values to the Specifications listed in this Chapter.

9   Install an injector test light ("noid" light) into each injector electrical connector, one at a time **(see illustrations)**. Crank the engine

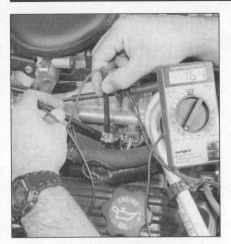

10.8b Measuring the resistance of an injector on an MPFI system

10.9a Install the "noid" light into each TB injector electrical connector and confirm that it blinks when the engine is cranking or running

10.9b Injector test light (noid light) in an MPFI electrical connector

over. The light should flash evenly on each connector. This will test for voltage to the injectors.

10   The remainder of the system checks can be found in the following Sections.

11   Throttle Body Injection (TBI) - component check and replacement

Warning: *Gasoline is extremely flammable, so take extra precautions when you work on any part of the fuel system. Don't smoke or allow open flames or bare light bulbs near the work area, and don't work in a garage where a natural gas-type appliance (such as a water heater or a clothes dryer) is present. Since gasoline is carcinogenic, wear latex gloves when there's a possibility of being exposed to fuel, and, if you spill any fuel on your skin, rinse it off immediately with soap and water. Mop up any spills immediately and do not store fuel-soaked rags where they could ignite. The fuel system is under constant pressure, so, if any fuel lines are to be disconnected, the fuel pressure in the system must be relieved first (see Section 2 for more information). When you perform any kind of work on the fuel system, wear safety glasses and have a Class B type fire extinguisher on hand.* **Note:** *All of the following procedures will require removal of the engine cover (see Chapter 2B, Section 2).*

Fuel injector(s)
Check

1   The fuel injectors are electric solenoids driven by the computer (SMEC). With the engine running, listen for sounds from the injectors with an automotive type stethoscope and verify that each injector sounds as if it is operating normally. If you don't have a stethoscope you can use a long screwdriver, or you can touch the area of the throttle body immediately above the fuel injector with your finger and try to determine if the injector feels

like it is operating smoothly. There should be a noticeable clicking sound, uniform in timing, that rises and falls according to engine speed. If the injector is not operating, or sounds and/or feels erratic, check the injector electrical connector and the wire harness connector. If the connectors are snug, check for voltage to the injector using a special injector harness test light (noid light) (see Section 10). If there is voltage to the injector and it is not operating correctly, replace the injector. Refer to Section 3 for additional testing procedures.

Replacement
Refer to illustrations 11.5, 11.7, 11.8 and 11.9

2   Remove the air cleaner (see Section 7).
3   Relieve the fuel system pressure (see Section 2).
4   Disconnect the negative cable from the battery.
5   Remove the Torx screw holding the injector hold-down clamp. Be careful not to lose the spacer under the clamp **(see illustration)**.

11.5 Remove the Torx screw from the center of the injector hold-down clamp

6   Using a small screwdriver, lift the top off the injector(s).
7   Using a small screwdriver, gently pry the fuel injector(s) from the throttle body **(see illustration)**.

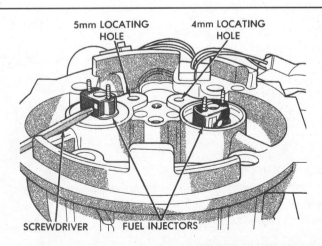

5mm LOCATING HOLE   4mm LOCATING HOLE

SCREWDRIVER   FUEL INJECTORS

11.7 Carefully pry the injector from the throttle body using a small screwdriver

4B

11.8 Make sure the lower O-ring remains attached to the injector

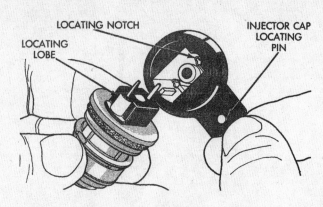

11.9 Align the locating notch in the injector cap with the locating lobe on the injector

8   Make sure the lower O-ring came out of the throttle body with the injector(s) **(see illustration)**. If not, remove it from the injector bore.

9   To reinstall an injector, align the injector cap with the locating notch on the injector **(see illustration)**.

10   Press the injector into the cap until the upper O-ring flange is flush with the lower surface of the cap.

11   Lightly lubricate the upper and lower O-rings with petroleum jelly.

12   Insert the injector and cap into the throttle body and align the cap locating pin with the locating hole in the casting. Note that the left and right injectors have different size locating pins and matching locating holes and will not interchange.

13   Press firmly on the injector caps until the injectors are flush with the casting surface.

14   While aligning the holes in the clamp with the pins on the caps, place the injector hold-down clamp and spacer on the rear of the caps.

15   Since squeezing the O-rings may cause the caps to lift, press firmly on both caps with one hand while installing the clamp screw.

16   The remainder of installation is the reverse of the removal procedure.

Fuel pressure regulator

Refer to illustration 11.23

Check

17   The fuel pressure regulator is a mechanical device that maintains a constant amount of fuel pressure (14.5 psi) across the fuel injector tip. The regulator uses a spring loaded rubber diaphragm to uncover a fuel return port. When fuel pressure reaches the constant, the spring compresses and uncovers the fuel return port. Follow the fuel pressure checks in Section 3 to diagnose the fuel pressure regulator.

Replacement

18   Remove the air cleaner assembly (see Section 7).

19   Release the fuel system pressure (see Section 2).

20   Disconnect the negative cable at the battery.

21   Remove the screws mounting the pressure regulator to the throttle body.

22   Wrap a cloth around the fuel inlet chamber to catch any residual fuel.

23   Withdraw the pressure regulator from the throttle body **(see illustration)**.

24   Carefully remove the O-ring from the pressure regulator, followed by the gasket.

25   Place a new gasket in position on the pressure regulator and carefully install a new O-ring.

26   Place the pressure regulator in position on the throttle body, press it into position and install the retaining screws.

27   Tighten the screws securely.

28   Connect the negative battery cable.

29   Install the air cleaner assembly.

30   Start the engine and check for fuel leaks.

Idle Speed Control (ISC) actuator (motor)

Refer to illustrations 11.40 and 11.44

31   The idle speed is controlled by the idle speed control actuator. This valve changes the amount of air that will bypass into the intake manifold. The idle speed control actuator is activated by the SMEC depending upon the running conditions of the engine (air conditioning system, power steering, cold and warm running etc.).

Check

32   Chrysler recommends the use of an exerciser tool installed in series between the idle speed control actuator and the electrical connector for testing purposes. There are several tests the home mechanic can perform on the idle speed system to verify operation but they are limited and are useful only in the case of definite problems rather than intermittent failure.

33   Disconnect the electrical connector from the idle speed actuator and listen carefully for a change in the idle. Connect the electrical connector and turn the air conditioning on and listen for a change in idle rpm. When the engine is cold, the actuator motor should fluctuate the idle as the engine goes from cold to warm and also when the air conditioning compressor is turned ON. If there are no obvious signs that the motor is working, continue testing.

34   Use a voltmeter and test for voltage to the idle speed control actuator motor. Back-

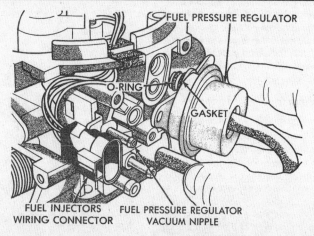

11.23 Be sure to replace the O-ring and the gasket with new parts

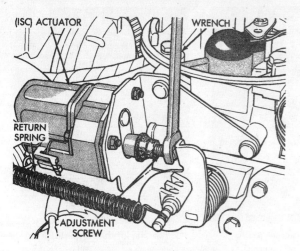

11.40 Adjusting the idle speed control actuator

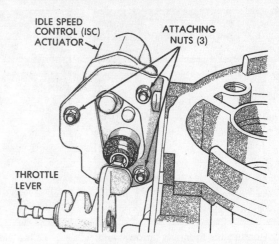

11.44 Remove the nuts (arrows) and separate the idle speed control actuator from the throttle body

4B

probe the motor electrical connector and check for voltage. Voltage should be present when the engine is running. As the engine rpm changes from high to low (or vice versa), the voltage will fluctuate between 2.0 volts and 6.0 volts. This test will be difficult to monitor through the various changes in engine conditions.

35   If there is no voltage present, have the electrical circuit for the motor diagnosed by a dealer service department or other qualified repair shop.

36   If the computer is delivering voltage to the motor, check that the actuator motor is not frozen or defective. Have an assistant inside the vehicle turn the ignition key ON (engine OFF). Carefully observe the actuator motor pintle. If the pintle retracts (pulls in) and also makes a very clear clicking sound, the actuator motor is operating.

Adjustment

37   Start the engine and allow it to run for two minutes or more. Shut off the engine and allow the throttle kicker to engage fully.

38   Disconnect the wire to the idle speed control actuator and the coolant temperature sensor (see Chapter 6).

39   Hook up a tachometer according to the manufacturer's specifications.

40   Start the engine. Adjust the extension screw **(see illustration)** on the ISC actuator until the rpm is correct:

    1) 3.9L engine - 2,500 to 2,600 rpm
    2) 5.2L and 5.9L engines - 2,750 to 2,850 rpm

41   Reconnect the ISC actuator electrical connector.

42   Reconnect the coolant temperature sensor.

Replacement

43   Disconnect the electrical connector from the IAC valve.

44   Remove the two mounting screws from the valve and lift it from the intake manifold **(see illustration)**.

45   Installation is the reverse of removal.

Throttle body temperature sensor

46   For information on the throttle body temperature sensor, see Chapter 6.

Throttle Position Sensor (TPS)

47   For information on the TPS see Chapter 6.

Throttle body unit

Removal

48   Remove the air cleaner.

49   Relieve the fuel system pressure (see Section 2).

50   Disconnect the negative cable from the battery.

51   Disconnect the vacuum hoses and wire harness connectors from the throttle body. Label them to ensure reinstallation in the same locations.

52   Remove the throttle return spring, cable and bracket, and, if equipped, the speed control cable and transmission kickdown linkage.

53   Disconnect the fuel supply and return hoses. Position rags under the fittings to catch any fuel spillage.

54   Remove the throttle body mounting bolts and lift the throttle body assembly off the intake manifold.

Installation

55   Installation is the reverse of the removal procedure. Be sure to use a new gasket between the throttle body and intake manifold. Tighten the mounting bolts, in a criss-cross pattern, to the torque listed in this Chapter's Specifications.

---

12   Multi Port Fuel Injection (MPFI) - component check and replacement

---

Warning: *Gasoline is extremely flammable, so take extra precautions when you work on any part of the fuel system. Don't smoke or allow open flames or bare light bulbs near the work area, and don't work in a garage where a natural gas-type appliance (such as a water heater or a clothes dryer) is present. Since gasoline is carcinogenic, wear latex gloves when there's a possibility of being exposed to fuel, and, if you spill any fuel on your skin, rinse it off immediately with soap and water. Mop up any spills immediately and do not store fuel-soaked rags where they could ignite. The fuel system is under constant pressure, so, if any fuel lines are to be disconnected, the fuel pressure in the system must be relieved first (see Section 2). When you perform any kind of work on the fuel system, wear safety glasses and have a Class B type fire extinguisher on hand.*

Throttle body

Refer to illustration 12.13

Check

1   On top of the throttle body, locate a vacuum hose that goes to the throttle body (port vacuum). Detach it from the throttle body and attach a vacuum gauge in its place.

2   Start the engine and warm it to its normal operating temperature (wait until the cooling fan comes on twice). Verify that the gauge indicates no vacuum.

3   Open the throttle slightly from idle and verify that the gauge indicates vacuum. If the gauge indicates no vacuum, check the port to make sure it is not clogged. Clean it with carburetor cleaner if necessary.

4   Stop the engine and verify the accelerator cable and throttle valve operate smoothly without binding or sticking.

5   If the accelerator cable or throttle valve binds or sticks, check for a build-up of sludge on the cable or throttle shaft.

6   If a build-up of sludge is evident, try removing it with carburetor cleaner or a similar solvent.

7   If this cleaning fails to remedy the problem, remove the throttle body and soak it in carburetor cleaner (be sure to remove all

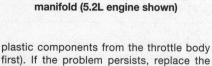

12.13 Remove the four nuts (arrows) and separate the throttle body from the intake manifold (5.2L engine shown)

12.16a Remove the fuel pressure regulator bolt (arrow) (5.2L engine shown) and . . .

12.16b . . . lift the regulator from the fuel rail

plastic components from the throttle body first). If the problem persists, replace the throttle body.

Removal and installation

Warning: *Wait until the engine is completely cool before beginning this procedure.*

8 Detach the cable from the negative battery terminal.

9 Remove the air duct that connects the air cleaner assembly to the throttle body.

10 Unplug the throttle position sensor connector from the throttle body. Also label and detach all vacuum hoses from the throttle body.

11 Detach the accelerator cable (see Section 8) and, if equipped, the transmission throttle valve cable (see Chapter 7B).

12 Detach the coolant hoses from the throttle body. Plug the lines to prevent coolant loss.

13 Unscrew the four mounting nuts **(see illustration)** and remove the throttle body and gasket. Remove all traces of old gasket material from the throttle body and intake manifold.

14 Installation is the reverse of removal. Be sure to use a new gasket. Adjust the accelerator cable (see Section 10) and, if equipped, the throttle valve cable (see Chapter 7B). Check the coolant level and add some, if necessary (see Chapter 1).

Fuel pressure regulator

Check

Refer to Section 3 for the fuel pressure checking procedure.

Replacement

1993 and earlier V6 and 5.2L V8 engines, 1992 and earlier 5.9L V8 engines

Refer to illustrations 12.16a and 12.16b

15 Relieve the fuel system pressure (see Section 2). Detach the cable from the negative battery terminal.

16 Detach the vacuum hose and fuel hose

from the pressure regulator, then unscrew the mounting bolt **(see illustrations)**. Remove the pressure regulator.

17 Installation is the reverse of removal. Be sure to use a new O-ring. Lubricate the O-ring with a light coat of clean engine oil before installation. Check for fuel leaks after installing the pressure regulator.

1993 and later 5.9L engines, all 1994 and later engines

Refer to illustration 12.19

18 Remove the fuel tank (see Chapter 4A). Mark the relationship of the outlet on the fuel filter/pressure regulator assembly to the fuel pump module, then pull the assembly out of its grommet in the fuel tank. **Caution:** *Don't pull the filter/regulator assembly out of the tank more than three inches before disconnecting the fuel line.*

19 On 1971 through 1997 models, remove the snap ring from the plastic cover **(see illustration)** and slide the cover to down to expose the clamp on the fuel tube. On 1998 and later models, carefully twist and pull up to expose the clamp on the fuel tube. Cut the old clamp off, being careful not to damage the fuel tube. Remove the fuel filter/regulator

from the fuel pump module.

20 Install a new clamp over the fuel tube and connect the fuel filter/regulator to the fuel tube. Turn the assembly so the outlet is aligned with the mark you made in Step 18, then tighten the clamp.

21 On 1971 through 1997 models, slide the cover up and fasten is with the snap ring. On all models, push the assembly back into its grommet, making sure the outlet is still aligned with the mark. Install the fuel tank, pressurize the fuel system by turning the ignition switch to the On position and check for fuel leaks.

Fuel injectors

Check

Refer to Section 10 for the fuel injector checking procedure.

Removal

Refer to illustrations 12.26a, 12.26b, 12.27a, 12.27b, 12.27c, 12.28a and 12.28b

22 Detach the cable from the negative battery terminal. Relieve the fuel pressure (see Section 2).

23 Unplug the injector connectors.

24 Detach the vacuum hose and fuel return hose from the fuel pressure regulator.

25 Detach any ground cables from the fuel rail.

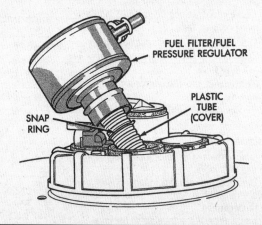

12.19 After pulling the fuel filter/pressure regulator out of its grommet in the fuel pump module, expand the snap-ring and slide the plastic tube down to expose the fuel tube clamp

12.26a To disconnect the fuel line from the fuel rail, remove the safety clip from the connector . . .

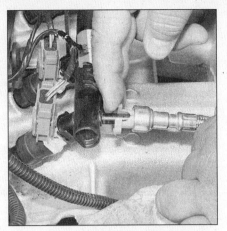

12.26b . . . then slide a plastic fuel disconnect tool onto the line and push the tool into the connector to force it (spring) off the fuel line lip

12.27a Remove the fuel rail mounting bracket located near the fuel pressure regulator (5.2L engine shown)

12.27b Remove the bolts (arrows) that secure the fuel rail to the intake manifold

12.27c Lift the fuel rail (with the fuel injectors attached) from the engine compartment

4B

26   Detach the fuel feed line from the fuel rail **(see illustrations)**.

27   Remove the mounting nuts **(see illustrations)** and detach the fuel rail and the injectors from the manifold.

28   Remove the retaining clips and pull the injector(s) from the bore(s) in the fuel rail and remove and discard the O-ring, cushion ring and seal ring **(see illustrations)**. **Note:** *Whether you're replacing an injector or a leaking O-ring, it's a good idea to remove all the injectors from the fuel rail and replace all the O-rings, seal rings and cushion rings.*

Installation

29   Coat the new cushion rings with clean engine oil and slide them onto the injectors.

30   Coat the new O-rings with clean engine oil and install them on the injector(s), then insert each injector into its corresponding bore in the fuel rail. Install the retainer clip.

31   Coat the new seal rings with clean engine oil and press them into the injector bore(s) in the intake manifold.

32   Install the injector and fuel rail assembly on the intake manifold. Tighten the fuel rail

mounting nuts to the torque listed in this Chapter's Specifications.

33   The remainder of installation is the reverse of removal.

34   After the injector/fuel rail assembly installation is complete, turn the ignition

switch to ON, but don't operate the starter (this activates the fuel pump for about two seconds, which builds up fuel pressure in the fuel lines and the fuel rail). Repeat this about two or three times, then check the fuel lines, rail and injectors for fuel leakage.

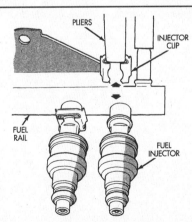

12.28a Pull the injector retaining clips from the fuel rail assembly

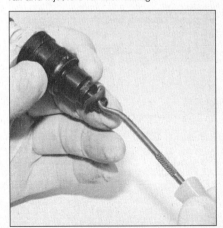

12.28b Carefully remove the O-rings from the injector

Idle Air Control (IAC) system

Refer to illustrations 12.38, 12.40 and 12.42

35   The idle speed is controlled by the IAC valve. This valve changes the amount of air that will bypass the throttle plate and flow into the intake manifold. The IAC valve is activated by the PCM depending upon the running conditions of the engine (air conditioning system, power steering, cold and warm running etc.).

Check

36   Chrysler recommends the use of an exerciser tool installed in series between the IAC motor and the harness electrical connector for testing purposes. There are several tests the home mechanic can perform on the IAC system to verify operation but they are limited and are useful only in the case of definite problems rather than intermittent failure.

37   Disconnect the electrical connector from the IAC valve and listen carefully for a change in the idle. Connect the IAC valve electrical connector and turn the air conditioning on and listen for a change in idle rpm. When the engine is cold, the IAC motor should fluctuate the idle as the engine goes from cold to warm and also when the air conditioning compressor is turned ON. If there are no obvious signs that the IAC motor is working, continue testing.

38   Use a voltmeter and test for voltage to the IAC motor. Backprobe the IAC motor electrical connector and check for voltage **(see illustration)**. Voltage should be present when the engine is running. As the engine rpm changes from high to low (or vice versa), the voltage will fluctuate between 2.0 volts and 6.0 volts. This test will be difficult to monitor through the various changes in engine conditions.

39   If there is no voltage present, have the electrical circuit for the IAC valve diagnosed by a dealer service department or other qualified repair shop.

40   If the computer is delivering voltage to the IAC motor, check that the IAC motor is not frozen or defective. Remove the IAC motor and leaving the IAC motor connected, turn the ignition key ON (engine not running) **(see illustration)**. If the pintle retracts (pulls in) start the engine and observe that the pintle extends to compensate for the high idle. It

12.38 Check for voltage on the brown/white wire (terminal B) and the purple/black wire (terminal A) with the engine idling (5.2L engine shown)

12.40 Remove the IAC valve from the throttle body but do not disconnect the electrical connector. Confirm that the pintle moves first with the ignition key ON and then with the engine running

12.42 Remove the screws from the IAC valve (5.2L engine shown)

will be necessary to plug the recess in the throttle body when the IAC motor is removed. **Caution:** *If the pintle extends, be sure to shut the engine off immediately before the pintle length exceeds 1/4-inch. If the pintle is extended more than this amount, the IAC motor may become damaged and separate from the body and must be replaced with a new part. Therefore, it is recommended that*

an assistant operate the ignition key while the IAC motor is being tested.

Replacement

41   Disconnect the electrical connector from the IAC valve.

42   Remove the two mounting screws from the valve and lift it from the intake manifold **(see illustration)**.

43   Installation is the reverse of removal. Be sure to install a new O-ring.

Chapter 5
Engine electrical systems

Contents

Specifications

Ignition system

Ignition coil resistance

Primary
  1971 through 1979
    Chrysler/Essex .... 1.41 to 1.55 ohms
    Chrysler/Prestolite .... 1.65 to 1.79 ohms
  1980 through 1984
    Chrysler/Essex .... 1.34 to 1.55 ohms
    Chrysler/Prestolite .... 1.60 to 1.79 ohms
  1985 through 1991 .... 1.34 to 1.55 ohms
  1992 on
    Diamond .... 0.96 to 1.18 ohms
    Toyodenso .... 0.95 to 1.20 ohms
Secondary
  1971 through 1979
    Chrysler/Essex .... 8,000 to 10,200 ohms
    Chrysler/Prestolite .... 9,400 to 11,700 ohms
  1980 though 1991 .... 8,000 to 12,000 ohms
    Chrysler/Essex .... 9,000 to 12,200 ohms
    Chrysler/UTC .... 9,000 to 12,200 ohms
    Chrysler/Prestolite .... 9,400 to 11,700 ohms
    Diamond .... 15,000 to 19,000 ohms
  1992 on
    Diamond .... 11,300 to 15,300 ohms
    Toyodenso .... 11,300 to 13,300 ohms

Pick-up coil resistance .... 150 to 900 ohms

Ballast resistor

Single type (points ignition)	0.5 to 0.6 ohm
Single type (electronic ignition)	1.1 to 1.8 ohms
Dual type (electronic ignition)	
Primary resistor (thermal)	0.5 to 0.6 ohm
Primary resistor (non-thermal)	1.12 to 1.38 ohms
Auxiliary resistor	4.75 to 5.75 ohms

Air gap

Single pick-up coil	
1974 through 1976	0.008 inch
1977 on	0.006 inch
Dual pick-up coils	
"START"	0.008 inch
"RUN"	
1977 through 1980	0.010 inch
1981 through 1985	0.012 inch

Charging system

Alternator brush length (minimum)	1/4-inch

1 General information and precautions

The engine electrical systems include all ignition, charging and starting components. Because of their engine-related functions, these components are discussed separately from chassis electrical devices such as the lights, the instruments, etc. (which are included in Chapter 12).

Always observe the following precautions when working on the electrical systems:

a) *Be extremely careful when servicing engine electrical components. They are easily damaged if checked, connected or handled improperly.*

b) *The alternator is driven by an engine drivebelt which could cause serious injury if your hands, hair or clothes become entangled in it with the engine running.*

c) *Both the alternator and the starter are connected directly to the battery and could arc or even cause a fire if mishandled, overloaded or shorted out.*

d) *Never leave the ignition switch on for long periods of time with the engine off.*

e) *Don't disconnect the battery cables while the engine is running.*

2.2 The battery hold-down clamp is retained by two nuts (arrows)

f) *Maintain correct polarity when connecting a battery cable from another source, such as a vehicle, during jump starting.*

g) *Always disconnect the negative cable first and hook it up last or the battery may be shorted by the tool being used to loosen the cable clamps.*

It's also a good idea to review the safety-related information regarding the engine electrical systems located in the *Safety First* section near the front of this manual before beginning any operation included in this Chapter.

2 Battery - removal and installation

Refer to illustration 2.2
Warning: *Hydrogen gas is produced by the battery, so keep open flames, bare light bulbs and lighted tobacco away from it at all times. Always wear eye protection when working around a battery. Rinse off spilled electrolyte immediately with large amounts of water.*

1   Disconnect both cables from the battery terminals. **Caution:** *Always disconnect the negative cable first and hook it up last or the battery may be shorted by the tool being used to loosen the cable clamps.*

2   Remove the battery hold-down clamp **(see illustration)**.

3   Lift out the battery. Use the proper lifting technique - the battery is heavy.

4   While the battery is out, inspect the battery carrier (tray) for corrosion (see Chapter 1).

5   If you are replacing the battery, make sure you get an identical battery, with the same dimensions, amperage rating, "cold cranking" rating, etc.

6   Installation is the reverse of removal.

3 Battery - emergency jump starting

Refer to the *Booster battery (jump) starting* procedure at the front of this manual.

4 Battery cables - check and replacement

1   Periodically inspect the entire length of each battery cable for damage, cracked or burned insulation and corrosion. Poor battery cable connections can cause starting problems and decreased engine performance.

2   Check the cable-to-terminal connections at the ends of the cables for cracks, loose wire strands and corrosion. The presence of white, fluffy deposits under the insulation at the cable terminal connection is a sign that the cable is corroded and should be replaced. Check the terminals for distortion, missing mounting bolts and corrosion.

3   When removing the cables, always disconnect the negative cable first and hook it up last or the battery may be shorted by the tool used to loosen the cable clamps. Even if only the positive cable is being replaced, be sure to disconnect the negative cable from the battery first (see Chapter 1 for further information regarding battery cable removal).

4   Disconnect the old cables from the battery, then trace each of them to their opposite ends and detach them from the starter solenoid and ground. Note the routing of each cable to insure correct installation.

5   If you are replacing either or both of the old cables, take them with you when buying new cables. It is vitally important that you replace the cables with identical parts. Cables have characteristics that make them easy to identify: positive cables are usually red, larger in cross-section and have a larger diameter battery post clamp; ground cables are usually black, smaller in cross-section and have a slightly smaller diameter clamp for the negative post.

6   Clean the threads of the solenoid or ground connection with a wire brush to remove rust and corrosion. Apply a light coat of battery terminal corrosion inhibitor, or petroleum jelly, to the threads to prevent future corrosion.

7   Attach the cable to the solenoid or

ground connection and tighten the mounting nut/bolt securely.

8   Before connecting a new cable to the battery, make sure that it reaches the battery post without having to be stretched.

9   Connect the positive cable first, followed by the negative cable.

5   Ignition system - general information

Refer to illustrations 5.4 and 5.6

1   The ignition system is designed to ignite the fuel/air charge entering each cylinder at just the right moment. It does this by producing a high voltage spark between the electrodes of each spark plug.

2   The types of systems installed on these vehicles evolved through the years to accommodate more strict emissions standards adapted by major automotive manufacturers. Some of the changes in the ignition systems overlapped in certain years and others changed with certain models while some did not. Here is a general listing of the various systems:

3   **Breaker point ignition systems:** (points and condenser type). This ignition system controls the ignition spark using the conventional points and condenser arrangement and timing is controlled by mechanical advance components. This system was originally installed on 1971 and 1972 models only.

4   **Electronic ignition systems (see illustration):** This breakerless ignition system basically controls the ignition spark without any electrical (computer-controlled) timing advance or retard capabilities. These sys-

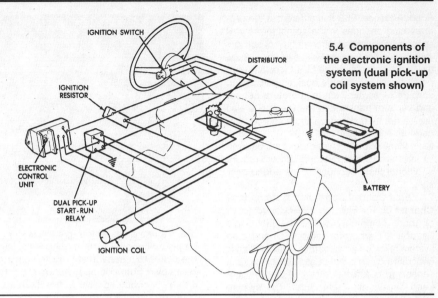

5.4 Components of the electronic ignition system (dual pick-up coil system shown)

tems were installed on certain models ranging from 1974 (select models only) until 1987. The components in these systems include a module (electronic control unit) usually mounted on the firewall, a coil and one or two magnetic pick-up coils mounted in the distributor. The dual pick-up coil systems were installed on models in 1981 through 1987. These systems also included a dual pick-up START-RUN relay that is used to transfer power to the operating pick-up coil. On 1974 through 1979 models, a five-pin connector is used on the module as well as a dual ballast resistor. On 1980 through 1987 models, a four-pin connector and a single ballast resistor was standard equipment. Timing is advanced using centrifugal or vacuum

advance components depending on the year and model.

5   **Spark Control Computer (SCC) systems:** On these systems, the spark timing is constantly adjusted by the computer in response to input from the various sensors located on the engine. Chrysler introduced two versions of this system in the early 1980's; ELB I and ELB II. Only ELB II is installed on models covered by this manual.

6   **Electronic Lean Burn (ELB) II system (see illustration):** This system was installed on most models from 1981 through 1989. This system is a redesigned version of the ELB I system. The system was renamed the ESC or Electronic Spark Control system. It was renamed again to ESA or Electronic

5

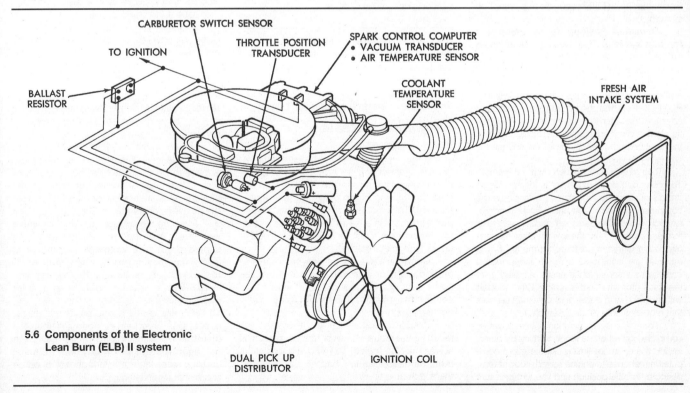

5.6 Components of the Electronic Lean Burn (ELB) II system

Spark Advance. ELB II is different in the programming changes in the computer as well as the additional and different type of sensors installed on the engine. Most systems include the same sensors as the ELB I (coolant temperature sensor, throttle position transducer, carburetor switch, air temperature sensor and vacuum transducer) but the ELB II also uses a detonation sensor and a charge temperature sensor (both are in the intake manifold). Many of the system checks will be similar for both systems but the ELB II will require additional tests because of the added sensors.

7   **Fuel Control Computer systems (see Chapter 6):** The Electronic Fuel Control (EFC) or more commonly called the Electronic Feedback Carburetor (also EFC) emission system relies on an electronic signal, which is generated by an exhaust gas sensor (oxygen sensor) to control a variety of devices and keep emissions within limits. The system works in conjunction with a three-way catalyst to control the levels of carbon monoxide, hydrocarbons and oxides of nitrogen. The EFC system also works in conjunction with the Spark Control Computer (SCC) system. The two systems share certain sensors and output actuators therefore diagnosing the EFC system will require a thorough check of all the SCC components.

8   **Hall Effect Electronic Ignition system:** This system was installed on models from 1988 through 1991. These systems incorporate sensors and output actuators along with a computer referred to as the Single Module Engine Controller (SMEC). This updated version of the SCC system controls ignition timing and fuel control with one computer. These systems also incorporate a Throttle Body Injection (TBI) electronic fuel injection system.

9   **Camshaft Position Sensor Electronic Ignition system:** This system was installed on 1992 and later models. The entire ignition system consists of the ignition coil, the camshaft position sensor, the crankshaft position sensor and the Powertrain Control Module (PCM). The PCM controls the ignition timing, spark and advance characteristics for the engine. The ignition timing is not adjustable, therefore, changing the position of the distributor will not change the timing in any way.

The distributor is driven by the camshaft. The camshaft position sensor is located inside the distributor and its purpose is to signal the computer concerning fuel and cylinder synchronization. **Note:** *The camshaft position sensor is formerly referred to as the* **switch plate assembly** *within Chrysler parts departments.* Some models are equipped with an independent camshaft sensor which is easily removed from the distributor while in the vehicle. Other models are equipped with dependent camshaft sensors that require distributor disassembly.

The computerized ignition system provides complete control of the ignition timing by determining the optimum timing using a micro computer in response to engine speed, coolant temperature, throttle position and vacuum pressure

6.1 To use a calibrated ignition tester, simply disconnect a spark plug wire, clip the tester to a convenient ground (like a valve cover bolt) and operate the starter - if there is enough power to fire the plug, sparks will be visible between the electrode tip and the tester body

6.11a Remove the retaining screw that secures the electrical connector to the ignition module

in the intake manifold. These parameters are relayed to the PCM by the camshaft position sensor, throttle position sensor (TPS), coolant temperature sensor and MAP Sensor. Ignition timing is altered during warm-up, idling and warm running conditions by the PCM. This electronic ignition system also consists of the ignition switch, battery, coil, distributor, spark plug wires and spark plugs. These ignition systems are equipped with the Multi Port Fuel Injection (MPFI) system.

Refer to a dealer parts department or auto parts store for any questions concerning the availability of the distributor parts and assemblies. Testing the crankshaft position sensor is covered in Chapter 6.

6   Ignition system - check

Caution: *Always disconnect the cable from the negative terminal of the battery before disconnecting the electrical connectors from the module or electronic control unit.*

All ignition systems

Refer to illustration 6.1
1   If the engine will not start even though it cranks over, check for spark at the spark plug by installing a calibrated ignition system tester to the end of the plug wire **(see illustration)**. This tool is available at most auto parts stores. Be sure to obtain the correct tool for your particular ignition system (breakerless type [electronic]).
2   Connect the clip on the tester to a ground such as a metal bracket or valve cover bolt, crank the engine and watch the end of the tester for a bright blue, well defined spark.
3   If sparks occur, sufficient voltage is reaching the plugs. However, the plugs themselves may be fouled, so remove and check them as described in Chapter 1 or replace them with new ones.

4   If no spark occurs, remove the distributor cap and check the cap and rotor as described in Chapter 1. If moisture is present, use WD-40 or something similar to dry out the cap and rotor, then reinstall the cap and repeat the spark test.
5   If there is still no spark, the tester should be attached to the wire from the coil and the test repeated again. If there is still no spark, the coil-to-cap wire may be bad (see Chapter 1).
6   If no spark occurs, check the primary wire connections at the coil to make sure they are clean and tight. Check the primary and secondary resistance of the ignition coil (see Section 7). Make any necessary repairs, then repeat the check again.
7   If sparks now occur, the distributor cap, rotor, plug wires or spark plugs may be defective.

Breaker type ignition systems (points)

Note: *Some of the following check(s) may require a voltmeter, ohmmeter and/or a jumper cable.*

8   If there is still no spark, check the ignition points (see Chapter 1). If the points appear to be in good condition (no pits or burned spots on the point surface) and the primary wires are hooked up correctly, adjust the points as described in Chapter 1.
9   Measure the voltage at the points with a voltmeter. With the ignition On, the points should produce a voltmeter reading of at least 10.5 volts. If not, the battery must be recharged or replaced. If the reading is 10.5 volts or more, record the reading for future reference.
10   If there is still no spark at the plugs, check for a grounded or open circuit in the distributor/points circuit. There may be a damaged resistor block or points terminal causing the ignition voltage to become shorted or diminished.

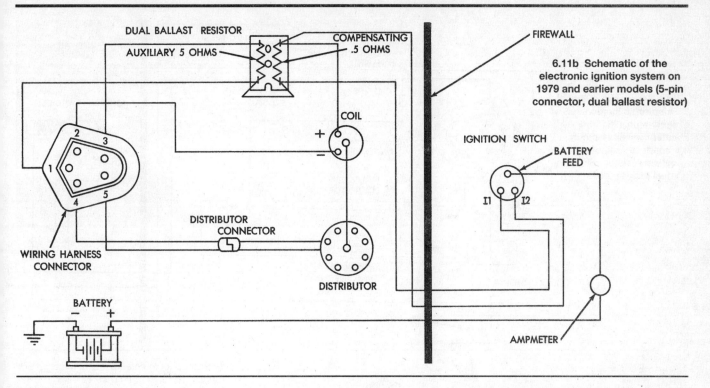

6.11b Schematic of the electronic ignition system on 1979 and earlier models (5-pin connector, dual ballast resistor)

Electronic type ignition systems

Refer to illustrations 6.11a, 6.11b, 6.11c, 6.11d, 6.11e, 6.12a, 6.12b, 6.14, 6.16a, 6.16b, 6.20, 6.22, 6.23 and 6.28

Note: These tests include checks for the early electronic ignition systems (starting in 1974) as well as for the Spark Computer Control (SCC) systems installed on later models (up to 1987).

11 If the battery voltage is satisfactory, turn the ignition off and disconnect the electronic control unit (ECU) at the wiring harness connector **(see illustrations)**. On SCC equipped ignition systems, disconnect the 10-way connector (ELB II and later SCC systems) **(see illustrations)**.

12 Check the voltage at cavity 2 of the ECU connector or cavity 1 of the 10-way connector with the ignition key ON (engine not running). It should be within one volt of battery voltage **(see illustrations)**.

13 If the reading is not within one volt of battery voltage, there is probably a short circuit between the ECU or 10-way connector and the coil negative terminal. Use a continu-

5

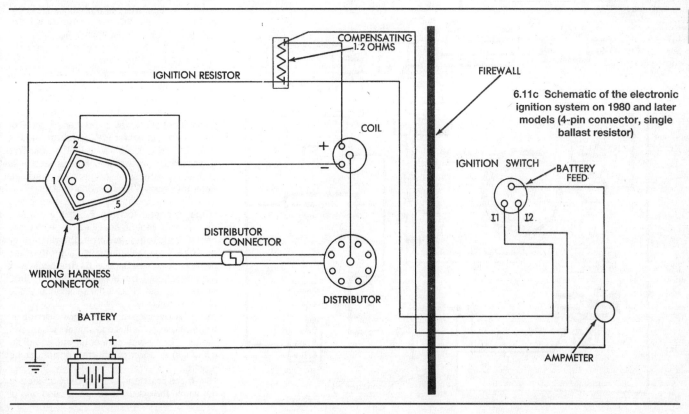

6.11c Schematic of the electronic ignition system on 1980 and later models (4-pin connector, single ballast resistor)

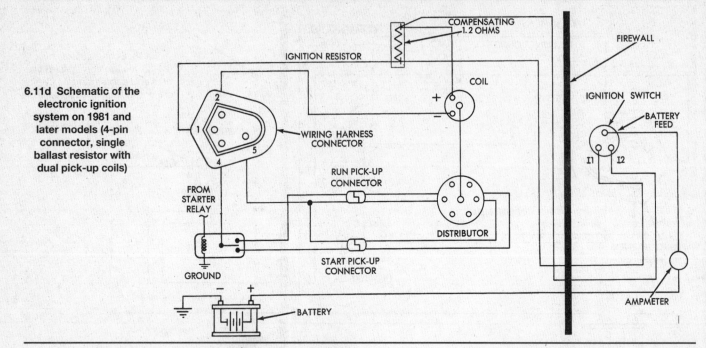

6.11d Schematic of the electronic ignition system on 1981 and later models (4-pin connector, single ballast resistor with dual pick-up coils)

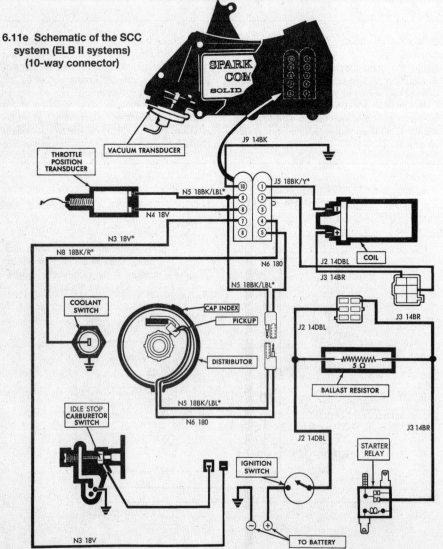

6.11e Schematic of the SCC system (ELB II systems) (10-way connector)

ity tester or ohmmeter to locate the short and repeat Step 9.

14 If the voltage reading is satisfactory, check the voltage at the carburetor switch on SCC equipped ignitions **(see illustration)**. Place a thin insulator between the idle adjusting screw and carburetor switch. The reading should be at least five volts. Measure the voltage at cavity 2 of the 10-way connector or cavity 1 of the ECU connector. It should be within one volt of battery voltage.

15 If the reading is not within one volt of battery voltage, there is probably a short circuit between the ECU connector or the 10-way connector and the ignition switch. Use a continuity tester or ohmmeter to locate the short and repeat Step 11.

16 If the voltage at cavities 1, 4 or 2 is as specified, check the resistance between the cavities with an ohmmeter. Measure between cavities 4 and 5 on the ECU connector **(see illustration)** or cavities 3 and 9 and 5 and 9 of the 10-way connector **(see illustration)**. The resistance should be 150 to 900 ohms. This test checks the resistance of the pick-up coils and wiring harness at the module/computer. Refer to Section 8 for additional checks.

17 If the resistance is within the specified range, proceed to Step 22. If the reading does not fall into the specified range, disconnect the pick-up leads from the distributor and check the resistance (ignition off) of the pick-up coils (see Section 8). The ohmmeter reading should be between 150 and 900 ohms. This test checks the resistance of the pick-up coils only.

18 If resistance at the pick-up leads is within the specified range, the wire between the ECU or 10-way connector and the pick-up coil(s) is open or shorted and must be repaired.

19 If the resistance at the pick-up lead is not within the specified range, ground one of

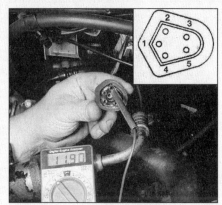

6.12a Check for battery voltage on cavity number 2 of the ignition module electrical (ECU) connector (1974 model shown)

6.12b Check for battery voltage on cavity number 1 of the 10-way electrical connector

6.14 Check the voltage of the carburetor switch - it should be greater than five volts

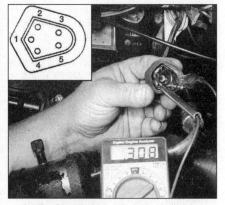

6.16a Check the resistance between cavity numbers 4 and 5 - it should be between 150 to 900 ohms

6.16b Check the resistance between cavity numbers 5 and number 9 on the 10-way connector - it should be between 150 to 900 ohms

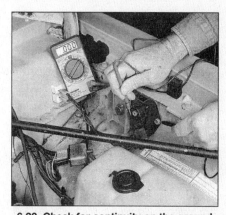

6.20 Check for continuity on the ground circuit (pin number 5) and ground on the ignition module (1974 model shown)

the ohmmeter leads and check for a ground or short circuit at each pick-up coil lead. A zero reading should be indicated.

20 If a zero reading is not indicated, check the pick-up leads for a ground or short circuit. If the pick-up leads are in good condition, test the ignition module. Working on the ignition module, check the ground circuit on ECU equipped ignitions (see illustration). Make sure there is a good ground at pin 5 of the ECU. Make sure the ECU connections are

clean and the mounting screws are tight. Refer to Section 9 for additional information on the ignition module.

21 If a zero reading was indicated and the resistance reading at the pick-up leads did not fall within the specified range (Step 19), one of the pick-up coils is defective and must be replaced.

22 On dual pick-up coil systems, check the START-RUN relay. Disconnect the electrical connectors from pin numbers 4 and 5 and

using an ohmmeter, measure the resistance of the relay between the same terminals (see illustration). Resistance should be between 20 and 30 ohms.

23 If there was no spark at the secondary coil wire while cranking the engine, a jumper cable will have to be fashioned to carry out further testing. Use two lengths of wire and three alligator clips to fabricate a three connector jumper cable. Splice a 0.33 MF capacitor into one of the wires (see illustration).

5

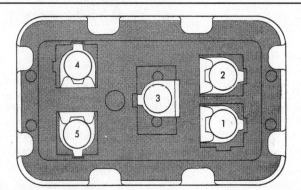

6.22 Check the resistance across terminal numbers 4 and 5 on the dual pick-up start relay - it should be between 20 and 30 ohms

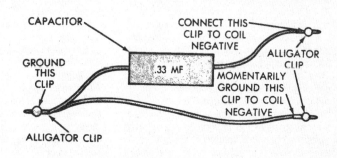

6.23 A homemade jumper cable used to test the primary ignition system

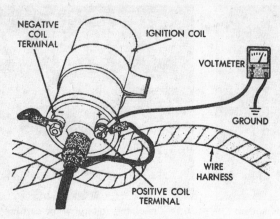

6.28 Checking the voltage at the coil positive terminal

6.50 Backprobe the coil positive terminal with a pin or clip and see if the test light flashes as the engine is cranked over

24   Disconnect the ECU or 10-way connector and connect one of the double leads to the coil negative terminal. Connect the opposite end (single lead) to a good ground.

25   Using the remaining wire lead, momentarily make contact with the coil negative terminal while holding the secondary coil wire about 1/4-inch away from the engine block.

26   If there is now spark at the secondary coil wire, the SCC or ECU is faulty and should be replaced.

27   If there is still no spark at the secondary coil wire, measure the voltage at the battery terminals with a voltmeter. Record the measurement.

28   Measure the voltage at the coil positive terminal. It should be within one volt of battery voltage **(see illustration)**. If the correct voltage is indicated, proceed to Step 32 for electronic ignition systems. Proceed to Step 34 for SCC equipped ignition systems. Also, check the resistance of the ballast resistor (see Section 10).

29   If sufficient voltage is not measured at the coil positive terminal, check the wiring between the battery and the coil terminal for an open or short circuit. Replace or repair any faulty wiring. On SCC equipped systems, proceed to Step 37.

30   If the wiring is not the problem, the next step is to replace the starter relay.

31   Repeat Step 29. If there is still insufficient voltage, the ballast resistor is defective and must be replaced (see Section 10).

32   If sufficient voltage was measured at the coil positive terminal, check the voltage at the coil negative terminal. It should be within one volt of the battery voltage.

33   If there is battery voltage at the coil negative terminal, but there is no spark when shorting the terminal (see Step 29), the coil is defective and must be replaced.

34   If the specified voltage was measured at the coil positive terminal on SCC equipped systems, check for voltage at the coil negative terminal. It should be within one volt of battery voltage.

35   If the voltage at the negative coil terminal is within this range, but no spark was produced when shorting the secondary coil lead

(Step 32), the ignition coil is faulty and should be replaced. Check the ignition coil (see Section 7).

36   If the voltage at the negative coil terminal is not within this range, the coil is faulty and should be replaced.

37   If voltage at the negative coil terminal is within the specified range and a spark was produced when shorting the secondary coil lead, the wiring should be checked further.

38   With the ignition off, use an ohmmeter to check for continuity between the carburetor switch and cavity 8 of the 10-way connector (ELB II and later SCC systems).

39   If the circuit is not continuous, isolate the open circuit with the ohmmeter and repair the wiring.

40   If the circuit is continuous, the spark control computer is probably faulty. It might be a good idea to have the computer diagnosed by a dealer service department before purchasing a replacement.

Computerized Ignition systems

Refer to illustration 6.50

Note: *These tests include checks for the Hall Effect Electronic ignition systems (1988 through 1991) as well as for the Camshaft Position Sensor ignition systems (1992 and later).*

41   With the ignition switch turned to the "ON" position, a "Battery" light or an "Oil Pressure" light is a basic check for ignition and battery supply to the SMEC (1988 through 1991) or PCM (1992 and later).

42   Check all ignition wiring connections for tightness, cuts, corrosion or any other signs of a bad connection.

43   Use a calibrated ignition tester to verify adequate secondary voltage (25,000 volts) at each spark plug **(see Illustration 6.1)**. A faulty or poor connection at that plug could also result in a misfire. Also, check for carbon deposits inside the spark plug boot.

44   Check for carbon tracking on the coil. If carbon tracking is evident, replace the coil and be sure the secondary wires related to that coil are clean and tight. Excessive wire resistance or faulty connections could cause damage to the coil.

45   Using an ohmmeter, check the resistance between the coil terminals. If an open is found (verified by an infinite reading), replace the coil (see Section 7).

46   Check the distributor cap for any obvious signs of carbon tracking, eroded terminals or cracks.

47   Using an ohmmeter, check the resistance of the spark plug wires. Each wire should measure less than 25,000 ohms.

48   Check the operation of the Automatic Shutdown Valve (ASD) (see Chapter 4).

49   On MPFI systems (1992 and later), check the operation of the camshaft position sensor (see Section 8) and the crankshaft position sensor (see Chapter 6).

50   If all the checks are correct, check the voltage signal from the computer. Using an LED-type test light, backprobe the coil power lead on the ignition coil **(see illustration)**. Remove the coil secondary wire and ground the terminal to the engine. Now have an assistant crank the engine over and observe that the test light pulses on and off. If there is no flashing from the test light, most likely the computer (SMEC or PCM) is damaged. Have the PCM diagnosed by a dealer service department.

51   Additional checks should be performed by a dealer service department or an automotive repair shop.

7   Ignition coil - check and replacement

Check

Refer to illustrations 7.4a, 7.4b, 7.5a and 7.5b

1   Remove the engine cover (see Chapter 2B). Mark the wires and terminals with pieces of numbered tape, then remove the primary wires and the high-tension lead from the coil.

2   Clean the outer case and check it for cracks and other damage.

3   Clean the coil primary terminals and check the coil tower terminal for corrosion. Clean it with a wire brush if any corrosion is found.

7.4a Checking the coil primary resistance on an early style coil

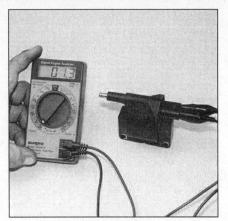

7.4b Checking the coil primary resistance on a late model type

7.5a Checking the coil secondary resistance on an early style coil

7.5b Checking the coil secondary resistance on a later style coil

7.8 Removing the ignition coil on the 5.2L engine

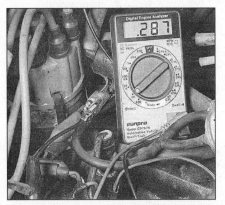

8.2a Disconnect the electrical connector and check the resistance of the pick-up coil at the distributor

8.2b On dual pick-up coil systems, the resistance should be the same for the START pick-up coil and the RUN pick-up coil

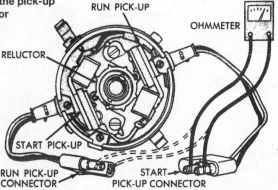

4   Check the coil primary resistance by attaching the leads of an ohmmeter to the positive and negative terminals. Compare the measured resistance to the value listed in this Chapter's Specifications **(see illustrations)**.

5   Check the coil secondary resistance by hooking one of the ohmmeter leads to one of the primary terminals and the other ohmmeter lead to the large center terminal **(see illustrations)**. Compare the measured resistance to the value listed in this Chapter's Specifications.

6   If the measured resistances are not as specified, the coil is defective and should be replaced with a new one.

7   It is essential for proper operation of the ignition system that all coil terminals and wires be kept clean and dry.

Removal and installation

Refer to illustration 7.8

8   Remove the engine cover (see Chapter 2B). To remove the coil, loosen the screw on the mounting clamp and slide the coil out of the clamp. On later models, remove the mounting bolts **(see illustration)**.

9   Installation is the reverse of removal.

8   Pick-up coil(s) - check and replacement

Note 1: *The pick-up coils changed design along with the various changes in ignitions systems. Early breakerless ignition systems (1974 through 1987 models) are equipped with a pick-up coil. 1988 through 1991 models are equipped with a Hall Effect switch. 1992 and later models are equipped with a camshaft position sensor. Each type of pick-up coil has different methods of testing and replacement procedures.*

Note 2: *The camshaft position sensor is formerly referred to as the **switch plate assembly** within Chrysler parts departments.*

1974 through 1987 pick-up coil(s)

Check

Refer to illustrations 8.2a and 8.2b

1   Before replacing the pick-up coil(s), check the ignition system thoroughly (see Section 6).

2   Disconnect the electrical connector(s) from the pick-up coil(s) at the distributor and using an ohmmeter, check for resistance **(see illustrations)**. There should be 150 to 900 ohms resistance.

3   If the readings are incorrect, replace the pick-up coil(s).

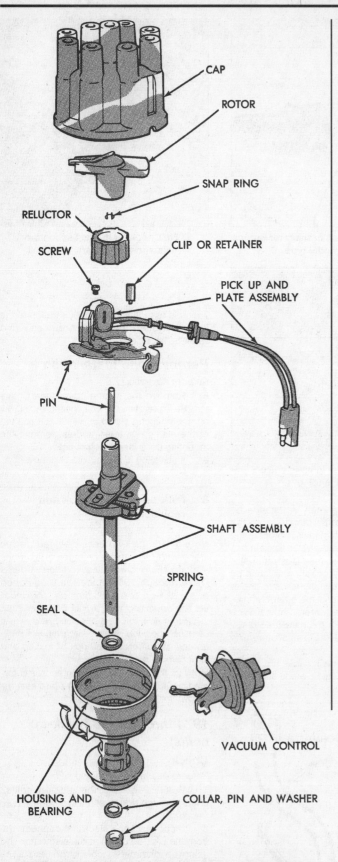

8.5 Exploded view of an electronic ignition distributor (this distributor is exactly the same for SCC systems except that the distributor weights are absent)

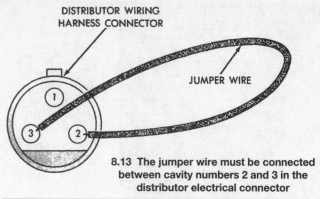

8.13 The jumper wire must be connected between cavity numbers 2 and 3 in the distributor electrical connector

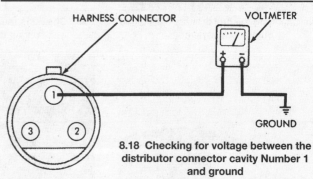

8.18 Checking for voltage between the distributor connector cavity Number 1 and ground

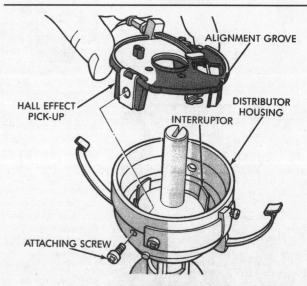

8.21 To remove the Hall Effect pick-up, remove the two attaching screws

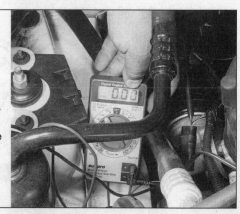

8.30 Connect the positive (+) probe of the voltmeter onto the tan/yellow wire on the camshaft position sensor connector and confirm that the voltage fluctuates between 0 and 5.0 volts

Replacement

Refer to illustration 8.5

4   Remove the engine cover (see Chapter 2B). Remove the distributor as described in Section 12 and carefully clamp it in a vise equipped with soft jaws (don't apply excessive pressure). **Note:** *On some early models, the pick-up coil can be detached after removing the mounting screw and separating the wires from the retainers on the upper plate and distributor housing.*

5   Using two screwdrivers not more than 7/16-inch wide, remove the reluctor by prying it up **(see illustration)**. Be careful not to damage the reluctor teeth.

6   Remove the two mounting screws from the vacuum advance unit, if equipped. Disconnect the control arm from the pick-up plate.

7   Remove the pick-up coil leads from the distributor housing.

8   Remove the two mounting screws from the pick-up plate.

9   Lift out (do not completely remove) the pick-up coil and plate as an assembly. Depress the retainer clip on the underside of the plate assembly and detach the assembly from the distributor. On most models, the pick-up coil and plate are replaced as an assembly and are not separable.

10   Installation is the reverse of removal, but be sure to place a small amount of distributor cam lubricant on the plate support pins before installing the plate assembly.

11   Set the air gap as described in Section 13.

1988 through 1991 Hall Effect switch

Check

Refer to illustrations 8.13 and 8.18

12   Disconnect the primary electrical connector from the distributor.

13   Jump cavity number 2 to cavity number 3 of the connector **(see illustration)**.

14   Turn the ignition switch to the ON (engine not running) position.

15   While holding the coil wire 1/4-inch from a good ground, make and break the connection at cavity two or cavity three several times.

16   If spark is not present at the coil wire in Step 4, the circuit connected to cavity number 2 and/or number 3 is open, or the coil is bad (see Section 10).

17   If spark is present at the coil wire in Step 4, go to the next Step.

18   Measure voltage at cavity number 1 of the distributor connector - it should be within one volt of battery voltage **(see illustration)**.

19   If battery voltage is within specification, replace the Hall Effect pick-up. If battery voltage is not present, repair open in the circuit connected to cavity number one.

Replacement

Refer to illustration 8.21

20   Disconnect the negative cable at the battery.

21   Remove the engine cover (see Chapter 2B). Remove the distributor cap and rotor (see Chapter 1). Unscrew the two Hall Effect pick-up screws (located on opposite sides of each other on the distributor housing) **(see illustration)**. Be careful not to drop them.

22   Carefully lift the Hall Effect pick-up assembly from the distributor housing.

23   Installation is the reverse of removal.

1992 and later camshaft position sensor

Check

Refer to illustration 8.30

24   Remove the distributor cap (see Chapter 1).

25   Install paper clips or pins into the backside of the distributor wire harness connector to make contact with the terminals. Be sure not to damage the wire harness when installing the clips.

26   Connect the positive (+) lead of the voltmeter onto the sensor output wire. Refer to the wiring diagrams in Chapter 12 for additional information on the harness schematic.

27   With the voltmeter installed, rotate the engine with the starter until the rotor is pointed toward the rear of the vehicle. The moveable pulse ring will be contained within the magnetic pick-up.

28   With the ignition key turned ON (engine not running), the voltmeter should read approximately 5 volts.

29   If there is no voltage, check for voltage

8.35 Removing the camshaft position sensor from the distributor

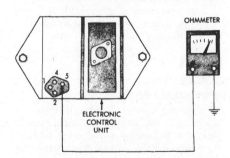

9.1 Checking for ground at the ECU (ignition module) connector (4 pin electronic ignition system shown)

on the supply wire (refer to the wiring diagrams at the end of Chapter 12). There should be approximately 8.0 volts with the ignition key ON (engine not running).

30   If voltage is present, install the probes of the voltmeter onto the signal wire (tan/yellow wire) and crank the engine over and observe the voltage reading. The meter should fluctuate between zero and 5.0 volts indicating that the ignition signal is pulsing properly **(see illustration)**.

31   If the camshaft position sensor fails either of the above checks, replace it.

Replacement

Refer to illustration 8.35

32   Disconnect the negative battery cable from the battery terminal.

33   Remove the engine cover (see Chapter 2B). Remove the distributor cap from the distributor (see Chapter 1).

34   Remove all the electrical connectors from the camshaft position sensor.

35   Remove the camshaft position sensor from the distributor **(see illustration)**.

36   Installation is the reverse of removal.

---

9   Electronic ignition system module - check and replacement

Refer to illustration 9.1

Check

1   Remove the engine cover (see Chapter 2B). Connect one ohmmeter lead to a good ground and the other lead to the control unit pin number 5 **(see illustration)**.

2   The ohmmeter should show continuity between ground and the control unit pin. If continuity does not exist, try removing the control unit (module), cleaning the back side of the housing and tightening the bolts a little tighter to make a better connection on the firewall.

3   Repeat step 1. If continuity still does not exist, replace the control unit.

Replacement

4   Remove the engine cover (see Chapter 2B). Remove the electrical connector from the control unit (module), remove the bolts and lift the assembly from the engine compartment.

---

10   Ballast resistor - check and replacement

Check

Refer to illustrations 10.3a, 10.3b, 10.3c and 10.3d

1   The ballast resistor limits voltage to the coil during low speed operation but allows it to increase as the engine speed increases. While the engine is cranking, the ballast resistor is bypassed to ensure adequate voltage

5

through the coil. During low speed operation when the primary circuit current flow is high, ballast resistor temperature rises, increasing resistance. This reduces current flow thereby prolonging ignition points (contact surface) longevity. At high speed operation, when primary current flow is low, the ballast resistance cools off allowing more current flow which is necessary for high speed operation.

2   On models equipped with a dual ballast resistor, voltage is limited by a pair of resistors; the primary resistor and the auxiliary resistor. Voltage to the electronic control unit is limited by the auxiliary side of the ballast resistor. Some earlier models were equipped with speed limiter circuitry to prevent engine damage from excessive rpms.

3   Disconnect the electrical leads from the ballast resistor and using an ohmmeter, check the resistance of the ballast resistor. On single ballast resistors, the resistance should be 1.1 to 1.8 ohms (electronic ignition systems) **(see illustration)**. On dual ballast resistors, first check the primary side **(see illustrations)**. The primary side is either a thermal type (wire resistor is openly exposed) (it should be 0.5 ohms), or the non-thermal type (wire resistor hidden in case) (it should be 1.12 to 1.38 ohms). Next check the auxiliary side of the dual ballast resistor **(see illustration)**. It should be 4.75 to 5.75 ohms.

Replacement

4   Remove the engine cover (see Chapter 2B). Disconnect the electrical leads from the ballast resistor.

5   Remove the screws from the ballast resistor. and lift it from the engine compartment.

6   Installation is the reverse of removal.

---

11 Spark Control Computer (SCC) system - check and component replacement

Note: *Refer to Chapter 6 for additional checks and replacement procedures for sensors and output actuators incorporated into the SCC system that are not covered in this Section*

General description

1   Some of the later model vehicles covered in this manual are equipped with an SCC ignition system **(see illustration 6.11e)**. This system delivers an infinitely variable spark advance. Some of these ignition advance modes occur only during cold engine operation, thereby reducing HC and nitrous oxide emissions.

2   The computer unit, mounted on the side of the air cleaner housing, constantly sends and receives signals to provide optimum ignition timing for current driving conditions.

3   Depending on the engine temperature and throttle position, vacuum is applied to the diaphragm in the distributor vacuum unit and the ignition timing is changed to reduce

10.3a Checking the resistance on the single type ballast resistor (1987 model - breakerless ignition system shown)

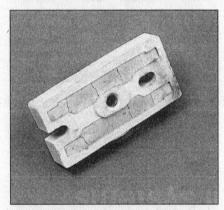

10.3c Remove the ballast resistor from the firewall and check the backside; non-thermal types will have the resistor coils sealed from the atmosphere (shown) while thermal types will have the resistor coils exposed

emissions and improve cold engine driveability. Here is a list of the various components, sensors and output actuators involved with these SCC systems:

> SCC computer
> Vacuum transducer
> Throttle control system
> Carburetor switch
> Coolant temperature sensor
> Charge air temperature sensor
> Detonation (knock) sensor
> Oxygen sensor
> Mixture control solenoid (also called duty cycle solenoid, oxygen feedback solenoid)

Check

SCC computer

Refer to Illustration 11.8

4   Visually check all vacuum hoses for cracks and kinks as well as correct routing. Make sure the engine is warmed up to operating temperature and all the sensors are working correctly (see Chapter 6).

5   Make sure the ignition timing is adjusted properly (see Chapter 1).

6   Make sure the curb idle adjusting screw

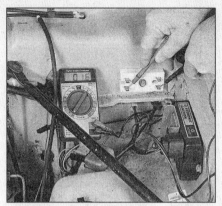

10.3b Checking the primary side of the ballast resistor on a dual ballast resistor type

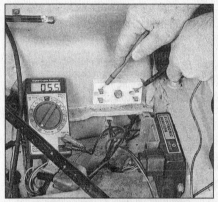

10.3d Check the resistance of the auxiliary side of the ballast resistor - it should be 4.75 to 5.75 ohms

is not touching the carburetor switch (see Chapter 4). If necessary, insert a piece of paper between them to prevent grounding.

7   Remove the vacuum line from the vacuum transducer and plug it with a suitable tool.

8   Connect a hand-held vacuum pump to the vacuum transducer **(see illustration)** and set the vacuum at 10 inches-Hg for six cylinder engines or 16 inches-Hg for eight cylinder engines.

11.8 Check the ignition timing while applying vacuum to the vacuum transducer on the SC computer - the ignition timing should advance 38 to 50-degrees with the engine running at 2,000 rpm

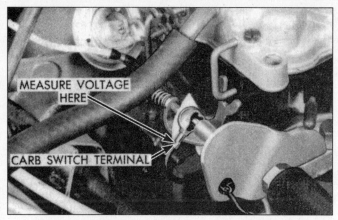

11.11 Install a piece of paper between the idle adjustment screw and the carburetor switch, then check for voltage at the switch terminal

11.14 Continuity should exist between terminal number 7 and the carburetor switch (ELB II systems)

9   Increase the engine speed to 2,000 rpm, wait one minute and confirm that the timing is advanced 38 to 50-degrees depending upon the year and model.

10   If the timing does not advance into the specified range, have the computer checked by a dealer service department.

Carburetor switch

Note: *Grounding the carburetor switch eliminates all spark advance on most of the SCC systems.*

ELB II systems level 4

Refer to illustrations 11.11 and 11.14

11   With the ignition key OFF, disconnect the 10-way connector from the SCC computer and install a piece of paper between the idle adjustment screw and the carburetor sensor switch **(see illustration)**.

12   With the ignition key ON, connect the positive probe of the voltmeter to the carburetor switch and the negative lead to ground.

a) *If the voltage was more than 5 volts but less than 10 volts, proceed to step number 16.*

b) *If the voltage was less than 5 volts, proceed to step number 14.*

c) *If the voltage is more than 10 volts, proceed to step number 13.*

13   Turn the ignition switch OFF and disconnect the 10-way connector on the SCC computer **(see illustration 11.14)**. Connect the probes of a voltmeter to terminal number 2 and ground. With the ignition key ON (engine not running), the voltmeter should indicate battery voltage.

14   If voltage is less than 5 volts, first check for voltage on terminal number 2 (see Step 13) and then with the ignition key OFF, check for continuity between terminal number 7 and the carburetor switch **(see illustration)**. Continuity should exist. **Note:** *On 1981 through 1987 models continuity should NOT exist with the throttle closed. Open the throttle and the circuit should indicate continuity.*

15   Next, check for continuity to the carburetor switch on terminal number 10 **(see illustration 11.14)**. There should be NO continuity. If there is continuity, check the wire harness for shorts or damage to the harness.

16   If the voltage is between 5 and 10 volts, check the pick-up coil resistance (see Section 8).

17   If the ohmmeter indicates continuity, battery voltage is reaching the SCC control unit, pick-up coil resistance is correct and all the circuits check out, replace the SC computer.

Throttle control system

Refer to illustration 11.23

18   The SCC incorporates into the computer an integral system called the throttle control system. A solenoid mounted on the carburetor is energized when the air conditioning switch, electronic backlite (EBL) or electronic timers are activated.

19   Connect a tachometer to the engine (see Chapter 1).

20   Start the engine and warm it up to operating temperature.

21   Depress the accelerator pedal and release it. The idle speed should remain slightly higher than the curb idle speed momentarily during deceleration.

22   On vehicles equipped with air conditioning or EBL, activate the system and confirm that the idle speed increases to compensate for the additional load. **Note:** *The air conditioning clutch will cycle ON and OFF during the test so do not be confused with the actual idle speed adjustments.*

23   If the idle speed does not increase, turn off the ignition, disconnect the electrical connector from the solenoid **(see illustration)** and check the resistance from the black wire to the ground. It should be between 15 and 35 ohms.

24   Start the vehicle and before the timer delay has elapsed, measure the voltage from the black wire of the 3-way connector. The voltmeter should read charging voltage.

25   If it does not, have the SCC computer checked by a dealer service department or other repair shop.

Replacement

Spark Control Computer

Refer to illustration 11.27

26   Remove the air cleaner from the carburetor (see Chapter 4).

27   Remove the screws from inside the air cleaner housing **(see illustration)** and separate the SC computer from the housing.

28   Installation is the reverse of removal.

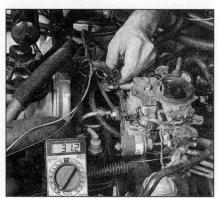

11.23 Disconnect the carburetor solenoid and check the resistance between the black wire and ground - it should read 15 to 35 ohms

11.27 Remove the screws (arrows) and separate the computer from the air cleaner housing

5

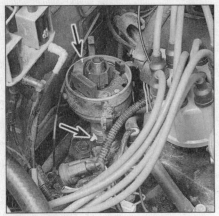

12.3a Be sure to mark the position of the distributor to the engine block (arrow) as well as the position of the rotor to the distributor body (arrow) before removing the assembly from the block

12 Distributor - removal and installation

Refer to illustrations 12.3a, 12.3b and 12.5

Removal

1   Remove the engine cover (see Chapter 2B). Unplug the primary electrical connector from the distributor.
2   Look for a raised "1" on the distributor cap. This marks the location for the number one cylinder spark plug wire terminal. If the cap doesn't have a mark for the number one terminal, locate the number one spark plug and trace the wire back to the terminal on the cap.
3   Mark the relationship of the distributor base to the engine block to ensure the distributor is installed correctly **(see illustration)**. Remove the distributor cap (see Chapter 1) and turn the engine over until the rotor is pointing toward the number one spark plug terminal **(see illustration)** (see locating TDC procedure in Chapter 2).
4   Make a mark on the edge of the distributor base directly below the rotor tip and in line with it.
5   Remove the distributor hold-down bolt and clamp, then pull the distributor straight up to remove it **(see illustration)**. **Caution:** *DO NOT turn the crankshaft while the distributor is out of the engine or the alignment marks will be useless.*

Installation

Note 1: *If the crankshaft has been turned while the distributor is out, the number one piston must be repositioned at TDC. This can be done by feeling for compression pressure at the number one plug hole as the crankshaft is turned. Once compression is felt, align the crankshaft timing mark with the zero on the engine timing scale.*
Note 2: *Be sure to inspect the distributor gear before installing it back into the engine block. Some models are equipped with a*

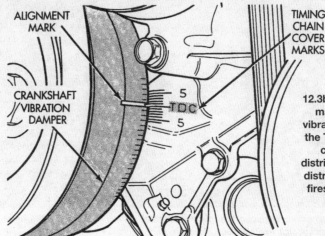

12.3b Be sure the alignment mark on the crankshaft vibration damper aligns with the TDC mark on the timing chain cover while the distributor rotor points to the distributor cap terminal that fires the number 1 cylinder on the engine

nylon drive gear that may need to be replaced because of wear and damage.
6   Insert the distributor into the engine in exactly the same relationship to the block that it was in when removed.
7   On inline six-cylinder models, to mesh the helical gears on the camshaft and distributor, you may have to turn the rotor slightly. Recheck the alignment marks between the distributor base and the block to verify the distributor is in the same position it was before removal. Also check the rotor to see if it's aligned with the mark you made on the edge of the distributor base.
8   On V6 and V8 models, engage the shaft with the slot in the driveshaft. Note that it's possible to install the distributor 180-degrees out of time.
9   Place the hold-down clamp (if equipped) in position and loosely install the bolt.
10   Install the distributor cap.
11   Plug in the primary electrical connector.
12   Reattach the spark plug wires to the plugs (if removed).
13   Connect the cable to the negative terminal of the battery.
14   Check the ignition timing (refer to Chapter 1) and tighten the distributor hold-down bolt securely.
15   If the engine was rotated with the distributor removed, perform the following:

a)   *Rotate the crankshaft until the number one piston is at top dead center (TDC) on the compression stroke. The mark on the damper should be aligned with the 0 degree (TDC) mark on the timing chain case. Do not confuse this setting with the companion cylinder TDC. Check the TDC setting procedure in Chapter 2 to verify valve position.*
b   *Install the distributor rotor onto the distributor shaft. Rotate the rotor approximately to the "CYL NO. 1" alignment mark on the distributor cap.*
c)   *Install the distributor into the engine (to its original position). Engage the distributor shaft with the slot in the oil pump drive gear (V6 and V8) or engage the gears (inline six-cylinder - see step 7).*

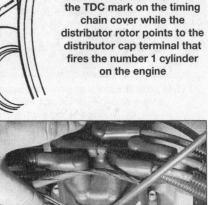

12.5 Location of the distributor hold-down clamp bolt

d)   *Install the distributor hold-down clamp and bolt. Do not tighten it at this time.*
e)   *Slowly rotate the distributor until the rotor is aligned with the "CYL. NO. 1" alignment mark on the distributor cap.*
f)   *Tighten the hold-down clamp bolt securely.*
g)   *Install the distributor cap.*

13 Air gap - check and adjustment

Single pick-up coil type

Refer to illustration 13.4

1   Release the spring clips or hold-down screws and remove the distributor cap. Secure the cap out of way.
2   Align one of the reluctor teeth attached to the distributor shaft with the pick-up coil tooth.
3   Loosen the pick-up coil adjusting screw.
4   Insert a non-magnetic type feeler gauge of the correct thickness between the reluctor and the pick-up coil **(see illustration)**. Refer to the Specifications listed in this Chapter for the correct gap and use a feeler gauge of that thickness.
5   Move the pick-up coil until light contact is made with the feeler gauge.
6   Tighten the pick-up coil adjusting screw.

13.4 Checking the air gap on a single pick-up coil distributor

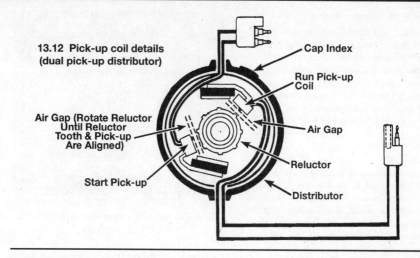

13.12 Pick-up coil details (dual pick-up distributor)

Cap Index

Run Pick-up Coil

Air Gap (Rotate Reluctor Until Reluctor Tooth & Pick-up Are Aligned)

Air Gap

Start Pick-up

Reluctor

Distributor

7   Remove the feeler gauge. No force should be required to remove the gauge.
8   Using a feeler gauge 0.002-inch larger than specified, try to check the air gap. The feeler gauge should not fit between the teeth (do not force it). Reset the air gap if necessary.
9   Replace the distributor cap.

Dual pick-up coil type

Refer to illustration 13.12
10   Some models equipped with Electronic Ignition systems from 1976 through 1987 are equipped with dual pick-up coils. These systems include ELB II, Combustion Computer, SC computer and Electronic Fuel Control computer systems. The ignition module or computer is separated into two integral units contained within the computer housing; the program schedule unit and the ignition control unit. The program schedule regulates the RUN pick-up coil (running condition) while the ignition control unit controls the START pick-up coil (starting condition) within the distributor. The air gap for each pick-up coil is set to a different specification designed for the different functions.
11   Release the spring clips or hold-down screws and remove the distributor cap. Secure the cap out of way.
12   Align one of the reluctor teeth attached to the distributor shaft with the pick-up coil tooth. **Note:** *First adjust the START pick-up coil and then the RUN pick-up coil* **(see illustration)**. *The procedure is the same for both but the air gap specifications are different. The START pick-up coil electrical connector is the larger of the two connectors.*
13   Loosen the pick-up coil adjusting screw.
14   Insert a non-magnetic type feeler gauge of the correct thickness between the reluctor and the pick-up coil. Refer to the Specifications listed in this Chapter for the correct gap and use a feeler gauge of that thickness.
15   Move the pick-up coil until light contact is made with the feeler gauge.
16   Tighten the pick-up coil adjusting screw.
17   Remove the feeler gauge. No force should be required to remove the gauge.
18   Using a feeler gauge 0.002-inch larger than specified, try to check the air gap. The

feeler gauge should not fit between the teeth (do not force it). Reset the air gap if necessary.
19   Reinstall the distributor cap.

14   Charging system - general information and precautions

The charging system on 1971 through 1987 models includes the alternator with an external voltage regulator regulating the charging system output. Also included in the system is a charge indicator light, the battery, a fusible link and the wiring between all the components. The charging system supplies electrical power for the ignition system, the lights, the radio, etc. The alternator is driven by a drivebelt at the front of the engine.

The fusible link is a short length of insulated wire integral with the engine compartment wiring harness. The link is four wire gauges smaller in diameter than the circuit it protects. Production fusible links and their identification flags are identified by the flag color. See Chapter 12 for additional information regarding fusible links.

The charging system on 1988 and later models is controlled by the computer (SMEC on 1988 through 1991 or PCM on 1992 and later). In the event of voltage regulator failure, the regulator contained within the SMEC/PCM must be replaced as a complete unit.

The alternator control system on later systems within the SMEC/PCM changes the voltage generated at the alternator in accordance with driving conditions. Depending upon electric load, vehicle speed, engine coolant temperature, accessories (A/C system, radio, cruise control etc.) and the intake air temperature, the system will adjust the amount of voltage generated, creating less load on the engine.

Starting in 1988, Chrysler also equipped these models with a self-diagnosis system for the charging system. The codes are accessed using the ignition key and observing the CHECK ENGINE light on the dashboard. Refer to Chapter 6 for the complete procedure on the self diagnosis system.

The purpose of the voltage regulator is to limit the alternator's voltage to a preset value. This prevents power surges, circuit overloads, etc., during peak voltage output.

The charging system doesn't ordinarily require periodic maintenance. However, the drivebelt, battery and wires and connections should be inspected at the intervals outlined in Chapter 1.

The dashboard charge warning light should come on when the ignition key is turned to Start, then go off immediately. If it remains on, there is a malfunction in the charging system (see Section 15). Some vehicles are also equipped with a voltmeter. If the voltmeter indicates abnormally high or low voltage, check the charging system

Be very careful when making electrical circuit connections to a vehicle equipped with an alternator and note the following:

a) *When reconnecting the wires to the alternator from the battery, be sure to note the polarity.*
b) *Before using arc welding equipment to repair any part of the vehicle, disconnect the wires from the alternator and the battery terminals.*
c) *Never start the engine with a battery charger connected.*
d) *Always disconnect both battery leads before using a battery charger.*
e) *The alternator is turned by an engine drivebelt which could cause serious injury if your hands, hair or clothes become entangled in it with the engine running.*
f) *Because the alternator is connected directly to the battery, it could arc or cause a fire if overloaded or shorted out.*
g) *Wrap a plastic bag over the alternator and secure it with rubber bands before steam cleaning the engine.*

15   Charging system - check

Refer to illustrations 15.7a and 15.7b
1   If a malfunction occurs in the charging circuit, don't automatically assume that the alternator is causing the problem.

5

First check the following items:

a) *Check the drivebelt tension and condition (see Chapter 1). Replace it if it's worn or deteriorated.*
b) *Make sure the alternator mounting and adjustment bolts are tight.*
c) *Inspect the alternator wiring harness and the electrical connectors at the alternator and voltage regulator (if equipped). They must be in good condition and tight.*
d) *Check the fusible link (if equipped) located between the starter solenoid and the alternator. If it's burned, determine the cause, repair the circuit and replace the link (the vehicle won't start and/or the accessories won't work if the fusible link blows). Sometimes a fusible link may look good, but still be bad. If in doubt, remove it and check for continuity.*
e) *Start the engine and check the alternator for abnormal noises (a shrieking or squealing sound indicates a bad bearing).*
f) *Check the specific gravity of the battery electrolyte. If it's low, charge the battery (doesn't apply to maintenance free batteries).*
g) *Make sure the battery is fully charged (one bad cell in a battery can cause overcharging by the alternator).*
h) *Disconnect the battery cables (negative first, then positive). Inspect the battery posts and the cable clamps for corrosion. Clean them thoroughly if necessary (see Chapter 1). Reconnect the cable to the positive terminal.*
i) *With the key OFF, connect a test light between the negative battery post and the disconnected negative cable clamp.*
   1) *If the test light does not come on, reattach the clamp and proceed to the next Step.*
   2) *If the test light comes on, there is a short (drain) in the electrical system of the vehicle. The short must be repaired before the charging system can be checked.*
   3) *Disconnect the alternator wiring harness.*
      (a) *If the light goes out, the alternator is bad.*
      (b) *If the light stays on, pull each fuse until the light goes out (this will tell you which component is shorted).*

2   Using a voltmeter, check the battery voltage with the engine off. If should be approximately 12-volts.
3   Start the engine and check the battery voltage again. It should now be approximately 14-to-15 volts.
4   Turn on the headlights. The voltage should drop, and then come back up, if the charging system is working properly.
5   If the voltage reading is more than the specified charging voltage, replace the voltage regulator (refer to Section 17). If the voltage is less, the alternator diode(s), stator or

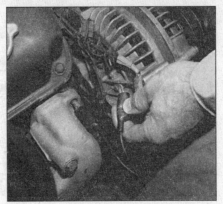

15.7a Disconnect the green wire from the FIELD terminal

15.7b Slowly raise the engine rpm and monitor the charging voltage. Do not allow the charging voltage to exceed 16 volts because of the danger of possible circuit and electrical component damage

rectifier may be bad or the voltage regulator may be malfunctioning. **Note:** *The charging system on 1988 and later models is controlled by the computer (SMEC on 1988 through 1991 or PCM on 1992 and later). In the event of voltage regulator failure, the regulator contained within the SMEC/PCM must be replaced as a complete unit.*
6   On 1974 through 1987 models, a good way to determine whether an undercharging problem is caused by the alternator or regulator is with a full-field test. Basically, the full-field test bypasses the regulator to send full battery voltage to the alternator's field (the rotor). If the charging voltage is normal, when the alternator is "full-fielded", you know the alternator is OK. If the voltage is still low, the problem is in the alternator. It's best to obtain wiring diagrams for the vehicle to determine the best way to send battery voltage to the field. However, this is the best method for the older style Delco alternators.
7   Be sure this is a Delco (Chrysler) alternator with an externally mounted voltage regulator. Disconnect the regulator connector and connect a jumper wire between the green wire terminal of the connector and ground **(see illustration)**. Make the connections with the ignition turned OFF, then start the engine and repeat the charging system check (see Step 3). The voltage reading should be high (above 15 to 16 volts) **(see illustration)**. If it is not the alternator is faulty. If is correct, the regulator is mostly likely defective. **Caution:** *Full-fielding sends high voltage through the vehicle's electrical system, which can damage components, particularly electronic components. Carefully monitor the charging system voltage during a full-fielding to be sure it does not exceed 16 volts. Operate it only long enough to take the voltage reading.*

16 Alternator - removal and installation

Refer to illustration 16.2
1   Detach the cable from the negative terminal of the battery.
2   Loosen the alternator adjustment and pivot bolts and detach the drivebelt **(see**

16.2 Loosen the pivot bolt (arrow) and the adjustment bolt on the alternator brackets

illustration).
3   Detach the wires from the alternator.
4   Remove the adjustment and pivot bolts and separate the alternator from the engine.
5   If you are replacing the alternator, take the old alternator with you when purchasing a replacement unit. Make sure the new/rebuilt unit is identical to the old alternator. Look at the terminals - they should be the same in number, size and location as the terminals on the old alternator. Finally, look at the identification markings - they will be stamped in the housing or printed on a tag or plaque affixed to the housing. Make sure these numbers are the same on both alternators.
6   Many new/rebuilt alternators do not have a pulley installed, so you may have to switch the pulley from the old unit to the new/rebuilt one. When buying an alternator, find out the shop's policy regarding pulleys - some shops will perform this service free of charge.
7   Installation is the reverse of removal.
8   After the alternator is installed, adjust the drivebelt tension (see Chapter 1).
9   Check the charging voltage to verify proper operation of the alternator (see Section 15).

17.1 The voltage regulator is mounted in the engine compartment and can be detached after unplugging the electrical connectors and removing the mounting bolts

17 Voltage regulator (external type) - replacement

Refer to illustration 17.1

1   Disconnect the electrical connector from the voltage regulator **(see illustration)**.

2   Remove the mounting screws and regulator from the firewall.

3   Installation is the reverse of removal. Before mounting the regulator to the firewall, make sure the regulator and the firewall are clean (bare metal) so a good ground will be achieved.

18 Alternator components - check and replacement

1   Remove the alternator (see Section 16) and place it on a clean workbench.

Nippondenso alternators

Brushes

Refer to illustrations 18.2a, 18.2b, 18.3, 18.4, 18.5, 18.13a, 18.13b, 18.18a, 18.18b, 18.19, 18.20a, 18.20b and 18.20c

2   Remove the three rear cover nuts, the nut and terminal insulator and the rear cover **(see Illustrations)**.

3   Remove the two brush holder retaining

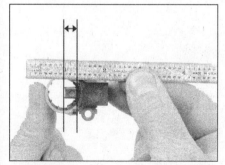

18.5 Measure the exposed length of the brushes and compare your measurements to the specified minimum length to determine if they should be replaced

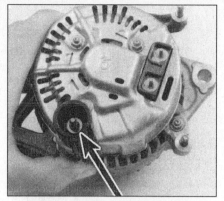

18.2a To replace the brushes on Nippondenso alternators, remove the B+ terminal insulator nut, washer and insulator

screws **(see illustration)**.

4   Remove the brush holder from the rear end frame **(see illustration)**.

5   Measure the exposed length of the brush **(see Illustration)** and compare it to the specified minimum length. If the length of the brush is less than the minimum listed in this Chapter's Specifications, replace the brush.

6   Make sure that each brush moves smoothly in the brush holder.

7   Install the brush holder by depressing the brush with a small screwdriver to clear the shaft.

8   Install the brush holder screws into the rear frame.

9   Install the rear cover and tighten the three nuts securely.

10 Install the terminal insulator and tighten it with the nut.

11 Install the alternator (see Section 16).

Voltage regulator

12 The voltage regulator is contained within the SMEC (1988 through 1991) or the PCM (1992 and later). Replace the computer in the event of voltage regulator charging problems (see Chapter 6).

Rotor, stator and diodes

13 Remove the field block and the rectifier assembly **(see illustrations)**. Remove the four rubber insulators and the seal plate.

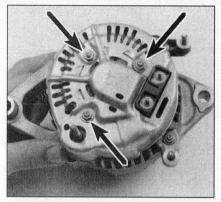

18.2b Remove the rear cover attaching nuts (arrows) and remove the rear cover

18.3 Remove the brush holder attaching screws (arrows) . . .

18.4 . . . and remove the brush holder assembly

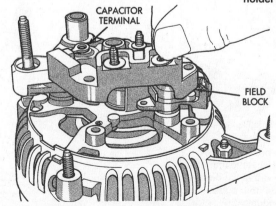

18.13a First remove the field block and . . .

5

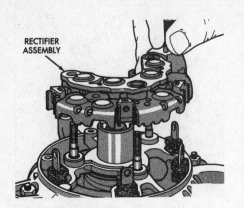

18.13b ... then remove the rectifier assembly

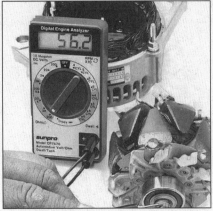

18.18a Continuity should exist between the rotor slip rings

18.18b Check the continuity between the rotor and the slip rings. There should be NO continuity

14   Scribe or paint marks on the front and rear end frame housings of the alternator to facilitate reassembly.

15   Remove the nut retaining the pulley to the rotor shaft and remove the pulley.

16   Remove the four nuts retaining the front and rear end frame together, then separate the rear end frame assembly from the front end frame.

17   Remove the thrust washer and separate the rotor from the end frame.

18   Check for an open between the two slip rings **(see illustration)**. There should be 2 to 4 ohms resistance between the slip rings. Check for grounds between each slip ring and the rotor **(see illustration)**. There should be no continuity (infinite resistance) between the rotor and either slip ring. If the rotor fails either test, or if the slip rings are excessively worn, the rotor is defective.

19   Check for opens between each end terminal of the stator windings **(see illustration)**. If either reading is high (infinite resistance), the stator is defective. Check for a grounded stator winding between each stator terminal and the frame. If there's continuity between any stator winding and the frame the stator is defective.

20   Start the checks on the diode assembly

by touching one probe (positive +) of the ohmmeter onto the diode terminal and the other probe (negative -) onto one of the other designated diode terminals **(see illustration)**. Then reverse the probes and check again **(see illustrations)**. The diode should have continuity with the ohmmeter one way and no continuity when the probes are reversed. Check each of the terminals in this manner. If any of the diodes fail the test (a total of sixteen tests), the diode assembly is defective.

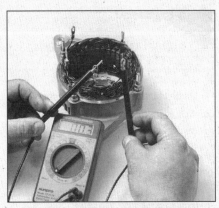

18.19 Check for continuity between the stator windings

Chrysler alternators

Alternator brushes

Refer to illustrations 18.21a, 18.21b, 18.21c and 18.22

21   Remove the brush holder mounting screws a little at a time to prevent distortion of the holder **(see illustrations)**.

22   Rotate the brush holder and separate it from the rear of the alternator **(see illustration)**.

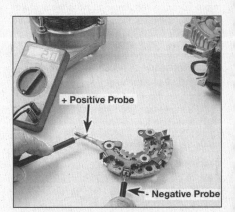

18.20a Position the positive probe of the ohmmeter onto the diode assembly positive post and the negative probe to ground. Continuity should exist

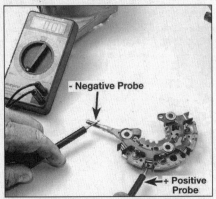

18.20b Switch the polarity of the ohmmeter probes and observe that now there is NO continuity within the diodes. Check each diode (four pairs total) individually

18.20c Checking the diodes on externally mounted rectifier assemblies on 1988 through 1991 models

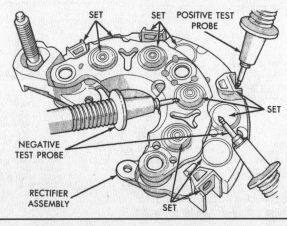

18.21a Typical mounting locations for the alternator brushes on early models

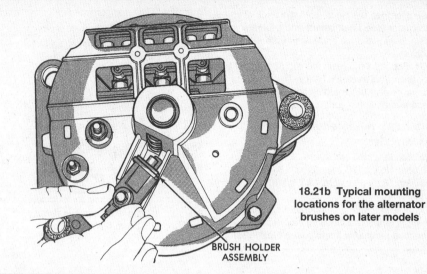

18.21b Typical mounting locations for the alternator brushes on later models

BRUSH HOLDER ASSEMBLY

23 If the brushes are worn badly, or if they do not move smoothly in the brush holder, replace the brush holder assembly with a new one.
24 Insert the holder into position, making sure the brushes seat correctly.
25 Hold the brush holder securely in place

and install the screws. Tighten them evenly, a little at a time, so the holder is not distorted. Once the screws are snug, tighten them securely.
26 Install the alternator.

Rotor, stator and diodes

Refer to illustrations 18.27, 18.28, 18.29a, 18.29b and 18.30

27 Remove the dust cover **(see illustration)** and the alternator brushes (see Steps 21 and 22).
28 On externally mounted rectifier assemblies, remove the rectifier from the alternator body **(see illustration)**.

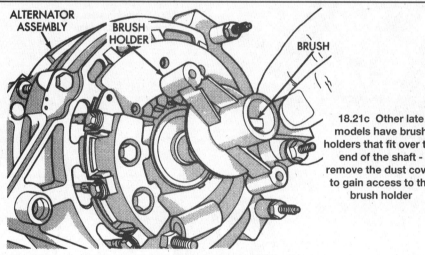

ALTERNATOR ASSEMBLY

BRUSH HOLDER

BRUSH

18.21c Other late models have brush holders that fit over the end of the shaft - remove the dust cover to gain access to the brush holder

18.22 The brushes on early model alternators slip out from under the rear cover

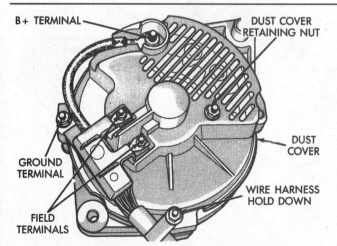

B+ TERMINAL

DUST COVER RETAINING NUT

DUST COVER

WIRE HARNESS HOLD DOWN

GROUND TERMINAL

FIELD TERMINALS

18.27 Remove the mounting nuts and separate the dust cover from the alternator body on late model alternators

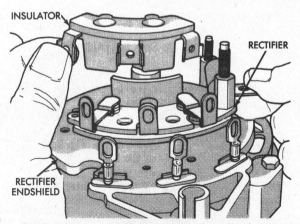

INSULATOR

RECTIFIER

RECTIFIER ENDSHIELD

18.28 On late model Chrysler alternators, first remove the insulator, then remove the rectifier assembly (1988 Chrysler 120HS model shown)

5

29   Remove the bolts from the back of the alternator **(see illustration)** and separate the rotor from the drive end shield **(see illustration)**.

30   Separate the rear housing from the stator **(see illustration)**. Follow the procedure for removing the rectifier assembly (below) before removing the stator.

Internally mounted rectifier assemblies

Refer to illustrations 18.31a, 18.31b 18.32, 18.33a, 18.33b, 18.33c, 18.34, 18.35a, 18.35b, 18.36a, 18.36b, 18.37a, 18.37b and 18.37c

31   Remove the four mounting screws from the top of the rear housing and separate the negative rectifier assembly **(see illustration)**.

18.29a Remove the three bolts (arrows) . . .

18.29b . . . and detach the rear cover from the alternator

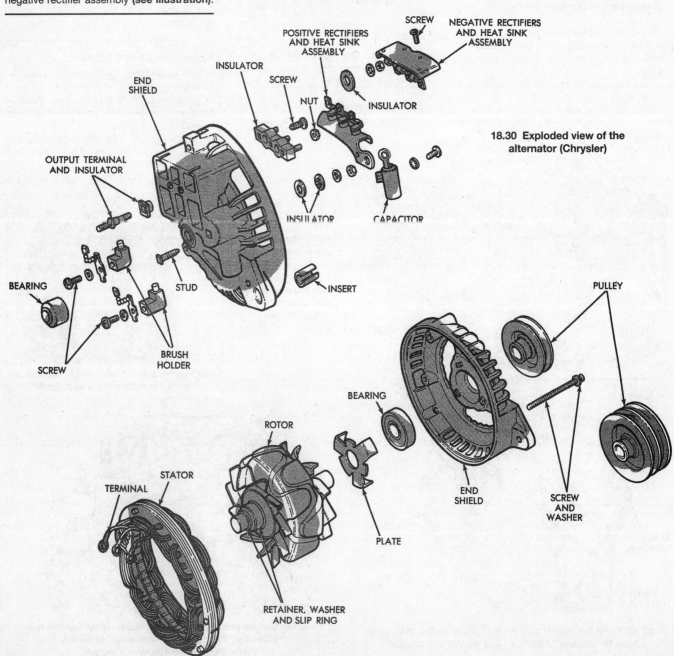

18.30 Exploded view of the alternator (Chrysler)

18.31a Remove the bolts (arrows) and separate the negative rectifier assembly from the end shield

18.31b Chrysler alternator rebuild kit

18.32 Remove the condenser mounting bolt

Note: *The following illustrations include parts and repair procedures commonly found in an alternator overhaul kit from a local parts distributor* (see illustration).

32   Remove the condenser mounting bolt and condenser (see illustration).

33   Remove the mounting nuts and separate the positive rectifier assembly (see illustration) from the rear housing, then the negative rectifier assembly and the stator (see illustrations).

34   Test the diodes.

a) *Test the positive rectifier assembly by placing the negative probe of the multimeter onto the terminal and the positive probe onto the body (see illustration). Continuity should exist. Next, reverse the position of the probes and check for continuity. Continuity should NOT exist. This checks the unidirectional capability of the operating diode. Next, repeat this test for each of the other two diodes (total six checks). Each diode should have the exact same characteristics. Note: It will be necessary to use a multi-*

18.33a Remove the nuts (arrows) that retain the positive rectifier assembly to the end shield

meter that includes a diode check function. This function allows slight current application to the diode to assist in opening the voltage gate.

b) *To test the negative rectifier assembly, follow the exact same procedure as*

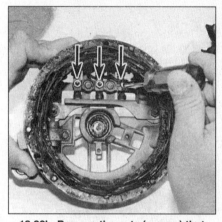

18.33b Remove the nuts (arrows) that retain the negative rectifier assembly to the end shield

detailed above. Be sure to use a multimeter with a diode check function. If the test results are incorrect for one of the diode terminals, replace the complete unit. The alternator overhaul kit contains both parts for the complete repair.

5

18.33c Lift the stator from the end shield

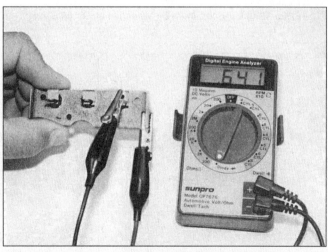

18.34 Position the negative probe (-) of the ohmmeter onto the terminal and the positive probe (+) onto the body and confirm that there is continuity - with the probes switched, there should be NO continuity

18.35a Continuity should exist between the rotor slip rings

18.35b Check the continuity between the rotor and the slip rings. There should be no continuity

35   Test the rotor. Check for an opening between the two slip rings **(see illustration)**. There should be 2 to 4 ohms resistance between the slip rings. Check for grounds between each slip ring and the rotor **(see illustration)**. There should be no continuity (infinite resistance) between the rotor and

either slip ring. If the rotor fails either test, or if the slip rings are excessively worn, the rotor is defective. **Note:** *If the rotor is determined to be defective, it will be necessary to use a special puller to remove the pulley from the rotor.*

36   Test the stator. Check for openings

between each end terminal of the stator windings **(see illustration)**. If either reading is high (infinite resistance), the stator is defective. Check for a grounded stator winding between each stator terminal and the frame **(see illustration)**. If there's continuity between any stator winding and the frame, the stator is defective.

37   Installation is the reverse of removal. Install the new brushes into the brush holders, placing an insulating washer between the mounting screw and the brush **(see illustrations)**.

Externally mounted rectifier assemblies

Refer to illustration 18.39

38   Remove the rectifier assembly from the alternator body (see Step 28).

39   Using an analog ohmmeter, test for continuity from the positive heat sink to each of the positive diode leads **(see illustration)**.

40   Reverse the test probes and repeat the test. Continuity should exist in one direction only.

41   Test the rotor and stator as detailed in Steps 35 and 36 above.

42   Installation is the reverse of removal.

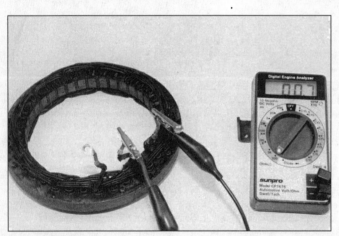

18.36a Check for continuity between the stator windings

18.36b Make sure NO continuity (open circuit) exists between the stator body and the windings

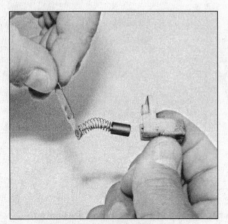

18.37a Install the new brush into the holder

18.37b Install new insulating washers into the end shield before installing the positive rectifier assembly

18.37c Apply a small amount of grease onto the bearing surface before assembly

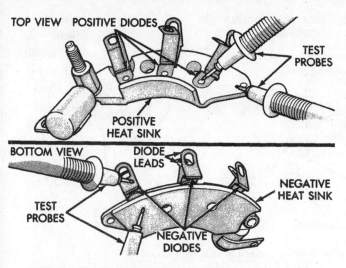

18.39 Position the probe of the ohmmeter onto the positive diodes and the other probe to heat sink, then reverse the leads and repeat the test - continuity should exist in one direction only

21.5 Starter motor assembly on a V8 engine - it's retained by two bolts (arrows)

19 Starting system - general information and precautions

The sole function of the starting system is to turn over the engine quickly enough to allow it to start.

The starting system consists of the battery, the starter motor, the starter relay, the starter solenoid and the wires connecting them. The solenoid is mounted directly on the starter motor or is a separate component located in the engine compartment.

The solenoid/starter motor assembly is installed on the lower part of the engine, next to the transmission bellhousing.

When the ignition key is turned to the START position, the starter solenoid is actuated through the starter relay and control circuit. The starter solenoid then connects the battery to the starter. The battery supplies the electrical energy to the starter motor, which does the actual work of cranking the engine.

The starter motor on a vehicle equipped with a manual transmission can only be operated when the clutch pedal is depressed; the starter on a vehicle equipped with an automatic transmission can only be operated when the transmission selector lever is in Park or Neutral.

Always observe the following precautions when working on the starting system:

a) *Excessive cranking of the starter motor can overheat it and cause serious damage. Never operate the starter motor for more than 15 seconds at a time without pausing to allow it to cool for at least two minutes.*

b) *The starter is connected directly to the battery and could arc or cause a fire if mishandled, overloaded or shorted out.*

c) *Always detach the cable from the negative terminal of the battery before working on the starting system.*

20 Starter motor - testing in vehicle

Note: *Before diagnosing starter problems, make sure the battery is fully charged.*

1   If the starter motor does not turn at all when the ignition switch is operated, make sure the shift lever is in Neutral or Park.

2   Make sure that the battery is charged and that all cables, both at the battery and starter solenoid terminals, are clean and secure.

3   If the starter motor spins but the engine is not cranking, the overrunning clutch in the starter motor is slipping and the starter motor must be replaced.

4   If, when the switch is actuated, the starter motor does not operate at all but the solenoid clicks, then the problem lies with either the battery, the main solenoid contacts or the starter motor itself, or the engine is seized.

5   If the solenoid plunger cannot be heard when the switch is actuated, the battery is bad, the fusible link is burned (the circuit is open), the starter relay is malfunctioning or the solenoid itself is defective.

6   To check the solenoid, connect a jumper lead between the battery (positive terminal) and the ignition switch terminal (the small terminal) on the solenoid. If the starter motor now operates, the solenoid is OK and the problem is in the ignition switch, neutral start switch, starter relay (see Section 23) or in the wiring.

7   If the starter motor still does not operate, remove the starter/solenoid unit for disassembly, testing and repair.

8   If the starter motor cranks the engine at an abnormally slow speed, first make sure that the battery is charged and that all terminal connections are tight. If the engine is partially seized, or has the wrong viscosity oil in it, it will crank slowly.

9   Run the engine until normal operating temperature is reached, then disconnect the coil wire from the distributor cap and ground it on the engine.

10   Connect a voltmeter positive lead to the battery positive post and then connect the negative lead to the negative post.

11   Crank the engine and take the voltmeter readings as soon as a steady figure is indicated. Do not allow the starter motor to turn for more than 15 seconds at a time. A reading of nine volts or more, with the starter motor turning at normal cranking speed, is normal. If the reading is nine volts or more but the cranking speed is slow, the motor is faulty. If the reading is less than nine volts and the cranking speed is slow, the solenoid contacts are probably burned, the starter motor is bad, the battery is discharged or there is a bad connection.

21 Starter motor - removal and installation

Refer to illustration 21.5

1   Disconnect the cable from the negative terminal of the battery.

2   Remove the engine cover (see Chapter 2B). Remove the heat shield (if equipped).

3   Detach the battery cable from the starter.

4   Disconnect the solenoid wire from the solenoid.

5   Remove the starter mounting nuts/bolts **(see illustration)**.

6   Remove the starter assembly.

7   Installation is the reverse of removal.

22 Starter motor brushes - replacement

Refer to illustrations 22.2, 22.5 and 22.19

1   Remove the starter motor assembly (see Section 21).

5

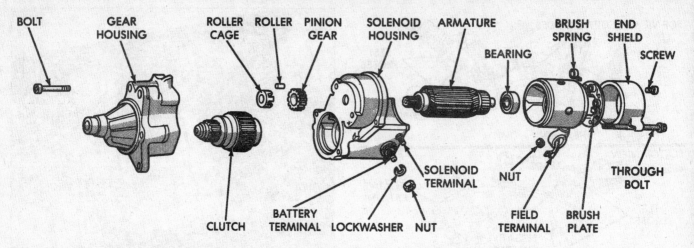

22.2 Exploded view of a Nippondenso starter/solenoid assembly on a 1988 model

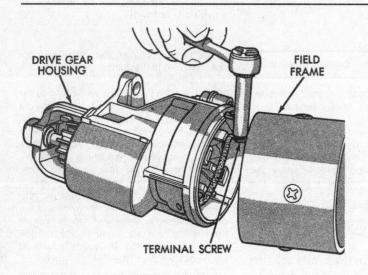
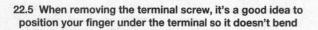

22.5 When removing the terminal screw, it's a good idea to position your finger under the terminal so it doesn't bend

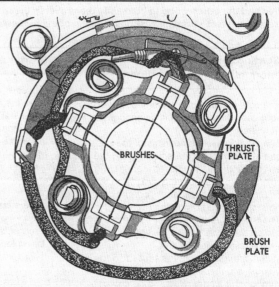

22.19 Correct positioning of the starter brushes on a Chrysler starter/solenoid assembly

2   Remove the through bolts and separate the end head assembly from the field frame assembly (see illustration).
3   Carefully pull the armature out of the gear housing and field frame assembly.
4   Pull the field frame assembly away from the gear housing just far enough to expose the terminal screw.
5   Support the brush terminal screw with a finger and remove the screw (see illustration).
6   Remove the field frame assembly.
7   Remove the mounting nuts from the solenoid and brush holder plate assembly and separate the assembly from the gear housing.
8   Remove the nut and washers from the solenoid terminal.
9   Unwind the solenoid wire from the starter brush terminal.
10   Separate the solenoid wire and the solenoid from the brush plate.
11   The brushes can now be removed from

the brush holders and new ones installed.
12   Check the condition of the contacts in the brush holder plate. If they are badly burned, replace the plate and brushes as an assembly.
13   Use a new gasket when attaching the brush holder plate to the solenoid. If a new gasket is not available, apply a thin bead of RTV sealant to the outside edge of the gasket.
14   Insert the solenoid wire through the hole in the brush holder.
15   Insert the brush holder plate mounting bolts and attach the plate to the solenoid.
16   Install the nut and washers on the solenoid terminal.
17   Wind the solenoid wire tightly around the brush terminal and solder it in place with a high temperature resin core solder and resin flux.
18   Carefully install the solenoid and brush plate assembly in the gear housing and install the bolts.

19   Position the brushes against the thrust plate (see illustration).
20   Install the brush terminal screw.
21   Carefully install the field frame assembly and armature. Rotate the armature shaft slightly to engage it with the reduction gear.
22   Install the thrust washer onto the armature shaft.
23   Install the end head assembly and through bolts. Tighten the through bolts securely.

23   Starter relay - check and replacement

Refer to illustrations 23.3, 23.4a, 23.4b and 23.7

Check

1   If the starter doesn't crank when the ignition key is turned to Start, but the battery

STARTER RELAY CONNECTIONS

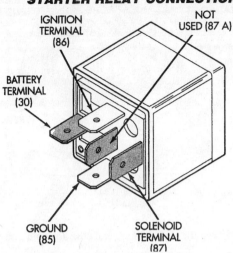

23.3 On 1992 and later starter relays, remove the relay and check for battery voltage on the electrical connector that coincides with terminal 30 on the relay

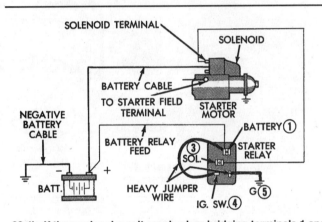

23.4b If the engine doesn't crank when bridging terminals 1 and 4, repeat the test with another jumper wire connected between terminal no. 5 and ground - if the engine cranks, the circuit between terminal no. 5 and ground is bad, the shift linkage is out of adjustment or the neutral start switch is out of adjustment or defective - if the engine doesn't crank, replace the relay

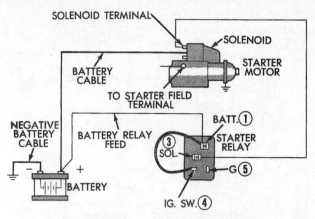

23.4a If the engine cranks when a jumper wire is connected between terminals 1 and 4, the relay is good

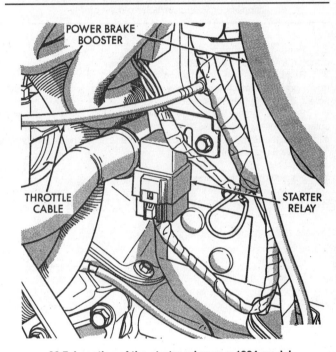

23.7 Location of the starter relay on a 1994 model

and ignition switch have checked out OK, check for power to the starter solenoid as an assistant turns the key. If there is no power present at the starter solenoid when the key is turned to start, check the starter relay.

2 Place the gear selector lever in the Neutral or Park position (automatic transmission) or the Neutral position (manual transmission). Set the parking brake.

3 Using a test light, check for battery voltage between the battery positive terminal of the starter relay (terminal no. 1) and ground (see illustration 23.4 for terminal locations). If there is no power present, repair the wire between the battery and the relay. Note: 1992 and later models are equipped with a slightly different style relay. The tests are the same but the terminal numbers are different (see illustration).

4 If there is power available on terminal no. 1, connect a jumper wire between terminals 1 (battery) and 4 (ignition) (see illustration). If the engine cranks, the starter relay is good. If the engine doesn't crank, connect another jumper wire between the ground terminal (no. 5) and a good ground (body) (see illustration).

5 Bridge terminals 1 and 4 again - if the engine now cranks, the relay is OK. Check the ground wire from the relay and the neutral safety switch or shift linkage adjustment. If these items check out OK, replace the starter relay.

Replacement

6 Disconnect the cable from the negative terminal of the battery.

7 The starter relay is located on the left

side of the firewall on some models. On others, it's located on the left inner fender panel (see illustration). If equipped, remove the cover from the relay.

8 Remove the relay mounting screw, remove the relay and unplug the electrical connector.

9 Installation is the reverse of removal.

24 Starter solenoid (Nippondenso starters) - removal and installation

Refer to illustrations 24.1, 24.3 and 24.4

1 Remove the starter from the vehicle (see Section 21), then remove the terminal cover (see illustration).

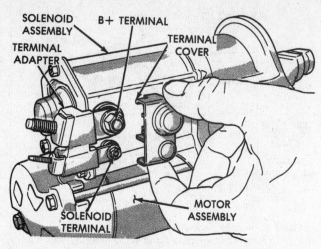

24.1 Remove the terminal cover to expose the terminals on a Nippondenso starter/solenoid assembly

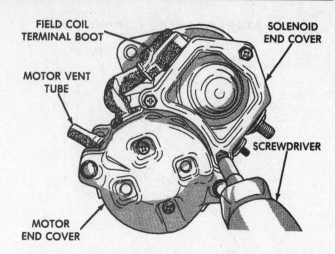

24.3 Remove the screws that retain the solenoid end cover

2   Remove the nuts mounting the terminal adapter, then remove the adapter.

3   Remove the screws retaining the end cover and separate the end cover from the solenoid housing **(see illustration)**.

4   Pull the solenoid plunger from the solenoid housing and inspect the contact and sliding surfaces for damage and excessive wear **(see illustration)**. Replace it if necessary.

5   Installation is the reverse of removal.

24.4 Remove the solenoid plunger and check for damage

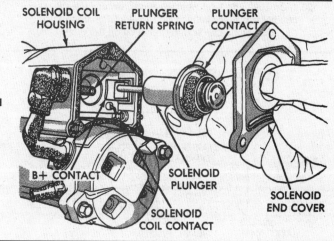

Chapter 6
Emissions and engine control systems

Contents

Specifications

Torque specifications

Crankshaft sensor bolts	70 in-lbs
Camshaft sensor retaining bolt	85 in-lbs

1   General information

Refer to illustrations 1.1 and 1.6a through 1.6as

Note: *1982 and 1983 models are not equipped with a factory vacuum schematic. Chrysler developed a vacuum harness that is not serviceable for those years. Consult a dealer parts department for additional information.*

1   To prevent pollution of the atmosphere from incompletely burned and evaporating gases, and to maintain good driveability and fuel economy, a number of emission control systems are incorporated. The combination of systems used depends on the year in which the vehicle was manufactured, the locality to which it was originally delivered (referred to in this Chapter as California, Canadian or Federal (which is any state except California) and the engine type. The systems incorporated on the

vehicles with which this manual is concerned include the **(see illustration):**

Air Injection (AIR) system
Air aspirator system
Catalytic converter (CAT)
Electric assist choke system
Exhaust Gas Recirculation (EGR) system
Feedback carburetor system (see Chapter 4A)
Fuel evaporative emission control (EVAP) system

6

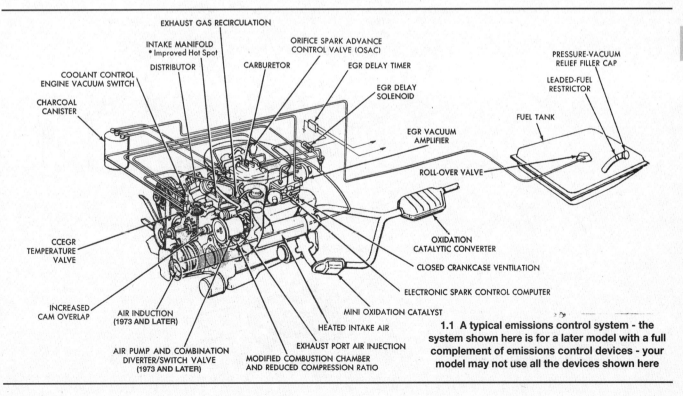

1.1 A typical emissions control system - the system shown here is for a later model with a full complement of emissions control devices - your model may not use all the devices shown here

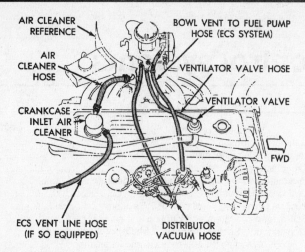

1.6a 1971 through 1973 crankcase ventilation system for inline six-cylinder engines

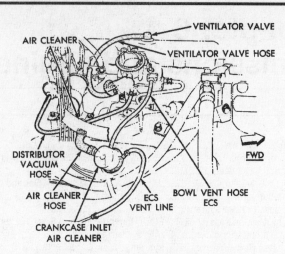

1.6b 1971 through 1973 crankcase ventilation system for V8 engines

Heated Inlet Air (HIA) system
Idle enrichment system
Orifice Spark Advance Control (OSAC)
Positive Crankcase Ventilation (PCV) system
Spark Control Computer (SCC) system
Vacuum throttle positioner

All of these systems are linked, directly or indirectly, to the emissions control system.
2   The Sections in this Chapter include general descriptions, checking procedures within the scope of the home mechanic and component replacement procedures (when possible).
3   Before assuming that an emissions con-trol system is malfunctioning, check the fuel and ignition systems carefully. The diagnosis of some emission control devices requires specialized tools, equipment and training. If checking and servicing become too difficult or if a procedure is beyond your ability, con-sult a dealer service department. Remember, the most frequent cause of emissions prob-lems is simply a loose or broken vacuum hose or wire, so always check the hose and wiring connections first.
4   This doesn't mean, however, that emis-sion control systems are particularly difficult to maintain and repair. You can quickly and easily perform many checks and do most of the regular maintenance at home with com-mon tune-up and hand tools. **Note:** *Because of a Federally mandated extended warranty (five years or 50,000 miles at the time this manual was written) which covers the emis-sion control system components, check with your dealer about warranty coverage before working on any emissions-related systems. Once the warranty has expired, you may wish to perform some of the component checks and/or replacement procedures in this Chap-ter to save money.*
5   Pay close attention to any special pre-cautions outlined in this Chapter. It should be noted that the illustrations of the various sys-

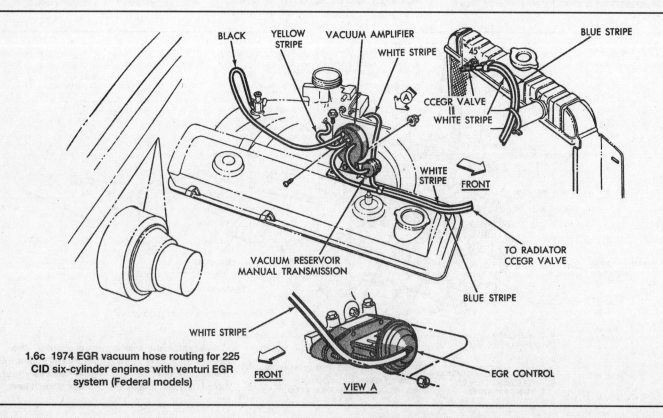

1.6c 1974 EGR vacuum hose routing for 225 CID six-cylinder engines with venturi EGR system (Federal models)

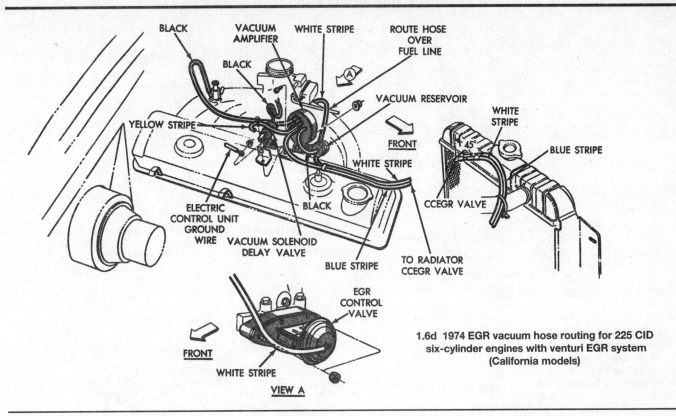

1.6d 1974 EGR vacuum hose routing for 225 CID six-cylinder engines with venturi EGR system (California models)

tems may not exactly match the system installed on your vehicle because of changes made by the manufacturer during production or from year-to-year.

6   A Vehicle Emissions Control Information (VECI) label is located in the engine compart-ment. It is usually located on the inner fender-well or on the radiator support. This label contains important emissions specifications and adjustment information, as well as a vac-uum hose schematic with emissions compo-nents identified **(see illustrations)**. When servicing the engine or emissions systems, the VECI label in your particular vehicle should always be checked for up-to-date information. **Note:** *In the event your VECI label is damaged or missing, contact a dealer parts department and order a replacement.*

6

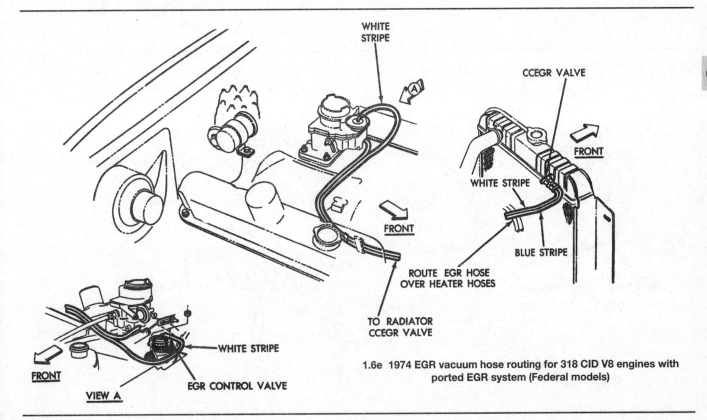

1.6e 1974 EGR vacuum hose routing for 318 CID V8 engines with ported EGR system (Federal models)

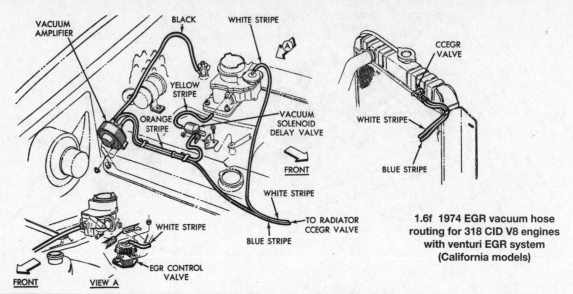

VACUUM AMPLIFIER
BLACK
WHITE STRIPE
YELLOW STRIPE
ORANGE STRIPE
VACUUM SOLENOID DELAY VALVE
FRONT
WHITE STRIPE
BLUE STRIPE
TO RADIATOR CCEGR VALVE
WHITE STRIPE
EGR CONTROL VALVE
FRONT
VIEW A

CCEGR VALVE
WHITE STRIPE
BLUE STRIPE

1.6f 1974 EGR vacuum hose routing for 318 CID V8 engines with venturi EGR system (California models)

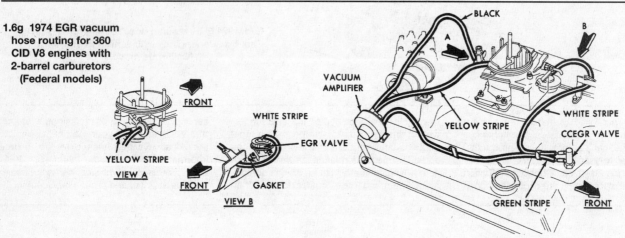

1.6g 1974 EGR vacuum hose routing for 360 CID V8 engines with 2-barrel carburetors (Federal models)

FRONT
YELLOW STRIPE
VIEW A
WHITE STRIPE
EGR VALVE
FRONT
GASKET
VIEW B

BLACK
A
B
VACUUM AMPLIFIER
YELLOW STRIPE
WHITE STRIPE
CCEGR VALVE
GREEN STRIPE
FRONT

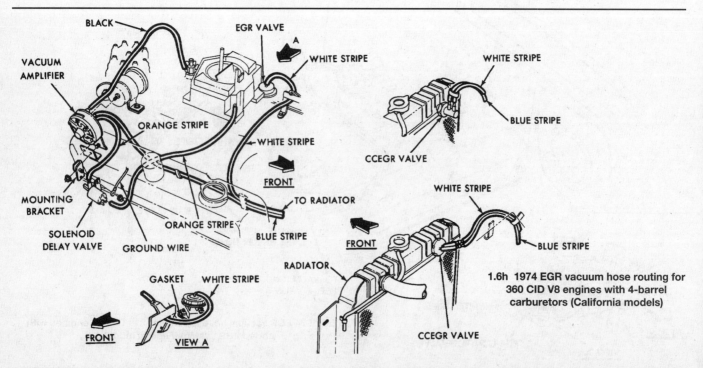

BLACK
EGR VALVE
A
WHITE STRIPE
VACUUM AMPLIFIER
ORANGE STRIPE
WHITE STRIPE
FRONT
TO RADIATOR
MOUNTING BRACKET
ORANGE STRIPE
BLUE STRIPE
FRONT
SOLENOID DELAY VALVE
GROUND WIRE
GASKET
WHITE STRIPE
FRONT
VIEW A
RADIATOR
CCEGR VALVE

WHITE STRIPE
BLUE STRIPE
CCEGR VALVE
WHITE STRIPE
BLUE STRIPE

1.6h 1974 EGR vacuum hose routing for 360 CID V8 engines with 4-barrel carburetors (California models)

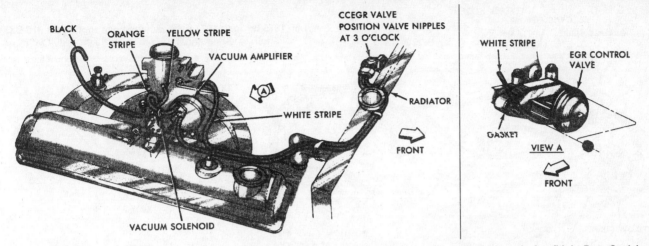

1.6i 1975 through 1977 EGR vacuum hose routing for 225 CID six-cylinder engines with manual transmission (Light Duty Cycle)

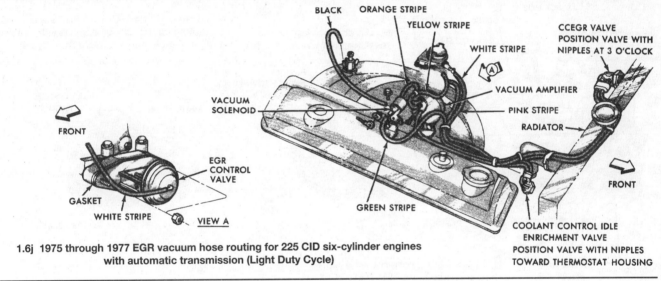

1.6j 1975 through 1977 EGR vacuum hose routing for 225 CID six-cylinder engines with automatic transmission (Light Duty Cycle)

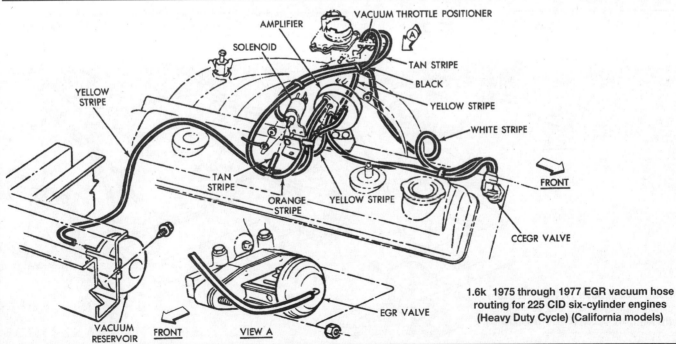

1.6k 1975 through 1977 EGR vacuum hose routing for 225 CID six-cylinder engines (Heavy Duty Cycle) (California models)

6

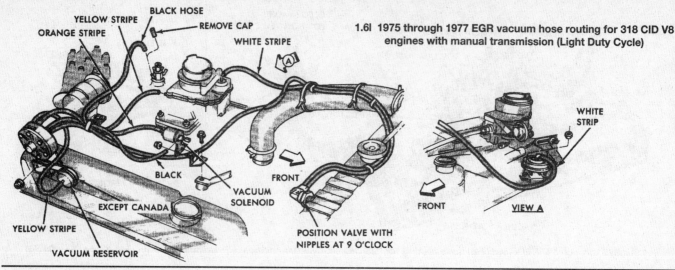

1.6l 1975 through 1977 EGR vacuum hose routing for 318 CID V8 engines with manual transmission (Light Duty Cycle)

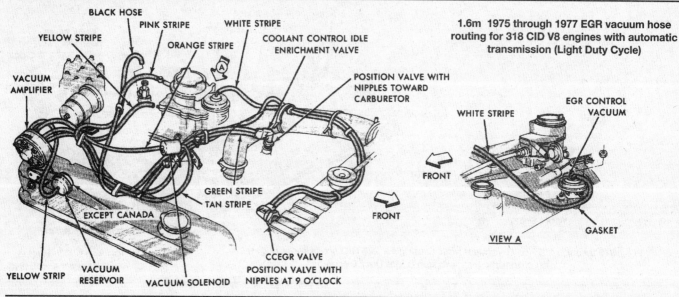

1.6m 1975 through 1977 EGR vacuum hose routing for 318 CID V8 engines with automatic transmission (Light Duty Cycle)

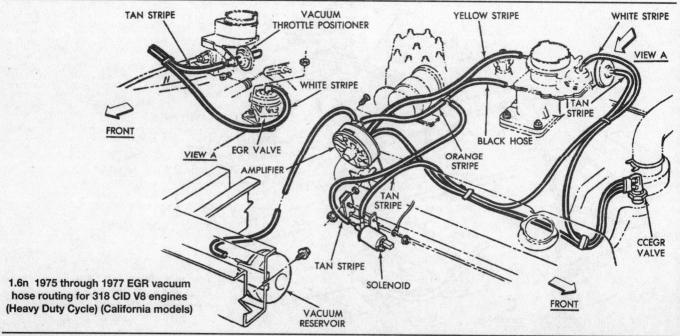

1.6n 1975 through 1977 EGR vacuum hose routing for 318 CID V8 engines (Heavy Duty Cycle) (California models)

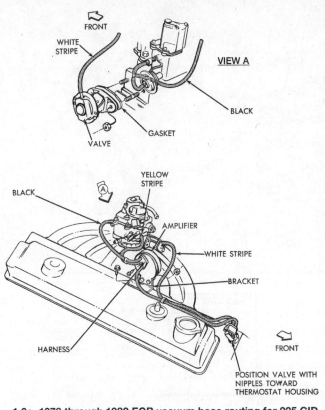

1.6o 1978 through 1980 EGR vacuum hose routing for 225 CID six-cylinder engines (Light Duty Cycle)

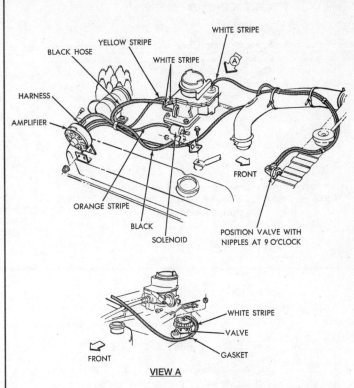

1.6p 1978 through 1980 EGR vacuum hose routing for 318 CID V8 engines with manual transmission (Light Duty Cycle)

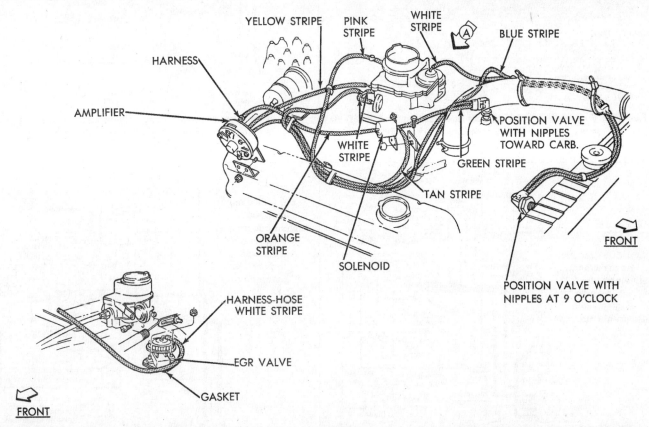

1.6q 1978 through 1980 EGR vacuum hose routing for 318 and 360 CID V8 engines with automatic transmission (Light Duty Cycle)

6

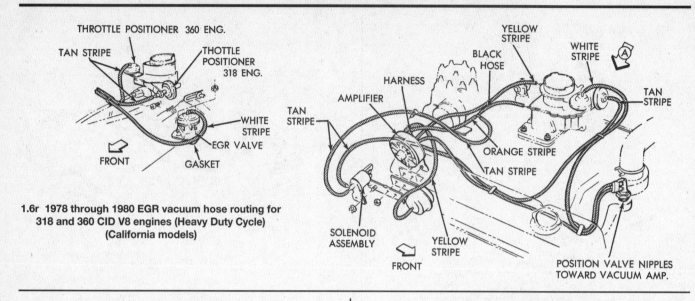

1.6r 1978 through 1980 EGR vacuum hose routing for 318 and 360 CID V8 engines (Heavy Duty Cycle) (California models)

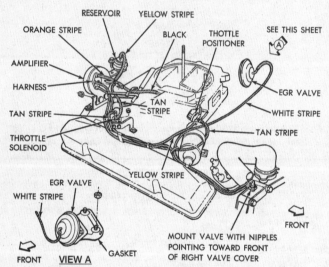

1.6s 1978 through 1980 EGR vacuum hose routing for 440 CID V8 engines (Heavy Duty Cycle) (California models)

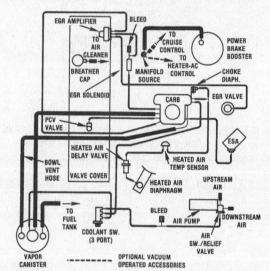

1.6t 1984 vacuum hose routing schematic for 3.7L six-cylinder engines with manual transmission, ESA with catalyst (Federal models)

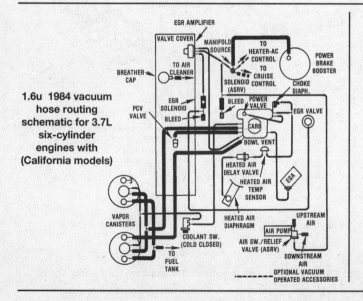

1.6u 1984 vacuum hose routing schematic for 3.7L six-cylinder engines with (California models)

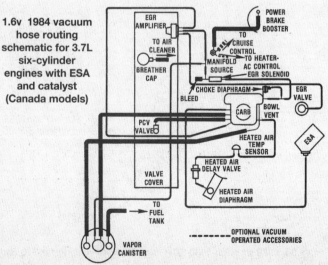

1.6v 1984 vacuum hose routing schematic for 3.7L six-cylinder engines with ESA and catalyst (Canada models)

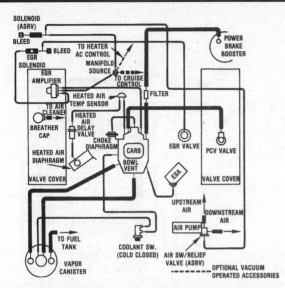

1.6w 1984 vacuum hose routing schematic for 5.2L V8 engines with High Altitude (California models)

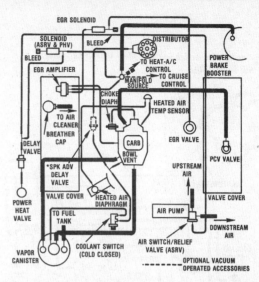

1.6x 1984 vacuum hose routing schematic for 5.2L V8 engines with manual transmissions, catalyst and no ESA (Federal and Canada models)

1.6y 1984 vacuum hose routing schematic for 5.2L with automatic transmission, catalyst, no ESA (Canada models)

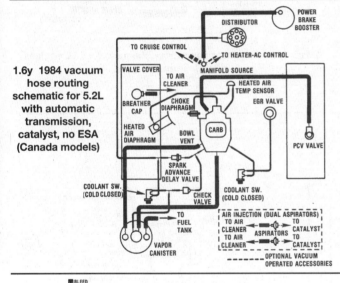

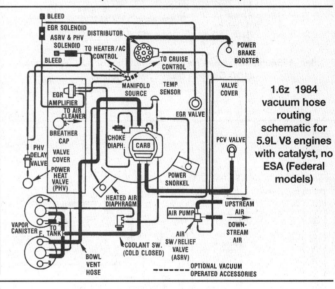

1.6z 1984 vacuum hose routing schematic for 5.9L V8 engines with catalyst, no ESA (Federal models)

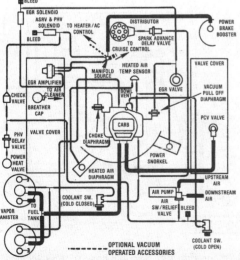

1.6aa 1984 vacuum hose routing schematic for 5.9L V8 engines with catalyst, High Altitude (Federal models)

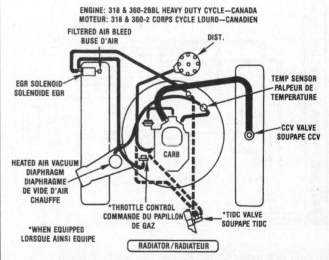

1.6ab 1984 vacuum hose routing schematic for 5.2L (318 CID) V8 engine (HDC Canada models)

6

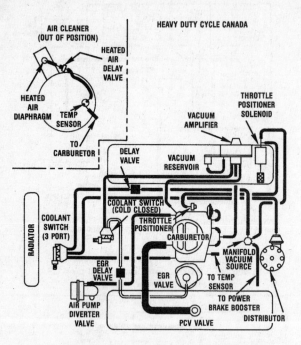

1.6ac 1984 vacuum hose routing schematic for 5.9L V8 engines (HDC Canada and Federal models)

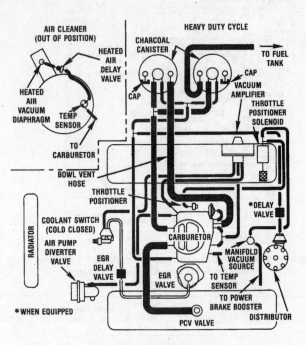

1.6ad 1984 vacuum hose routing schematic for 5.9L engines (HDC California models)

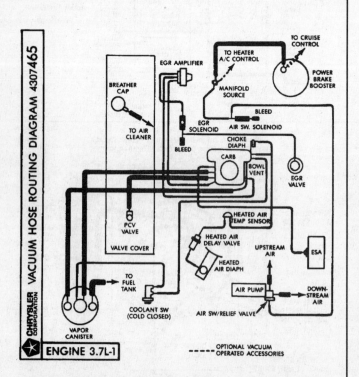

1.6ae 1985 through 1987 vacuum hose routing schematic for 3.7L six-cylinder engines with manual transmission ESA (Federal models)

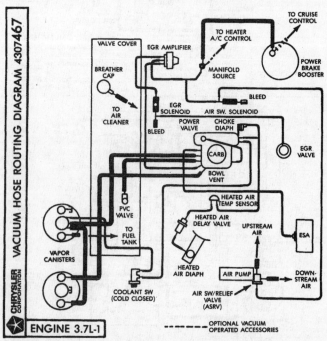

1.6af 1985 through 1987 vacuum hose routing schematic for 3.7L six-cylinder engines with (California models)

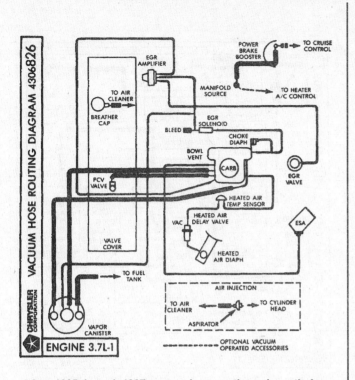

1.6ag 1985 through 1987 vacuum hose routing schematic for
3.7L six-cylinder engines with ESA and catalyst (Canada models)

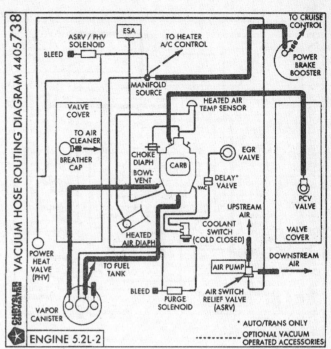

1.6ah 1985 through 1987 vacuum hose routing schematic for
5.2L V8 engines with catalyst and ESA (Federal and Canada
models) (1985 models are not equipped with ESA)

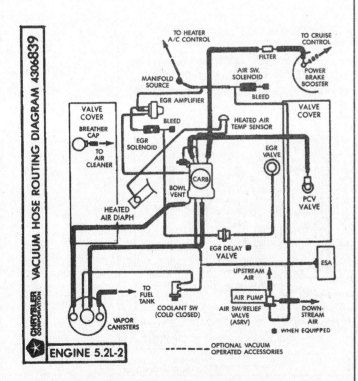

1.6ai 1985 through 1987 vacuum hose routing schematic for
5.2L engine (high-altitude California models)

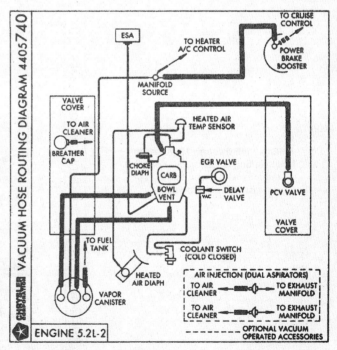

1.6aj 1985 through 1987 vacuum hose routing schematic for
5.2L V8 engines with catalyst, ESA and Aspirator (Canadian
models) (1985 models are not equipped with ESA)

6

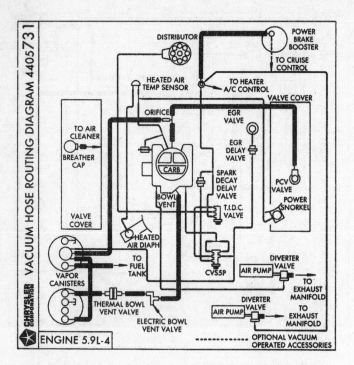

1.6ak 1985 through 1987 vacuum hose routing schematic for 5.9L V8 engines without catalyst and ESA (Federal models)

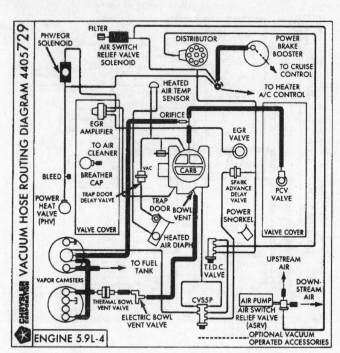

1.6al 1985 through 1987 vacuum hose routing schematic for 5.9L V8 engine (high-altitude Federal models)

1.6am 1985 through 1987 vacuum hose routing schematic for 5.9L V8 engines with air conditioning (California models)

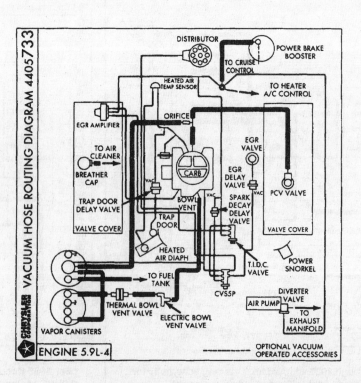

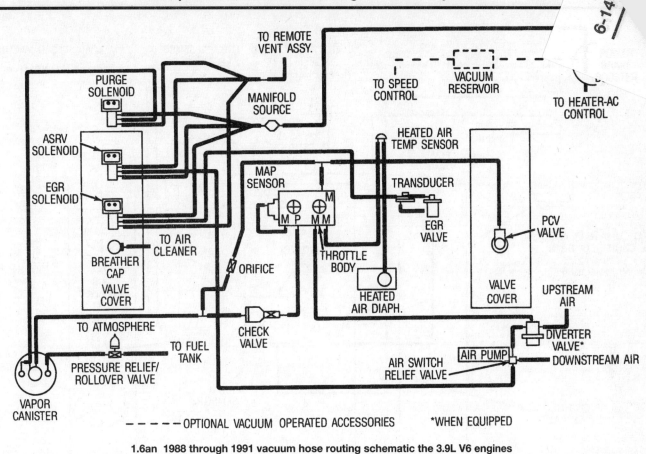

1.6an 1988 through 1991 vacuum hose routing schematic the 3.9L V6 engines

1.6ao 1988 through 1991 vacuum hose routing for 5.2L V8 and 3.9L V6 engines

6

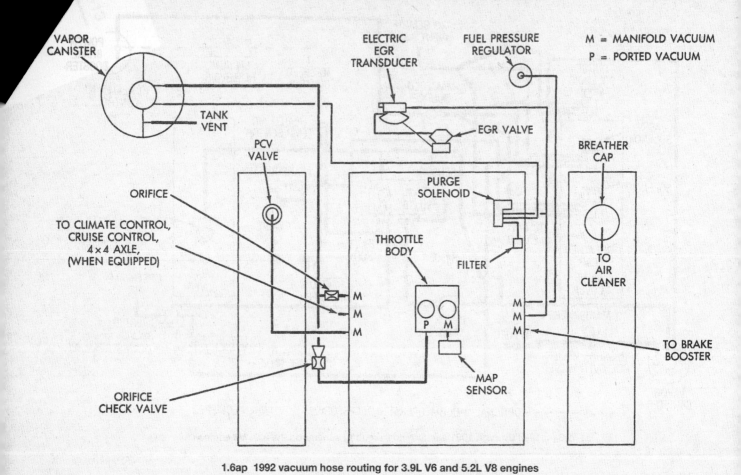

1.6ap 1992 vacuum hose routing for 3.9L V6 and 5.2L V8 engines

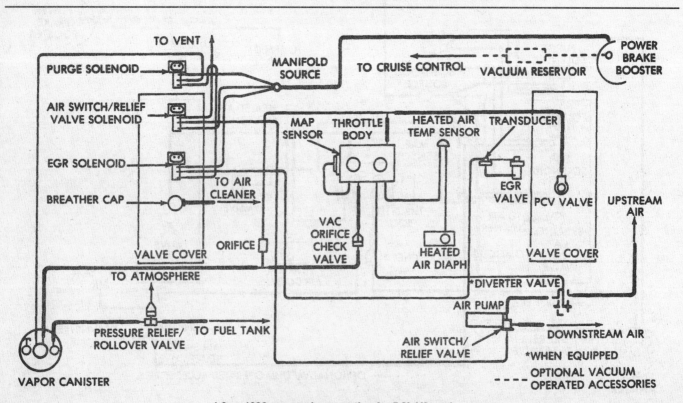

1.6aq 1992 vacuum hose routing for 5.9L V8 engines

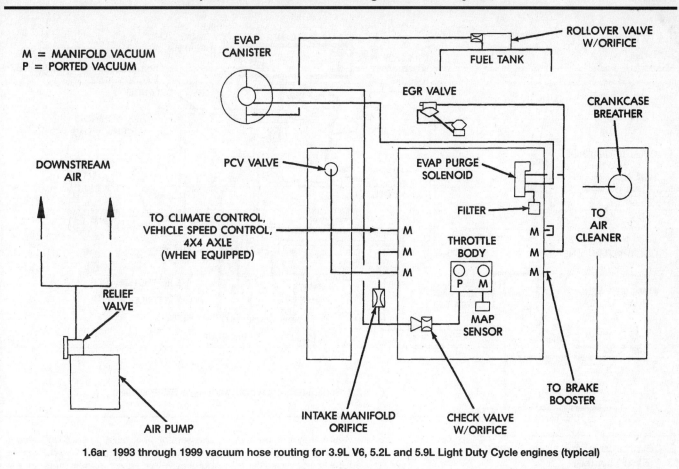

1.6ar 1993 through 1999 vacuum hose routing for 3.9L V6, 5.2L and 5.9L Light Duty Cycle engines (typical)

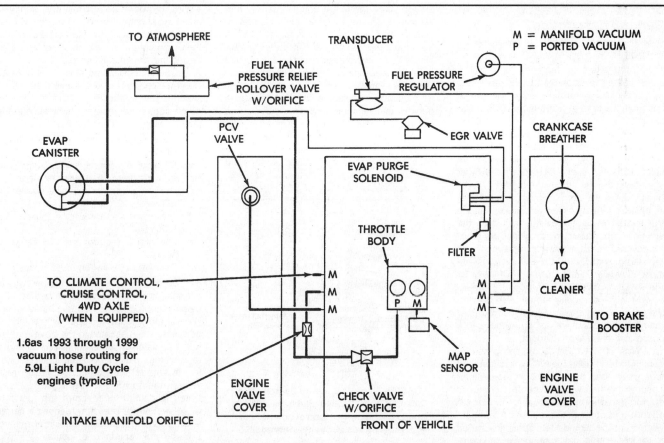

1.6as 1993 through 1999 vacuum hose routing for 5.9L Light Duty Cycle engines (typical)

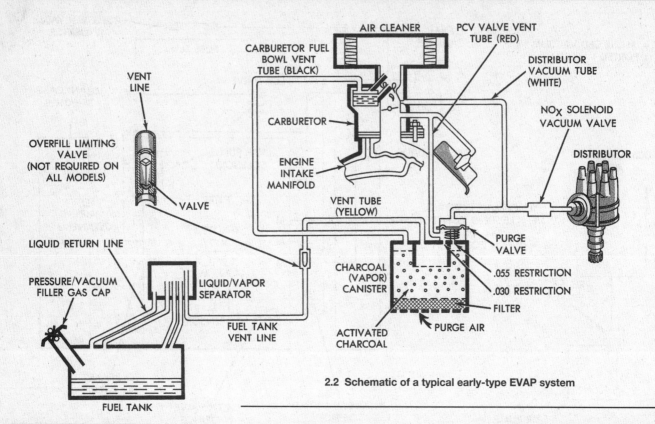

2.2 Schematic of a typical early-type EVAP system

2 Evaporative emissions control (EVAP) system

General description

Refer to illustrations 2.2, 2.6, 2.10a, 2.10b and 2.13

1   This system is designed to trap and store fuel that evaporates from the fuel system that would normally enter the atmosphere in the form of hydrocarbon (HC) emissions.

2   The system consists of a charcoal-filled vapor canister **(see illustration)**, a purge control valve (early systems), a purge control solenoid (later systems), a rollover valve (later systems), overfill-limiting valve (early systems), pressure vacuum filler cap (on tank inlet) and connecting lines and hoses (see the VECI label attached to the inside of the engine compartment). **Note:** *Not all the models covered in this manual are equipped with the same components. Early systems are more simplified. Refer to the VECI label for the exact location and component breakdown for the particular model. Also refer to the vacuum diagrams in Section 1. Note that many early models are not equipped with a purge control valve or solenoid.*

3   When the engine is off and pressure begins to build up in the fuel tank (caused by fuel expansion), the charcoal in the canister absorbs the fuel vapor. When the engine is started (cold), the charcoal continues to absorb and store fuel vapor. As the engine warms up, the stored fuel vapors are routed to the intake manifold or air cleaner and com-

bustion chambers where they are burned during normal vehicle operation.

4   On early models the charcoal canister is purged using the purge control valve. When vacuum reaches a specified level, the control diaphragm opens and allows vapor to pass from the canister into the carburetor venturi where they are burned during the normal combustion process. This is a mechanically operated system.

5   On later models, the charcoal canister is purged using engine vacuum via the purge control solenoid which is controlled by the Spark Control Computer (SCC). This type is an electrically operated system.

6   The purge control valve or solenoid operates in conjunction with the Coolant Vacuum Switch Cold Closed (CVSCC) valve **(see**

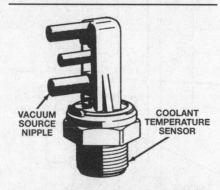

2.6 The Coolant Vacuum Switch Cold Closed (CVSCC) valve allows vacuum to pass to the purge control solenoid at temperatures above 98-degrees F

illustration). This valve is usually mounted in the thermostat housing. The CVSCC and the purge control valves remain shut at low engine temperatures. When the engine is operating at normal temperatures, the CVSCC valve opens, allowing vacuum to be applied to the purge control valve so fuel vapors can be vented into the intake manifold.

7   On all models equipped with Spark Control Computer or Electronic Fuel Control system, before the SCC will energize the purge control solenoid for canister purging, the following conditions must be met:

a) *The coolant temperature must be above 90-degrees F.*
b) *The engine must have been running for approximately 1.5 minutes or more.*
c) *The vehicle speed must be above 5 to 15 mph.*
d) *The engine speed must be above 800 to 1200 rpm.*
e) *The engine vacuum must be below 30 inches.*
f) *The carburetor switch must be open.*

8   Some models are also equipped with an external bowl vent valve which is controlled by air pump operation. Some Canadian and Federal models are equipped with a thermal bowl vent valve (located in the bowl vent line to the canister) which is controlled by engine temperature to aid in cold starting.

9   All models are equipped with a relief valve, mounted in the fuel tank filler cap, which is calibrated to open the fuel tank when vacuum or pressure reaches a certain level. This vents the fuel tank and relieves extreme tank vacuum or pressure.

2.10a With the ignition key on, check for voltage to the purge-control solenoid

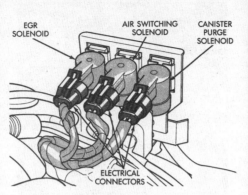

2.10b Location of the canister purge solenoid on later TBI systems (typical)

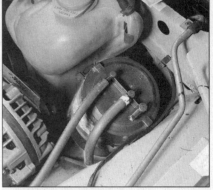

2.13 Check all the hoses on the charcoal canister for leaks, breaks or misrouting - note that the right hose in this photo is disconnected (typical)

Check

Purge control solenoid

10   Check the power to the solenoid. With the ignition key on (engine not running), check for voltage at the electrical connector (see illustrations). There should be battery voltage (approximately 12 volts) present. If not, check the wiring harness (see Chapter 12). If the solenoid is suspect, use jumper wires and apply battery voltage to the solenoid and check for solenoid activation.

11   Also check for vacuum to the solenoid. With the engine running, install a vacuum gauge and check for a vacuum signal. If vacuum is not present, check all the lines for damage or improper vacuum hose routing.

Charcoal canister

12   Remove the filter at the bottom of the canister and inspect it for damage and saturation. If necessary, replace it with a new one.

13   Check all the hoses, valves and connectors for damage or cracked hoses (see illustration).

14   Replace any component that is damaged.

Purge control valve

15   Disconnect the upper (smaller) vacuum hose from the purge control valve and attach a hand vacuum pump.

16   Disconnect the lower (larger) hose from the purge control valve and attach a short length of clean hose to the fitting on the valve.

17   Try to blow through the hose connected to the lower hose fitting; air should not pass through the valve.

18   Apply 15 inches of vacuum to the upper hose fitting with the vacuum pump, then try again to blow through the lower hose fitting; air should not pass through the valve. If the valve does not operate as specified, the valve is faulty and should be replaced.

3   Exhaust Gas Recirculation (EGR) system

General description

Refer to illustrations 3.1 and 3.4

1   This system recirculates a portion of the exhaust gases into the intake manifold or carburetor in order to reduce the combustion temperatures and decrease the amount of nitrogen oxide (NOx) produced (see illustration). The main component in the system is the EGR valve.

2   Early models are equipped with either a Ported Vacuum Control system or a Venturi Vacuum Control system. The Venturi Vacuum system utilizes a vacuum tap at the throat of the carburetor venturi to provide the control signal. This signal is a low amplitude signal that requires a vacuum amplifier to increase the level of vacuum needed to operate the EGR valve. The ported vacuum control system uses a slot in the carburetor body which is exposed to an increasing percentage of manifold vacuum as the throttle valve is opened up during acceleration. Ported systems do not require a vacuum amplifier.

3   The early models are also equipped with a Coolant Control Exhaust Gas Recirculation (CCEGR) valve mounted on the radiator. When the coolant temperature in the top radiator tank reaches 65-degrees F, the valve opens so that vacuum is applied to the EGR valve controlling exhaust gas recirculation.

4   Some models are equipped with an EGR delay system that uses an electrical timer mounted on the firewall in the engine compartment which controls an engine-mounted solenoid (see illustration). The timer prevents any exhaust gas recirculation for at least 35 seconds after the engine has started.

6

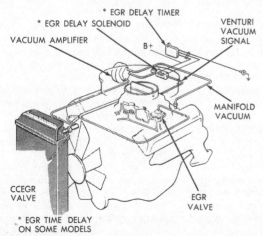

3.1 A typical EGR system with Venturi Vacuum EGR System

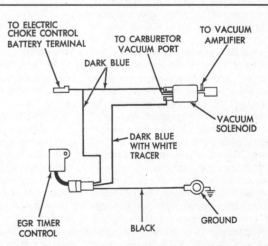

3.4 Schematic of the EGR time delay system

3.9 Apply vacuum to the EGR valve and put your finger on the underside of the valve, feeling for movement - the valve diaphragm should move up (small-block V8 engine shown)

3.15 Remove the rubber junction block along with the vacuum hoses from the vacuum amplifier

3.16 Check to make sure engine vacuum is reaching the vacuum amplifier by using a vacuum gauge to check for vacuum on the corresponding line

5   Engines equipped with SCC (see Chapter 5) and/or EFC (see Chapter 4A) systems are equipped with a charge temperature switch that is activated above 60-degrees F, thus allowing EGR timer to control exhaust gas recirculation. The CTS is located in the number 6 intake manifold runner on six-cylinder engines or the number 8 intake manifold runner on eight-cylinder engines. Refer to Section 14 for the charge temperature switch diagnostic procedures.

6   These later models are also equipped with an EGR solenoid. This solenoid is electrically operated by the computer and it is designed to control vacuum to the EGR valve.

Check

Refer to illustrations 3.9, 3.15, 3.16, 3.17, 3.18 and 3.19

7   Remove the engine cover (see Chapter 2B, Section 2). Check all hoses for cracks, kinks, broken sections and proper connection. Inspect all system connections for damage, cracks and leaks.

8   To check the EGR system operation, bring the engine up to operating temperature and, with the transmission in Neutral (parking brake set and tires blocked to prevent movement), allow it to idle for 70 seconds. Open the throttle abruptly so the engine speed is between 2000 and 3000 rpm and then allow it to close. The EGR valve stem should move if the control system is working properly. The test should be repeated several times. Movement of the stem indicates the control system is functioning correctly.

9   If the EGR valve stem does not move, check all of the hose connections to make sure they are not leaking or clogged. Disconnect the vacuum hose and apply ten inches of vacuum with a hand pump **(see illustration)**. If the stem still does not move, replace the EGR valve with a new one. If the valve does open, measure the valve travel to make sure it is approximately 1/8-inch. Also, the engine should run roughly when the valve is

3.17 Also check to make sure vacuum is reaching the CCEGR valve and the EGR valve from the amplifier. Replace any damaged or broken hoses.

open. If it doesn't, the passages are probably clogged.

10   Apply vacuum with the pump and then clamp the hose shut. The valve should stay open for 30 seconds or longer. If it does not, the diaphragm is leaking and the valve should be replaced with a new one.

11   If the EGR valve does not operate with the CCEGR or CVSCC bypassed, the carburetor must be removed to check and clean the slotted port in the throttle bore and the vacuum passages and orifices in the throttle body. Use solvent to remove deposits and check for flow with light air pressure.

12   If the engine idles roughly and it is suspected the EGR valve is not closing, remove the EGR valve and inspect the poppet and seat area for deposits.

13   If the deposits are more than a thin film of carbon, the valve should be cleaned. To clean the valve, apply solvent and allow it to penetrate and soften the deposits, making sure that none gets on the valve diaphragm, as it could be damaged.

14   Use a vacuum pump to hold the valve open and carefully scrape the deposits from

3.18 Watch carefully for a change in vacuum supply as battery voltage is applied to the solenoid - when voltage is applied, vacuum should drop

the seat and poppet area with a tool. Inspect the poppet and stem for wear and replace the valve with a new one if wear is found.

15   Locate the vacuum amplifier (if equipped) and remove the rubber junction block along with the vacuum hoses **(see illustration)**.

16   With the engine running, check for a vacuum signal from the intake manifold **(see illustration)**. This test will determine if manifold vacuum is reaching the vacuum amplifier from the intake manifold.

17   Apply vacuum to the port (vacuum hose) that is routed to the EGR valve or the CCEGR valve **(see illustration)**. The gauge should hold vacuum. If not, check for broken hoses or ruptured diaphragms in the EGR valve or CCEGR valve.

18   Check the EGR solenoid (if equipped). Using a hand-held vacuum pump, apply vacuum and then apply battery voltage. The solenoid should activate and allow vacuum to pass through the solenoid **(see illustration)**.

19   Also remove the vacuum hose(s) from the solenoid and check for vacuum to the EGR valve **(see illustration)**.

3.19 Also check to make sure vacuum is reaching the EGR valve from the EGR solenoid (1986 5.2L shown). Replace any damaged or broken hoses

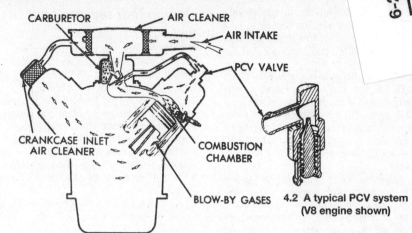

4.2 A typical PCV system (V8 engine shown)

4.3 Place the tip of your finger over the end of the PCV valve and check for engine vacuum. There should be a strong vacuum signal. This indicates that the system is drawing crankcase vapors from the engine

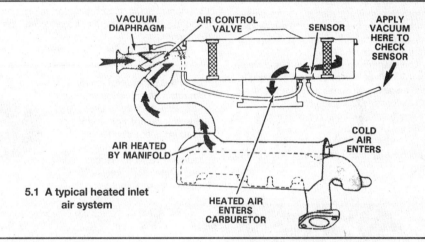

5.1 A typical heated inlet air system

EGR valve replacement

20 Unscrew the fitting attaching the EGR valve tube to the intake manifold (if equipped). Remove the two nuts retaining the valve and remove the assembly.
21 Check the tube for signs of leaking or cracks.
22 Installation is the reverse of removal.

4 Positive Crankcase Ventilation (PCV) system

General description

Refer to illustrations 4.2 and 4.3

1 This system is designed to reduce hydrocarbon emissions (HC) by routing blow-by gases (fuel/air mixture that escapes from the combustion chamber past the piston rings into the crankcase) from the crankcase to the intake manifold and combustion chambers, where they are burned during engine operation.
2 The system is very simple and consists of rubber hoses and a small, replaceable metering valve (PCV valve) **(see illustration)**.

Check and component replacement

3 Remove the engine cover (see Chapter 2B, Section 2). With the engine running at idle, pull the PCV valve out of the valve cover and place your finger over the valve inlet **(see illustration)**. A strong vacuum will be felt if the valve is operating properly. Replace the valve with a new one if it is not functioning as described. Do not attempt to clean the old valve.
4 With the engine off, shake the valve and make sure you hear a rattling noise. If not, the valve is defective and should be replaced.
5 To replace the PCV valve, simply pull it from the hose or vent module and install a new one.
6 Inspect the hose prior to installation to ensure that it isn't plugged or damaged. Compare the new valve with the old one to make sure they are the same. Chapter 1 contains additional PCV valve information.

5 Heated Inlet Air (HIA) system

Refer to illustrations 5.1, 5.6, 5.9, 5.11 and 5.14

General description

1 The heated inlet air system is used to keep the air entering the carburetor at a constant temperature. The carburetor can then be calibrated much leaner for emissions reductions, faster warm up and improved driveability **(see illustration)**.
2 Two air flow circuits are used. They are controlled by various carburetor vacuum and coolant temperature sensing valves. A vacuum diaphragm operates an air control valve in the air cleaner snorkel. When the underhood temperature is cold, air is drawn through the shroud which fits over the exhaust manifold, up through the air pipe and into the air cleaner. On dual-snorkel air cleaners, both air control valves open during hard acceleration (when intake manifold vacuum drops below 5-inches).
3 As the underhood temperature rises, the air control valve will be closed by the vacuum diaphragm and the air that enters the air cleaner will be drawn through the cold air snorkel or duct.

Check

4 Checking of the system should be done while the engine is cold. If the engine is equipped with a cold air inlet snorkel duct, remove the clamp and the duct so you can view the air control valve through the air inlet snorkel. Start the engine and see if the air control valve moves up to allow hot air from the manifold shroud to enter the air cleaner. As the engine warms up, the vacuum

6

diaphragm should allow the air control valve to close, eventually allowing cool air to enter through the snorkel. If this does not occur, any one of the many components in the system could be defective.

5   Remove the engine cover (see Chapter 2B, Section 2). Remove the air cleaner assembly and allow it to cool down completely. **Note:** *Early models can be checked with the air cleaner just under 90-degrees F, but later models must be cooled to below 60-degrees F (50-degrees F on 1982 and later models).*

6   Apply 20-inches of vacuum to the sensor in the air cleaner housing with a hand pump. The air control valve should move to the up or closed position. If it doesn't, the vacuum should be applied directly to the vacuum diaphragm **(see illustration)**. If the air control valve now moves up, to the closed position, replace the sensor with a new one. If it doesn't, replace the vacuum diaphragm and repeat the check.

7   The diaphragm should not bleed down more than 10-inches of vacuum in five minutes. The air control valve should begin to move at 5 to 5-1/2 inches of vacuum minimum and be fully open at no more than 8-1/2 to 9 inches of vacuum.

8   On dual-snorkel air cleaners, check the second air control valve as described in Step 7.

Component replacement

Sensor

9   To replace the sensor, remove the air cleaner, detach the vacuum hoses from the sensor ports and pry up the retaining clips with a screwdriver **(see illustration)**.

10   Discard the old sensor and the gasket.

11   Position the new gasket and sensor on the housing, support the sensor at the outer edge and push the new retaining clips onto the hose ports **(see illustration)**. **Note:** *Make sure the gasket is compressed slightly to form an air-tight seal. Do not attempt to adjust the sensor.*

12   Hook up the vacuum hoses and install the air cleaner.

Vacuum diaphragm

13   Remove the air cleaner housing and disconnect the vacuum hose from the diaphragm.

14   On later models, drill through the welded tab on the air cleaner snorkel **(see illustration)**.

15   Tip the diaphragm slightly forward and rotate it counterclockwise.

16   Slide the diaphragm assembly to one side and disengage the operating rod from the heat control valve.

17   Check the heat control valve for free movement.

18   Installation is the reverse of removal. On later models, use a pop rivet tool to install a new rivet in the vacuum diaphragm tab. Do not manually operate the heat control valve - use vacuum only.

6   Air Injection (AI) system

Refer to illustrations 6.1a and 6.1b

General description

1   The air injection system is employed to reduce carbon monoxide and hydrocarbon

5.6 If the air control valve operates when vacuum is applied directly to the diaphragm, but not when vacuum is applied to the sensor, the sensor may be faulty - check the hoses and connections carefully for leaks before replacing it

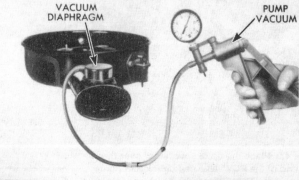

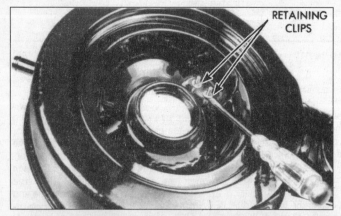

5.9 The sensor is held in place with two retaining clips - discard them and use new ones when the new sensor is installed

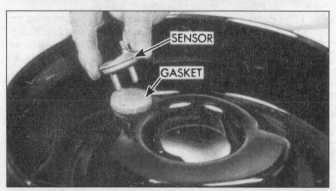

5.11 Make sure a new gasket is used and that it is compressed slightly when the sensor is fastened in place

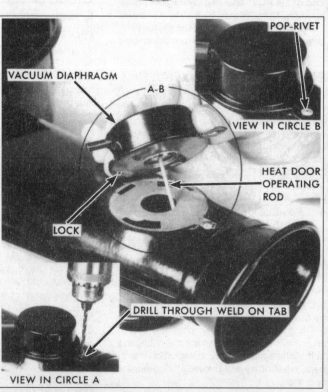

5.14 Removing the air control valve vacuum diaphragm (later model diaphragms are attached with a rivet)

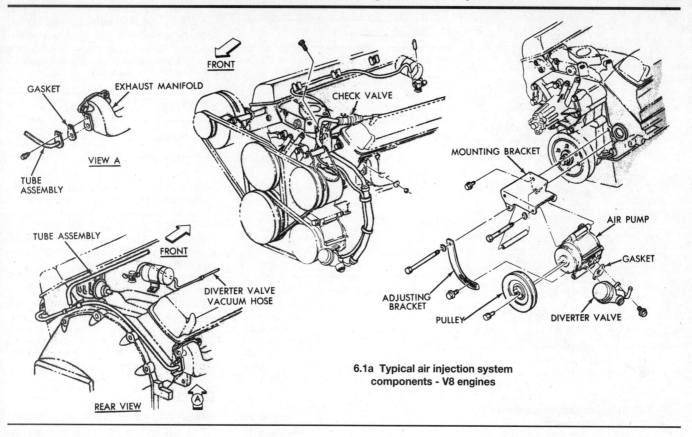

6.1a Typical air injection system
components - V8 engines

emissions, The system consists of an air supply pipe, an air control valve, a check valve, injection tubes and connecting hoses **(see illustrations)**.

2   The air injection system continues combustion of unburned gases after they leave the combustion chamber by injecting fresh air into the hot exhaust stream leaving the exhaust ports. At this point, the fresh air mixes with hot exhaust gases to promote further oxidation of both hydrocarbons and carbon monoxide, thereby reducing their concentration and converting some of them into harmless carbon dioxide and water. During some modes of operation, such as higher engine speeds, the air is dumped into the atmosphere by the air control valve to prevent overheating of the exhaust system.

3   There are three types of air control valves used on the different air injection systems:

a) *The relief valve diverts excessive pump output at higher engine speeds to the atmosphere.*
b) *The diverter valve incorporates a similar relief valve, but also uses a vacuum sensitive valve, which dumps air into the atmosphere upon sudden deceleration to prevent backfiring.*
c) *The switch/relief valve, besides releasing excessive pressure, directs air injection flow to either the exhaust port location or to the downstream injection at the catalytic converter, depending on engine temperature.*

Check

Air supply pump

4   Check and adjust the drivebelt tension (refer to Chapter 1).
5   Remove the engine cover (see Chapter 2B, Section 2). Disconnect the air supply hose at the air bypass valve inlet.
6   The pump is operating satisfactorily if air flow is felt at the pump outlet with the engine running at idle, increasing as the engine speed is increased.
7   If the air pump does not successfully pass the above tests, replace it with a new or rebuilt unit.

Relief valve and diverter valve

8   Failure of the relief valve or diverter valve will cause excessive noise and air pump output at the valve.
9   There should be little or no air escaping from the silencer of the valve body at engine speed.
10   If these tests indicate a faulty air control valve, the unit should be replaced.

Switch/relief valve

11   If air injection flow is being directed to both the upstream and downstream inlets simultaneously, the switch/relief valve is malfunctioning.

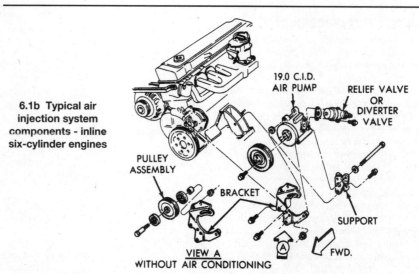

6.1b Typical air
injection system
components - inline
six-cylinder engines

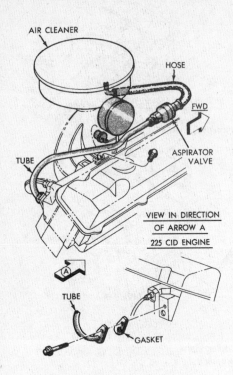

AIR CLEANER

HOSE

FWD

ASPIRATOR VALVE

TUBE

VIEW IN DIRECTION
OF ARROW A
225 CID ENGINE

TUBE

GASKET

**7.1 Typical air aspirator system
components**

12   There should be no air escaping from
the relief valve silencer at idle speed.
13   If these tests indicate a faulty
switch/relief valve the unit should be replaced.

Check valve

14   The check valve, located on the injection tube assembly, can be tested by removing the air hose from the valve inlet tube.
15   Start the engine and check for exhaust gas escaping from the valve inlet tube.
16   If there is an exhaust leak, the check valve is faulty and should be replaced.

Component replacement

17   The check valve and air control valve may be replaced by disconnecting the hoses leading to them (be sure to label the hoses as they are disconnected to facilitate reconnection), replacing the faulty element with a new one and reconnecting the hoses to the proper ports. Be sure to use new gaskets and remove all traces of the old gasket, where applicable. Make sure that the hoses are in good condition. If not, replace them with new ones.
18   To replace the air supply pump, first loosen the appropriate engine drivebelt(s) (refer to Chapter 1), then remove the faulty pump from the mounting bracket, labeling all wires and hoses as they are removed to facilitate installation of the new unit.
19   After the new pump is installed, adjust the drivebelt(s) (refer to Chapter 1).

7   Air aspirator system

Refer to illustration 7.1

General description

1   Later model Canadian vehicles with a catalytic converter are equipped with an air aspirator system to reduce carbon monoxide and hydrocarbon emissions. The system draws fresh air from the air cleaner, past a one-way diaphragm, allowing fresh air to mix with the exhaust gases during negative pressure pulses in the exhaust system **(see illustration)**.

Check

2   Failure of the aspirator valve results in excessive exhaust noise in the engine compartment at idle and hardening of the rubber hose from the valve to the air cleaner.
3   Remove the engine cover (see Chapter 2B, Section 2.) To check the aspirator valve, located midway in the hose between the air cleaner and the cylinder head, first disconnect the hose from the aspirator valve.
4   With the engine at idle and the transmission in Neutral, the negative (vacuum) exhaust pulses should be felt at the aspirator inlet. If hot exhaust gases escape from the inlet, the valve is faulty and should be replaced.
5   If the aspirator tube is leaking at the exhaust manifold joint, the aspirator tube-to-cylinder head gasket should be replaced.
6   Inspect the rubber hose and replace it, if it has cracked or hardened.

Component replacement

7   The aspirator valve can be replaced by disconnecting the air hose and unscrewing the valve from the tube assembly.
8   The aspirator tube-to-cylinder head gasket can be replaced by disconnecting the aspirator valve from the air hose and removing the aspirator tube mounting bolts. Be sure to scrape off all traces of the old gasket.

8   Electric assist choke system

Refer to illustration 8.1

General description

1   Most vehicles covered in this manual are equipped with a choke assist device to help reduce hydrocarbon and carbon monoxide emissions during engine warm up. An electric heat element, located next to the thermostatic coil spring in the choke well, assists engine heat to shorten choke duration **(see illustration)**.
2   Some models may employ a dual stage choke assist, which provides the choke heater with partial power below 58-degrees F and full power above 68-degrees F. The single stage system energizes the control switch above 68-degrees F. On both systems, the heater shuts off after the control switch reaches a temperature of 110-degrees F.

Check

Single stage choke control switch

3   A test light will be needed to check the control switch.
4   Remove the engine cover (see Chapter 2B, Section 2). With the engine off, remove the battery lead wire connector from the control switch.
5   Connect a test light to the load (small) terminal of the control switch.
6   Start the engine and allow it to warm up to operating temperature.

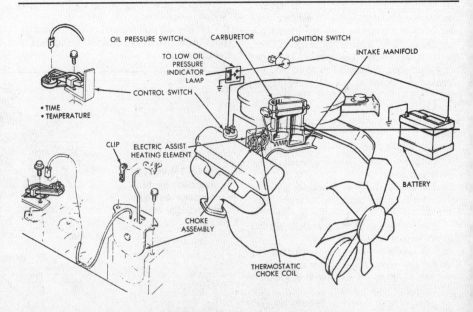

OIL PRESSURE SWITCH   CARBURETOR   IGNITION SWITCH

TO LOW OIL
PRESSURE
INDICATOR
LAMP

INTAKE MANIFOLD

CONTROL SWITCH

TIME
TEMPERATURE

CLIP

ELECTRIC ASSIST
HEATING ELEMENT

BATTERY

CHOKE
ASSEMBLY

THERMOSTATIC
CHOKE COIL

8.1 Typical electric assist choke system

7   Apply 12 volts to the battery lead terminal of the control switch.

8   The test light should glow for a few seconds to several minutes. However, if the test light stays on for more than five minutes, the control switch is defective and must be replaced.

Dual stage choke control switch

9   The same procedure as above is used to check the dual stage control switch, except that two levels of bulb intensity should be observed.

10   Connect the test light directly to the battery terminals. Observe the intensity of the bulb.

11   Repeat Steps 4 through 8. When applying voltage to the circuit as described in Step 8, the test light bulb should initially show the same intensity as when connected directly to the battery.

Choke heating element

12   To test the choke heating element, an ohmmeter will be needed. Remove the battery lead connector from the control switch terminal (B+ terminal).

13   Connect one ohmmeter lead to the choke housing or choke retainer screw.

14   Touch the other ohmmeter lead to the metal portion of the choke wire connector at the control switch (not the B+ terminal).

15   A resistance of 4 to 12 ohms should be indicated. Meter readings indicating an open or short circuit are cause for installation of a new heating element.

Component replacement

16   The electric assist choke switch is mounted on the right side of the engine block, adjacent to the carburetor.

17   Disconnect the electrical lead and remove the two mounting bolts to remove the choke switch. Installation is the reverse of removal.

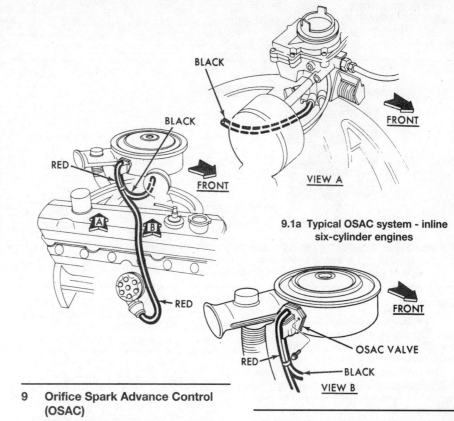

9.1a Typical OSAC system - inline six-cylinder engines

9   Orifice Spark Advance Control (OSAC)

Refer to illustrations 9.1a, 9.1b and 9.3

General description

1   The OSAC system was installed on some engines to aid in the control of NOx (oxides of nitrogen). This system controls the vacuum to the vacuum advance diaphragm of the distributor. The OSAC valve delays the change in ported vacuum to the distributor when transferring from idle to part throttle, and so helps to reduce the quantity of oxides of nitrogen produced by the engine (see illustrations).

9.1b Typical OSAC systems - V8 engines

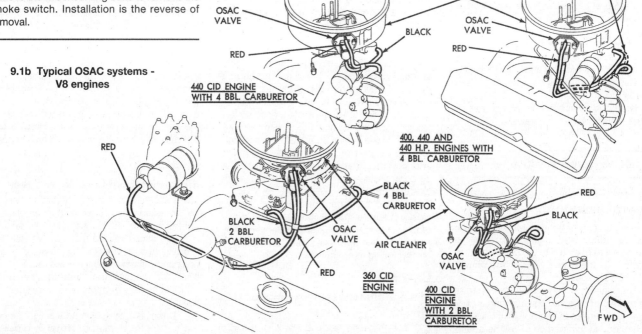

440 CID ENGINE WITH 4 BBL. CARBURETOR

400, 440 AND 440 H.P. ENGINES WITH 4 BBL. CARBURETOR

360 CID ENGINE

400 CID ENGINE WITH 2 BBL. CARBURETOR

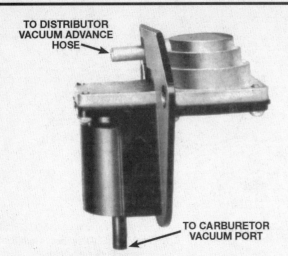

9.3 When checking OSAC valve operation, connect the vacuum gauge T-fitting to the hose between the upper OSAC valve port and the distributor vacuum advance unit

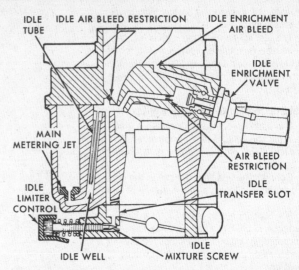

10.3 Idle enrichment control components (not all models)

Check

2   Maintenance consists of periodically inspecting the condition and security of the vacuum hoses. Replace any that are hardened, cracked or loose fitting.

3   Remove the engine cover (see Chapter 2B, Section 2). To check the operation of the OSAC valve, run the engine to normal operating temperature, then connect a vacuum gauge to the vacuum hose between the OSAC valve and the distributor **(see illustration)**. Use a T-fitting so the hose can remain connected at both components.

4   Set the parking brake and put the transmission in Neutral. Run the engine at 2000 rpm and note the vacuum reading on the gauge. The valve is operating correctly if a very gradual increase in vacuum occurs. If the vacuum increases suddenly or if there is no increase at all, replace the valve with a new one.

Component replacement

5   The vacuum hoses are simple to replace - just pull off the old one and install the new one. Replace one hose at a time to avoid mix-ups.

6   The OSAC valve is attached to the air cleaner housing with two screws.

10 Idle enrichment system

Refer to illustration 10.3

General description

1   The purpose of the idle enrichment system is to reduce cold engine stalling with a metering system related to the basic carburetor instead of the choke. It was used on mid to late 70's vehicles.

2   The system enriches carburetor mixtures in the curb idle and fast idle ranges. The carburetor will have the complete idle system enriched during that portion of vehicle operation related to cold or semi-cold operation.

3   A small vacuum-controlled diaphragm mounted near the top of the carburetor controls idle system air. When control vacuum is applied to the diaphragm, idle system air is reduced. Air losses strengthen the small vacuum signal within the idle system and fuel flow increases. As a result of more fuel and less air, the fuel/air mixture is richer **(see illustration)**.

4   The vacuum signal to the carburetor diaphragm is controlled by a thermal switch threaded into a coolant passage on the engine. When the engine is cold, the switch is open and passes the vacuum signal to the carburetor diaphragm. During warm-up, the switch closes to eliminate the vacuum signal and return carburetor metering to normal, lean levels. In one type system, the coolant vacuum switch also controls idle system enrichment duration.

5   In a second system, a similar coolant vacuum switch is used, but it receives its vacuum signal from a solenoid valve operated by an electric timer. Enrichment duration is approximately 35 seconds after the engine starts. The switch will prevent additional cycles of idle system enrichment after the engine reaches normal operating temperature, but while the engine is cold, each restart results in another 35 seconds of enrichment.

6   When the engine cools below the switch opening temperature, enriched idle will occur after restart. Duration of the enrichment will be controlled as described above for each type system.

7   The solenoid controlled system switch closes at approximately 150-degrees F, while others close at about 98-degrees F. The switches open approximately 12-degrees F below closing temperatures.

Check

Idle enrichment system functional check

8   The following test is a check of the idle enrichment system within the carburetor.

9   Remove the engine cover (see Chapter 2B, Section 2). Start the engine and allow it to reach normal operating temperature.

10   Remove the air cleaner but DO NOT cap any vacuum ports opened by hose removal (leakage is needed for this check).

11   Disconnect the hose to the idle enrichment valve diaphragm at the plastic connector (the connector must be removed from the carburetor hose because it is a filtered bleed which will interfere with the check).

12   Start the engine and place the fast idle speed screw on the slowest speed step of the fast idle cam.

13   Connect 3 or 4 feet of hose to the enrichment valve hose with a splice connector.

14   Apply vacuum to the other end of the hose with a hand vacuum pump and listen for an engine speed change. If speed can be controlled by vacuum, the diaphragm and idle enrichment valve are operating properly. If engine speed cannot be controlled, proceed to the next step (Carter carburetor) or replace the valve (Holley carburetor).

15   Place your finger over the idle air bleed passage and listen for an engine speed change. If engine speed can be controlled, the diaphragm is leaking or the valve is stuck open. If engine speed cannot be controlled, the valve is stuck closed. In either case, clean the valve and repeat Step 14. If engine speed still cannot be controlled, replace the idle enrichment valve.

Idle enrichment time delay check

16   In some applications, the timing module and solenoid valve serve the dual function of controlling both EGR delay and idle enrichment duration. Prior to testing the timer, the wires should be checked for proper connections and repairs made if necessary.

17   With the ignition switch off, detach the wiring connector from the time delay solenoid valve.

18   Connect a self-powered test light across the connector terminals. **Note:** *In order to avoid timer overload, test light current draw*

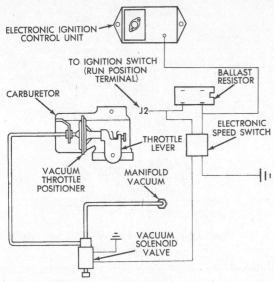

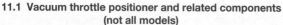

11.1 Vacuum throttle positioner and related components (not all models)

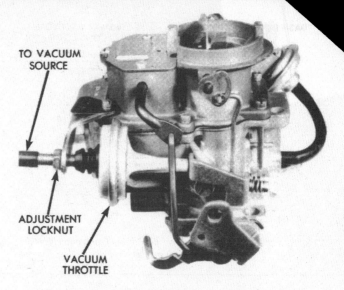

11.11 The vacuum throttle positioner can be turned after loosening the adjustment locknut

should not exceed 0.5 amps. A typical automotive 6-12 volt test light is satisfactory. If the light is shop built, instrument panel size bulbs or smaller should be used.

19   Start the engine. The light should come on and remain on for 60 seconds after the engine starts (30 to 40 seconds on 1976 models).

20   If the light does not come on, remains on indefinitely or does not conform to delay times described in Step 19, the timer should be replaced.

21   If the timer is replaced, repeat the check on the new one.

Coolant Control Idle Enrichment (CCIE) valve check

22   Check all vacuum hoses for proper routing and valve installation. Check engine coolant level (Chapter 1).

23   With the valve installed properly, disconnect the molded connector from the valve. Attach a 1/8-inch ID hose to the bottom port of the valve.

24   With the radiator top tank warm to the touch (no warmer than 75-degrees F), blow through the hose. If it is not possible to blow through the hose/valve, it is defective and must be replaced.

25   Bring the engine to normal operating temperature. Attach a vacuum pump and gauge to the bottom port of the valve. Apply a 10 in-Hg vacuum signal. If the vacuum level drops more than one in-Hg in 15 seconds, replace the valve.

11   Vacuum throttle positioner

Refer to illustrations 11.1 and 11.11

General description

1   Some carburetors are equipped with a throttle positioner system which consists of an electronic speed switch, an electronically

controlled vacuum solenoid valve and a vacuum-actuated throttle positioner **(see illustration)**.

2   The system's function is to prevent unburned hydrocarbons from entering the atmosphere through the vehicle's exhaust system when the throttle is closed after running at high rpm. The electronic speed switch receives ignition pulses from the five ohm ballast resistor terminal connected to the electronic ignition control unit. It then senses when the engine speed exceeds 2000 rpm and allows vacuum to energize the throttle positioner. When the positioner is energized, it prevents the throttle from returning to the idle position. When the throttle is released, it will return to the new stop position (positioner energized) calibrated for a steady-state engine speed of 1750 rpm. As the engine is decelerating, the electronic speed switch senses when the engine speed drops below 2000 rpm and de-energizes the throttle positioner. This allows the throttle to return to the normal idle stop position and the engine will continue to decelerate to idle speed. This operation positions the throttle partially open (1750 rpm) whenever the engine decelerates from a speed above 2000 rpm to a speed just below 2000 rpm. This provides sufficient air flow through the engine to adequately dilute the fuel/air mixture.

Check

3   Check all wiring harness and hose connections to make sure they are secure.

4   Apply vacuum from an external source to the positioner port. If the plunger does not extend, replace the positioner. If it does extend, pinch off the vacuum hose and watch the plunger. If it remains extended for one minute or more, the positioner is in good condition.

5   Apply vacuum from an external source to the manifold vacuum hose connection on the solenoid valve. Disconnect the wiring har-

ness from the solenoid and ground one solenoid terminal with a jumper wire. Apply battery voltage to the other terminal and watch the positioner plunger. If the plunger does not extend, replace the solenoid. If operation is normal, replace the speed switch.

Adjustment

6   It is essential that the curb idle speed adjustment be made using only the curb idle speed screw on the carburetor. Curb idle speed adjustment must not be made using the adjustment for the vacuum throttle positioner. Improper adjustment of the throttle positioner may result in excessive curb idle speed.

7   When finished with idle speed or throttle positioner adjustments, the engine speed must be increased to above 2500 rpm to check for proper idle return.

8   Start the engine and allow it to idle in Neutral, then increase engine speed to above 2000 rpm and verify that the vacuum positioner unit plunger extends. Try to retract it by hand - resistance should be encountered. If the positioner plunger does not extend, refer to the checking procedure above.

9   If the plunger operates normally, adjust the positioner as follows:

10   Increase engine speed to approximately 2500 rpm.

11   Loosen the positioner adjustment locknut and rotate the throttle positioner assembly until the plunger just contacts the throttle lever **(see illustration)**.

12   Release the throttle, then slowly turn the positioner in the opposite direction (away from the throttle lever) until a sudden drop in speed occurs (over 1000 rpm). At this point, turn the positioner an additional 1/4-turn and tighten the locknut. Increase the engine speed to approximately 2500 rpm and release the throttle; the engine should return to normal idle.

6

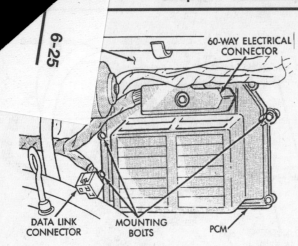

12.3a Location of the Single Module Engine Controller (SMEC) on a 1989 model

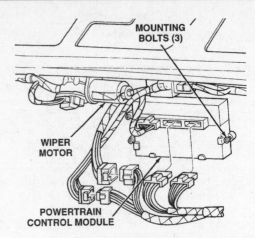

12.3b Location of the Powertrain Control Module (SMEC/PCM) on a 1992 model

12 Single Module Engine Controller (SMEC) or Powertrain Control Module (PCM) - removal and installation

Refer to illustrations 12.3a and 12.3b
Note: Avoid any static electricity damage to the SMEC/PCM by grounding yourself to the body before touching the PCM and using a special anti-static pad to store the SMEC/PCM on, once it is removed.
1 Disconnect the negative battery cable from the battery terminal.
2 Remove the coolant reserve overflow tank (see Chapter 3).
3 Remove the harness retaining bolt for the SMEC/PCM electrical connector **(see illustrations)**.
4 Remove the bolts that retain the SMEC/PCM to the engine compartment and lift the assembly from the vehicle.
5 Installation is the reverse of removal.

13 Self-diagnosis system - description and code access

Note 1: Before outputting the trouble codes, thoroughly inspect ALL electrical connectors and vacuum hoses. Make sure all electrical connections are tight, clean and free of corrosion; make sure all hoses are properly connected, fit tightly and are in good condition (no cracks or tears).

Obtaining diagnostic trouble codes

Refer to illustration 13.4
1 On 1985 through 1997 models, the computer can be placed in the diagnostic mode and the trouble codes flashed on the Check Engine light. On 1998 and later models, a scan tool must be used to tap into the PCM's memory.
2 The PCM will illuminate the CHECK ENGINE light (also known as the Malfunction Indicator Lamp) on the dash if it recognizes certain faults in the system. The light will remain illuminated until the problem is repaired and the code is cleared or the PCM does not detect any malfunction for several consecutive drive cycles. However, not all trouble codes will turn on the light.
3 On 1985 through 1997 models, the system can be accessed to flash the trouble codes on the CHECK ENGINE light. This method will not allow you to obtain all possible codes - only the two-digit trouble codes. To extract the diagnostic trouble codes on these models using this method, proceed as follows:
a) Without starting the engine, turn the ignition key ON, OFF, ON, OFF and finally ON. The CHECK ENGINE light on the dash will begin to flash.
b) If any trouble codes are stored in the PCM memory, the CHECK ENGINE light will flash the number of the first digit, pause and flash the number of the second digit. For example: Code 23, air temperature sensor circuit, would be indicated by two flashes, pause, then three flashes. A long pause will appear between individual codes if more than one code is present. Carefully observe the flashes and record the exact code number(s) onto paper.
c) After the stored codes have been indicated (or if everything in the self diagnosis system is functioning properly), the CHECK ENGINE light will flash a Code 55. Refer to the "Check Engine Light Flash Code" column of the trouble code chart for trouble code identification.
d) If the ignition key is turned OFF during the code extraction process and possibly turned back ON, the self diagnostic system will automatically invalidate the procedure. Restart the procedure to extract the codes. **Note:** The self diagnostic system cannot be accessed with the engine running.
4 The preferred code extraction method requires a special SCAN tool that is programmed to interface with the OBD system by plugging into the diagnostic connector **(see illustration)**. This method will work on 1996 and later models. On 1998 and later models this is the only way to obtain trouble codes. When used, the scan tool has the ability to diagnose in-depth driveability problems and it allows freeze frame data to be retrieved from the PCM stored memory. Freeze frame data is an OBD II PCM feature that records all related sensor and actuator activity on the PCM data stream whenever an engine control or emissions fault is detected and a trou-

13.4 Scanners like the Actron OBD II Diagnostic Tester, Actron Scantool and the AutoXray XP240 are powerful diagnostic aids - programmed with comprehensive diagnostic information, they can tell you just about anything you want to know about your engine management system

Check Engine Light Flash Code	Generic Scan Tool Code	DRB Scan Tool Display	Description of Diagnostic Trouble Code
12			Battery power to PCM was disconnected
54	P0340	No Cam Signal at PCM	No camshaft signal detected during engine cranking.
53	P0601	Internal Controller Failure	PCM Internal fault condition detected.
47	P0162	Charging System Voltage Too Low	Battery voltage sense input below target charging during engine operation. Also, no significant change detected in battery voltage during active test of generator output circuit.
46	P1594	Charging System Voltage Too High	Battery voltage sense input above target charging voltage during engine operation.
42	P1388	Auto Shutdown Relay Control Circuit	An open or shorted condition detected in the auto shutdown relay circuit.
41	P0622	Generator Field Not Switching Properly	An open or shorted condition detected in the generator field control circuit.
37	P0743	Torque Converter Clutch Soleniod/Trans Relay Circuits	An open or shorted condition detected in the torque converter part throttle unlock solenoid control circuit (3 speed auto RH trans. only).
35	P1491	Rad Fan Control Relay Circuit	An open or shorted condition detected in the low speed radiator fan relay control circuit (2.5L only).
34	P1595	Speed Control Solenoid Circuits	An open or shorted condition detected in the Speed Control vacuum or vent solenoid circuits.
33	P0645	A/C Clutch Relay Circuit	An open or shorted condition detected in the A/C clutch relay circuit.
31	P0443	EVAP Purge Solenoid Circuit	An open or shorted condition detected in the duty cycle purge solenoid circuit.
27	P0203 or P0202 or P0201	Injector #3 Control Circuit / Injector #2 Control Circuit / Injector #1 Control Circuit	Injector #3 output driver does not respond properly to the control signal. / Injector #2 output driver does not respond properly to the control signal. / Injector #1 output driver does not respond properly to the control signal.
25	P0505	Idle Air Control Motor Circuits	A shorted or open condition detected in one or more of the idle air control motor circuits.
24	P0122 or P0123	Throttle Position Sensor Voltage Low / Throttle Position Sensor Voltage High	Throttle position sensor input below the minimum acceptable voltage / Throttle position sensor input above the maximum acceptable voltage.
22	P0117	ECT Sensor Voltage Too Low	Engine coolant temperature sensor input below minimum acceptable voltage.

Diagnostic Trouble Code Identification Chart (continued on the following pages)

Check Engine Light Flash Code	Generic Scan Tool Code	DRB Scan Tool Display	Description of Diagnostic Trouble Code
22	or P0118	ECT Sensor Voltage Too High	Engine coolant temperature sensor input above maximum acceptable voltage.
17	1281	Engine Is Cold Too Long	Engine did not reach operating temperature within acceptable limits.
14	P0107	MAP Sensor Voltage Too Low	MAP sensor input below minimum acceptable voltage.
	or P0108	MAP Sensor Voltage Too High	MAP sensor input above maximum acceptable voltage.
13	P1297	No Change in MAP From Start to Run	No difference recognized between the engine MAP reading and the barometric (atmospheric) pressure reading from start-up.
11	P0320	No Crank Reference Signal at PCM	No crank reference signal detected during engine cranking.
	P0351	Ignition Coil #1 Primary Circuit	Peak primary circuit current not achieved with maximum dwell time.
42	P1389	No ASD Relay Output Voltage at PCM	An Open condition Detected In The ASD Relay Output Circuit.
63	P1696	PCM Failure EEPROM Write Denied	Unsuccessful attempt to write to an EEPROM location by the PCM.
37	P0753	Trans 3-4 Shift Sol/Trans Relay Circuits	Current state of output port for the solenoid is different from expected state.
23	P0112	Intake Air Temp Sensor Voltage Low	Intake air temperature sensor input below the maximum acceptable voltage.
	or P0113	Intake Air Temp Sensor Voltage High	Intake air temperature sensor input above the minimum acceptable voltage.
27	P0204	Injector #4 Control Circuit	Injector #4 output driver does not respond properly to the control signal.
21	P0132	Left Upstream O2S Shorted to Voltage	Oxygen sensor input voltage maintained above the normal operating range.
	P0154	O2 2/1 Signal Inactive	No signal at O2 2/1 sensor
	P0152	O2 2/1 Shorted High	Oxygen sensor input voltage sustained above the normal operating range.
53	PO600	PCM Failure SPI Communications	PCM internal fault condition detected
27	P0205	Injector #5 Control Circuit	Injector #5 output driver does not respond properly to the control signal.
	or P0206	Injector #6 Control Circuit	Injector #6 output driver does not respond properly to the control signal.
45	P0712	Trans Temp Sensor Voltage Too Low	Voltage less than 1.55 volts.
	or		

Diagnostic Trouble Code Identification Chart (continued)

Check Engine Light Flash Code	Generic Scan Tool Code	DRB Scan Tool Display	Description of Diagnostic Trouble Code
45	P0713	Trans Temp Sensor Voltage Too High	Voltage greater than 3.76 volts.
27	P0207 or P0208	Injector #7 Control Circuit Injector #8 Control Circuit	Injector #7 output driver does not respond properly to the control signal. Injector #8 output driver does not respond properly to the control signal.
77	P1683	SPD CTRL PWR RLY; or S/C 12V Driver CKT	Malfuntion detected with power feed to speed control servo solenoids
34	P1596 or P1597	MUX S/C Switch High MUX S/C Switch Low	Speed control switch input above the maximum acceptable voltage. Speed control switch input below the minimum acceptable voltage.
42	P1282	Fuel Pump Relay Control Circuit	An open or shorted condition detected in the fuel pump relay control circuit.
21	P0133 or P0152 or P0135	O2 1/1 Slow Response O2 1/1 Heater Circuit	Oxygen sensor response slower than minimum required switching frequency. Upstream oxygen sensor heating element circuit malfunction
	P0139	O2 1/1 Slow Response	Oxygen sensor response slower than minimum required switching frequency.
	P0141	O2 1/2 Heater Circuit	Oxygen sensor heating element circuit malfunction.
43	P0300 or P0301 or P0302 or P0303 or P0304	Multiple Cylinder Mis-fire Cylinder #1 Mis-fire Cylinder #2 Mis-fire Cylinder #3 Mis-fire Cylinder #4 Mis-fire	Misfire detected in multiple cylinders. Misfire detected in cylinder #1. Misfire detected in cylinder #2. Misfire detected in cylinder #3. Misfire detected in cylinder #4.
72	P0420	Catalyst 1/1 Effic	Catalyst efficiency below required level.
31	P0441	Incorrect Purge Flow	Insufficient or excessive vapor flow detected during evaporative emission system operation.
37	P1899	P/N Switch Stuck in Park or in Gear	Incorrect input state detected for the Park/Neutral switch, auto. trans. only.
	P0551	Pwr Steering Sw Perf	Power steering high pressure seen at high speed (2.5L only).
52	P0172	Left Bank or Fuel System Rich	A rich air/fuel mixture has been indicated by an abnormally lean correction factor.
51	P0171	Right Rear (or just) Fuel System Lean	A lean air/fuel mixture has been indicated by an abnormally rich correction factor.

Diagnostic Trouble Code Identification Chart (continued)

Check Engine Light Flash Code	Generic Scan Tool Code	DRB Scan Tool Display	Description of Diagnostic Trouble Code
	P0175	Fuel System 2/1 Rich	A rich air/fuel mixture has been indicated by an abnormally lean correction factor.
	P0174	Fuel System 2/1 Lean	A lean air/fuel mixture has been indicated by an abnormally lean correction factor.
	P0153	O2 2/1 Slow Response	Oxygen sensor response slower than minimum required switching frequency.
	P0159	O2 2/1 Slow Response	Oxygen sensor response slower than minimum required switching frequency.
	P0155	O2 2/1 Heater circuit	Oxygen sensor heater element malfunction.
	P0161	O2 2/1 Heater circuit	Oxygen sensor heater element malfunction.
21	P0138	Left Bank Downstream or Downstream and Pre-Catalyst O2S Shorted to Voltage	Oxygen sensor input voltage maintained above the normal operating range.
	P0158	O2 2/2 Shorted High	Oxygen sensor input voltage maintained above the normal operating range.
17	P0125	Closed Loop Temp Not Reached	Engine does not reach 20°F within 5 minutes with a vehicle speed signal.
24	P0121	TPS Voltage Does Not Agree With MAP	TPS signal does not correlate to MAP sensor
14	P1296	No 5 Volts To MAP Sensor	5 Volt output to MAP sensor open.
25	P1294	Target Idle Not Reached	Actual idle speed does not equal target idle speed.
37	P1756 or P1757	Governor Pressure Not Equal to Target @ 15-20 PSI Governor Pressure Above 3 PSI In Gear With 0 MPH	Governor sensor input not between 10 and 25 psi when requested. Governor pressure greater than 3 psi when requested to be 0 psi.
37	P0740	Torq Conv Clu, No RPM Drop At Lockup	Relationship between engine speed and vehicle speed indicates no torque converter clutch engagement (auto. trans. only).
42	P0462 or P0463 or P0460	Fuel Level Sending Unit Volts Too Low Fuel Level Sending Unit Volts Too High Fuel Level Unit No Change Over Miles	Open circuit between PCM and fuel gauge sending unit. Circuit shorted to voltage between PCM and fuel gauge sending unit. No movement of fuel level sender detected.
44	P1493 or P1492	Ambient/Batt Temp Sen VoltsToo Low Ambient/Batt Temp Sensor VoltsToo High	Battery temperature sensor input voltage below an acceptable range. Battery temperature sensor input voltage above an acceptable range.

Diagnostic Trouble Code Identification Chart (continued)

Check Engine Light Flash Code	Generic Scan Tool Code	DRB Scan Tool Display	Description of Diagnostic Trouble Code
21	P0131 or P0137	Left Bank and Upstream O2S Shorted to Ground Downstream, Left Bank Downstream and Pre-Catalyst O2S Shorted to Ground	O2 sensor voltage too low, tested after cold start. O2 sensor voltage too low, tested after cold start.
11	P1391	Intermittent Loss of CMP or CKP	Intermittent loss of either camshaft or crankshaft position sensor
31	P0442 or P0455	Evap Leak Monitor Small Leak Detected Evap Leak Monitor Large Leak Detected	A small leak has been detected by the leak detection monitor The leak detection monitor is unable to pressurize Evap system, indicating a large leak.
45	P0711	Trans Temp Sensor, No Rise After Start	Sump temp did not rise more than 16°F within 10 minutes when starting temp is below 40°F or sump temp is above 260°F with coolant below 100°F.
37	P0783	3-4 Shift Sol, No RPM Drop @ 3-4 Shift	The ratio of engine rpm/output shaft speed did not change beyond on the minimum required.
15	P0720	Low Ouput Spd Sensor RPM Above 15 mph	Output shaft speed is less than 60 rpm with vehicle speed above 15 mph.
45	P1764 or P1763 or P1762	Governor Pessure Sensor Volts Too Low Governor Pressure Sensor Volts Too HI Governor Press Sen Offset Volts Too Lo or High	Voltage less than .10 volts. Voltage greater than 4.89 volts. Sensor input greater or less than calibration for 3 consecutive Neutral/Park occurances.
37	P0748	Governor Pressure Sol Control/Trans Relay Circuits	Current state of solenoid output port is different than expected.
37	P1765	Trans 12 Volt Supply Relay Ctrl Circuit	Current state of solenoid output port is different than expeted.
43	P0305 or P0306 or P0307 or P0308	Cylinder #5 Mis-fire Cylinder #6 Mis-fire Cylinder #7 Mis-fire Cylinder #8 Mis-fire	Misfire detected in cylinder #5. Misfire detected in cylinder #6. Misfire detected in cylinder #7. Misfire detected in cylinder #8.
	P0432	Catalyst 2/1 EFFIC	Catalyst 2/1 efficiency below required level
	P0151	O2 2/1 Voltage Low	Oxygen sensor input voltage maintained below normal operating range.

6

Diagnostic Trouble Code Identification Chart (continued)

Check Engine Light Flash Code	Generic Scan Tool Code	DRB Scan Tool Display	Description of Diagnostic Trouble Code
	P0157	O2 2/1 Voltage Low	Oxygen sensor input voltage maintained below normal operating range.
31	P1495 or P1494	Leak Detection Pump Solenoid Circuit Leak detection pump SW or mechanical fault	Leak detection pump solenoid circuit fault (open or short) Leak detection pump switch does not respond to input.
11	P1398	Mis-fire Adaptive Numerator at Limit	CKP sensor target windows have too much variation
31	P1486	Evap leak monitor pinched hose found	Plug or pinch detected between purge solenoid and fuel tank
45	P0751	O/D Switch Pressed (LO) More Than 5 Min	Overdrive Off switch input too low for more than 5 minutes.
	P0147	O2 1/3 Heater Circuit	Oxygen sensor heater element malfunction.
21	P0133 or P1195	Cat Mon slow O2 1/1	A slow switching oxygen sensor has been detected in bank 1/1 during catalyst monitor test.
	P0153 or P1196	Cat Mon slow O2 2/1	A slow switching oxygen sensor has been detected in bank 2/1 during catalyst monitor test.
	P0129 or P1197	Cat Mon slow O2 1/2	A slow switching oxygen sensor has been detected in bank 1/2 during catalyst monitor test.
15		No vehicle speed sensor signal	No Vehicle speed sensor signal detected during driving conditions.

Diagnostic Trouble Code Identification Chart (continued)

ble code is set. This ability to look at the circuit conditions and values when the malfunction occurs provides a valuable tool when trying to diagnose intermittent driveability problems. If the tool is not available and intermittent driveability problems exist, have the vehicle checked at a dealer service department or other qualified repair shop. **Note:** *The scan tool trouble code designation is referred to as a P0 or P1 code. Refer to the "Generic Scan Tool Code" column of the trouble code chart for the code identification.*

Clearing diagnostic trouble codes

5   After the system has been repaired, the codes must be cleared from the PCM memory. On 1997 and earlier models the codes can be erased by disconnecting the cable from the negative terminal of the battery for at least ten seconds. On 1998 and later models the codes must be erased with the scan tool (follow the directions on the tool display).
6   Always clear the codes from the PCM before starting the engine after a new electronic emission control component is installed onto the engine. The PCM stores the operating parameters of each sensor. The PCM may set a trouble code if a new sensor

is allowed to operate before the parameters from the old sensor have been erased.

Diagnostic trouble code identification

7   The accompanying list of diagnostic trouble codes is a compilation of all the codes that may be encountered. Not all codes pertain to all models and not all codes will illuminate the Check Engine light when set. The codes listed under the "Check Engine Light Flash Code" column are codes that may be displayed by the Check Engine light on 1985 through 1997 models only (see Step 3). All other models require a SCAN tool to access the diagnostic trouble codes.

14   Information sensors (1988 and later models)

Oxygen sensor

General description

1   The oxygen sensor, which is located in the exhaust manifold, monitors the oxygen content of the exhaust gas stream. The oxygen content in the exhaust reacts with the

oxygen sensor to produce a voltage output which varies from 0.1-volt (high oxygen, lean mixture) to 0.9-volts (low oxygen, rich mixture). The SMEC/PCM constantly monitors this variable voltage output to determine the ratio of oxygen to fuel in the mixture. The SMEC/PCM alters the air/fuel mixture ratio by controlling the pulse width (open time) of the fuel injectors. A mixture ratio of 14.7 parts air to 1 part fuel is the ideal mixture ratio for minimizing exhaust emissions, thus allowing the catalytic converter to operate at maximum efficiency. It is this ratio of 14.7 to 1 which the SMEC/PCM and the oxygen sensor attempt to maintain at all times.
2   The oxygen sensor produces no voltage when it is below its normal operating temperature of about 600-degrees F. During this initial period before warm-up, the SMEC/PCM operates in OPEN LOOP mode.
3   If the engine reaches normal operating temperature and/or has been running for two or more minutes, and if the oxygen sensor is producing a steady signal voltage below 0.45-volts at 1,500 rpm or greater, the computer will set a trouble code.
4   When there is a problem with the oxygen sensor or its circuit, the SMEC/PCM operates in the open loop mode - that is, it

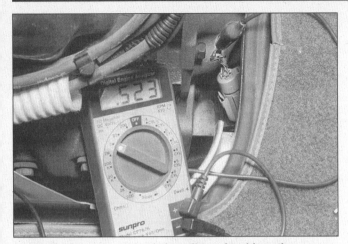

14.7 On TBI systems, check the voltage signal from the oxygen sensor on the black/green wire

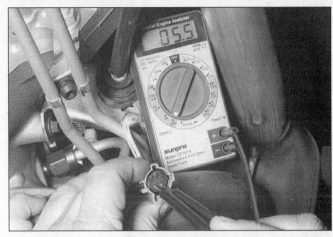

14.10 Measure the resistance of the oxygen sensor heater on the white wires (lower terminals) and check the resistance. It should be between 5 and 7 ohms

controls fuel delivery in accordance with a programmed default value instead of feedback information from the oxygen sensor.

5   The proper operation of the oxygen sensor depends on four conditions:

a *Electrical - The low voltages generated by the sensor depend upon good, clean connections which should be checked whenever a malfunction of the sensor is suspected or indicated.*

b *Outside air supply - The sensor is designed to allow air circulation to the internal portion of the sensor. Whenever the sensor is removed and installed or replaced, make sure the air passages are not restricted.*

c *Proper operating temperature - The SMEC/PCM will not react to the sensor signal until the sensor reaches approximately 600-degrees F. This factor must be taken into consideration when evaluating the performance of the sensor.*

d *Unleaded fuel - The use of unleaded fuel is essential for proper operation of the sensor. Make sure the fuel you are using is of this type.*

6   In addition to observing the above conditions, special care must be taken whenever the sensor is serviced.

a) *The oxygen sensor has a permanently attached pigtail and electrical connector which should not be removed from the sensor. Damage or removal of the pigtail or electrical connector can adversely affect operation of the sensor.*

b) *Grease, dirt and other contaminants should be kept away from the electrical connector and the louvered end of the sensor.*

c) *Do not use cleaning solvents of any kind on the oxygen sensor.*

d) *Do not drop or roughly handle the sensor.*

e) *The silicone boot must be installed in the correct position to prevent the boot from being melted and to allow the sensor to operate properly.*

Check

Refer to illustrations 14.7 and 14.10

7   Remove the engine cover (see Chapter 2B, Section 2). Locate the oxygen sensor electrical connector and backprobe the backside of the connector green/black wire with the positive probe of a digital voltmeter **(see illustration)**. Attach the negative probe to a good ground. **Note:** *Consult the wiring diagrams at the end of Chapter 12 for additional information on the oxygen sensor electrical connector wire color designations.*

8   Monitor the voltage signal (millivolts) as the engine goes from cold to warm.

9   The oxygen sensor will produce a steady voltage signal at first (open loop) of approximately 0.1 to 0.2 volts with the engine cold. After a period of approximately two minutes, the engine will reach operating temperature and the oxygen sensor will start to fluctuate between 0.1 to 0.9 volts (closed loop). If the oxygen sensor fails to reach the closed loop mode or there is a very long period of time until it does switch into closed loop mode, replace the oxygen sensor with a new part.

10   Also inspect the oxygen sensor heater. Disconnect the oxygen sensor electrical connector and connect an ohmmeter between the two white wires **(see illustration)**. It should measure approximately 5 to 7 ohms.

11   Check for proper supply voltage to the heater. Measure the voltage on the oxygen sensor electrical connector between the light blue/black wire (+) and black/green wire (-) (on a TBI system) or the orange/green wire (+) and black/light blue wire (-) (on a MPFI system). There should be battery voltage with the ignition key ON (engine not running). If there is no voltage, check the circuit between the main relay, the SMEC/PCM and the sensor. **Note:** *It is important to remember that supply voltage will only last approximately 2 seconds because the system uses the air conditioning relay to divert the voltage.*

12   If the oxygen sensor fails any of these tests, replace it with a new part.

Replacement

Note: *Because it is installed in the exhaust manifold or pipe, which contracts when cool, the oxygen sensor may be very difficult to loosen when the engine is cold. Rather than risk damage to the sensor (assuming you are planning to reuse it in another manifold or pipe), start and run the engine for a minute or two, then shut it off. Be careful not to burn yourself during the following procedure.*

13   Disconnect the cable from the negative terminal of the battery.

14   Raise the vehicle and place it securely on jackstands.

15   Carefully disconnect the electrical connector from the sensor.

16   Carefully unscrew the sensor from the exhaust manifold.

17   Anti-seize compound must be used on the threads of the sensor to facilitate future removal. The threads of new sensors will already be coated with this compound, but if an old sensor is removed and reinstalled, recoat the threads.

18   Install the sensor and tighten it securely.

19   Reconnect the electrical connector of the pigtail lead to the main engine wiring harness.

20   Lower the vehicle, take it on a test drive and check to see that no trouble codes set.

Manifold Absolute Pressure (MAP) sensor

General description

21   The Manifold Absolute Pressure (MAP) sensor monitors the intake manifold pressure changes resulting from changes in engine load and speed and converts the information into a voltage output. The SMEC/PCM uses the MAP sensor to control fuel delivery and ignition timing. The SMEC/PCM will receive information as a voltage signal that will vary from 1.5 to 2.5 volts at closed throttle (high vacuum) and 4.0 to 5.0 volts at wide open throttle (low vacuum). The MAP sensor is located inside the control box which is attached to the firewall.

6

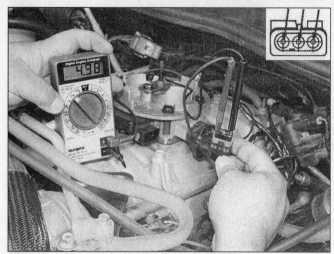

14.25 Using a voltmeter, check for reference voltage to the MAP sensor (purple/white (+) wire). It should be approximately 5.0 volts

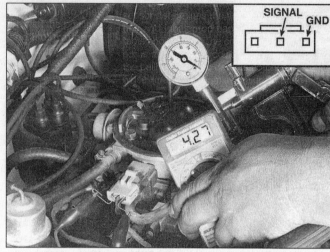

14.26 With no vacuum applied to the MAP sensor there should be approximately 4.5 to 5.0 volts on the green/red wire

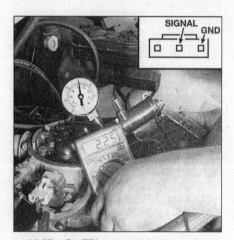

14.27a On TBI systems, increase the vacuum using the hand-held pump and confirm that the voltage decreases

14.27b On MPFI systems, with the engine idling (high manifold vacuum), check the signal voltage on the green/red wire. It should decrease to approximately 1.5 to 3.0 volts

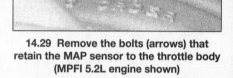

14.29 Remove the bolts (arrows) that retain the MAP sensor to the throttle body (MPFI 5.2L engine shown)

22   A failure in the MAP sensor circuit should set a trouble code.

Check

Refer to illustrations 14.25, 14.26, 14.27a and 14.27b

23   The MAP sensor draws manifold air from a port in the throttle body housing. Check to make sure that the O-ring does not leak. Carefully spray a small amount of aerosol carburetor cleaner around the seal and listen for any changes in rpm. If the engine suddenly speeds up, the MAP sensor is not sealed properly.

24   Check the electrical connector at the sensor for a snug fit. Check the terminals in the connector and the wires leading to it for looseness and breaks. Repair as required.

25   Disconnect the MAP sensor connector, turn the ignition key ON (engine not running) and check for voltage on the purple/white wire **(see illustration)**. There should be approximately 5.0 volts. This checks for ref-

erence voltage to the MAP sensor.

26   Check for voltage on the signal wire. Connect the electrical connector and back-probe the connector using pins or paper clips into the green/red (signal) wire and black/blue (ground) wire terminals. With the ignition key on (engine not running), there should be approximately 5.0 volts **(see illustration)**.

27   On TBI systems, install a hand-held vacuum pump to the MAP sensor to simulate engine vacuum conditions. **Note:** *On MPFI engines, the vacuum connection on the MAP sensor is difficult to reach, so it will be necessary to run the engine to obtain test results.* Apply vacuum to the MAP sensor (or on MPFI engines start the engine and allow it to idle) while probing the signal wire with the positive probe of the voltmeter **(see illustrations)**. Voltage should decrease to approximately 1.5 to 2.5 volts as vacuum increases to full manifold vacuum. If the readings are incorrect, replace the MAP sensor with a new part.

Replacement

Refer to illustration 14.29

28   Remove the engine cover (see Chapter 2B, Section 2). Disconnect the electrical connector from the MAP sensor.

29   Remove the bolts that retain the MAP sensor and remove the MAP sensor **(see illustration)**.

30   Installation is the reverse of removal.

Crankshaft position sensor (MPFI systems only)

General description

Refer to illustrations 14.31a and 14.31b

31   On these models, the crankshaft position sensor determines the timing for the fuel injection and ignition on each cylinder. It also detects engine rpm. The crankshaft position sensor is a Hall-Effect device that is mounted on the bellhousing and detects notches in the flywheel (manual transmission) or flexplate (auto-

matic transmission) **(see illustrations)**. The engine will not operate if the SMEC/PCM does not receive a crankshaft position sensor input.

Check

Refer to illustration 14.32

32   Position a digital ohmmeter across terminals B and C **(see illustration)** of the crankshaft position sensor electrical connector. Set the meter on the 1K to 10K scale, there should be zero resistance.

33   If there is a low resistance reading, replace the sensor with a new part.

Replacement

34   Remove the engine cover (see Chapter 2B, Section 2). Remove the spark plug wires from the right bank spark plugs. Mark the correct cylinder location of each wire as it is removed. Detach the wires from the mounting stud at the rear of the valve cover and position the wires out of the way.

35   Remove the right exhaust manifold heat shield (see Chapter 2B).

36   Remove the EGR valve and tube **(see illustration 14.31a)**. Remove the oil pressure sending unit for access to the tube nut if necessary.

37   Disconnect the crankshaft sensor wiring harness connector, remove the mounting bolts and lift the sensor out.

38   Installation is the reverse of removal. Tighten the bolts to the torque listed in this Chapter's Specifications.

Coolant temperature sensor

General description

39   The coolant temperature sensor is a thermistor (a resistor which varies the value of its voltage output in accordance with temperature changes). The change in the resistance values will directly affect the voltage signal from the water thermosensor. As the sensor temperature DECREASES, the resistance values will INCREASE. As the sensor temperature INCREASES, the resistance val-

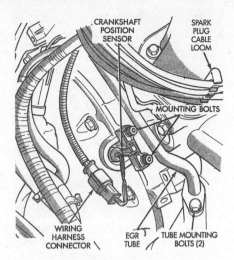

14.31a Location of the crankshaft position sensor on a late model MPFI system

ues will DECREASE. A failure in this sensor circuit should set a trouble code. This code indicates a failure in the coolant temperature sensor circuit, so in most cases the appropriate solution to the problem will be either repair of a wire or replacement of the sensor.

Check

Refer to illustrations 14.40 and 14.41

40   To check the sensor, disconnect the electrical connector and check the resistance values of the coolant temperature sensor while it is completely cold (50 to 80-degrees F = 17,900 to 10,800 ohms). Next, start the engine and warm it up until it reaches operating temperature. The resistance should be lower (180 to 200-degrees F = 1,170 to 820 ohms) **(see illustration)**. **Note:** *Access to the coolant temperature sensor makes it difficult to position probes of the meter on the terminals. If necessary, remove the sensor and perform the tests in a pan of heated water to simulate the conditions.*

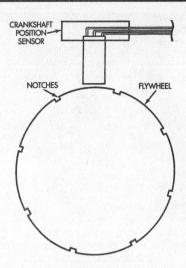

14.31b Crankshaft sensor operation on the 5.2L engine

14.32 Position the probes of the ohmmeter across terminals B and C - there should be infinite resistance

41   Probe the harness connector with a voltmeter and check the reference voltage with the ignition key ON (engine not running). It should be approximately 5.0 volts **(see illustration)**.

6

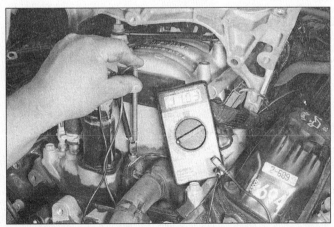

14.40 The coolant temperature sensor on the 5.2L engine is very difficult to access. First remove the alternator and then the A/C compressor to check the resistance on the CTS (MPFI 5.2L engine shown)

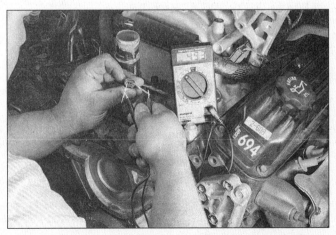

14.41 Check for reference voltage to the coolant temperature sensor with an voltmeter. It should be approximately 5.0 volts.

14.43 First remove the alternator and the A/C compressor to gain access to the coolant temperature sensor on the 5.2L MPFI engines

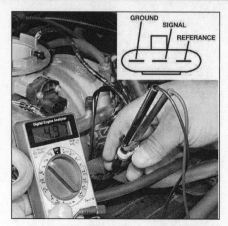

14.47 On MPFI systems, check for reference voltage to the TPS sensor on the purple/white wire (reference) and the black/blue wire (ground). It should be approximately 5.0 volts

14.48a Backprobe the orange/blue wire (SIGNAL) and the purple (GROUND) and monitor the signal voltage with the TPS closed - it should be approximately 0.5 volts (TBI system shown)

Replacement

Refer to illustration 14.43

Warning: *Wait until the engine has cooled completely before beginning this procedure.*

42 Before installing the new sensor, wrap the threads with Teflon sealing tape to prevent leakage and thread corrosion.

43 To remove the sensor, depress the locking tab, unplug the electrical connector, then carefully unscrew the sensor **(see illustration)**. Coolant will leak out when the sensor is removed, so install the new sensor as quickly as possible. **Caution:** *Handle the coolant sensor with care. Damage to this sensor will affect the operation of the entire fuel injection system.*

44 Installation is the reverse of removal. Check the coolant level and add some, if necessary (see Chapter 1).

Throttle Position Sensor (TPS)

General description

45 The Throttle Position Sensor (TPS) is located on the end of the throttle shaft on the throttle body. By monitoring the output voltage from the TPS, the SMEC/PCM can determine fuel delivery based on throttle valve angle (driver demand). A broken or loose TPS can cause intermittent bursts of fuel from the injector and an unstable idle because the SMEC/PCM thinks the throttle is moving.

Check

Refer to illustrations 14.47, 14.48a and 14.48b

46 Remove the engine cover (see Chapter 2B, Section 2). Locate the Throttle Position Sensor (TPS) on the throttle body.

47 Using a voltmeter, check the reference voltage from the SMEC/PCM. Install the positive probe (+) onto the purple/white wire and the negative probe (-) onto the black/blue wire **(see illustration)**. The voltage should read approximately 5.0 volts.

48 Next, check the TPS signal voltage. With the throttle fully closed, install the positive probe (+) of the voltmeter onto the

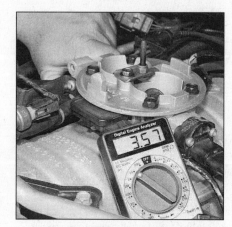

14.48b On MPFI systems, check the voltage signal from the signal wire (orange/blue) and ground wire (black/blue). First, check the voltage with the throttle completely closed (see illustration 14.48a), then at full throttle - the voltage should increase to approximately 3.5 to 4.5 volts

orange/blue wire and the negative probe (-) onto the black/blue wire **(see illustrations)**. Gradually open the throttle valve and observe the TPS sensor voltage. With the throttle valve fully closed, the voltage should read approximately 0.5 to 1.5 volts. Slowly open the throttle valve and observe a gradual change in the voltage values as the sensor travels from idle to full throttle. The voltage should increase to approximately 3.5 to 4.5 volts at wide open throttle. If the readings are incorrect, replace the TPS sensor.

49 A problem in any of the TPS circuits will set a trouble code. Once a trouble code is set, the SMEC/PCM will use an artificial default value for TPS and some vehicle performance will return.

Replacement

Refer to illustrations 14.51 and 14.52

50 Disconnect the electrical connector

14.51 Remove the Torx bolts (arrows) from the TPS

from the TPS.

51 Remove the Torx drive bolts from the TPS **(see illustration)** and remove the TPS from the throttle body.

52 When installing the TPS, be sure to align the socket locating tangs on the TPS with the throttle shaft in the throttle body **(see illustration)**.

53 Installation is the reverse of removal.

Air temperature sensor (MPFI systems only)

General information

54 The air temperature sensor is located in the intake manifold. This sensor is also referred to as the "charge air temperature sensor" or the "intake air temperature (IAT) sensor". This sensor operates as a negative temperature coefficient (NTC) device. As the sensor temperature INCREASES, the resistance values will DECREASE. Most cases, the appropriate solution to the problem will be either repair of a wire or replacement of the sensor.

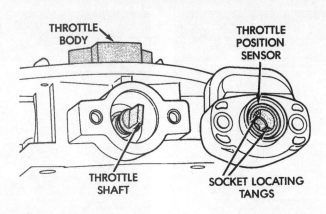

14.52 TPS installation details on the 5.2L engine (others similar)

14.55 Checking the resistance of the air temperature sensor on the 5.2L engine

Check

Refer to illustrations 14.55 and 14.56

55   Remove the engine cover (see Chapter 2B, Section 2). To check the sensor, check the resistance values of the air temperature sensor while it is completely cold (50 to 80-degrees F = 17,900 to 10,800 ohms). Next, start the engine and warm it up until it reaches operating temperature. The resistance should be lower (180 to 200-degrees F = 1,170 to 820 ohms). With the ignition switch ON, disconnect the electrical connector from the air temperature sensor, which is located on the intake manifold. Using an ohmmeter, measure the resistance between the two terminals on the sensor **(see illustration)**.

56   With the ignition key ON (engine not running), probe the wiring harness connector with a volt meter and check for reference voltage to the sensor **(see illustration)**. It should be approximately 5.0 volts.

57   If the test results are incorrect, replace the air temperature sensor.

58   If the sensor checks out okay but there is still a problem, have the vehicle checked at a dealer service department or other qualified repair shop, as the SMEC/PCM may be malfunctioning.

Replacement

59   Unplug the electrical connector from the air temperature sensor.

60   Unscrew the sensor from the intake manifold.

61   Installation is the reverse of removal.

Throttle body temperature sensor (5.2L and 5.9L TBI systems only)

62   The throttle body temperature sensor is located in the throttle body. This sensor monitors throttle body temperature or fuel temperature (located near fuel inlet). This sensor operates as a negative temperature coefficient (NTC) device. As the sensor temperature INCREASES, the resistance values will DECREASE. Most cases, the appropriate solution to the problem will be either repair of a wire or replacement of the sensor.

Check

Refer to illustration 14.63

63   Remove the engine cover (see Chapter 2B, Section 2). To check the sensor, check the resistance values of the throttle body temperature sensor while it is completely cold (50 to 80-degrees F = 17,900 to 10,800 ohms). Next, start the engine and warm it up until it reaches operating temperature. The resistance should be lower (180 to 200-degrees F = 1,170 to 820 ohms). With the ignition switch ON, disconnect the electrical connector from the sensor. Using an ohmmeter, measure the resistance between the two terminals on the sensor **(see illustration)**.

64   With the ignition key ON (engine not running), probe the wiring harness connector with a volt meter and check for reference voltage to the sensor. It should be approximately 5.0 volts.

65   If the test results are incorrect, replace the throttle body temperature sensor.

66   If the sensor checks out okay but there is still a problem, have the vehicle checked at a dealer service department or other qualified repair shop, as the SMEC/PCM may be malfunctioning.

14.56 Checking the reference voltage to the air temperature sensor on the 5.2L engine

Replacement

67   Unplug the electrical connector from the throttle body temperature sensor.

68   Unscrew the sensor from the throttle body.

69   Installation is the reverse of removal.

Vehicle Speed Sensor (VSS)

Refer to illustrations 14.71 and 14.73

General description

70   The Vehicle Speed Sensor (VSS) is located on the transmission. This sensor is a permanent magnetic variable reluctance sensor that produces a pulsing voltage whenever vehicle speed is over three mph. These pulses are translated by the SMEC/PCM and provided for other systems for fuel and transmission shift control.

Check

71   To check the vehicle speed sensor, remove the electrical connector in the wiring harness near the sensor. Using a voltmeter, check for reference voltage to the sensor

14.63 Checking the resistance of the throttle body temperature sensor on a TBI system

6

14.71 Connect the positive probe of the voltmeter onto the reference wire (orange) and the negative probe onto the ground wire (black/light blue) (MPFI system shown)

14.73 Removing the VSS mounting bolt (TBI system shown)

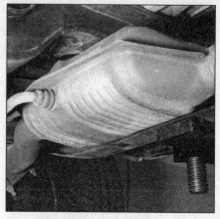

15.1 Catalytic converter on a 1990 TBI system

(see illustration). The reference wire should have approximately 5.0 volts or more available. If there is no voltage available, have the SMEC/PCM diagnosed by a dealership service department.

Replacement

72   To replace the VSS, disconnect the electrical connector from the VSS. Disconnect the speedometer cable, if equipped.
73   Loosen the large nut retaining the sensor to the adapter and remove the VSS from the transmission **(see illustration)**.
74   Installation is the reverse of removal. Be sure to use a new O-ring.

Lock-up control solenoid

General Description

75   The Lock-up Control Solenoid is a computer controlled output actuator that is used to activate the lock-up torque converter on vehicles equipped with an automatic transmission. Refer to Chapter 7B for the check and replacement procedures.

Camshaft position sensor

76   Refer to Chapter 5 for the checks and replacement procedures on the camshaft position sensor.

Neutral start switch

77   Refer to Chapter 7B for the checks and replacement procedures for the Neutral start switch.

---

15   Catalytic converter

General description

Refer to illustration 15.1
1   The catalytic converter **(see illustration)** is an emission control device added to the exhaust system to reduce pollutants from the exhaust gas stream. There are two types of converters. The conventional oxidation catalyst reduces the levels of hydrocarbon (HC) and carbon monoxide (CO). The three-way catalyst lowers the levels of nitrogen oxides (NOx) as well as hydrocarbons (HC) and carbon monoxide (CO).

Check

2   The test equipment for a catalytic converter is expensive and highly sophisticated. If you suspect that the converter on your vehicle is malfunctioning, take it to a dealer or authorized emissions inspection facility for diagnosis and repair.

3   Whenever the vehicle is raised for servicing of underbody components, check the converter for leaks, corrosion, dents and other damage. Check the welds/flange bolts that attach the front and rear ends of the converter to the exhaust system. If damage is discovered, the converter should be replaced.
4   Although catalytic converters don't break too often, they do become plugged. The easiest way to check for a restricted converter is to use a vacuum gauge to diagnose the effect of a blocked exhaust on intake vacuum.

 a) *Open the throttle until the engine speed is about 2000 rpm.*
 b) *Release the throttle quickly.*
 c) *If there is no restriction, the gauge will quickly drop to not more than 2 in-Hg or more above its normal reading.*
 d) *If the gauge does not show 5 in-Hg or more above its normal reading, or seems to momentarily hover around its highest reading for a moment before it returns, the exhaust system, or the converter, is plugged (or an exhaust pipe is bent or dented, or the core inside the muffler has shifted).*

Replacement

5   Refer to Chapter 4A.

Chapter 7 Part A
Manual transmission

Contents

Specifications

General

Transmission lubricant type	See Chapter 1

Torque specifications

Ft-lbs (unless otherwise indicated)

Shift rod swivel bolts	
A-250/A-230	100 in-lbs
A-390	125 in-lbs
Transmission-to-clutch housing bolts	
A230/250/390	50
Overdrive-4	50
NP2500	50
AX-15 (see illustration 4.13)	
Clutch housing-to-transmission bolts	28
Clutch housing-to-engine bolts	
B bolts	50
C bolts	45
D bolts	30

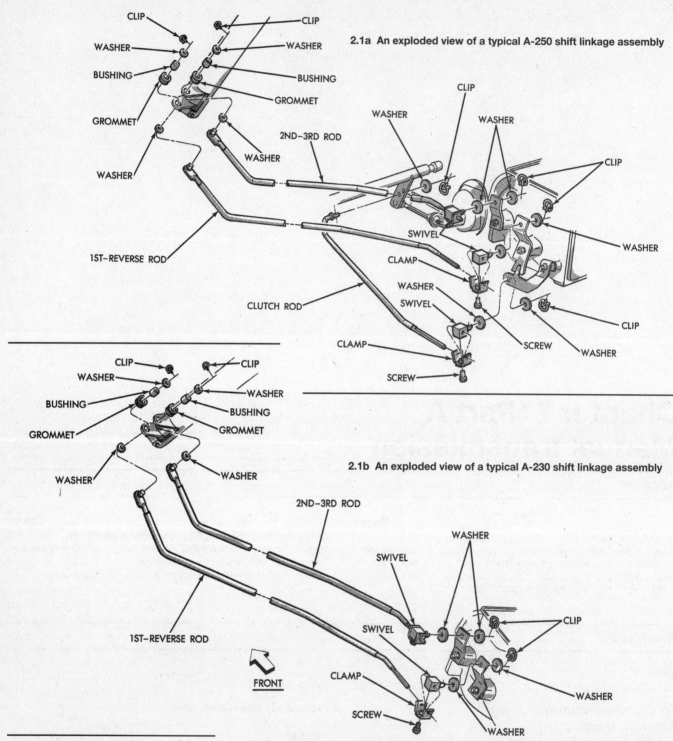

2.1a An exploded view of a typical A-250 shift linkage assembly

2.1b An exploded view of a typical A-230 shift linkage assembly

1   General information

Vehicles covered by this manual are equipped with a three-, four- or five-speed manual or a three- or four-speed automatic transmission. Information on the manual transmission is included in this Part of Chapter 7. Information on the automatic transmission can be found in Part B of this Chapter.

The models covered by this Chapter were equipped with a variety of manual trans-missions that featured either external or internal shift mechanisms.

Models with external shift linkage include the A-230, A-250 and A-390 three-speeds, and the four-speed Overdrive-4. The Overdrive-4 is similar in design to the three-speed transmissions, except that it has an overdrive fourth gear. Units with an internal-shift mechanism include the NP2500 and the AX-15, both of which are five-speeds.

Depending on the expense involved in having a transmission overhauled, it may be a better idea to consider replacing it with either a used or rebuilt one. Your local dealer or transmission shop should be able to supply information concerning cost, availability and exchange policy. Regardless of how you decide to remedy a transmission problem, you can still save a lot of money by removing and installing the unit yourself.

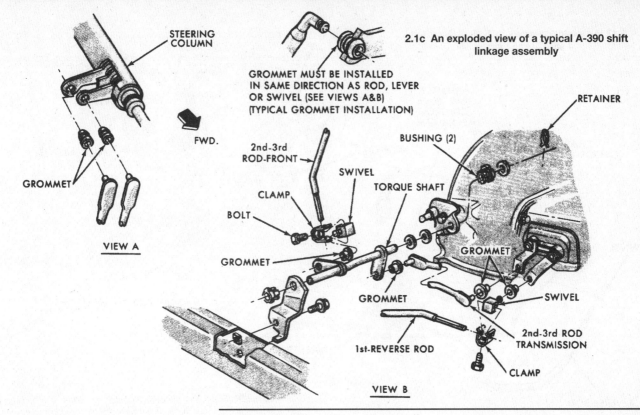

2.1c An exploded view of a typical A-390 shift linkage assembly

2 Shift linkage - removal, installation and adjustment

Column-shift models (A-250/A-230/A-390)

Removal and installation

Refer to illustrations 2.1a, 2.1b and 2.1c

1   The shifter assembly is integral with the steering column on column-shift models. The mechanism in the column normally requires no service. However, the bushings at the ends of the shift rods can wear enough to cause difficult shifting. To replace the bushings, remove the clips (if equipped) and pull or pry the rods out of the levers at the steering column or transmission **(see illustrations)**. Anytime you disassemble the shift linkage assembly, be sure to adjust it after reassembly (see below).

Adjustment

Refer to illustrations 2.4a and 2.4b

2   Remove both shift rod swivels from the transmission shift levers **(see illustration 2.1a, 2.1b or 2.1c)**.

3   Make sure the transmission shift levers are in the neutral (middle detent) position. Move the shift lever so that it lines up with the locating slots in the bottom of the steering column shift housing and bearing housing.

4   Insert a screwdriver between the crossover blade and the 2nd/3rd lever at the steering column so that both lever pins are engaged by the crossover blade **(see illustrations)**.

7A

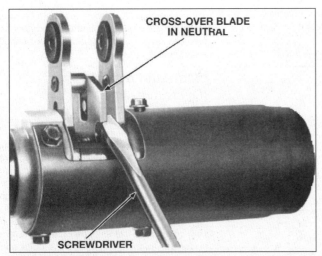

2.4a On A-250/A-230 units, insert a screwdriver between the crossover blade and the 2nd/3rd lever at the steering column so that both lever pins are engaged by the crossover blade as shown

2.4b The setup on A-390 units is similar to A-250/A-230s, but note that the lever pins are reversed, i.e. the right pin - not the left - is on top (the left pin is on top on an A-250/A-230)

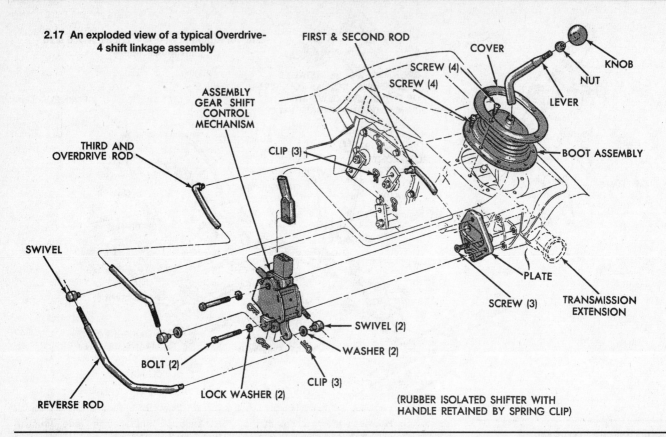

2.17 An exploded view of a typical Overdrive-4 shift linkage assembly

5   Set the 1st/Reverse lever on the transmission to the Reverse position. Adjust the 1st/Reverse rod swivel by loosening the clamp bolt and sliding the swivel along so it will enter the 1st/Reverse lever at the transmission. Install the washers and clip and tighten the swivel bolt to the torque listed in this Chapter's Specifications.
6   Remove the gearshift housing locating tool and shift into the Neutral position.
7   On A-250 and A-230 units, adjust the 2nd/3rd swivel rod by loosening the clamp bolt and sliding the swivel along the rod so that it will enter the 2nd/3rd lever at the transmission. Install the washers and clip and tighten the swivel bolt to the torque listed in this Chapter's Specifications.
8   Remove the tool from the crossover

blade at the steering column and shift it through all gears to check the adjustment and the smoothness of the crossover during operation.
9   Check for proper operation of the steering column lock in Reverse. If the linkage adjustment is correct, the ignition should lock in the Reverse position.
10   Remove the jackstands and lower the vehicle. Test drive the vehicle to make sure that the adjustment is correct.

A250 gearshift interlock

11   Disconnect the clutch rod swivel from the interlock pawl **(see illustration 2.1a)**, then adjust the clutch pedal freeplay (see Chapter 1).
12   With the 1st/reverse lever in Neutral

(middle detent), the interlock pawl will enter the slot in the 1st/reverse lever.
13   Loosen the swivel clamp bolt and slide the swivel on the rod to enter the pawl. Install the washers and the clip.
14   While holding the interlock pawl forward, tighten the swivel clamp bolt but make sure the clutch pedal is returned completely during this adjustment. *Don't pull the clutch rod to the rear to engage the swivel pawl during these steps.*
15   Shift from Neutral to 1st, then from Neutral to Reverse, using the clutch in the normal manner. The clutch action should be normal.
16   Disengage the clutch, and shift halfway to 1st or Reverse. The clutch should now be held down by the interlock to prevent clutch engagement.

Floor-shift models (Overdrive-4)

Removal and installation

Refer to illustrations 2.17 and 2.18

17   Remove the retaining screws from the floor pan boot **(see illustration)** and remove the boot.
18   Insert a 0.10-inch feeler gauge into the shift lever base **(see illustration)**, then pull up on the lever to detach it.
19   Remove the retaining clips, washers and shift rods from the shift mechanism levers under the floor pan. Remove the bolts and washers which secure the shift lever assembly to the transmission extension housing mounting plate and remove the shift lever unit.

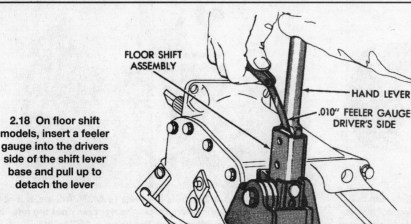

2.18 On floor shift models, insert a feeler gauge into the drivers side of the shift lever base and pull up to detach the lever

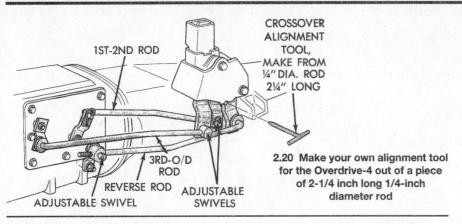

1ST-2ND ROD

CROSSOVER
ALIGNMENT
TOOL,
MAKE FROM
¼" DIA. ROD
2¼" LONG

3RD-O/D
ROD

REVERSE ROD

ADJUSTABLE SWIVEL

ADJUSTABLE
SWIVELS

2.20 Make your own alignment tool for the Overdrive-4 out of a piece of 2-1/4 inch long 1/4-inch diameter rod

Adjustment

Refer to illustration 2.20

20   Fabricate a lever alignment tool out of 1/4-inch diameter rod **(see illustration)**.
21   Insert the lever alignment tool into the holes in the levers and the mechanism frame to hold the levers in the Neutral crossover position.
22   With all rods removed from the transmission shift levers, place the levers in the Neutral detent position.
23   Rotate the threaded shift rods until their length is correct for entering the transmission levers. Start with the 1-2 shift rod (you may have to pull the clip at the shifter end to rotate this rod).
24   Replace the washers and the clips.
25   Remove the aligning tool and test the shifting action.
26   Attach the shift lever to the mechanism.
27   Slide the boot and retainer over the shift lever and fasten it to the floor with the screws.

28   Connect the negative battery cable.
29   Check the shifting action for smoothness.

3   Shift lever (NP2500 and AX-15) - removal and installation

NP2500

Refer to illustration 3.1

1   Remove the bezel screws and pull off the bezel and shifter boot, then unscrew the gearshift lever **(see illustration)**.
2   Installation is the reverse of removal.

AX-15

Refer to illustration 3.5

3   Shift the transmission into Neutral. Raise the front of the vehicle and place it securely on jackstands.
4   Support the transmission with a floor jack and remove the rear crossmember.

5   Lower the transmission about three inches, reach up on top of the transmission, unseat the shift lever dust boot from the transmission shift tower **(see illustration)**, move the boot up on the lever for better access to the lever retainer and disengage the shift lever from the transmission as follows:

a) *Reach up and around the transmission case and press the shift lever retainer down with your fingers.*
b) *Turn the retainer counterclockwise to release it.*
c) *Lift the lever and retainer out of the shift tower.*

Note: *If you're detaching the shift lever to remove the transmission rather than to replace the lever itself, it isn't necessary to remove the shift lever from the boot. Simply leave the lever in place for later reassembly.*

6   To install the shift lever, reach up and around the transmission and insert the lever into the shift tower. Press the lever retainer down and turn it clockwise to lock it into place. Install the shift lever dust boot on the shift tower. Raise the transmission back into place. Install the rear crossmember, remove the jackstands and lower the vehicle.

4   Transmission - removal and installation

Removal

Refer to illustration 4.13

1   Disconnect the negative cable from the battery,
2   Raise the front end of the vehicle on a jack. Raise the vehicle sufficiently to provide clearance to easily remove the transmission. Support the vehicle securely on jackstands.
3   Disconnect the external shift levers or rods from the transmission levers (see Section 2) or remove the shift lever (see Section 3).

7A

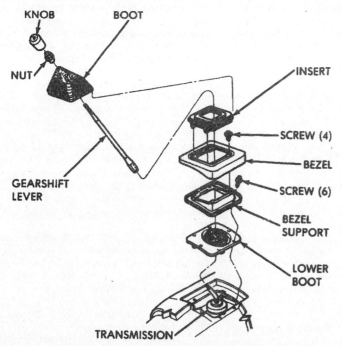

KNOB

BOOT

NUT

INSERT

SCREW (4)

BEZEL

SCREW (6)

BEZEL
SUPPORT

GEARSHIFT
LEVER

LOWER
BOOT

TRANSMISSION

3.1 The shift lever on the NP 2500 transmission simply unscrews

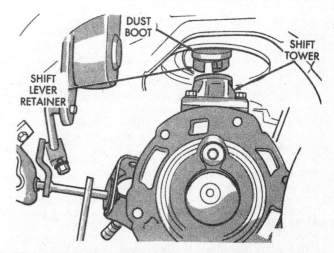

DUST
BOOT

SHIFT
TOWER

SHIFT
LEVER
RETAINER

3.5 To remove the shift lever from an AX-15 transmission, lower the transmission about three inches, reach up and press the shift lever retainer down, turn the retainer counterclockwise, and lift the lever and retainer out of the tower

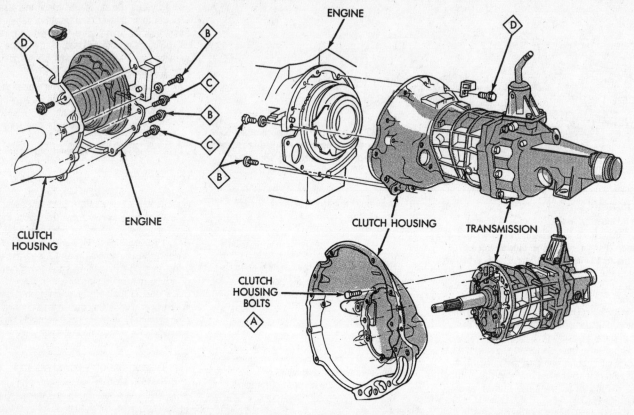

4.13 Fastener torque guide for the AX-15 transmission

4   Disconnect the clutch linkage (1971 through 1987 models) or the hydraulic release cylinder (1988 through 1993 models) (see Chapter 8).

5   Disconnect the speedometer cable (see Chapter 7B), then unplug the back-up light switch connector.

6   Drain the lubricant from the transmission (see Chapter 1).

7   Remove the driveshaft (see Chapter 8). Use a plastic bag to cover the end of the transmission to prevent fluid loss and contamination.

8   Remove the exhaust system components as necessary for clearance (see Chapter 4).

9   Support the engine. This can be done from above with an engine hoist, or by placing a jack (with a block of wood as an insulator) under the engine oil pan. The engine should remain supported at all times while the transmission is out of the vehicle.

10   Support the transmission with a jack - preferably a special jack made for this purpose. Safety chains will help steady the transmission on the jack.

11   Raise the engine slightly and disconnect the extension housing from the center crossmember.

12   Raise the transmission slightly and remove the center crossmember.

13   On all transmissions except AX-15 units, remove the bolts securing the transmission to the clutch housing. On AX-15s, remove the clutch housing-to-engine bolts **(see illustration)**. (The bolts securing the transmission to the bellhousing on AX-15s are *inside* the bellhousing.)

14   Make a final check that all wires and hoses have been disconnected from the transmission, then move the transmission and jack toward the rear of the vehicle until the transmission drive pinion shaft (input shaft) clears the splined hub in the clutch disc. Keep the transmission level as this is done.

15   Once the input shaft is clear, lower the transmission and remove it from under the vehicle.

16   The clutch components can be inspected by removing the clutch housing from the engine (see Chapter 8). In most cases, new clutch components should be routinely installed if the transmission is removed.

Installation

17   Insert a small amount of multi-purpose grease into the pilot bushing in the crankshaft and lubricate the inner surface of the bushing. Make sure no grease gets on the input shaft, clutch disc splines or the release lever.

18   If removed, install the clutch components (see Chapter 8).

19   If removed, attach the clutch housing to the engine and tighten the bolts to the torque listed in the Chapter 8 Specifications.

20   With the transmission secured to the jack as on removal, raise the transmission into position behind the clutch housing and then carefully slide it forward, engaging the input shaft with the clutch plate hub. Do not use excessive force to install the transmission - if the input shaft does not slide into place, readjust the angle of the transmission so it is level and/or turn the input shaft so the splines engage properly with the clutch.

21   Install the transmission-to-clutch housing bolts. Tighten the bolts to the torque listed in this Chapter's Specifications.

22   Install the crossmember and attach it to the transmission housing. Tighten all nuts and bolts securely.

23   Remove the jacks supporting the transmission and the engine.

24   Install the various items removed previously, referring to Chapter 8 for the installation of the driveshaft and Chapter 4 for information regarding the exhaust system components.

25   Make a final check that all wires, hoses and the speedometer cable have been connected and that the transmission has been filled with lubricant to the proper level (see Chapter 1). Lower the vehicle.

26   Connect the shift levers or rods to the transmission levers (see Section 2) or install the shift lever (see Section 3).

27   Connect the negative battery cable. Road test the vehicle for proper operation and check for leakage.

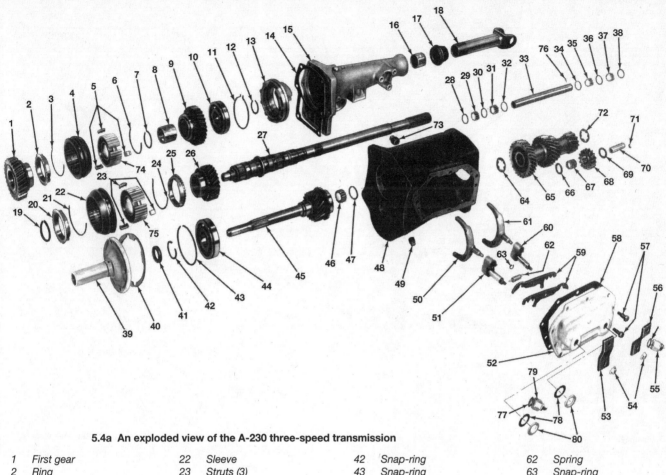

5.4a An exploded view of the A-230 three-speed transmission

1	First gear	22	Sleeve	42	Snap-ring	62	Spring
2	Ring	23	Struts (3)	43	Snap-ring	63	Snap-ring
3	Spring	24	Spring	44	Bearing	64	Washer
4	Sleeve	25	Ring	45	Drive pinion	65	Countershaft gear
5	Struts (3)	26	Second gear	46	Roller	66	Washer
6	Spring	27	Output shaft	47	Snap-ring	67	Roller
7	Snap-ring	28	Washer	48	Case	68	Idler gear
8	Bushing	29	Roller	49	Drain plug	69	Washer
9	Reverse gear	30	Washer	50	Fork	70	Shaft
10	Bearing	31	Roller	51	Lever	71	Key
11	Snap-ring	32	Washer	52	Housing	72	Washer
12	Snap-ring	33	Countershaft	53	Lever	73	Filler plug
13	Retainer	34	Washer	54	Locking nut	74	Clutch gear
14	Gasket	35	Roller	55	Switch	75	Clutch gear
15	Extension housing	36	Washer	56	Lever	76	Key
16	Bushing	37	Roller	57	Bolt	77	Exhaust emission switch gasket
17	Seal	38	Washer	58	Gasket		
18	Yoke	39	Retainer	59	Interlock lever	78	O-rings (2)
19	Snap-ring	40	Gasket	60	Lever	79	Exhaust emission switch
20	Ring	41	Seal	61	Fork	80	O-ring retainers (2)
21	Spring						

5   Transmission overhaul - general information

Refer to illustrations 5.4a, 5.4b, 5.4c and 5.4d

Overhauling a manual transmission is a difficult job for the do-it-yourselfer. It involves the disassembly and reassembly of many small parts. Numerous clearances must be precisely measured and, if necessary, changed with select fit spacers and snap-rings. As a result, if transmission problems arise, it can be removed and installed by a competent do-it-yourselfer, but overhaul should be left to a transmission repair shop. Rebuilt transmissions may be available - check with your dealer parts department and auto parts stores. At any rate, the time and money involved in an overhaul is almost sure to exceed the cost of a rebuilt unit.

Nevertheless, it's not impossible for an inexperienced mechanic to rebuild a transmission if the special tools are available and the job is done in a deliberate step-by-step manner so nothing is overlooked.

The tools necessary for an overhaul include internal and external snap-ring pliers, a bearing puller, a slide hammer, a set of pin punches, a dial indicator and possibly a hydraulic press. In addition, a large, sturdy workbench and a vise or transmission stand will be required.

During disassembly of the transmission, make careful notes of how each piece comes off, where it fits in relation to other pieces and what holds it in place. Exploded views are included **(see illustrations)** to show where the parts go - but actually noting how they

7A

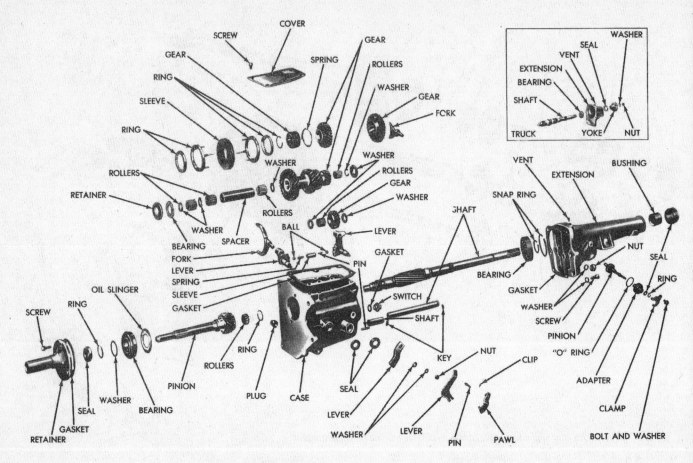

5.4b An exploded view of the A-250 three-speed transmission

are installed when remove the parts will make it much easier to get the transmission back together.

Before taking the transmission apart for repair, it will help if you have some idea what area of the transmission is malfunctioning. Certain problems can be closely tied to specific areas in the transmission, which can make component examination and replacement easier. Refer to the Troubleshooting Section at the front of this manual for information regarding possible sources of trouble.

5.4c An exploded view of the A-390 three-speed transmission

1	Case cover	21	Extension housing screw and lockwasher	41	2nd-speed gear
2	Case cover screw	22	Output shaft bearing retainer	42	Low-speed gear thrust washer snap-ring
3	Case cover gasket	23	Extension gasket	43	Low-speed gear thrust washer
4	Countershaft bearing roller	24	Gearshift 1st and Reverse rail	44	Low-speed gear
5	Countershaft bearing washer	25	Fork set screw	45	Synchronizer low stop ring
6	Countershaft thrust washer	26	Gearshift 1st and Reverse fork	46	Synchronizer low and Reverse clutch-gear snap-ring
7	Reverse idler thrust washer	27	Gearshift lever shaft oil seal	47	Low and Reverse synchronizer assembly
8	Reverse idler bushing	28	Gearshift lever	48	Output shaft
9	Countershaft	29	Transmission case	49	Output shaft pilot roller
10	Countershaft roll pin	30	Plug	50	Input shaft
11	Reverse idler gear	31	Gearshift 2nd and 3rd rail	51	Input shaft bearing
12	Output shaft bearing	32	Gearshift detent-pin spring	52	Bearing outer snap-ring
13	Reverse idler shaft	33	Gearshift 2nd and 3rd fork	53	Bearing inner snap-ring
14	Reverse idler stop pin	34	Gearshift detent pin	54	Bearing retainer oil seal
15	Output shaft bearing outer snap-ring	35	Gearshift detent-pin spring	55	Bearing retainer gasket
16	Output shaft bearing inner snap-ring	36	Plug	56	Bearing retainer
17	Extension housing	37	Case filler plug	57	Bearing retainer screw
18	Extension housing seal	38	Countershaft gear		
19	Back-up light switch	39	2nd and 3rd synchronizer ring		
20	Back-up light switch gasket	40	Synchronizer 2nd and 3rd stop ring		

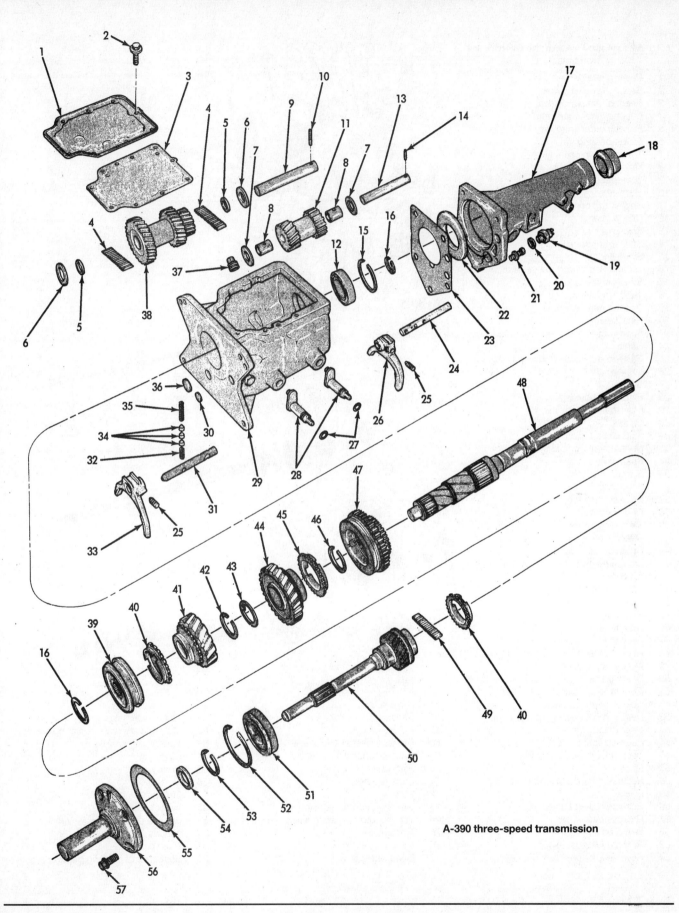

A-390 three-speed transmission

5.4d An exploded view of the Overdrive-4 transmission

1   Bearing retainer
2   Bearing retainer gasket
3   Bearing retainer oil seal
4   Bearing inner snap-ring
5   Bearing outer snap-ring
6   Pinion bearing
7   Transmission case
8   Filler plug
9   2nd-speed gear
10 Snap-ring
11 Shift strut springs
12 Clutch gear
13 Shift struts (3)
14 Shift strut spring
15 Snap-ring
16 1st and 2nd clutch sleeve gear
17 Synchronizer ring
18 1st-speed gear
19 Bearing retainer ring
20 Rear bearing
21 Snap-ring
24 Baffle
25 Case-to-extension housing gasket
26 Lock-washer
27 Bolt
28 Extension housing
29 Mainshaft yoke bushing
30 Oil seal
31 Main drive pinion
33 Needle bearing rollers
34 Snap-ring
35 Synchronizer ring
36 Snap-ring
37 Shift-strut spring
38 Clutch gear
39 Shift-strut spring
40 Clutch sleeve
41 Synchronizer ring
42 3rd speed gear (A-833), Overdrive gear (Overdrive-4)
43 Mainshaft (output shaft)
44 Shift struts (3)
45 Woodruff key
46 Countershaft
47 Gear thrust washer
48 Needle roller bearing spacer rings
49 Needle roller bearing
50 Bearing spacer
51 Countershaft gear cluster
52 Needle bearing rollers
53 Needle roller bearing spacer rings
54 Gear thrust washer
55 Back-up light switch
56 Back-up light switch gasket
57 Plug
58 Reverse detent ball spring retainer
59 Gasket
60 Reverse detent ball spring
61 Reverse detent ball
62 Woodruff key
63 Reverse idler gear shaft
64 Reverse idler gear bushing

65 Reverse idler gear
66 Reverse shifter fork
67 Reverse lever
68 Reverse lever shaft oil seal
69 Reverse operating lever
70 Flat washer
71 Lock-washer
72 Nut
73 Gearshift control housing
74 1st and 2nd operating lever
75 Flat washer
76 Lock-washer lever
77 Lever nut

78 Lock-washer lever
79 Flat-washer lever
80 3rd and Overdrive operating lever
83 Interlock lever (2)
84 E-ring
85 Spring
86 Oil seal (2)
87 3rd and Overdrive lever
88 1st and 2nd
89 3rd and Overdrive-speed fork
90 1st and 2nd-speed fork
91 Drain plug
92 Shift control housing gasket

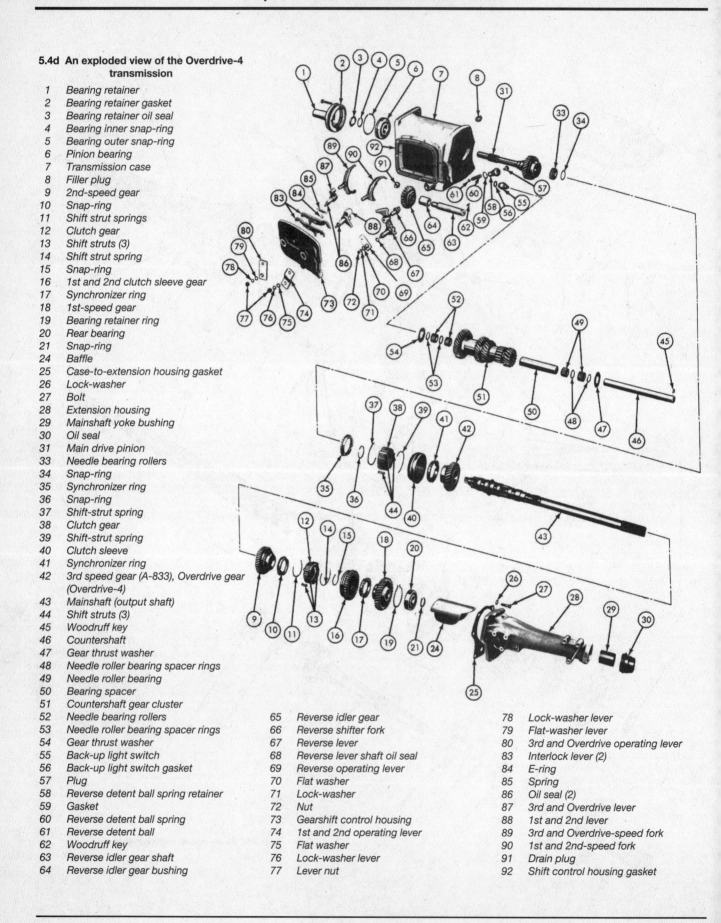

Chapter 7 Part B
Automatic transmission

Contents

Specifications

Torque specifications

Ft-lbs (unless otherwise indicated)

Neutral start/backup light switch	24
Transmission fluid pan bolts	See Chapter 1
Transmission-to-engine bolts	
1971 through 1994	30
1995 on **(see illustration 8.26)**	
A	20 to 40
B	80 to 120 in-lbs
C	30 to 50 in-lbs
D	35 to 65
Torque converter-to-driveplate bolts	
9.5- inch, 3-lug bolts	40
9.5- inch, 4-lug bolts	55
10.0-inch, 4-lug bolts	55
10.75-inch, 4-lug bolts	22.5

1 General information

All vehicles covered in this manual come equipped with a three-, four- or five-speed manual transmission or an automatic transmission. All information on the automatic transmission is included in this Part of Chapter 7. Information on the manual transmission can be found in Part A of this Chapter. You'll also find certain procedures common to both automatic and manual transmissions - such as oil seal replacement - here in Part B.

All automatic transmissions are three- or four-speed units based on the basic Torque-flite design. The A-727 is the most common of the three-speed types along with several variants; the A-904T, A-998, A-999, 32RH and 36RH. Four-speed versions of this transmission also go by several names; the A-500, A-518, 42RH and 46RH. The four-speed transmissions are basically the three-speed

unit with an overdrive unit attached to the rear.

Many 1978 and later transmissions are equipped with a lock-up torque converter that engages in high gear. The lock-up torque converter provides a direct connection between the engine and the drive wheels for improved efficiency and economy. The lock-up converter consists of a solenoid-controlled clutch on the torque converter that engages to lock up the converter in high gear.

Due to the complexity of the automatic transmissions covered in this manual and the need for specialized equipment to perform most service operations, this Chapter contains only general diagnosis, routine maintenance, adjustment and removal and installation procedures.

If the transmission requires major repair work, it should be left to a dealer service department or an automotive or transmission

repair shop. You can, however, remove and install the transmission yourself and save the expense, even if the repair work is done by a transmission shop.

2 Diagnosis - general

Note: *Automatic transmission malfunctions may be caused by five general conditions: poor engine performance, improper adjustments, hydraulic malfunctions, mechanical malfunctions or malfunctions in the computer or its signal network (later models). Diagnosis of these problems should always begin with a check of the easily repaired items: fluid level and condition (see Chapter 1), shift linkage adjustment and throttle rod linkage adjustment. Next, perform a road test to determine if the problem has been corrected or if more diagnosis is necessary. If the problem per-*

sists after the preliminary tests and corrections are completed, additional diagnosis should be done by a dealer service department or transmission repair shop. Refer to the Troubleshooting section at the front of this manual for information on symptoms of transmission problems.

Preliminary checks

1   Drive the vehicle to warm the transmission to normal operating temperature.
2   Check the fluid level as described in Chapter 1:
   a) If the fluid level is unusually low, add enough fluid to bring the level within the designated area of the dipstick, then check for external leaks (see below).
   b) If the fluid level is abnormally high, drain off the excess, then check the drained fluid for contamination by coolant. The presence of engine coolant in the automatic transmission fluid indicates that a failure has occurred in the internal radiator walls that separate the coolant from the transmission fluid (see Chapter 3).
   c) If the fluid is foaming, drain it and refill the transmission, then check for coolant in the fluid or a high fluid level.

3   Check the engine idle speed. **Note:** If the engine is malfunctioning, do not proceed with the preliminary checks until it has been repaired and runs normally.
4   Check the throttle rod for freedom of movement. Adjust it if necessary (see Section 5). **Note:** The throttle rod may function properly when the engine is shut off and cold, but it may malfunction once the engine is hot. Check it cold and at normal engine operating temperature.
5   Inspect the shift linkage (see Section 4). Make sure it's properly adjusted and that the linkage operates smoothly.

Fluid leak diagnosis

6   Most fluid leaks are easy to locate visually. Repair usually consists of replacing a seal or gasket. If a leak is difficult to find, the following procedure may help.
7   Identify the fluid. Make sure it's transmission fluid and not engine oil or brake fluid (automatic transmission fluid is a deep red color).
8   Try to pinpoint the source of the leak. Drive the vehicle several miles, then park it over a large sheet of cardboard. After a minute or two, you should be able to locate the leak by determining the source of the fluid dripping onto the cardboard.
9   Make a careful visual inspection of the suspected component and the area immediately around it. Pay particular attention to gasket mating surfaces. A mirror is often helpful for finding leaks in areas that are hard to see.
10   If the leak still cannot be found, clean the suspected area thoroughly with a degreaser or solvent, then dry it.
11   Drive the vehicle for several miles at normal operating temperature and varying

3.4 Use a hammer and chisel to dislodge the rear seal

3.6 A large socket can be used to drive the new seal evenly into the bore

speeds. After driving the vehicle, visually inspect the suspected component again.
12   Once the leak has been located, the cause must be determined before it can be properly repaired. If a gasket is replaced but the sealing flange is bent, the new gasket will not stop the leak. The bent flange must be straightened.
13   Before attempting to repair a leak, check to make sure that the following conditions are corrected or they may cause another leak. **Note:** Some of the following conditions cannot be fixed without highly specialized tools and expertise. Such problems must be referred to a transmission repair shop or a dealer service department.

Gasket leaks

14   Check the pan periodically. Make sure the bolts are tight, no bolts are missing, the gasket is in good condition and the pan is flat (dents in the pan may indicate damage to the valve body inside).
15   If the pan gasket is leaking, the fluid level or the fluid pressure may be too high, the vent may be plugged, the pan bolts may be too tight, the pan sealing flange may be warped, the sealing surface of the transmission housing may be damaged, the gasket may be damaged or the transmission casting may be cracked or porous. If sealant instead of gasket material has been used to form a seal between the pan and the transmission housing, it may be the wrong type sealant.

Seal leaks

16   If a transmission seal is leaking, the fluid level or pressure may be too high, the vent may be plugged, the seal bore may be damaged, the seal itself may be damaged or improperly installed, the surface of the shaft protruding through the seal may be damaged or a loose bearing may be causing excessive shaft movement.
17   Make sure the dipstick tube seal is in good condition and the tube is properly seated. Periodically check the area around the speedometer gear or sensor for leakage. If transmission fluid is evident, check the O-ring for damage.

Case leaks

18   If the case itself appears to be leaking, the casting is porous and will have to be repaired or replaced.
19   Make sure the oil cooler hose fittings are tight and in good condition.

Fluid comes out vent pipe or fill tube

20   If this condition occurs, the transmission is overfilled, there is coolant in the fluid, the case is porous, the dipstick is incorrect, the vent is plugged or the drain-back holes are plugged.

---

3   Oil seal replacement

1   Oil leaks frequently occur due to wear of the extension housing oil seal, and/or the speedometer drive gear oil seal and O-ring. Replacement of these seals is relatively easy, since the repairs can usually be performed without removing the transmission from the vehicle.

Extension housing oil seal

Refer to illustrations 3.4 and 3.6

2   The extension housing oil seal is located at the extreme rear of the transmission, where the driveshaft is attached. If leakage at the seal is suspected, raise the vehicle and support it securely on jackstands. If the seal is leaking, transmission fluid will be built up on the front of the driveshaft and may be dripping from the rear of the transmission.
3   Refer to Chapter 8 and remove the driveshaft.
4   Using a chisel and hammer, carefully pry the oil seal out of the rear of the transmission **(see illustration)**. Be careful not to damage the splines on the transmission output shaft.
5   If the oil seal and bushing cannot be removed with a chisel, a special oil seal removal tool (available at auto parts stores) will be required.
6   Using a large section of pipe or a very large deep socket as a drift, install the new oil seal **(see illustration)**. Drive it into the bore

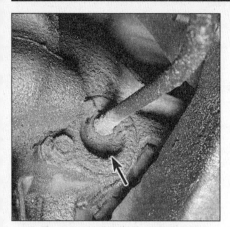

3.8 The adapter (arrow), or driven gear housing, as it's sometimes called, is located on the left side of the extension housing, right in front of the crossmember; it can be hard to find if it's coated with oil and dirt like this one

3.9 Unscrew the speedometer cable collar with a wrench (1977 and earlier models)

3.10a To remove the driven gear housing, remove the retainer bolt . . .

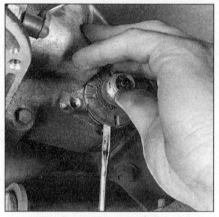

3.10b . . . and pry the housing out with a screwdriver

3.11 Use a hooked tool to remove the O-ring

squarely and make sure it's completely seated.

7   Lubricate the splines of the transmission output shaft and the outside of the driveshaft sleeve yoke with multi-purpose grease, then install the driveshaft. Be careful not to damage the lip of the new seal.

Speedometer driven gear seals

Refer to illustrations 3.8, 3.9, 3.10a, 3.10b, 3.11, 3.12, 3.13 and 3.15

Note: *Any time the speedometer pinion adapter is removed, you MUST install a new O-ring.*

8   The speedometer cable and driven gear housing **(see illustration)**. is located on the left (driver's) side of the extension housing. Look for transmission fluid around the cable and housing to determine if the seal and O-ring are leaking.

9   On 1977 and earlier models, disconnect the speedometer cable **(see illustration)**.

10   Mark the relationship of the adapter to the transmission, then remove the bolt and retainer and pull out the adapter **(see illustrations)**. **Caution:** *Every combination of rear axle gear ratio and tire size requires a different pinion (driven gear), and each pinion is a dif-*

ferent diameter, so the position of the (eccentric) adapter is critical. If you try to install the adapter in some position other than the one it was in before you removed it, the pinion will be stripped or will seize as soon as the vehicle is driven. That's why it's important to carefully mark the position of the adapter in relation to the transmission housing so that the adapter is installed in exactly the same orientation it was in prior to disassembly.

11   Remove the old O-ring **(see illustration)** and install a new O-ring in the adapter.

12   On 1977 and earlier models, remove the

retainer ring with a small screwdriver, then, using a hook, remove the seal **(see illustration)**.

13   On 1978 and later models, remove the adapter clip and pinion clip, then remove the pinion gear and speedometer cable from the adapter **(see illustration)**.

14   Using a small socket as a drift, install the new seal.

15   Installation is the reverse of removal. Make SURE you align the marks you made so the adapter will be in the same position. If you can't find the mark, there should be a

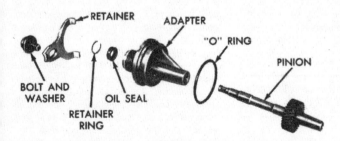

3.12 An exploded view of a typical speedometer pinion and adapter assembly (1977 and earlier models)

RETAINER — ADAPTER — "O" RING — PINION — BOLT AND WASHER — RETAINER RING — OIL SEAL

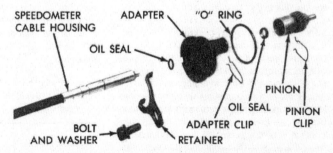

3.13 An exploded view of a typical speedometer pinion and adapter assembly (1978 and later models)

SPEEDOMETER CABLE HOUSING — ADAPTER — "O" RING — OIL SEAL — BOLT AND WASHER — RETAINER — ADAPTER CLIP — OIL SEAL — PINION — PINION CLIP

7B

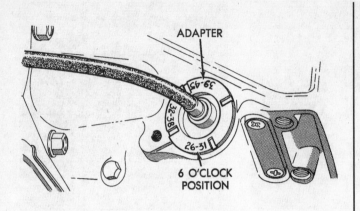

3.15 If you forgot to mark the adapter, look at the number molded into the end of the pinion gear, then find the number range it falls into on the adapter - position the correct range in the six O'clock position

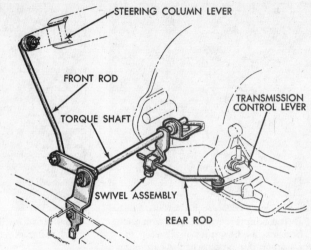

4.4a Typical shift linkage assembly for column-shift models (1978 and earlier models)

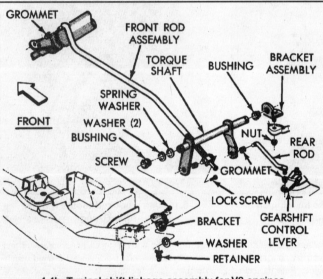

4.4b Typical shift linkage assembly for V8 engines (1979 through 1994 models)

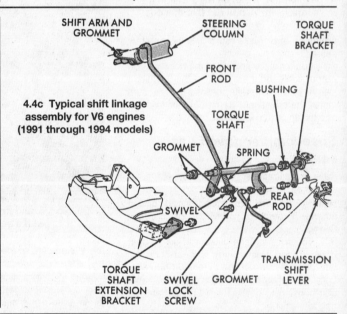

4.4c Typical shift linkage assembly for V6 engines (1991 through 1994 models)

number molded into the end of the plastic pinion gear. This number will fall into one of four ranges cast into the outside flange of the adapter. When reinstalling the adapter, rotate it until the correct range is in the six O'clock position **(see illustration)**. For example, if the number on the end of the pinion gear is 28, position the 26-31 range down.

16   Reinstall the retainer and bolt, tightening the bolt securely.

Front seal

17   If you find transmission fluid leaking from the front of the transmission, coming out from around the torque converter cover, the transmission front seal is probably leaking. Remove the torque converter cover **(see illustration 8.6)** and check for transmission fluid in the cover area, which will normally confirm your diagnosis. However, oil leaking from the engine rear seal can also accumulate in the torque converter housing area and lead to confusion. Transmission fluid is a deep red color.

18   Replacing the front seal is a difficult job, since the transmission and torque converter must be removed first (see Section 8). Also, the torque converter bushing in the transmission front pump should be replaced at this time, since front seal leaks are frequently caused by a worn bushing which causes increased movement of the torque converter hub and subsequent seal failure. We recommend taking the removed transmission to an automotive transmission shop for replacement of the front seal and bushing - this way you'll make sure the seal will not fail on you again soon.

4   Shift linkage/cable - check and adjustment

Shift linkage

Refer to illustrations 4.4a, 4.4b and 4.4c
Note: *Should it be necessary to disassemble any linkage rods from levers which use plastic*

grommets as retainers, always replace the grommets.

1   To check the shift linkage adjustment, verify that the engine starts only in the Park or Neutral position. If the engine won't start at all, the Neutral start/backup light switch may be faulty. If it starts in any other gear, adjust the linkage as follows:

2   Disconnect the negative battery cable.

3   Raise the front of the vehicle and support it securely on jackstands. In the passenger compartment, place the shift lever in Park.

4   Before proceeding with the adjustment, loosen the adjustment swivel lockscrew **(see illustrations)** and make sure the adjustable swivel block is free to turn on the shift rod. If the swivel block binds at all on the shift rod, clean off any corrosion, dirt or grease with solvent and a wire brush before proceeding.

5   Place the shift lever (the one inside the vehicle) in the Park position and lock the steering column with the ignition key.

6   With all linkage assembled and the

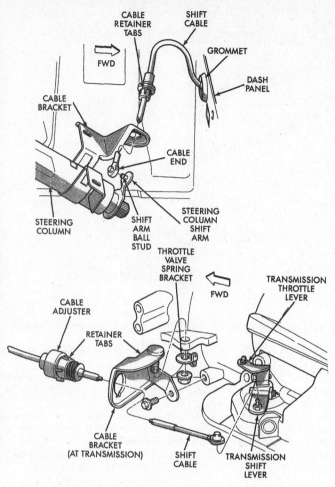

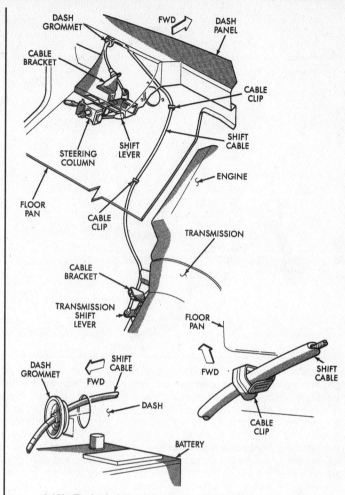

4.10a Typical shift cable assembly connections at steering column and at transmission (1995 and later models)

4.10b Typical shift cable routing (1995 and later models)

adjustment swivel lockscrew still loose, move the shift lever on the transmission all the way to its rear detent (the Park position).

7   With the shift lever inside the vehicle and the shift lever on the transmission in their Park positions, tighten the adjustment swivel lockscrew securely.

8   Lower the vehicle and check the shift linkage operation, again making sure the engine starts only in Park or Neutral. The detent position for Neutral and Drive should be within the limits of the shift lever stops.

Shift cable

Refer to illustrations 4.10a and 4.10b

9   To check the shift cable adjustment, verify that the engine starts only in the Park or Neutral position. If the engine won't start at all, the Neutral start/backup light switch may be faulty. If it starts in any gear other than Park or Neutral, adjust the shift cable as follows:

Description

10   The upper end of the shift cable is attached to, and operated by, an arm on the steering column shaft **(see illustration)** and the lower end is attached to the transmission shift lever. Brackets support and anchor the

cable at each end and clips are used to attach the cable to the floorpan **(see illustration)**. The cable has a built-in adjuster device at the transmission end. This device consists of an internal clamp and release button that allows the cable to be adjusted by extending or shortening it.

Cable adjustment check

11   You should be able to rotate the ignition key from Off to Lock only when the shift lever is in Park.

12   You should be able to shift out of Park only when the ignition key is in the Off, Run or Start positions. You should not be able to shift out of Park when the key is the Accessory or Lock position.

13   The engine should start only in Park or Neutral.

Cable adjustment

14   Put the shift lever in Park.

15   Raise the vehicle and place it securely on jackstands.

16   Unsnap the cable from the transmission shift lever **(see illustration 4.10a)**.

17   Move the transmission shift lever all the way to the rear to the Park detent.

18   Release the cable adjuster (at the trans-

mission end of the cable) to unlock the cable.

19   Snap the cable back onto the pin on the transmission shift lever.

20   To secure the cable, press the cable lock in.

21   Lower the vehicle and verify that the cable is adjusted correctly.

5   Throttle rod and cable adjustment

Caution: *This adjustment is critical and must be correct for the transmission to function properly. Also, incorrect adjustment can lead to early transmission failure. When the adjustment is complete, the transmission should shift into third (high) gear at 20 to 30 mph under light throttle pressure. If it does not, take the vehicle to a dealer service department or qualified transmission shop to have the transmission checked.*

Throttle rod

1992 and earlier models

1   On these models the throttle rod (which is the only transmission shift control, since the Torqueflite has no vacuum modulator)

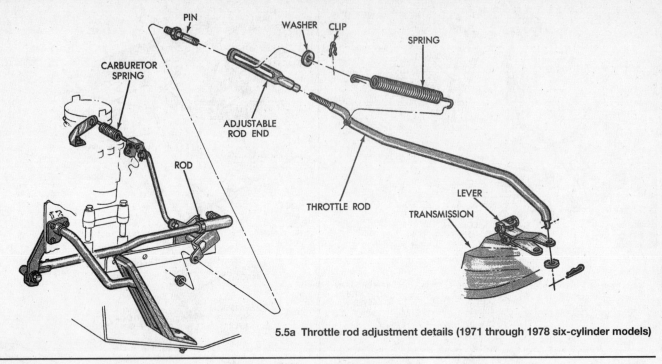

5.5a Throttle rod adjustment details (1971 through 1978 six-cylinder models)

controls a valve in the transmission which governs shift quality and speed. Under light throttle pressure, the transmission should shift into third gear at 20 to 30 mph. If shifting is harsh, erratic or not at the right speed, the throttle rod should be adjusted. You can quickly check the throttle rod's operation by moving the rod (near the carburetor or fuel injection throttle body) rearward, then slowly releasing it. If properly adjusted and in good condition, the throttle rod will return to its full forward position when released. If it doesn't, it should be adjusted and/or lubricated so it does.

2   Before performing the following adjustment, start the engine and warm it up to its normal operating temperature. Remove the air cleaner assembly and make sure the carburetor is **off** the fast-idle cam (see Chapter 4) and the idle speed is correctly adjusted (see Chapter 1). Also, make sure all linkage is cleaned and lubricated, especially bellcranks and carburetor and throttle rod levers.

1971 through 1978 models

Refer to illustrations 5.5a, 5.5b, 5.7a and 5.7b
3   Disconnect the choke at the carburetor or block the choke valve in its fully-open

position.
4   Open the throttle slightly to release the fast-idle cam, then return the carburetor to curb idle.
5   Remove the return spring from the carburetor pin and tab on the transmission throttle rod **(see illustrations)**.
6   Remove the clip, washer and slotted throttle rod end from the carburetor pin.
7   The following adjustment must be made with the throttle lever on the transmission all the way forward against its stop. To insure correct adjustment, it is essential that the transmission lever remain firmly against the

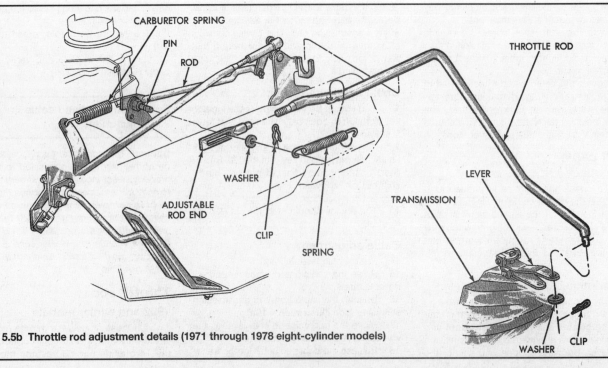

5.5b Throttle rod adjustment details (1971 through 1978 eight-cylinder models)

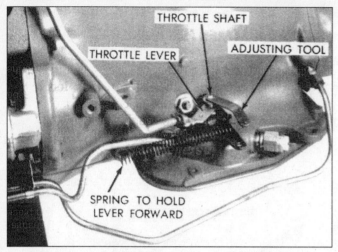

5.7a On 1971 through 1978 models, obtain or fabricate an adjusting tool and use a spring to hold the throttle lever on the transmission all the way forward; to install the tool, attach the spring to the hole in the tool and hook the spring to the front end of the transmission case

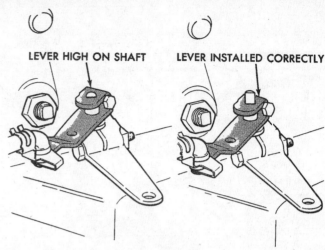

5.7b If the throttle lever is positioned too high on the shaft to install the special tool, loosen the throttle lever, move it down and retighten it

stop, so the factory recommends using a special adjusting tool and spring **(see illustration)**; however, if the tool is not available or cannot be fabricated, have an assistant hold the throttle lever all the way forward during the adjustment. While adjusting the linkage, make sure that no other force is exerted against the lower end of the rod. **Note:** *If the lever is too high on the throttle shaft, you may have to loosen the throttle lever, slide it down the shaft and retighten it before you can install the special tool* **(see illustration).**

8   Rotate the threaded rod end so that the rear edge of the slot contacts the carburetor pin when the transmission throttle lever is held forward against its internal stop.

9   Install the flat washer and clip to retain the throttle rod on the carburetor pin.
10   Install the throttle rod return spring to the tab on the throttle rod and to the carburetor pin.
11   Remove the special tool and spring from the transmission throttle lever.
12   To check the linkage for freedom of operation, move the slotted adjuster link to its full rearward position, then allow it to return slowly, making sure it returns to the full forward position.
13   Connect the choke rod or remove the blocking fixture.
14   Reinstall the air cleaner.

1979 through 1991 models and 1992 models with a throttle rod

Refer to illustrations 5.16a, 5.16b, 5.16c and 5.16d

15   Refer to Steps 1 and 2, then raise the vehicle and place it securely on jackstands.

16   Loosen the adjustment swivel lockscrew **(see illustrations)**. To insure proper adjustment, the swivel must be free to slide along the flat end of the throttle rod so that the preload spring action is unrestricted. If necessary, disassemble, clean and lubricate all parts before proceeding.
17   On 1990 through 1992 models, the ISC actuator must be fully retracted before beginning the adjustment procedure. With the engine off, unplug the ISC electrical connector and connect two jumper wires to the battery. Connect the negative jumper wire to the ISC actuator top pin and the positive wire to the actuator second pin for no longer than five seconds and then remove them to retract the actuator.
18   Push the transmission lever forward as far as it will go and hold it there while tightening the swivel lockscrew securely. That's all there is to it! Linkage backlash is automatically removed by the preload spring.
19   Lower the vehicle, reattach the choke (earlier models) and verify that the linkage operates freely - move the throttle rod to the rear, slowly release it and verify that it returns fully to its forward position.

7B

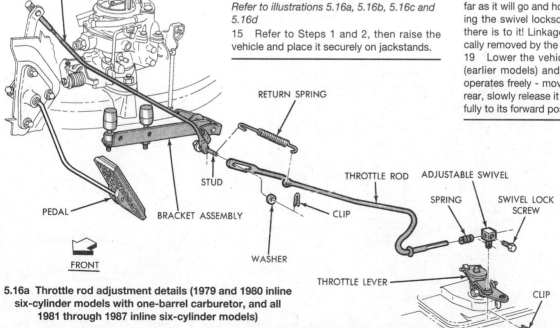

5.16a Throttle rod adjustment details (1979 and 1980 inline six-cylinder models with one-barrel carburetor, and all 1981 through 1987 inline six-cylinder models)

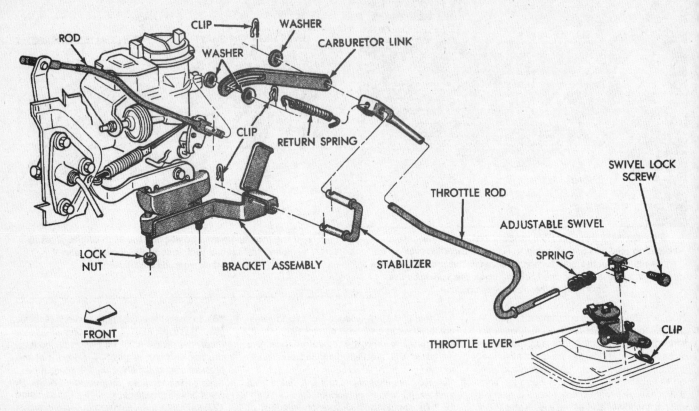

5.16b Throttle rod adjustment details (1979 and 1980 inline six-cylinder models with two-barrel carburetor)

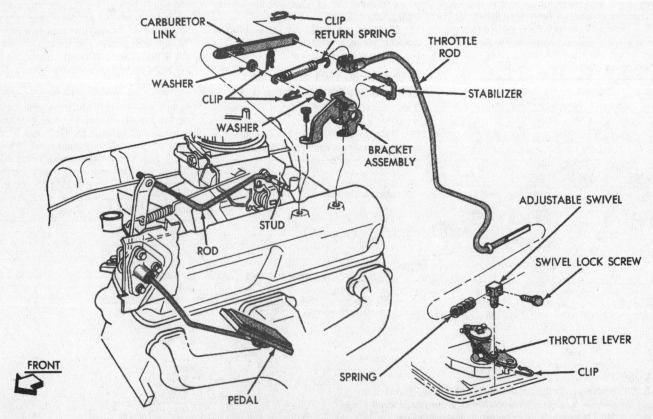

5.16c Throttle rod adjustment details (1979 through 1989 V6 and V8 models)

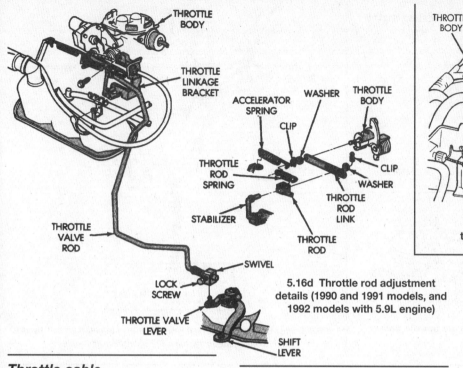

5.16d Throttle rod adjustment details (1990 and 1991 models, and 1992 models with 5.9L engine)

5.22 Insert the feeler gauge between the throttle lever and the idle stop (1993 models)

Throttle cable

1993 models

Refer to illustrations 5.22, 5.24 and 5.26

20   These models use a cable to control the transmission throttle valve.

21   Remove the air cleaner (see Chapter 4).

22   Insert a 0.110 to 0.120-inch feeler gauge spacer between the idle stop and the throttle lever located on the fuel injection throttle body (see illustration).

23   Depress the cable lock button to release the cable.

24   Pull the cable rearward until the wiper contracts completely, then release the cable lock button (see illustration).

25   Raise the vehicle and place it securely on jackstands.

26   Rotate the throttle valve lever forward until the cable adjuster ratcheting sound stops (see illustration).

27   Remove the spacer and check the cable adjustment. Make sure the transmission throttle lever and the lever on the throttle body move simultaneously.

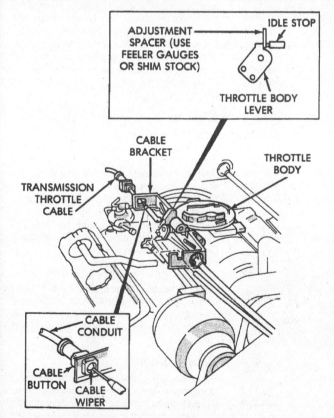

5.24 Pull the throttle cable rearward until the wiper is completely contracted (1993 models)

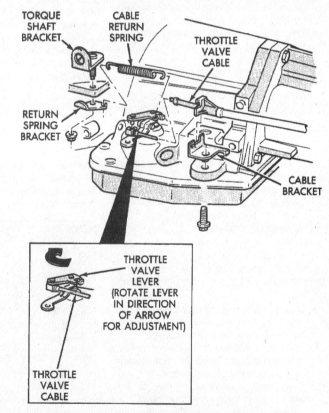

5.26 Throttle cable adjustment details at the transmission (1993 models)

7B

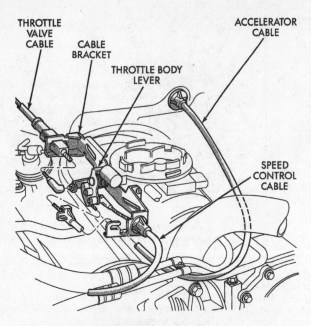

5.30 Throttle cable adjustment details at the throttle body (1994 and later models)

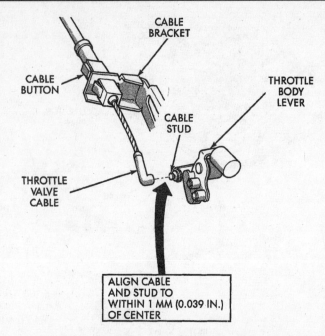

ALIGN CABLE AND STUD TO WITHIN 1 MM (0.039 IN.) OF CENTER

5.32 Compare the position of the cable end in relation to the stud on the throttle body lever (1994 and later models)

1994 and later models

Throttle cable adjustment check

Refer to illustrations 5.30 and 5.32

28   Turn the ignition to the Off position.

29   Remove the air cleaner.

30   Verify that the throttle body lever **(see illustration)** is at the curb idle position, then verify that the transmission throttle lever **(see illustration 5.26)** is also at the idle (full forward) position.

31   Slide the cable off the attachment stud on the throttle body lever.

32   Compare the position of the cable end to the attachment stud on the throttle body lever **(see illustration)**.

a) *The cable end and attachment stud should be aligned (or centered on one another) to within 0.039 inch in either direction.*

b) *If the cable end and attachment stud are misaligned (off center), the cable will have to be adjusted as described in the following procedure.*

33   Reconnect the cable end to the attachment stud. Then with the aid of a helper, observe the movement of the transmission throttle lever and the lever on the throttle body.

a) *If both levers move simultaneously from idle to half-throttle and back to the idle position, the adjustment is correct.*

b) *If the transmission throttle lever moves ahead of or lags behind the throttle body lever, adjust the cable (see below).*

c) *If the throttle body lever prevents the transmission lever from returning to the closed position, adjust the cable (see below).*

Throttle cable adjustment

34   Turn the ignition switch to the Off position and shift the transmission into Park.

35   Remove the air cleaner.

36   Disconnect the cable end from the attachment stud on the throttle body.

37   Verify that the transmission throttle lever is in the idle (full forward) position. Then verify that the lever on the throttle body is at the curb idle position.

38   Press the cable button in to release the cable **(see illustration 5.32)**. The button only has to move about 0.070 inch to release the cable in the adjuster head.

39   Center the cable end on the attachment stud to within 0.039 inch and release the cable button.

40   Verify that the transmission throttle lever and the lever on the throttle body move simultaneously and as described in the cable adjustment checking procedure (Steps 28 through 33).

6.3 Unplug the electrical connector from the switch for access to the terminals (arrow)

6   Neutral start/backup light switch - check and replacement

1   The Neutral start/backup switch is threaded into the lower left front edge of the transmission case. The Neutral start and backup light switch functions are combined into one unit, with the center terminal of the switch grounding the starter solenoid circuit when the transmission is in Park or Neutral, allowing the engine to start. The outer terminals make up the backup light switch circuit.

Check

Refer to illustration 6.3

2   Prior to checking the switch, make sure the shift linkage is properly adjusted (see Section 4). Raise the vehicle and support it securely on jackstands.

3   Unplug the connector and use an ohmmeter or self-powered test light to check for continuity between the center terminal and the transmission case **(see illustration)**. Continuity should exist only when the transmission is in Park or Neutral.

4   Check for continuity between the two outer terminals. There should be continuity only when the transmission is in Reverse. There should be no continuity between either of the outer terminals and the transmission case.

Replacement

5   Place a container under the transmission to catch the fluid which will be released, then use a six-point socket to remove the switch.

6   Move the shift lever from Park to Neutral while checking to see that the switch operating fingers are centered in the opening. If

they aren't, the shift linkage is incorrectly adjusted or there is an internal problem with transmission.

7   Install the new switch and O-ring (be sure to use a new O-ring, even if for some reason you end up putting the same switch back in), tighten it to the torque listed in this Chapter's Specifications, then repeat the checks before plugging in the connector.

7   Transmission mount - check and replacement

Check

Refer to illustration 7.1

1   Raise the vehicle and support it securely on jackstands. Insert a large screwdriver or prybar into the space between the transmission extension housing and the crossmember and try to pry the transmission up slightly **(see illustration)**.
2   The transmission should not move much at all and the rubber in the center of the mount should fully insulate the center of the mount from the mount bracket around it.

Replacement

3   To replace the mount, remove the through-bolt attaching the mount to the crossmember and the bolts attaching the mount to the transmission.
4   Raise the transmission slightly with a jack and remove the mount.
5   Installation is the reverse of the removal procedure. Be sure to tighten the bolts/nut securely.

8   Automatic transmission - removal and installation

Removal

Refer to illustrations 8.6 and 8.8

1   The transmission and converter must be removed as a single assembly. If you try to

7.1 To check a transmission mount, insert a prybar or large screwdriver between the mount and its bracket, then try to lever the mount sideways; it should move very little - excessive movement indicates that the mount is worn out and needs to be replaced

leave the torque converter attached to the driveplate, the converter driveplate, pump bushing and oil seal will be damaged. The driveplate is not designed to support the load, so none of the weight of the transmission should be allowed to rest on the plate during removal.
2   Disconnect the negative cable from the battery.
3   Raise the vehicle and support it securely on jackstands, then remove the starter motor (see Chapter 5). **Note:** *On inline six-cylinder models, it will probably be easier to remove the starter before you raise the vehicle.*
4   Drain the transmission fluid (see Chapter 1), then reinstall the pan.
5   Unbolt and remove the engine-to-transmission struts, if so equipped.
6   Remove the torque converter cover **(see illustration)**.
7   Mark the relationship of the torque converter and driveplate with white paint so they can be installed in the same position.
8   Remove the torque converter-to-driveplate bolts **(see illustration)**. Turn the crankshaft for access to each bolt. Turn the crankshaft in a clockwise direction only (as viewed from the front).
9   Mark the yokes and remove the drive-

shaft (see Chapter 8).
10   Disconnect the speedometer cable (see Section 3). On later models, you can remove the adapter assembly from the transmission and use wire to tie it out of the way, with the speedometer cable still attached. Just make sure you tie a plastic bag around the adapter to prevent it from getting dirty.
11   Detach all wire harness connectors from the transmission.
12   Remove any exhaust components which will interfere with transmission removal (see Chapter 4).
13   Disconnect the throttle rod or cable at the transmission (see Section 5).
14   Disconnect the shift linkage or cable (see Section 4).
15   Support the engine with a jack. Use a block of wood under the oil pan to spread the load.
16   Support the transmission with a jack - preferably a jack made for this purpose (often available from rental yards). Safety chains will help steady the transmission on the jack.
17   Remove the transmission mount through-bolt, then remove the bolts holding the crossmember to the chassis.
18   Remove the bolts securing the transmission to the engine and also remove the trans-

7B

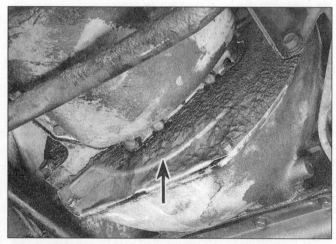

8.6 Unbolt and remove the torque converter cover . . .

8.8 . . . then remove the four torque converter-to-driveplate bolts - be sure to mark the relationship of the torque converter to the driveplate for reassembly reference

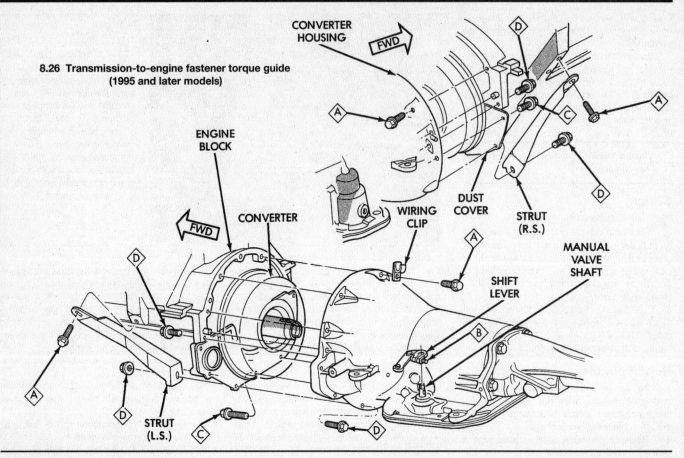

8.26 Transmission-to-engine fastener torque guide (1995 and later models)

mission dipstick tube. **Note:** *The upper bolts are often easier to remove from above, in the engine compartment. Also, on many V8 models, you'll have to remove the oil filter (see Chapter 1) before you can remove the lower right (passenger's side) bolt.*

19 Lower the transmission slightly and disconnect the transmission cooler lines at the transmission, using a flare-nut wrench. **Note 1:** *You'll often need to use a back-up wrench on the transmission cooler line fitting so it doesn't unscrew and cause the line to twist.* **Note 2:** *Fluid will leak out of the lines after you disconnect them. To prevent this, connect the ends of the disconnected lines together with a piece of 3/8-inch diameter fuel hose.*

20 Remove the transmission dipstick tube.

21 Clamp a pair of locking pliers onto the lower portion of the transmission case, just in front of the torque converter (but make sure they're not clamped in front of one of the torque converter-to-driveplate bolt locations, or the driveplate will hang up on the pliers). The pliers will prevent the torque converter from falling out while you're removing the transmission. Move the transmission to the rear to disengage it from the engine block dowel pins and make sure the torque converter is detached from the driveplate. Lower the transmission with the jack.

Installation

Refer to illustration 8.26

22 Prior to installation, make sure the torque converter hub is securely engaged in

the pump. If you've removed the converter, spread transmission fluid on the torque converter rear hub, where the transmission front seal rides. With the front of the transmission facing up, rotate the converter back and forth. It should drop down into the transmission front pump in stages. To make sure the converter is fully engaged, lay a straightedge across the transmission-to-engine mating surface and make sure the converter hub is at least 1/2-inch below the straightedge. Reinstall the locking pliers to hold the converter in this position.

23 With the transmission secured to the jack, raise it into position. Connect the transmission fluid cooler lines.

24 Turn the torque converter to line up the holes with the holes in the driveplate. The white paint mark on the torque converter and driveplate made in Step 7 must line up.

25 Move the transmission forward carefully until the dowel pins and the torque converter are engaged. Make sure the transmission mates with the engine with no gap. If there's a gap, make sure there are no wires or other objects pinched between the engine and transmission and also make sure the torque converter is completely engaged in the transmission front pump. Try to rotate the converter - if it doesn't rotate easily, it's probably not fully engaged in the pump. If necessary, lower the transmission and install the converter fully.

26 Install the transmission-to-engine bolts and tighten them to the torque listed in this

Chapter's Specifications **(see illustration)**. As you're tightening the bolts, check again that the engine and transmission mate completely at all points. If not, find out why. Never try to force the engine and transmission together with the bolts or you'll break the transmission case!

27 Install the torque converter-to-driveplate bolts. Tighten them to the torque listed in this Chapter's Specifications.

28 Install the transmission mount and crossmember.

29 Remove the jacks supporting the transmission and the engine.

30 Install the dipstick tube and engine-to-transmission bolts.

31 Install the starter motor (see Chapter 5).

32 Connect the shift and throttle rod/cable linkage (see Sections 4 and 5, respectively).

33 Plug in the transmission wire harness connectors.

34 Install the torque converter cover and tighten the bolts securely.

35 Install the driveshaft (see Chapter 8).

36 Connect the speedometer cable (see Section 3).

37 Adjust the shift linkage.

38 Install any exhaust system components that were removed or disconnected (see Chapter 4).

39 Remove the jackstands and lower the vehicle.

40 Fill the transmission with the specified fluid (see Chapter 1), run the engine and check for fluid leaks.

Chapter 8
Clutch and driveline

Contents

Specifications

General
Clutch disc lining thickness .... 1/16 inch (above rivet)

Torque specifications
Ft-lbs (unless otherwise indicated)

Clutch
Pressure plate-to-flywheel bolts
 5/16-inch bolts .... 200 in-lbs
 3/8-inch bolts 30
Release fork pivot bolts .... 200 in-lbs
Clutch housing-to-engine bolts
 All except NP2500, Overdrive-4 and AX-15
  3/8-inch bolts 30
  7/16-inch bolts 50
 NP2500 and Overdrive-4 **(see illustration 3.24)**
  A .... 30 to 50 in-lbs
  B .... 20 to 40
  C .... 35 to 65
 AX-15 .... See Chapter 2A Specifications

Driveshaft
Clamp bolts .... 170 in-lbs

Rear axle
Pinion shaft lock bolt .... 100 in-lbs
Differential cover bolts .... See Chapter 1
Pinion shaft-to-companion flange nut (Spicer model 60/70) .... 250 to 270

1 General information

The information in this Chapter deals with the components from the rear of the engine to the rear wheels, except for the transmission, which is dealt with in the previous Chapter. For the purposes of this Chapter, these components are grouped into three categories: clutch, driveshaft and axles. Separate Sections within this Chapter offer general descriptions and checking procedures for components in each of the three groups.

Since nearly all the procedures covered in this Chapter involve working under the vehicle, make sure it's securely supported on sturdy jackstands or on a hoist where the vehicle can be easily raised and lowered.

2 Clutch - description and check

Refer to illustration 2.1

1 All vehicles with a manual transmission use a single dry-plate type clutch **(see illustration)**. The clutch disc has a splined hub which allows it to slide along the splines of the transmission input shaft. On 1971 through 1987

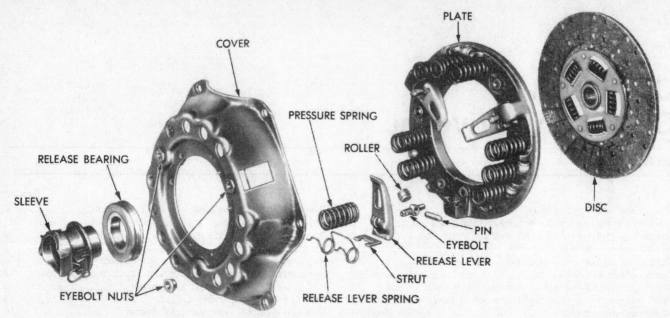

2.1 An exploded view of a typical clutch assembly

models, the clutch and pressure plate are held in contact by spring pressure exerted by nine coil springs installed between the cover and the pressure plate. The clutch is a diaphragm spring design on 1988 and later models.

2   On 1971 through 1987 models, the clutch release system consists of the clutch pedal, a rod connecting the pedal to the torque shaft, the torque shaft itself, another rod connecting the torque shaft to the clutch release lever, and the release, or throwout, bearing. A hydraulic clutch release system is used on 1988 and later models. It consists of the clutch pedal, the clutch master cylinder and remote fluid reservoir, the clutch release cylinder, and a metal line connecting the master cylinder to the release cylinder, a release arm and a release bearing.

3   When the clutch pedal is depressed on a 1971 through 1987 model, the pedal rod pushes against the upper lever on the torque shaft, rotating the torque shaft the same direction as forward wheel rotation. This motion moves another lever on the bottom of the torque shaft to the rear, pushing against the rod connected to the outer end of the release lever. When the outer end of the release lever is pushed to the rear, its inner end slides the release bearing forward on the input shaft. The bearing pushes against the fingers of the release levers mounted on the pressure plate, which in turn relieves coil spring pressure against the clutch disc. When the clutch pedal is depressed on a 1988 or later model with a hydraulic release system, the pedal pushes against the pushrod, which pushes against a piston inside the master cylinder, which raises the hydraulic pressure inside the fluid line and inside the release cylinder. This pressure inside the release cylinder pushes against another piston which pushes against another pushrod that moves

the release arm, which slides the release bearing forward on the input shaft. The release bearing pushes against the diaphragm spring, releasing the clutch.

4   Terminology can be a problem when discussing the clutch components because common names are in some cases different from those used by the manufacturer. For example, the clutch housing is sometimes called the bellhousing, the driven plate is also called the clutch plate or disc, the pressure plate is sometimes called the clutch cover, the clutch release bearing is sometimes called a throwout bearing, the release fork is also called a release lever, the release cylinder is sometimes referred to as a slave cylinder, and so on.

5   Before replacing any components with obvious damage, some preliminary checks should be performed to diagnose clutch problems.

Mechanical system

a) *The first check should be "clutch spin down time." Run the engine at normal idle speed with the transmission in Neutral (clutch pedal up - engaged). Disengage the clutch (pedal down), wait several seconds and shift the transmission into Reverse. No grinding noise should be heard. A grinding noise would most likely indicate a problem in the pressure plate or the clutch disc.*

b) *To check for complete clutch release, run the engine (with the parking brake applied to prevent movement) and hold the clutch pedal approximately 1/2-inch from the floor. Shift the transmission between First gear and Reverse several times. If the shift is hard or the transmission grinds, component failure is indicated. Make sure clutch pedal freeplay is adjusted correctly (see Chapter 1).*

c) *Visually inspect the pivot bushing at the top of the clutch pedal to make sure there is no binding or excessive play.*

d) *Crawl under the vehicle and make sure the clutch release lever is solidly mounted on the pivot.*

Hydraulic system

a) *The first check should be of the fluid level in the clutch master cylinder. If the fluid level is low, add fluid as necessary and inspect the hydraulic system for leaks. If the master cylinder reservoir has run dry, the hydraulic clutch assembly will have to be replaced as a unit. Retest the clutch operation.*

b) *To check "clutch spin down time," run the engine at normal idle speed with the transmission in Neutral (clutch pedal up - engaged). Disengage the clutch (pedal down), wait several seconds and shift the transmission into Reverse. No grinding noise should be heard. A grinding noise would most likely indicate a problem in the pressure plate or the clutch disc (assuming the transmission is in good condition).*

c) *To check for complete clutch release, run the engine (with the parking brake applied to prevent movement) and hold the clutch pedal approximately 1/2-inch from the floor. Shift the transmission between 1st gear and Reverse several times. If the shift is rough, component failure is indicated. Check the release cylinder pushrod travel. With the clutch pedal depressed completely, the release cylinder pushrod should extend substantially. If it doesn't, check the fluid level in the clutch master cylinder.*

d) *Visually inspect the pivot bushing at the top of the clutch pedal to make sure there is no binding or excessive play.*

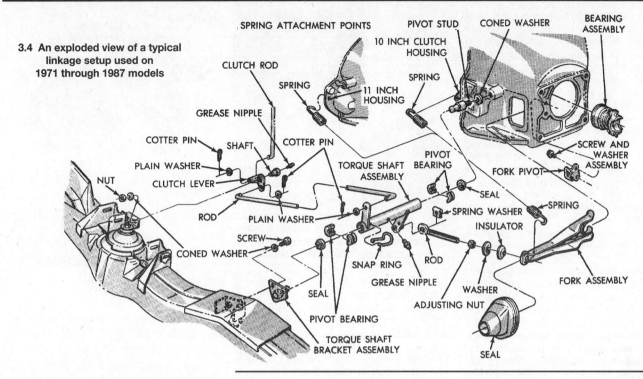

3.4 An exploded view of a typical linkage setup used on 1971 through 1987 models

SPRING ATTACHMENT POINTS
PIVOT STUD
CONED WASHER
BEARING ASSEMBLY
10 INCH CLUTCH HOUSING
CLUTCH ROD
SPRING
SPRING
11 INCH HOUSING
GREASE NIPPLE
COTTER PIN
SHAFT
COTTER PIN
PLAIN WASHER
CLUTCH LEVER
TORQUE SHAFT ASSEMBLY
PIVOT BEARING
NUT
ROD
PLAIN WASHER
FORK PIVOT
SEAL
SPRING
SCREW AND WASHER ASSEMBLY
SPRING WASHER
INSULATOR
SPRING
SCREW
CONED WASHER
SNAP RING
ROD
FORK ASSEMBLY
SEAL
GREASE NIPPLE
WASHER
PIVOT BEARING
ADJUSTING NUT
TORQUE SHAFT BRACKET ASSEMBLY
SEAL

---

3 Clutch components - removal, inspection and installation

Warning: *Dust produced by clutch wear and deposited on clutch components may contain asbestos, which is hazardous to your health. DO NOT blow it out with compressed air and DO NOT inhale it. DO NOT use gasoline or petroleum-based solvents to remove the dust. Brake system cleaner should be used to flush the dust into a drain pan. After the clutch components are wiped clean with a rag, dispose of the contaminated rags and cleaner in a covered, marked container.*

Removal

Refer to illustrations 3.4 and 3.10

1   Access to the clutch components is normally accomplished by removing the transmission and bellhousing, leaving the engine in the vehicle. If, of course, the engine is being removed for major overhaul, then check the clutch for wear and replace worn components as necessary. However, the relatively low cost of the clutch components compared to the time and trouble spent gaining access to them warrants their replacement anytime the engine or transmission is removed, unless they are new or in near-perfect condition. The following procedures are based on the assumption the engine will stay in place.

2   Referring to Chapter 7 Part A, remove the transmission from the vehicle. **Note:** *The transmission is removed separately from the bellhousing on all models except those equipped with an AX-15, on which the transmission and the clutch housing must be removed as a single assembly (the bellhousing-to-transmission bolts are inside the bell-*

housing). Support the engine while the transmission is out. Preferably, an engine hoist should be used to support it from above. However, if a jack is used underneath the engine, make sure a piece of wood is positioned between the jack and oil pan to spread the load. **Caution:** *The pick-up for the oil pump is very close to the bottom of the oil pan. If the pan is bent or distorted in any way, engine oil starvation could occur.*

3   Remove the bellhousing access cover.

4   On 1971 through 1987 models, remove the return spring from the clutch release fork and the clutch housing, or from the torque shaft lever **(see illustration).** On 1988 and later models, detach the clutch release cylinder from the transmission (see Section 6).

5   On 1971 through 1987 models, remove the spring washer securing the fork rod to the torque shaft lever and remove the rod from the torque shaft and release fork.

6   On six-cylinder models with an A-903 or A-250 transmission, remove the clip and plain washer securing the interlock rod to the torque shaft lever and remove the spring washer, the plain washer and the rod from the torque shaft **(see illustration 3.4).**

7   On all models except those equipped with an AX-15 transmission, remove the bellhousing-to-engine bolts and then detach the housing. It may have to be gently pried off the alignment dowels with a screwdriver or prybar.

8   The clutch fork and release bearing can remain attached to the bellhousing for the time being.

9   To support the clutch disc during removal, install a clutch alignment tool through the clutch disc hub.

10   Carefully inspect the flywheel and pressure plate for indexing marks. The marks are usually an X, an O or a white letter. If they

cannot be found, scribe marks yourself so the pressure plate and the flywheel will be in the same alignment during installation **(see illustration).**

11   Turning each bolt only 1/4-turn at a time, loosen the pressure plate-to-flywheel bolts. Work in a criss-cross pattern until all spring pressure is relieved. Then hold the pressure plate securely and completely remove the bolts, followed by the pressure plate and clutch disc.

Inspection

Refer to illustration 3.13

12   Ordinarily, when a problem occurs in the clutch, it can be attributed to wear of the clutch driven plate assembly (clutch disc). However, all components should be inspected at this time.

3.10 Be sure to mark the pressure plate and flywheel in order to insure proper alignment during installation (this won't be necessary if a new pressure plate is to be installed)

13   Inspect the flywheel for cracks, heat checking, grooves and other obvious defects **(see illustration)**. If the imperfections are slight, a machine shop can machine the surface flat and smooth, which is highly recommended regardless of the surface appearance. See Chapter 2 for the flywheel removal and installation procedure.

14   Inspect the pilot bushing (see Section 5).

15   Inspect the lining on the clutch disc. There should be at least 1/16-inch of lining above the rivet heads. Check for loose rivets, distortion, cracks, broken springs and other obvious damage. As mentioned above, ordinarily the clutch disc is routinely replaced, so if in doubt about the condition, replace it with a new one.

16   The release bearing should also be replaced along with the clutch disc (see Section 4).

17   Check the coil springs for bluing (which could indicate overheating, and loss of temper) and look for damage such as cracks. Check the ends of the release levers (where they contact the throwout bearing) for signs of excessive wear. Finally, be sure to inspect the pressure plate friction surface itself for obvious damage, distortion, cracking, etc. The friction surface of the pressure plate must be clean, dry, lightly sanded and free of scoring, warpage and heavy checking. Light glazing can be removed with sandpaper or emery cloth. If a new pressure plate is required, new and factory-rebuilt units are available.

Installation

Refer to illustration 3.24

18   Before installation, clean the flywheel and pressure plate machined surfaces with brake system cleaner, lacquer thinner or acetone. It's important that no oil or grease is on these surfaces or the lining of the clutch disc. Handle the parts only with clean hands.

19   Position the clutch disc and pressure plate against the flywheel with the clutch held in place with an alignment tool (available from auto parts stores). Make sure it's installed properly (most replacement clutch plates will be marked "flywheel side" or something similar - if not marked, install the clutch disc with the damper springs toward the transmission).

20   Tighten the pressure plate-to-flywheel bolts only finger tight, working around the pressure plate.

21   Center the clutch disc by ensuring the alignment tool extends through the splined hub and into the pilot bushing in the crankshaft. Wiggle the tool up, down or side-to-side as needed to bottom the tool in the pilot bushing. Tighten the pressure plate-to-flywheel bolts a little at a time, working in a criss-cross pattern to prevent distorting the cover. After all of the bolts are snug, tighten them to the torque listed in this Chapter's Specifications. Remove the alignment tool.

22   Using high-temperature grease, lubricate the inner groove of the release bearing (see Section 4). Also place grease on the

3.13 Check the flywheel for cracks, hot spots and other obvious defects (slight imperfections can be removed by resurfacing) - this flywheel should be resurfaced

release lever contact areas and the transmission input shaft bearing retainer.

23   Install the clutch release bearing as described in Section 4.

24   Install the clutch housing and tighten the bolts to the torque listed in this Chapter's Specifications **(see illustration)**.

25   Install the transmission, linkage (1971 through 1987 models) or release cylinder (1988 and later models) and any other components previously removed. Tighten all fasteners to the proper torque specifications.

4   Clutch release bearing - removal, inspection and installation

Warning: *Dust produced by clutch wear and deposited on clutch components may contain asbestos, which is hazardous to your health. DO NOT blow it out with compressed air and DO NOT inhale it. DO NOT use gasoline or petroleum-based solvents to remove the dust. Brake system cleaner should be used to flush the dust into a drain pan. After the clutch components are wiped clean with a rag, dispose of the contaminated rags and cleaner in a covered, marked container.*

Removal

Refer to illustrations 4.4a and 4.4b

1   Disconnect the negative cable from the battery.

2   Remove the transmission (see Chapter 7 Part A).

3   Remove the access cover from the bellhousing, disconnect the linkage (1971 through 1987 models) or detach the clutch release cylinder (1988 through 1988 models) and remove the clutch housing from the engine (see Section 3).

4   Remove the clutch release fork from the pivot, then remove the bearing and sleeve from the fork **(see illustrations)**.

Inspection

5   Hold the center of the bearing and

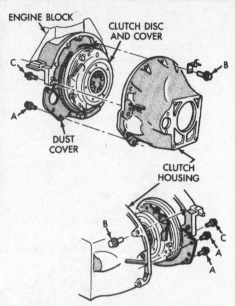

3.24 Fastener torque guide for NP2500 and Overdrive-4 clutch housings

rotate the outer portion while applying pressure. If the bearing doesn't turn smoothly or if it's noisy, remove it from the sleeve and replace it with a new one. Wipe the bearing with a clean rag and inspect it for damage, wear and cracks. Don't immerse the bearing in solvent - it's sealed for life and to do so would ruin it. Also check the release fork for cracks and other damage. **Note:** *Because of the relative low expense of the release bearing compared with the amount of work it takes to remove it, we recommend replacing the bearing routinely whenever it is removed.*

Installation

6   Lightly lubricate the clutch fork crown and spring retention crown where they contact the bearing sleeve with high-temperature grease. Fill the inner groove of the bearing with the same grease.

7   Attach the release bearing sleeve to the clutch release fork.

8   Lubricate the fork pivot socket with high-temperature grease and push the fork onto the pivot until it's firmly seated. **Note:** *Be sure the retaining spring is positioned behind the pivot.*

9   Apply a light coat of high-temperature grease to the face of the release bearing, where it contacts the pressure plate diaphragm fingers.

10   Install the clutch housing and tighten the bolts to the torque listed in this Chapter's Specifications. Reattach the clutch release linkage (see Section 3).

11   Prior to installing the transmission, apply a light coat of grease to the transmission front bearing retainer.

12   The remainder of installation is the reverse of the removal procedure.

13   Be sure to adjust clutch pedal freeplay when you're done (see Chapter 1).

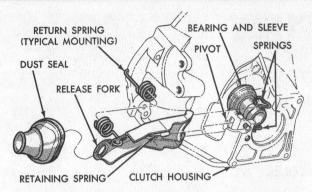

4.4a An exploded view of a typical release fork and release bearing assembly (mechanical release system)

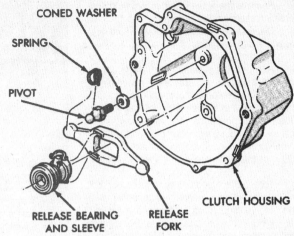

4.4b An exploded view of a typical release fork and release bearing assembly (hydraulic release system)

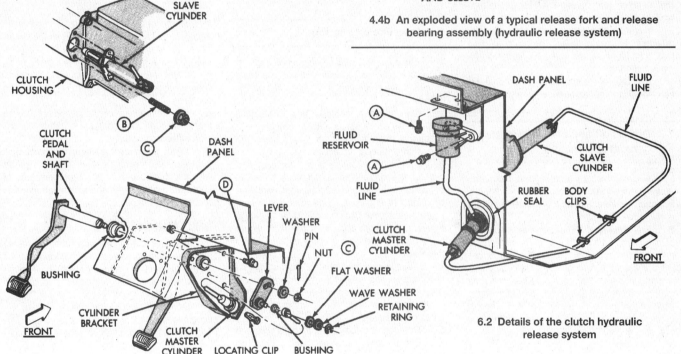

6.2 Details of the clutch hydraulic release system

5 Pilot bushing - inspection and replacement

1   The clutch pilot bushing is an oil-impregnated bronze bushing which is pressed into the rear of the crankshaft. It is greased at the factory and does not require additional lubrication. Its primary purpose is to support the front of the transmission input shaft. The pilot bushing should be inspected whenever the clutch components are removed from the engine. Due to its inaccessibility, if you are in doubt as to its condition, replace it with a new one. **Note:** *If the engine has been removed from the vehicle, disregard the following steps which do not apply.*
2   Remove the transmission (see Chapter 7 Part A).
3   Remove the clutch components (see Section 3).
4   Inspect for any excessive wear, scoring, lack of grease, dryness or obvious damage. If

any of these conditions are noted, the bushing should be replaced. A flashlight will be helpful to direct light into the recess.
5   The bushing must be pulled from its hole in the crankshaft, gripping the bushing at the rear. Special tools are available, but the do-it-yourselfer may be able to get by with a an alternative tool or fabricated tool. One method that works well is to use a slide-hammer with a tip that has two adjustable hooks, 180 degrees apart. Such tips are commonly available for slide-hammers, and are often included with better-quality slide-hammer kits. A slide-hammer is a tool with many uses, such as pulling dents from body parts and removing seals; you'll use it later for more things than just removing bushings. If a slide-hammer with the correct hooked tip is not available, try to find a hooked tool that will fit into the bushing hole and hook behind the bushing, then clamp a large pair of Vise-Grip locking pliers to the tool and strike the pliers, near the jaws, to pull the bushing out.

6   To install the new bushing, lightly lubricate the outside surface with multi-purpose grease, then drive it into the recess with a soft-face hammer. Most new bushings come already lubricated, but if it is dry, apply a thin coat of high-temperature grease to it.
7   Install the clutch components, transmission and all other components removed previously, tightening all fasteners properly.

6 Clutch hydraulic release system - removal and installation

Refer to illustration 6.2
Note: *The clutch hydraulic release system is serviced as an assembly. If the system is leaking or has air in it, replace the entire assembly.*
1   Raise the vehicle and place it securely on jackstands.
2   Remove the nuts attaching the slave cylinder to the clutch housing **(see illustration)**.

8

3   Remove the slave cylinder and fluid line retaining clip from the housing.
4   Disengage the clutch hydraulic fluid line from the retaining clips on the body.
5   Lower the vehicle.
6   Remove the retainer ring, flat washer and wave washer that attach the clutch master cylinder pushrod to the clutch pedal.
7   Unplug the electrical connector for the clutch pedal position switch wires, then slide the clutch master cylinder pushrod off the clutch pedal-to-pushrod pin. Inspect the condition of the pin bushing. If it's worn, replace it.
8   To avoid spillage, verify that the cap on the master cylinder reservoir is tight. Remove the reservoir mounting screws and the master cylinder mounting stud nuts that attach the reservoir and master cylinder to the firewall.
9   Remove the entire assembly. Do not attempt to disconnect the clutch fluid hydraulic line from the master cylinder or the slave cylinder.
10   Installation is the reverse of removal. Be sure to tighten all fasteners securely.

7   Driveshaft and universal joints - general information

1   A driveshaft is a tube that transmits power between the transmission and the differential. Universal joints are located at either end of the driveshaft.
2   The driveshaft employs a splined yoke at the front, which slips into the extension housing of the transmission. This arrangement allows the driveshaft to alter its length during vehicle operation. An oil seal prevents leakage of fluid at this point and keeps dirt from entering the transmission. If leakage is evident at the front of the driveshaft, replace the oil seal (see Chapter 7 Part B).
3   The driveshaft assembly requires very little service. The original equipment universal joints are lubricated for life and must be replaced if problems develop. The driveshaft must be removed from the vehicle for this procedure.
4   Since the driveshaft is a balanced unit, it's important that no undercoating, mud, etc. be allowed to stay on it. When the vehicle is raised for service it's a good idea to clean the driveshaft and inspect it for any obvious damage. Also, make sure the small weights used to originally balance the driveshaft are in place and securely attached. Whenever the driveshaft is removed it must be reinstalled in the same relative position to preserve the balance.
5   Problems with the driveshaft are usually indicated by a noise or vibration while driving the vehicle. A road test should verify if the problem is the driveshaft or another vehicle component. Refer to the *Troubleshooting* Section at the front of this manual. If you suspect trouble, inspect the driveline (see the next Section).

9.3 Always make reference marks on the driveshaft and pinion flange to insure that the driveshaft's balance is maintained

8   Driveline inspection

1   Raise the rear of the vehicle and support it securely on jackstands. Block the front wheels to keep the vehicle from rolling off the stands.
2   Crawl under the vehicle and visually inspect the driveshaft. Look for any dents or cracks in the tubing. If any are found, the driveshaft must be replaced.
3   Check for oil leakage at the front and rear of the driveshaft. Leakage where the driveshaft enters the transmission extension housing indicates a defective extension housing seal (see Chapter 7 Part B). Leakage where the driveshaft enters the differential indicates a defective pinion seal (see Section 18).
4   While under the vehicle, have an assistant rotate a rear wheel so the driveshaft will rotate. As it does, make sure the universal joints are operating properly without binding, noise or looseness.
5   The universal joints can also be checked with the driveshaft motionless, by gripping your hands on either side of a joint and attempting to twist the joint. Any movement at all in the joint is a sign of considerable wear. Lifting up on the shaft will also indicate movement in the universal joints.
6   Finally, check the driveshaft mounting bolts at the ends to make sure they're tight.

9   Driveshaft - removal and installation

Removal

Refer to illustrations 9.3 and 9.4
1   Disconnect the negative cable from the battery.
2   Raise the vehicle and support it securely on jackstands. Place the transmission in Neutral with the parking brake off.
3   Make reference marks on the driveshaft and the pinion flange in line with each other **(see illustration)**. This is to make sure the driveshaft is reinstalled in the same position to preserve the balance.

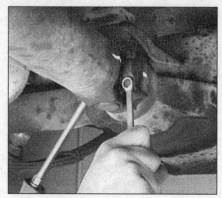

9.4 Use a large screwdriver to hold the flange while breaking the U-joint bolts loose

4   Remove the rear universal joint bolts and clamps **(see illustration)**. Turn the driveshaft (or wheels) as necessary to bring the bolts into the most accessible position.
5   Tape the bearing caps to the cross to prevent the caps from coming off during removal.
6   Lower the rear of the driveshaft. Slide the front of the driveshaft out of the transmission extension housing.
7   Wrap a plastic bag over the transmission extension housing and hold it in place with a rubber band. This will prevent loss of fluid and protect against contamination while the driveshaft is out.

Installation

8   Remove the plastic bag from the transmission extension housing and wipe the area clean. Inspect the oil seal carefully. Procedures for replacement of this seal can be found in Chapter 7.
9   Slide the front of the driveshaft into the transmission.
10   Raise the rear of the driveshaft into position, checking to be sure the marks are in alignment. If not, turn the rear wheels to match the pinion flange and the driveshaft.
11   Remove the tape securing the bearing caps and install the clamps and bolts. Tighten all clamp bolts to the torque listed in this Chapter's Specifications. Lower the vehicle and connect the negative battery cable.

10   Universal joints - replacement

Refer to illustrations 10.3a, 10.3b, 10.3c, 10.5a, 10.5b, 10.9a and 10.9b
Note: *A press or large vise will be required for this procedure. It may be a good idea to take the driveshaft to a repair or machine shop where the universal joints can be replaced for you, normally at a reasonable charge.*
1   Remove the driveshaft (see Section 9).
2   Apply penetrating oil to the bearing caps.
3   Use a screwdriver and hammer to tap off the bearing cap retainers **(see illustrations)**.

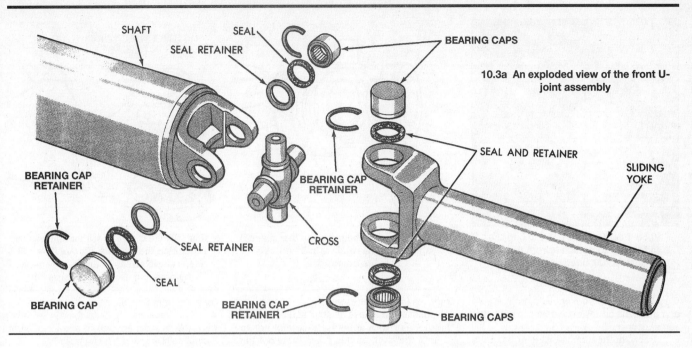

10.3a An exploded view of the front U-joint assembly

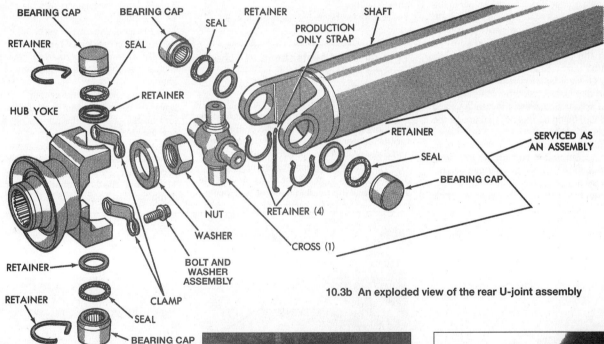

10.3b An exploded view of the rear U-joint assembly

8

4   Supporting the driveshaft, place it in position on either an arbor press or on a workbench equipped with a vise.

5   Place a piece of pipe or a large socket, having an inside diameter slightly larger than the outside diameter of the bearing caps, over one of the bearing caps. Position a socket with an outside diameter slightly smaller than that of the opposite bearing cap against the cap **(see illustration)** and use the vise or press to force the bearing cap out (inside the pipe or large socket), stopping just before it comes completely out of the yoke. Use the vise or large pliers to work the bearing cap the rest of the way out **(see illustration)**.

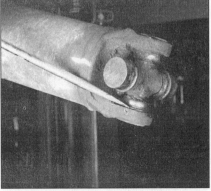

10.3c Remove the inner retainers from the U-joint by tapping them off with a screwdriver and hammer

10.5a Press the bearing caps from the yoke with sockets and a large vise

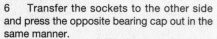

10.5b Pliers can be used to grip the bearing cap to detach it from the yoke after it has been pushed out

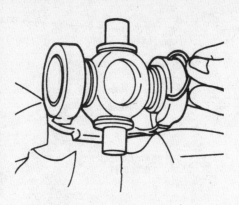

10.9a Press the bearing cap into place and install the retainer in its groove in the bearing cap

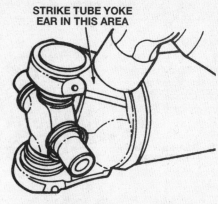

10.9b Strike the yoke with a hammer to reposition the retainer groove (this will also relieve stress in the joint caused by pressing the bearing caps in)

6   Transfer the sockets to the other side and press the opposite bearing cap out in the same manner.

7   Pack the new universal joint bearings with chassis grease. Position the spider in the yoke and partially install one bearing cap in the yoke. If the replacement spider is equipped with a grease fitting, be sure it's offset in the proper direction (toward the driveshaft).

8   Start the spider into the bearing cap and then partially install the other bearing cap. Align the spider and press the bearing caps into position, being careful not to damage the dust seals.

9   Install the bearing cap retainers. If difficulty is encountered in seating the retainers, strike the driveshaft yoke sharply with a hammer. This will spring the yoke ears slightly and allow the retainers to seat in their grooves **(see illustrations)**.

10   Install the grease fitting (if equipped) and fill the joint with grease. Be careful not to overfill the joint, as this could blow out the grease seals.

11   Install the driveshaft (see Section 9).

11   Driveshaft center bearing - check and replacement

Refer to illustration 11.1

1   The center bearing can be checked in the same manner as the U-joints. Inspect the rubber center bearing mounts for damage and deterioration **(see illustration)**.

2   Remove the driveshaft assembly.

3   Mount the front driveshaft assembly in a vise and pull the bearing support and insula-

tor away from the bearing.

4   Use a hammer to bend the slinger away from the bearing enough to allow sufficient clearance to install a bearing puller.

5   Remove the bearing with the puller.

6   Remove the slinger.

7   Always use an overhaul kit when reassembling the center bearing assembly.

8   To assemble the bearing, position the slinger, bearing assembly and retainer on the shaft.

9   Use a piece of the proper size pipe to drive the parts onto the shaft until they bottom on the shoulder (be careful not to damage the shaft splines).

10   Place the slip yoke seal cap and seal on the center bearing splines for installation purposes.

11   Install the driveshaft assembly and crimp the seal cap tabs to the slip yoke.

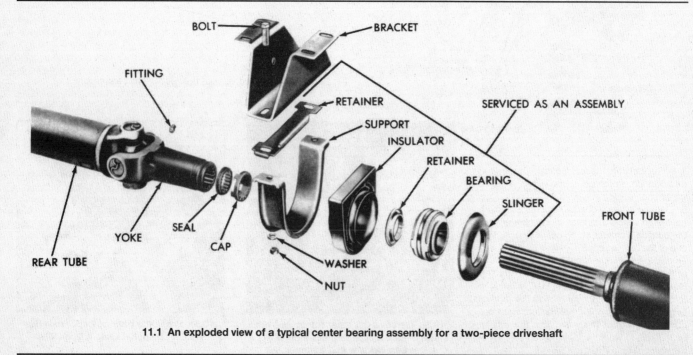

11.1 An exploded view of a typical center bearing assembly for a two-piece driveshaft

12 Rear axle - description and check

Description

1   The rear axle assemblies on 1/2-ton and light duty 3/4-ton models are a *semi-floating* design, i.e. the axle supports the weight of the vehicle on the axleshaft in addition to transmitting driving forces to the rear wheels. These axles are referred to by the diameter (in inches) of the differential ring gear - "8-3/8," "8-3/4" or "9-1/4". Of these, two basic types of axles are used: On 8-3/4 inch units, the bearing is pressed onto the axleshaft; on all others, the bearing is pressed into the axle tube. Although you can't actually see how the axle bearings are installed in either type, you can quickly identify which type you're servicing by looking for a retainer bolted to the inside of the brake backing plate. 8-3/4 inch axles use flange-type retainers at the brake backing plate (so you know the bearing is pressed onto the axleshaft). 8-3/8 inch and 9-1/4 inch axles use "C-washer" locks to secure the axles to the differential (there are no retainers, so you know the bearing is pressed into the axle tubes on these units).

2   Heavy duty 3/4-ton and 1-ton models use *full-floating* axles and these are identified by the name of their manufacturer; Spicer 60 or Spicer 70. A full floating axleshaft doesn't carry any of the vehicle's weight; the weight is supported on the axle housing itself by roller bearings. Full-floating axles can be identified by the large hub projecting from the center of the wheel. The axle flange is secured by bolts on the end of the hub.

3   All axles also use a *hypoid*-type pinion gear, i.e. the pinion gear teeth are cut in a spiral configuration to allow the centerline of the pinion to be lower than the centerline of the ring gear, so the vehicle floor can be lower. All differential housings are cast iron and the axle tubes are made of steel, pressed and welded into the differential housing. Except for the 8-3/4 inch axle, all other differentials can be inspected and serviced by removing a pressed-steel access cover from the back side of the differential housing. The 8-3/4 inch axle uses a "drop-out" style differential: The carrier, which is bolted to the front side of the axle housing, is also the differential housing; the entire differential assembly can be removed as a unit by removing the axleshafts, then unbolting and "dropping out" the differential from the front of the axle housing.

4   Most axles use a conventional differential. But a "Sure-Grip" model is an available option. A conventional differential allows the driving wheels to rotate at different speeds during cornering (the outside wheel must rotate faster than the inside wheel), while equally dividing the driving torque between the two wheels. In most normal driving situations, this function of the differential is adequate. However, the total driving torque can't be more than double the torque at the lower-traction wheel. When traction conditions aren't the same for both driving wheels, some

of the available traction is lost. An optional "Sure-Grip" differential is available for poor traction conditions - i.e. mud, ice or snow. During normal cornering and turning conditions, its controlled internal friction is easily overcome, allowing the driving wheels to turn at different speeds, just like a conventional differential. However, during extremely slippery conditions, the Sure-Grip differential allows the driving wheel with the better traction condition to develop more driving torque than the other wheel. Thus, the total driving torque can be significantly greater than with a conventional differential.

5   Obviously, with all these variations in axle size and type, proper identification is critical. If you're not sure what axle the vehicle has, make the following inspection:

a) *Does the axle have a bolt-on pressed-steel cover on the back side of the differential housing? If not, it must be an 8-3/4 inch axle.*

b) *Does it use flange-type axle retainers? If not, it could only be an 8-3/8 inch or 9-1/4 inch axle.*

c) *Ring gear diameter can be determined by counting the number of cover screws and measuring the diameter of the axle tubes: 8-3/8 inch axles have a 10-screw cover and 3-inch axle tubes; 9-1/4 inch axles have 12-screw covers.*

d) *Does it have a Sure-Grip differential? One easy way to tell is to raise the rear of the vehicle, place it securely on jackstands and rotate the rear wheels. If they both turn in the same direction, the axle is equipped with a Sure-Grip; if they turn in opposite directions, it's a conventional differential. Another way to tell - without raising the vehicle - is to shine a flashlight into the filler plug hole and look at the differential case. When you look into the "windows" of a Sure-Grip differential, you'll see coil springs.*

e) *Need to know the axle ratio? It's usually stamped onto a small metal tag attached to one of the bolts that secures the differential access cover.*

Check

6   Many times, a problem is suspected in an axle area when, in fact, it lies elsewhere. For this reason, a thorough check should be performed before assuming an axle problem.

7   The following noises are those commonly associated with axle diagnosis procedures:

a) *Road noise is often mistaken for mechanical faults. Driving the vehicle on different surfaces will show whether the road surface is the cause of the noise. Road noise will remain the same if the vehicle is under power or coasting.*

b) *Tire noise is sometimes mistaken for mechanical problems. Tires which are worn or low on pressure are particularly susceptible to emitting vibrations and noises. Tire noise will remain about the same during varying driving situations,*

where axle noise will change during coasting, acceleration, etc.

c) *Engine and transmission noise can be deceiving because it will travel along the driveline. To isolate engine and transmission noises, make a note of the engine speed at which the noise is most pronounced. Stop the vehicle and place the transmission in Neutral and run the engine to the same speed. If the noise is the same, the axle is not at fault.*

8   Overhaul and general repair of the front or rear axle differential are beyond the scope of the home mechanic due to the many special tools and critical measurements required. Thus, the procedures listed here will involve axleshaft removal and installation, axleshaft oil seal replacement, axleshaft bearing replacement and removal of the entire unit for repair or replacement.

13 Axleshaft (semi-floating axles) - removal and installation

8-3/8 inch and 9-1/4 inch axles

Refer to illustrations 13.2, 13.3a, 13.3b and 13.4

1   The usual reason for axleshaft removal is that the bearing is worn or the seal is leaking. To check the bearing, loosen the wheel lug nuts, raise the vehicle, support it securely on jackstands and remove the wheel and brake drum (see Chapter 9). Try to move the axle flange up and down. If the axleshaft moves up and down, bearing wear is excessive. Also check for differential lubricant leaking out from below the axleshaft - this indicates the seal is leaking.

2   Remove the cover from the differential carrier and allow the oil to drain into a suitable container **(see illustration)**.

3   Remove the lock bolt from the differential pinion shaft **(see illustration)**. Remove the pinion shaft **(see illustration)**.

4   Push the outer (flanged) end of the axleshaft in and remove the C-lock from the inner end of the shaft **(see illustration)**.

5   Withdraw the axleshaft, taking care not to damage the oil seal in the end of the axle housing as the splined end of the axleshaft passes through it.

6   Installation is the reverse of removal. Tighten the pinion shaft lock bolt to the torque listed in this Chapter's Specifications.

7   Always use a new cover gasket and tighten the cover bolts to the torque listed in the Chapter 1 Specifications.

8   Refill the axle with the correct quantity and grade of lubricant (see Chapter 1).

8-3/4 inch axles

Refer to illustrations 13.10 and 13.11

9   The usual reason for axleshaft removal is that the bearing is worn or the seal is leaking. To check the bearing, raise the vehicle, support it securely on jackstands and remove the wheel and brake drum. Move the axle flange

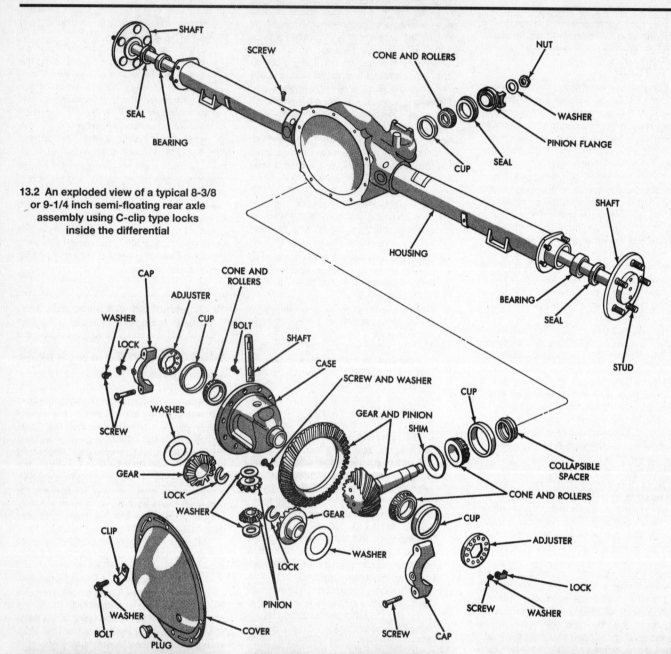

13.2 An exploded view of a typical 8-3/8 or 9-1/4 inch semi-floating rear axle assembly using C-clip type locks inside the differential

13.3a Remove the pinion shaft lock bolt . . .

13.3b . . .then carefully remove the pinion shaft from the differential carrier (don't turn the wheels or the carrier after the shaft has been removed, or the pinion gears may fall out)

13.4 Push the axle flange in, then remove the C-lock from the inner end of the axleshaft

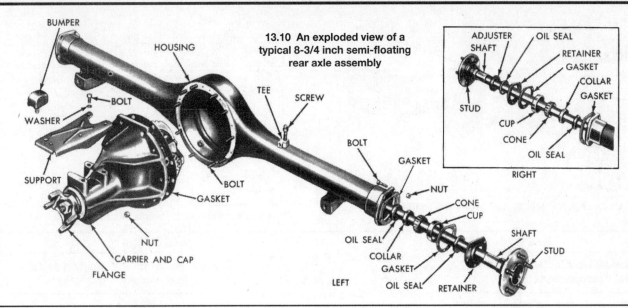

13.10 An exploded view of a typical 8-3/4 inch semi-floating rear axle assembly

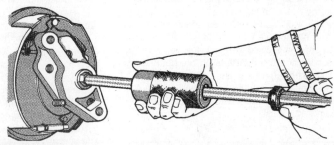

13.11 Attach a slide hammer to the wheel mounting studs and withdraw the axleshaft

14.2 The end of the axleshaft can be used to pry the old seal out of the axle housing

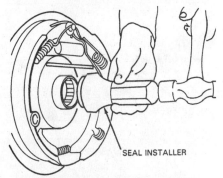

14.3 Use a seal installer tool, a large socket, a piece of pipe or a block of wood to tap the seal evenly into place

up and down. If the axleshaft moves up and down, bearing wear is excessive. Also check for differential lubricant leaking out from below the axleshaft - this indicates the seal is leaking.

10   Unscrew the nuts which attach the retainer to the brake backing plate **(see illustration)**. A large hole is provided in the axle flange (where the brake drum attaches) for access to the nuts with a socket, ratchet and extension. Rotate the axle to align this hole with each nut.

11   Attach a slide hammer to the wheel mounting studs and withdraw the axleshaft **(see illustration)**. Sometimes, if the axleshaft is not too tight, you can pull it out by hand, but be very careful not to pull too hard or you might pull the vehicle off the jackstands.

12   As the axleshaft is removed, it is possible that the bearing will come apart. This indicates the bearing has failed and will need to be replaced; be sure to remove all bearing parts from the axle housing. Take the axleshaft to an automotive machine shop to have the bearing replaced, since replacement requires pressing a new bearing and collar onto the shaft with a hydraulic press.

13   Considering the relative low cost of axle seals compared with the amount of work involved in removing the axle, we recommend replacing the axle seal whenever the axleshaft is removed. To replace the seal, use a slide-hammer-type tool or seal removal tool to pull the oil seal out of the axle housing. Clean the seal bore so there's no oil where the new seal will seat. Coat the outside diam-

eter of the new seal with RTV sealant, then drive the seal into the seal bore with the lip of the seal facing in, using a hammer and seal-installer tool **(see illustration 13.3)**. Note: *If a seal removal tool is not available, a large-diameter socket (slightly smaller than the seal bore) and an extension will also work. Coat the rubber seal lip with differential lubricant so the seal won't be damaged.*

14   Before installing the axleshaft assembly, smear wheel bearing grease onto the bearing end and in the bearing recess in the axle housing tube.

15   Apply differential lubricant to the axleshaft splines. Also, apply differential lubricant to the oil seal in the axle housing.

16   Hold the axleshaft horizontal and insert it into the axle housing. Feel when the shaft splines have picked up those in the differential side gears and then push the shaft fully into position, using a soft-faced hammer on the end flange as necessary.

17   Bolt the retainer to the brake backplate, install the brake drum and wheel and lower the vehicle to the ground.

---

14   Axleshaft oil seal (semi-floating axles) - replacement

Refer to illustrations 14.2 and 14.3
Note: *Oil seal replacement on 8-3/4 axles is covered as part of the axleshaft replacement procedure in Section 12.*

1   Remove the axleshaft (see Section 13).
2   Pry the oil seal out of the end of the axle housing with a large screwdriver or the inner end of the axleshaft **(see illustration)**.
3   Apply high-temperature grease to the oil seal recess and tap the new seal evenly into place with a hammer and seal installation tool **(see illustration)**, large socket or piece of pipe so the lips are facing in and the metal face is visible from the end of the axle housing. When correctly installed, the face of the oil seal should be flush with the end of the axle housing.

8

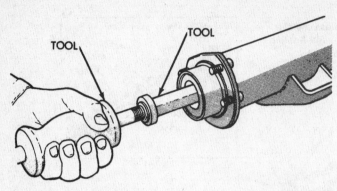

15.3 Use a slide hammer to remove the axleshaft bearing

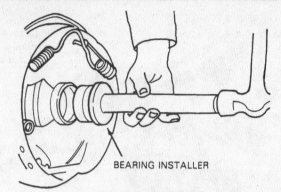

15.4 Use a special bearing installer tool, large socket or piece of pipe to tap the bearing evenly into the axle housing

15 Axleshaft bearing (semi-floating axles) - replacement

Refer to illustrations 15.3 and 15.4
Note: *This procedure does not apply to models with 8-3/4 inch axles. If you're working on a model with an 8-3/4 inch rear axle, remove the axle (see Section 13) and take it to an automotive machine shop to have the old bearing pressed off and a new one pressed on.*
1   Remove the axleshaft (see Section 13) and the oil seal (see Section 14).
2   A slide-hammer-type puller which grips the bearing from behind will be required for this job.
3   Attach the puller and extract the bearing from the axle housing **(see illustration)**.
4   Clean out the bearing recess and drive in the new bearing with a bearing installer or a piece of pipe positioned against the outer bearing race **(see illustration)**. Make sure the bearing is tapped in to the full depth of the recess and the numbers on the bearing are visible from the outer end of the axle housing.
5   Install a new oil seal (see Section 13), then install the axleshaft (see Section 12).

16 Axleshaft (full-floating axles) - removal and installation

Refer to illustrations 16.3a, 16.3b and 16.6
1   Remove the axleshaft flange bolts or nuts and lock washers.
2   Rap the axleshaft sharply in the center of the flange with a hammer to loosen the tapered dowels, if equipped.
3   Remove the dowels and pull out the axleshaft **(see illustrations)**.
4   When installing the axleshaft, clean the gasket sealing area and place a new gasket

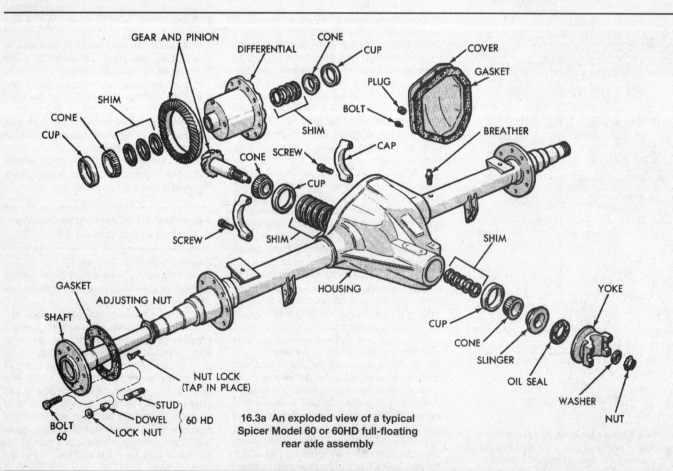

16.3a An exploded view of a typical Spicer Model 60 or 60HD full-floating rear axle assembly

16.3b Remove the axleshaft flange nuts and tapered dowels and slide out the axle

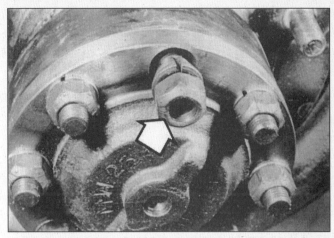

16.6 Install the tapered dowels and seat them with the flange nuts (arrow)

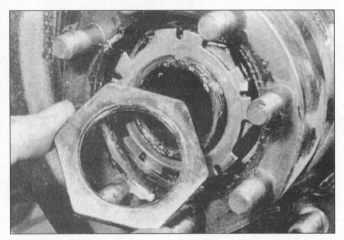

17.3 After flattening the lockring tab, remove the outer nut . . .

17.4 . . . followed by the washer, inner nut and bearing

on the flange.

5   Slide the axleshaft into the housing, engaging the axleshaft splines into the differential.

6   Install the bolts or tapered dowels, lock washers and nuts (if equipped) **(see illustration)**. Tighten the bolts/nuts to the torque specified at the beginning of this Chapter.

17 Hub/drum assembly and wheel bearings (full-floating axles) - removal, installation and adjustment

Removal

Refer to illustrations 17.3, 17.4 and 17.6

1   Raise the vehicle, support it securely on jackstands and remove the rear wheels.

2   Remove the axleshafts (see Section 16).

3   Straighten the locking tabs on the lock-ring and remove the locknut, lock-ring and adjusting nut **(see illustration)**.

4   Slide the hub off the axle slightly and remove the outer wheel bearing **(see illustration)**.

5   Pull the hub/drum assembly straight off the axle tube and place it on a clean flat surface with the inside of the drum facing up.

6   Use a large screwdriver, prybar or seal removal tool to pry out the oil seal **(see illustration)**.

7   Remove the inner wheel bearing from the hub.

8   Use solvent to wash the bearings, hub and axle tube. A small brush may prove useful; make sure no bristles from the brush embed themselves in the bearing rollers. Allow the parts to air dry.

9   Carefully inspect the bearings for cracks, wear and damage. Check the axle tube flange, studs, and hub splines for damage and corrosion. Check the bearing cups (races) for pitting or scoring. Worn or damaged components must be replaced with new ones. If necessary, tap the bearing cups from the hub with a hammer and brass drift and install new ones with the appropriate size bearing driver.

10   Inspect the brake drum for scoring or damage (see Chapter 9).

Installation

11   Lubricate the bearings and the axle tube contact areas with wheel bearing grease.

Work the grease completely into the bearings, forcing it between the rollers, cone and cage. Place the inner bearing into the hub and install the seal (with the seal lip facing into the hub). Using a seal driver or block of wood, drive the seal in until it's flush with the hub. Lubricate the seal with gear oil or grease.

17.6 Check the inner oil seal and replace it if there is any sign of leakage or damage

8

17.16 Bend over the lockring tabs to keep both the inner and outer nuts from turning

18.4 Use an inch-pound torque wrench to check the torque necessary to move the pinion shaft

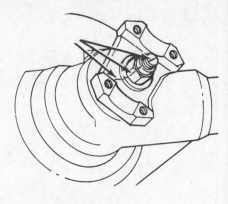

18.5 Mark the relative positions of the pinion, nut and flange (arrows) before removing the nut

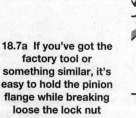

18.7a If you've got the factory tool or something similar, it's easy to hold the pinion flange while breaking loose the lock nut

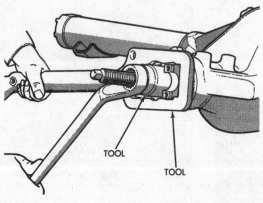

TOOL

TOOL

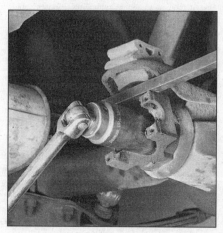

18.7b If you don't have the factory tool or one like it, try holding the flange with a prybar or large screwdriver while you loosen the locknut

12   Place the hub assembly on the axle tube, taking care not to damage the oil seals. Place the outer bearing into the hub.

13   Install the adjusting nut and adjust the bearings as described below.

Adjustment

Refer to illustration 17.16

14   Spin the wheel, and while it is spinning, tighten the inner nut until there is a slight drag, then back off the nut one flat.

15   Install the lock-ring and outer nut. make sure the inner nut doesn't turn while tightening the outer nut.

16   Bend over two tabs of the lock-ring to make sure that neither the inner or outer nuts can turn **(see illustration)**.

17   Install the axleshafts (see Section 16). Install the rear wheels and lower the vehicle.

18   Pinion oil seal - replacement

Refer to illustrations 18.4, 18.5, 18.7a, 18.7b and 18.11

1   Disconnect the cable from the negative battery terminal.

2   Raise the rear of the vehicle and support it securely on jackstands. Block the front wheels to keep the vehicle from rolling off the stands.

3   Disconnect the driveshaft and fasten it out of the way.

8-3/8, 8-3/4 and 9-1/4 inch rear axles

4   Use an inch-pound torque wrench to check the torque required to rotate the pinion. Record it for use later **(see illustration)**.

5   Scribe or punch alignment marks on the pinion shaft, nut and flange **(see illustration)**.

6   Count the number of threads visible between the end of the nut and the end of the pinion shaft and record it for use later.

7   A special tool can be used to keep the companion flange from moving while the self-locking pinion nut is loosened **(see illustration)**. If the special tool isn't available, try locking the flange with a prybar or a large screwdriver **(see illustration)**.

8   Remove the pinion nut.

9   Withdraw the companion flange. It may be necessary to use a two or three-jaw puller engaged behind the flange to draw it out. Do not attempt to pry behind the flange or hammer on the end of the pinion shaft.

10   Pry out the old seal and discard it.

11   Lubricate the lips of the new seal with high-temperature grease and tap it evenly into position with a seal installation tool or a large socket. Make sure it enters the housing squarely and is tapped in to its full depth **(see illustration)**.

12   Align the mating marks made before disassembly and install the companion flange. If necessary, tighten the pinion nut to draw the flange into place. Do not try to hammer the flange into position.

13   Apply non-hardening sealant to the ends of the splines visible in the center of the flange so oil will be sealed in.

14   Install the washer (if equipped) and pinion nut. Tighten the nut carefully until the original number of threads are exposed and the marks are aligned.

15   Measure the torque required to rotate the pinion and tighten the nut in small increments until it matches the figure recorded in Step 4. In order to compensate for the drag of the new oil seal, the nut should be tightened more until the rotational torque of the pinion exceeds earlier recording by 10 in-lbs. with a torque reading of at least 210 ft-lbs. on the nut.

16   Connect the driveshaft and lower the vehicle.

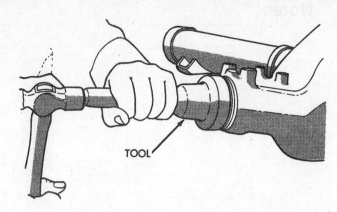

18.11 Use the proper-size driver to install the pinion seal

Spicer model 60 and 70 rear axles

17   Remove the pinion nut. A special tool, available at most auto parts stores, can be used to keep the companion flange from moving while the self-locking pinion nut is loosened **(see illustration 18.7a)**.

18   Mark the relationship of the companion flange to the pinion shaft and withdraw the companion flange. It may be necessary to use a two or three-jaw puller engaged behind the flange to draw it out. Do not attempt to pry behind the flange or hammer on the end of the pinion shaft.

19   Pry out the old seal and discard it.

20   Lubricate the lips of the new seal with high-temperature grease and tap it evenly into position with a seal installation tool or a large socket. Make sure it enters the housing squarely and is tapped in to its full depth **(see illustration 18.11)**.

21   Align the mating marks made before disassembly and install the companion flange. If necessary, tighten the pinion nut to draw the flange into place. Do not try to hammer the flange into position.

22   Install the washer (if equipped) and pinion nut. Tighten the nut to the torque listed in this Chapter's Specifications.

23   Connect the driveshaft, add gear oil to the differential housing, if necessary (see Chapter 1) and lower the vehicle.

19   Rear axle assembly - removal and installation

Removal

1   Loosen the rear wheel lug nuts, raise the rear of the vehicle and support it securely on jackstands. Block the front wheels to keep the vehicle from rolling off the stands. Remove the rear wheels.

2   Position a jack under the rear axle differential case.

3   Disconnect the driveshaft from the rear axle companion flange. Fasten the driveshaft out of the way with a piece of wire to the underbody.

4   Disconnect the shock absorbers at the lower mounts, then compress them to get them out of the way.

5   If equipped, disconnect the vent hose from the fitting on the axle housing and fasten it out of the way.

6   Disconnect the brake hose from the junction block on the axle housing, then plug the hose to prevent fluid leakage.

7   Remove the brake drums (see Chapter 9).

8   Disconnect the parking brake cables from the actuating levers and the backing plate (see Chapter 9).

9   Disconnect the spring U-bolts (see Chapter 10). Remove the spacers and clamp plates.

10   Disconnect any wiring interfering with removal.

11   Lower the jack under the differential, then remove the rear axle assembly from under the vehicle.

Installation

12   Installation is the reverse of removal. Lower the vehicle weight onto the wheels before tightening the U-bolt nuts completely.

13   Bleed the brakes (see Chapter 9).

8

Notes

Chapter 9   Brakes

Contents

Specifications

General

Brake fluid type	See Chapter 1

Disc brakes

Brake pad minimum thickness	See Chapter 1
Disc lateral runout limit	0.004 inch
Disc minimum thickness	Cast into disc
Thickness variation (parallelism)	0.0005 inch

Drum brakes

Minimum brake lining thickness	See Chapter 1
Maximum drum diameter	Cast into drum

Torque specifications

Ft-lbs (unless otherwise indicated)

Brake caliper	
Retaining clip/anti-rattle spring bolts	
1971 through 1975	220 in-lbs
1976 through 1997	15
Caliper mounting bolts (1998 on)	24
Caliper adapter-to-steering knuckle bolts	
1997 and earlier	
1/2-inch bolts	110
5/8-inch bolts	160
1998 on	
1500 series	130
2500/3500 series	210
Brake hose-to-caliper banjo bolt	
1971 through 1975	25
1976 through 1987	19 to 29
1988 through 1997	30 to 40
1998 on	18
Master cylinder mounting nuts	
1971 through 1975	80 to 200 in-lbs
1976 on	170 to 260 in-lbs
Brake booster mounting nuts	
1971	200 in-lbs
1972 through 1995	20
1996 and 1997	
In-line booster	18
Transverse booster	30
1998 on	156 in-lbs

1 General information

General

All models covered by this manual are equipped with hydraulically operated, four wheel brake systems. Except for 1971 models, which had front drum brakes, all vans covered by this manual use front disc brakes and rear drums.

All brakes are self-adjusting. The front disc brakes automatically compensate for pad wear, while the rear drum brakes (and the fronts on 1971 models) incorporate an adjustment mechanism which is activated as the brakes are applied.

The hydraulic system has separate circuits for the front and rear brakes. If one circuit fails, the other circuit will remain functional and a warning indicator will light up on the dashboard when a substantial amount of brake fluid is lost, showing that a failure has occurred.

Master cylinder

Refer to illustration 1.9

The master cylinder is located under the hood on the driver's side, and can be identified by the large fluid reservoir on top. The master cylinder has separate primary and secondary piston assemblies for the front and rear circuits.

Two basic types of master cylinders are used on the vehicles covered by this book. A one-piece cast-iron unit with an integral reservoir was used on all 1971 through 1978 models and on 1979 through 1983 models with Hydro-Boost. A two-piece unit with an aluminum body and a separate plastic reservoir was used on 1979 through 1983 models without Hydro-Boost, and on all 1984 and later models. All cast iron units are rebuildable (although you may have some difficulty obtaining rebuild kits for certain older cast-iron units).

Early (1979 through 1982) aluminum master cylinders are rebuildable, but you will probably have difficulty tracking down a rebuild kit. Although Chrysler originally supplied rebuildable aluminum master cylinder units on 1979 through 1982 models, rebuild kits are no longer available. Some parts departments may be able to direct you to an "old parts supplier" near you, but because properly rebuilt brake components are a safety issue, we feel that your best bet is to simply replace one of these older aluminum master cylinders with a new or rebuilt unit.

Aluminum units manufactured from 1983 on are *not* rebuildable. If your vehicle falls into this category, or if you have an older van equipped with a master cylinder for which a rebuild kit is no longer available, you'll have to obtain a new or remanufactured master cylinder unit.

When you go to a dealer parts department or a an auto parts store to purchase a new or rebuilt master cylinder unit, be sure to specify the make, model, type of master

1.9 A typical transverse master cylinder/power brake booster installation

cylinder (cast-iron, Hydro-Boost, aluminum, etc.) in your vehicle, etc. Be as specific as you can. It's a good idea to take the unit you're rebuilding or replacing with you so you can compare parts. If that's not possible, be sure to take the serial numbers off the casting to assist the parts people in providing you with the correct unit.

There are also two basic types of master cylinder installations: The conventional fore-and-aft type, with the master cylinder directly in front of, and bolted to, the power brake booster assembly, and the side-by-side type, which uses special right-angle linkage to swing the booster and master cylinder 90-degrees to the right so that they're parallel with the firewall instead of perpendicular to it. This setup is used to allow the installation of a bigger power brake booster **(see illustration)**. Removing the master cylinder on one of these vehicles is no different from removing a conventionally installed master cylinder unit.

Power brake booster

The power brake booster uses engine manifold vacuum to provide assistance to the brakes. It is mounted on the firewall in the engine compartment, directly behind the master cylinder.

Most booster units are bolted directly to the firewall. However, larger boosters used on heavier vehicles are mounted sideways; these units use a special right-angle linkage arrangement which transmits movement from the brake pushrod **(see illustration 1.9)**.

Rear Wheel Anti-Lock (RWAL) brake system

A rear wheel anti-lock brake system is used on 1990 and later models to improve directional stability and control during hard braking, regardless of load.

Anti-lock Brake System (ABS)

Four-wheel ABS was introduced as an option on 1994 models. In 1995, it became standard equipment on all wagons and B2500 conversion vans, and optional on the heaviest GVWRs of the 3500 van.

Parking brake

The parking brake mechanically operates the rear brakes only. The parking brake cables pull on a lever attached to the brake shoe assembly, causing the shoes to expand against the drum.

Precautions

There are some general precautions and warnings related to the brake system:

a) Use only brake fluid conforming to DOT 3 specifications.
b) The brake pads and linings may contain asbestos fibers which are hazardous to your health if inhaled. Whenever you work on brake system components, clean all parts with brake system cleaner. Do not allow the fine dust to become airborne.
c) Safety should be paramount whenever any servicing of the brake components is performed. Do not use parts or fasteners which are not in perfect condition, and be sure all clearances and torque specifications are adhered to. If you are at all unsure about a certain procedure, seek professional advice. Upon completion of any brake system work, test the brakes carefully in a controlled area before driving the vehicle in traffic.

If a problem is suspected in the brake system, don't drive the vehicle until it's fixed.

2 Disc brake pads - replacement

Refer to illustrations 2.3, 2.4a through 2.4i and 2.5a through 2.5e

Warning: *Disc brake pads must be replaced on both front wheels at the same time - never replace the pads on only one wheel. Also, the dust created by the brake system may contain asbestos, which is harmful to your health. Never blow it out with compressed air and don't inhale any of it. An approved filtering mask should be worn when working on the brakes. Do not, under any circumstances, use petroleum-based solvents to clean brake parts. Use brake system cleaner only!*

2.3 Using a large C-clamp, push the piston back into the caliper bore - note that one end of the clamp is positioned against the flat end of the piston housing and the screw is pushing against the outer pad

2.4a Remove these two bolts (arrows), then remove the upper and lower retainer clips and springs

2.4b Remove the caliper housing and detach the outer brake pad from the caliper; after the outer pad has been removed from the caliper, suspend the caliper from the coil spring with a piece of wire to protect the brake hose

2.4c Remove the inner brake pad from the caliper adapter

1   Loosen the front wheel lug nuts, raise the front of the vehicle and support it securely on jackstands. Apply the parking brake. Remove the front wheels.

2   Remove about two-thirds of the fluid from the master cylinder reservoir and discard it. **Caution:** *Do not spill brake fluid on painted surfaces. Position a drain pan under the brake assembly and clean the caliper and surrounding area with brake system cleaner.*

3   Position a large C-clamp over the caliper as shown and squeeze the piston back into in its bore **(see illustration)** to provide room for the new brake pads. As the piston is depressed to the bottom of its caliper bore, the fluid in the master cylinder will rise. Make sure it does not overflow. If necessary, siphon off some of the fluid.

4   If you're working on a 1971 through 1997 model, follow the accompanying illustrations, beginning with **illustration 2.4a,** for the pad replacement procedure. Be sure to

stay in order and read the caption under each illustration. Work on one brake assembly at a time, using the other side for reference, if necessary.

5   If you're working on a 1998 or later model, follow the accompanying illustrations, beginning with **illustration 2.5a,** for the pad removal procedure. Be sure to stay in order and read the caption under each illustration. Work on one brake assembly at a time, using the other side for reference, if necessary.

6   While the pads are removed, inspect the caliper for seal leaks (evidenced by fluid moisture around the cavity) and for any damage to the piston dust boots. If excessive moisture is evident, install a new piston dust boot and piston seal (see Section 3). When you're done, make sure you tighten the bolts that secure the anti-rattle springs and retainers to the torque listed in this Chapter's Specifications.

7   Install the brake pads on the opposite

wheel, then install the wheels and lower the vehicle. Tighten the lug nuts to the torque listed in the Chapter 1 Specifications. Add brake fluid to the reservoir until it's full (see Chapter 1).

8   Pump the brakes several times to seat the pads against the disc, then check the fluid level again.

9   Check the operation of the brakes before driving the vehicle in traffic. Try to avoid heavy brake applications until the brakes have been applied lightly several times to seat the pads.

3   Disc brake caliper - removal, overhaul and installation

Note: *If an overhaul is indicated (usually because of fluid leaks, a stuck piston or broken bleeder screw) explore all options before beginning this procedure. New and factory*

9

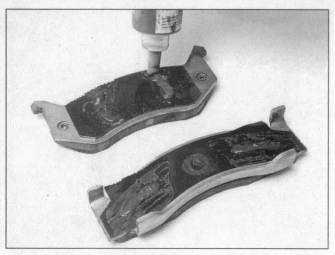

2.4d Apply anti-squeal compound to the backing plates of the new pads - let the compound dry a few minutes before installing the pads

2.4e When installing the new outer pad, make sure the "ears" are fully seated against the caliper as shown

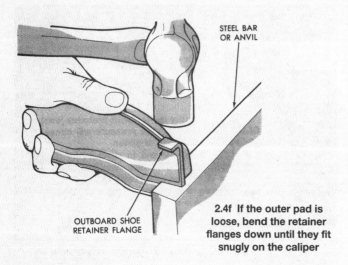

STEEL BAR
OR ANVIL

OUTBOARD SHOE
RETAINER FLANGE

2.4f If the outer pad is loose, bend the retainer flanges down until they fit snugly on the caliper

2.4g Clean the "ways" of the machined surfaces on the adapter, then lubricate them with high temperature brake grease

2.4h When you install the new inner brake pad, make sure the "ears" at either end of the backing plate are properly seated onto the "ways" of the machined surfaces on the adapter

2.4i Install the caliper, then install the upper and lower retainer clips and anti-rattle springs as shown, with the anti-rattle spring on top of the retainer clip, and tighten the bolts to the torque listed in this Chapter's Specifications

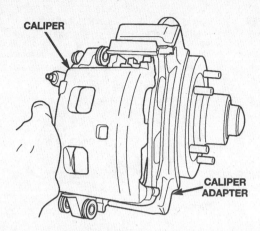

2.5a Remove the caliper mounting bolts, then lift the caliper up and off the caliper adapter

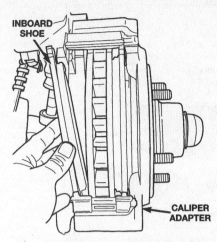

2.5b Remove the inner brake pad . . .

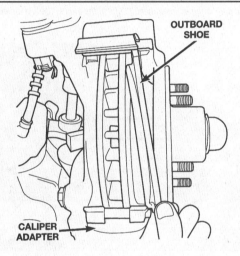

2.5c . . . then the outer brake pad from the caliper adapter

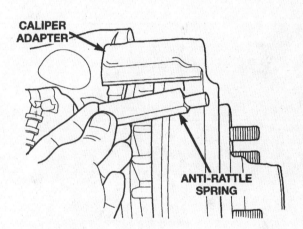

2.5d Remove the upper anti-rattle spring and inspect it . . .

rebuilt calipers are available on an exchange basis, which makes this job quite easy. If you decide to rebuild the calipers, make sure rebuild kits are available before proceeding. Always rebuild or replace the calipers in pairs - never rebuild just one of them.

Removal

1   Loosen the front wheel lug nuts, raise the vehicle and support it securely on jackstands. Remove the front wheels.
2   Remove the brake hose-to-caliper banjo bolt and detach the hose from the caliper. (If the caliper is only being removed for access to other components, don't disconnect the hose.) Discard the two copper sealing washers on each side of the fitting and use new ones during installation. Wrap a plastic bag around the end of the hose to prevent fluid loss and contamination.
3   Remove the caliper following the first few steps of Section 2 (it's part of the brake pad replacement procedure), remove the brake pads, then clean the caliper with brake

system cleaner. DO NOT use kerosene, gasoline or petroleum-based solvents.

Overhaul

Refer to illustrations 3.4, 3.5a, 3.5b, 3.6, 3.11a, 3.11b and 3.12
Note: This procedure is shown on a 1971

through 1997 model with a single piston caliper. 1998 and later calipers are equipped with two pistons, two dust boots and two piston seals. The service procedure steps shown apply to both caliper assemblies.
4   Place several shop towels or a block of wood in the center of the caliper to act as a cushion, then use compressed air, directed

2.5e . . . then the lower anti-rattle spring. Refer to illustration 2.4d and apply anti-squeal compound to the backs of the new pads, then reverse the removal procedure to install them (be sure to tighten the caliper mounting bolts to the torque listed in this Chapter's Specifications)

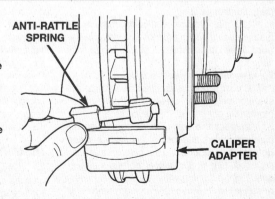

9

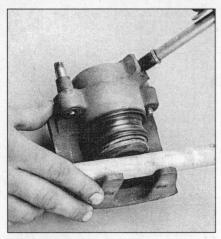

3.4 With a block of wood placed between the piston and caliper frame, use compressed air to ease the piston out of the bore

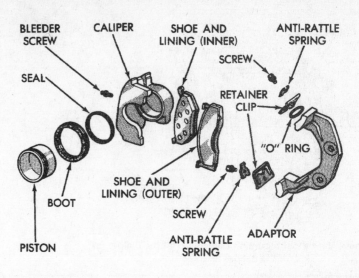

3.5a An exploded view of the caliper assembly (1971 through 1997 models)

BLEEDER SCREW

CALIPER

SHOE AND LINING (INNER)

ANTI-RATTLE SPRING

SCREW

SEAL

RETAINER CLIP

"O" RING

BOOT

SHOE AND LINING (OUTER)

SCREW

ANTI-RATTLE SPRING

ADAPTOR

PISTON

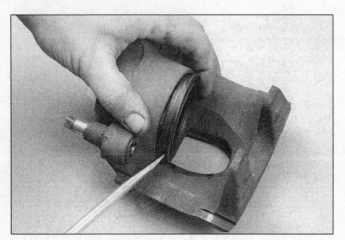

3.5b Carefully pry the dust boot out of the caliper

3.6 The piston seal should be removed with a plastic or wooden tool to avoid damage to the bore and seal groove (a pencil will do the job)

into the fluid inlet, to remove the piston **(see illustration)**. Use only enough air pressure to ease the piston out of the bore. If the piston is blown out, even with the cushion in place, it may be damaged. **Warning:** *Never place your fingers in front of the piston in an attempt to catch or protect it when applying compressed air, as serious injury could occur.*

5   Pry the dust boot from the caliper bore **(see illustrations)**.

6   Using a wood or plastic tool, remove the piston seal from the groove in the caliper bore **(see illustration)**. Metal tools may cause bore damage.

7   Remove the bleeder screw.

8   Clean the remaining parts with brake system cleaner or clean brake fluid, then blow them dry with compressed air.

9   Inspect the surfaces of the piston for nicks and burrs and loss of plating. If surface defects are present, the piston must be replaced. Check the caliper bore in a similar way. Light polishing with crocus cloth is permissible to remove slight corrosion and

stains. **Note:** *On later-model vans, the caliper pistons are made of phenolic resin, not metal. These should never be polished or sanded, and do not interchange metal pistons for plastic or vice-versa.*

10   Lubricate the new piston seal with clean brake fluid and position the seal in the cylinder groove using your fingers only. Make sure it isn't twisted.

11   Install the new dust boot in the groove in the end of the piston **(see illustration)**. Dip the piston in clean brake fluid and insert it squarely into the cylinder. Depress the piston to the bottom of the cylinder bore **(see illustration)**.

12   Seat the boot in the caliper counterbore using a boot installation tool or a blunt punch **(see illustration)**.

13   Install the bleeder screw.

Installation

14   Clean the sliding surfaces of the caliper and the caliper adapter and lube lightly with high-temperature grease.

15   Install the caliper and brake pads as described in Section 2.

16   Using new sealing washers, connect the brake hose to the inlet fitting. Tighten the

3.11a Slip the dust boot over the piston

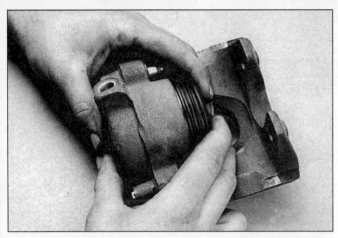

3.11b Push the piston straight into the caliper - make sure it doesn't become cocked in the bore

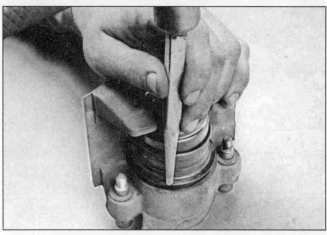

3.12 If you don't have a boot installation tool, gently seat the boot with a drift punch, going evenly around the boot

brake hose-to-caliper banjo bolt to the torque listed in this Chapter's Specifications.

17 Bleed the brakes as outlined in Section 10.

18 Install the wheels and lug nuts. Lower the vehicle and tighten the lug nuts to the torque listed in the Chapter 1 Specifications.

19 After the job has been completed, firmly depress the brake pedal a few times to bring the pads into contact with the disc.

20 Check the operation of the brakes before driving the vehicle in traffic. Check the calipers, bleeder fittings and hose connections for leaks.

4 Brake disc - inspection, removal and installation

Inspection

Refer to illustrations 4.3, 4.4a and 4.4b

1 Loosen the wheel lug nuts, raise the front of the vehicle and support it securely on jackstands. Apply the parking brake. Remove the front wheels.

2 Remove the brake caliper as described in Section 2, and support it with a piece of heavy wire so that it won't hang by the brake hose. Visually inspect the disc surface for score marks, hard spots and other damage. Light scratches and shallow grooves are normal after use and won't affect brake operation. Deep grooves - over 0.015-inch (0.38 mm) deep - require disc removal and refinishing by an automotive machine shop. Be sure to check both sides of the disc.

3 To check disc runout, place a dial indicator at a point about 1/2-inch from the outer edge of the disc **(see illustration)**. Set the indicator to zero and turn the disc. The indicator reading should not exceed the runout limit listed in this Chapter's Specifications. If it does, the disc should be refinished by an automotive machine shop. **Note:** *Professionals recommend resurfacing the brake discs regardless of the dial indicator reading (to produce a smooth, flat surface that will elimi-*

4.3 Use a dial indicator to check disc runout - if the reading exceeds the specified runout limit, the disc will have to be machined or replaced

4.4a On some models, the minimum thickness is cast into the inside of the disc - on others, it's located on the outside of the disc (typical)

nate brake pedal pulsations and other undesirable symptoms related to questionable discs). At the very least, if you elect not to have the discs resurfaced, deglaze them with sandpaper or emery cloth.

4 The disc must not be machined to a thickness less than the specified minimum thickness. The minimum (or discard) thickness is cast into the disc **(see illustration)**. The disc thickness can be checked with a micrometer **(see illustration)**, at several points around the circumference of the disc.

4.4b Use a micrometer to measure the thickness of the disc at several points

9

5.4a If you can't pull off the brake drum, apply some penetrating oil at the hub-to-drum joint and allow it to soak in, lightly tap the drum to break it loose . . .

5.4b . . . then carefully tap around the outer edge of the drum to drive it off the studs - don't use excessive force!

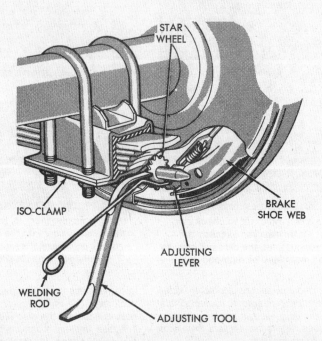

5.4c If you still can't pull off the drum, the shoes have worn into the drum and will have to be retracted; insert a screwdriver or piece of stiff wire into the slot in the backing plate to hold the adjusting lever away from the star wheel, then turn the star wheel with another screwdriver or a brake adjusting tool

5.4d Before beginning work, wash away all traces of dust with brake system cleaner - DO NOT use compressed air

Removal and installation

5   Remove the dust cap, cotter pin, nut and outer wheel bearing (see Chapter 1), then pull off the hub and disc.

6   Refer to Chapter 1 and service the wheel bearings, then install the hub/disc assembly and adjust the wheel bearings (also in Chapter 1).

5   Drum brake shoes - replacement

Refer to illustrations 5.4a through 5.4r, 5.4s through 5.4vv and 5.5

Warning: *Drum brake shoes must be replaced on both wheels at the same time - never replace the shoes on only one wheel. Also, the dust created by the brake system may contain asbestos, which is harmful to your health. Never blow it out with compressed air and don't inhale any of it. An approved filtering mask should be worn when working on the brakes. Do not, under any circumstances, use petroleum-based solvents to clean brake parts. Use brake system cleaner only!*

Caution: *Whenever the brake shoes are replaced, the retractor and hold-down springs should also be replaced. Due to the continuous heating/cooling cycle that the springs are subjected to, they lose their tension over a period of time and may allow the shoes to drag on the drum and wear at a much faster rate than normal.*

1   There are two types of drum rear brakes on the vehicles covered by this book, the 11-inch-diameter Chrysler brakes used on most models, and the Bendix 12-inch units used on heavy-duty (1-ton) models.

2   Loosen the wheel lug nuts, raise the rear of the vehicle and support it securely on jackstands. Block the front wheels to keep the vehicle from rolling and release the parking brake.

3   Remove the wheel. **Note:** *All four rear*

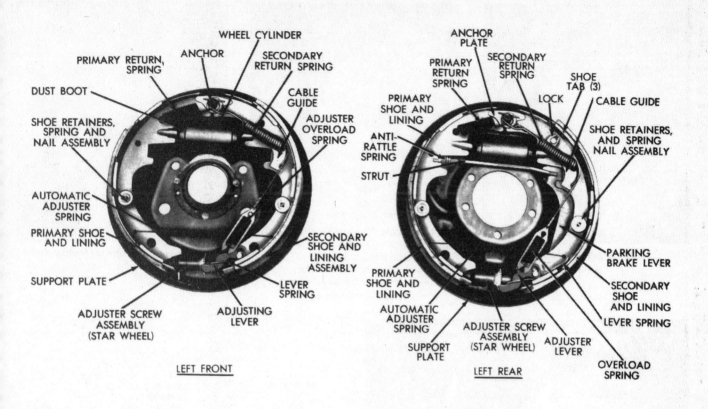

PRIMARY RETURN, SPRING
ANCHOR
WHEEL CYLINDER
SECONDARY RETURN SPRING
DUST BOOT
CABLE GUIDE
ADJUSTER OVERLOAD SPRING
SHOE RETAINERS, SPRING AND NAIL ASSEMBLY
AUTOMATIC ADJUSTER SPRING
PRIMARY SHOE AND LINING
SUPPORT PLATE
ADJUSTER SCREW ASSEMBLY (STAR WHEEL)
ADJUSTING LEVER
LEVER SPRING
SECONDARY SHOE AND LINING ASSEMBLY

LEFT FRONT

ANCHOR PLATE
PRIMARY RETURN SPRING
SECONDARY RETURN SPRING
SHOE TAB (3)
PRIMARY SHOE AND LINING
LOCK
CABLE GUIDE
ANTI-RATTLE SPRING
SHOE RETAINERS, AND SPRING NAIL ASSEMBLY
STRUT
PARKING BRAKE LEVER
PRIMARY SHOE AND LINING
SECONDARY SHOE AND LINING
AUTOMATIC ADJUSTER SPRING
LEVER SPRING
SUPPORT PLATE
ADJUSTER SCREW ASSEMBLY (STAR WHEEL)
ADJUSTER LEVER
OVERLOAD SPRING

LEFT REAR

5.4e Details of the 11-inch Chrysler drum brake assembly (left side shown)

brake shoes must be replaced at the same time, but to avoid mixing up parts, work on only one side's brake assembly at a time.

4   Follow the accompanying illustrations **(5.4a through 5.4r** for 11-inch brakes; **5.4s through 5.4vv** for 12-inch Bendix brakes) for the inspection and replacement of the brake shoes. Be sure to stay in order and read the caption under each illustration. **Note 1:** *If the brake drum cannot be easily pulled off the axle and shoe assembly, make sure that the parking brake is completely released, then apply some penetrating oil at the hub-to-drum joint. Allow the oil to soak in and try to pull the drum off. If the drum still cannot be pulled off, the brake shoes will have to be retracted. This is accomplished by first removing the plug from the backing plate. With the plug removed, pull the lever off the adjusting star wheel with one narrow screw-driver while turning the adjusting wheel with another narrow screwdriver, moving the shoes away from the drum* **(see illustration 5.4c)**. *The drum should now come off.* **Note 2:** *On models with Bendix rear brakes, remove the rear axle before you start (see Chapter 8). To remove the brake drum, refer to Chapter 8, Hub/drum assembly and wheel bearings (full-floating axle) - removal, installation and adjustment.*

5.4f Remove the shoe return springs - the spring removal tool shown here, which can be purchased at most auto parts stores, greatly simplifies this step

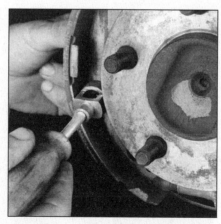

5.4g Remove the hold-down springs by pushing the retainers in and turning them 90-degrees

5.4h Remove the adjuster cable

9

5.4i Remove the cable guide
and spring

5.4j Separate the shoes at the top, lift
them away from the backing plate, then
detach the parking brake lever (rear
brakes only) from the secondary shoe -
the shoes, adjuster and spring can
now be separated

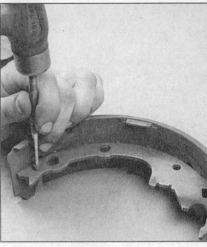

5.4k Drive a new adjuster lever pin into
the new shoe

5.4l Apply high-temperature grease to the areas on the backing
plate that support the shoes (don't use too much grease
or get it on the shoes or drum)

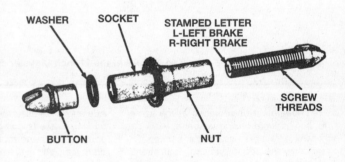

5.4m Clean the adjuster screw, then lubricate the threads and the
sliding surface of the button with high-temperature grease

5.4n Install the shoe-to-shoe spring and
adjuster assembly

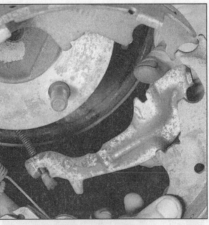

5.4o Raise the shoe assembly up to the
backing plate and engage the tang on
the parking brake lever with its slot
in the secondary shoe

5.4p Fit the shoes over the axle flange,
then install the hold-down pins, springs
and retainers; make sure the shoes
engage with the wheel cylinder pistons
(or, on some models, the wheel
cylinder pushrods)

5.4q Install the parking brake strut (rear brakes only)
and anti-rattle spring

5.4r Install the anchor plate and adjuster cable over the anchor
pin, then install the return springs, route the adjuster cable
around the cable guide and hook it to the adjuster lever;
the brake drum can now be installed

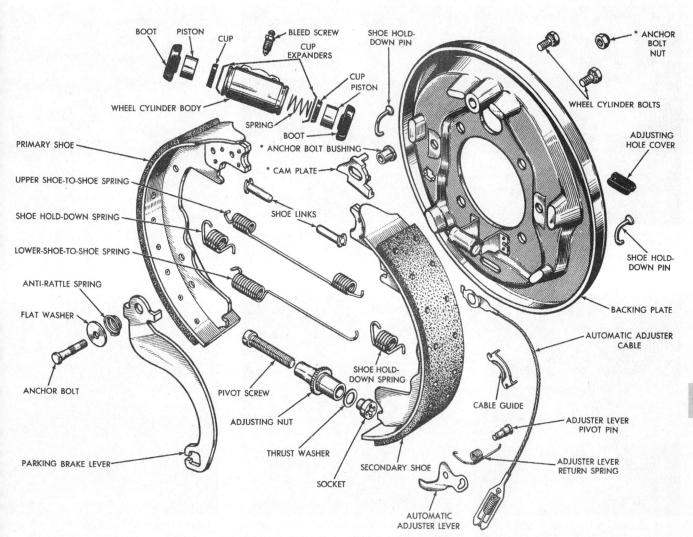

5.4s An exploded view of a 12-inch Bendix rear drum brake assembly

9

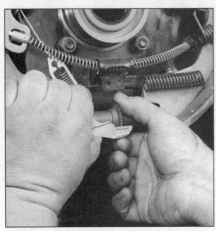

5.4t Clean the brake assembly with brake system cleaner, then lift the automatic adjuster lever off the adjusting ratchet and back off the adjuster to retract the brake shoes

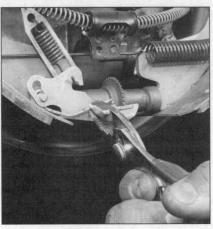

5.4u Pull the adjuster lever return spring off the lever

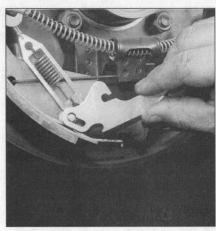

5.4v Remove the automatic adjuster lever

5.4w Remove the adjuster lever return spring

5.4x Disengage the front shoe hold-down spring from the front hold-down pin

5.4y Pull forward on the front shoe and remove the adjuster assembly

5.4z Disengage the lower shoe-to-shoe spring from both shoes and remove it

5.4aa Disengage the front end of the upper shoe-to-shoe spring from the front shoe (shown) and remove the front shoe, then unhook the spring from the rear shoe and remove the spring

5.4bb Pull back the spring with a pair of pliers as shown and disengage the parking brake cable from the parking brake lever

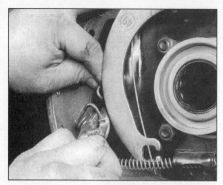

5.4cc Disengage the rear shoe hold-down spring from the rear hold-down pin, then remove the rear shoe

5.4dd Knock out the adjuster lever pivot pin from the old rear shoe with a drift punch and drive it into the new rear shoe

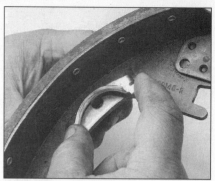

5.4ee The cable guide is attached to the rear shoe by a pair of tangs that are inserted through holes in the shoe and bent over on the other side; straighten the tangs, remove the guide and install it the same way on the new rear shoe

5.4ff Lubricate the friction surfaces of the backing plate with high-temperature grease

5.4gg Place the new rear shoe in position, install the hold-down spring and hook it over the shoe hold-down pin

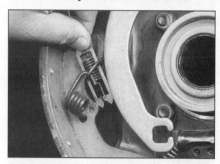

5.4hh Route the automatic adjuster cable behind the hold-down spring and pin

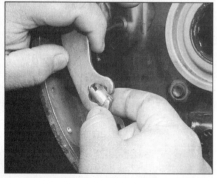

5.4ii Place the parking brake cable retainer in position in the parking brake lever . . .

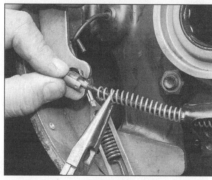

5.4jj . . . and reattach the parking brake cable to the parking brake lever

5.4kk Hook the rear end of the upper shoe-to-shoe spring through its hole in the rear shoe . . .

5.4ll . . . then hook it through the hole in the front shoe and place the shoe in position

5.4mm Hook the front end of the lower shoe-to-shoe spring into its hole in the front shoe . . .

5.4nn . . . then pull it back and hook the rear end of the spring into its hole in the rear shoe

9

5.4oo Hook the front hold-down spring through its hole in the front shoe . . .

5   Before reinstalling the drum, check it for cracks, score marks, deep scratches and hard spots, which will appear as small discolored areas. If the hard spots cannot be removed with fine emery cloth or if any of the other conditions listed above exist, the drum must be taken to an automotive machine shop to have it resurfaced. **Note:** *Professionals recommend resurfacing the drums whenever a brake job is done. Resurfacing will eliminate the possibility of out-of-round*

5.4pp . . . and - using the special hold-down spring tool, if you've got it - reattach the hold-down spring to the hold-down pin (if you don't have the special tool, you can use a pair of pliers, but it's easier, and safer, with the special tool)

drums. If the drums are worn so much that they can't be resurfaced without exceeding the maximum allowable diameter (stamped into the drum) **(see illustration)**, then new ones will be required. At the very least, if you elect not to have the drums resurfaced, remove the glazing from the surface with emery cloth or sandpaper using a swirling motion.

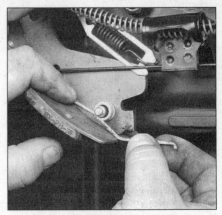

5.4qq Install the adjuster lever return spring

6   Install the brake drum on the axle flange (if you're working on a Bendix brake, refer to Chapter 8, *Hub/drum assembly and wheel bearings (full-floating axle) - removal, installation and adjustment.* Turn the brake adjuster until the shoes rub on the drum, then back-off the adjuster until the shoes don't rub.
7   Mount the wheel, install the lug nuts, then lower the vehicle. On models with Bendix brakes, reinstall the rear axles (see Chapter 8).

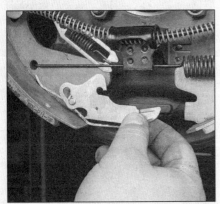

5.4rr Install the adjuster lever

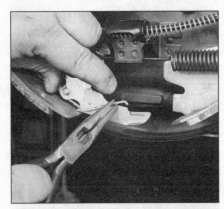

5.4ss Lift the end of the spring up and hook it over the adjuster lever as shown

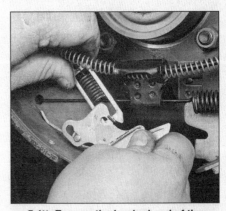

5.4tt Engage the hooked end of the automatic adjuster cable with its slot in the adjuster lever

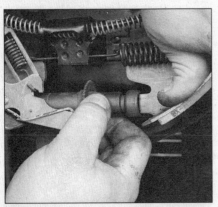

5.4uu Pull out on the adjuster lever and install the adjuster assembly as shown; make sure both ends are properly engaged with the lower ends of the shoes

5.4vv This is how it should look when you're done

5.5 The maximum allowable inside diameter is cast into the outer edge of the drum (typical)

8   Make a number of forward and reverse stops to adjust the brakes until satisfactory pedal action is obtained.

9   Check brake operation before driving the vehicle in traffic.

6   Wheel cylinder - removal, overhaul and installation

Note: *If an overhaul is indicated (usually because of fluid leakage or sticky operation) explore all options before beginning the job. New wheel cylinders are available, which makes this job quite easy. If you decide to rebuild the wheel cylinder, make sure a rebuild kit is available before proceeding. Never overhaul only one wheel cylinder. Always rebuild both of them at the same time.*

Removal

Refer to illustration 6.2

1   Refer to Section 5 and remove the brake shoes.

2   Unscrew the brake line fitting from the rear of the wheel cylinder **(see illustration)**. If available, use a flare-nut wrench to avoid rounding off the corners on the fitting. Don't pull the metal line out of the wheel cylinder - it could bend, making installation difficult.

3   Remove the two bolts securing the wheel cylinder to the brake backing plate.

4   Remove the wheel cylinder.

5   Plug the end of the brake line to prevent the loss of brake fluid and the entry of dirt.

Overhaul

Refer to illustrations 6.6a, 6.6b, 6.11, 6.12a, 6.12b, 6.14 and 6.15

6   To disassemble the wheel cylinder, first remove the rubber boot from each end of the cylinder and push out the two pistons, cups (seals) and spring expander **(see illustrations)**. Discard the rubber parts and use new ones from the rebuild kit when reassembling the wheel cylinder.

7   Inspect the pistons for scoring and scuff marks. If defects are present, the pistons should be replaced with new ones. **Note:** *Most wheel cylinder overhaul kits include all components except the pistons and the pushrods, all other original parts are discarded.*

8   Examine the inside of the cylinder bore for score marks and corrosion. If these conditions exist, the cylinder can be polished slightly with crocus cloth to restore it, but replacement is recommended.

9   If the cylinder is in good condition, clean it with brake system cleaner or brake fluid. **Warning:** *DO NOT, under any circumstances, use gasoline or petroleum-based solvents to clean brake parts!*

10   Remove the bleeder screw and make sure the hole at its inner end is clean.

11   Lubricate the cylinder bore with clean brake fluid, then insert one of the new rubber cups into the bore. Make sure the lip on the rubber cup faces in **(see illustration)**.

12   Attach the two cup expanders to the new expander spring and place the assembly in the opposite end of the bore and push it in until it contacts the rear of the rubber cup **(see illustrations)**.

13   Install the remaining cup in the cylinder bore.

14   Install both pistons **(see illustration)**.

15   Insert the pushrods into the boots **(see illustration)**, then install the boots and pushrods.

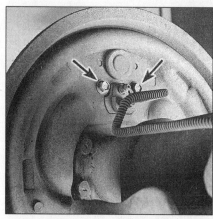

6.2 To remove the wheel cylinder from the brake backing plate, disconnect the brake line fitting with a flare-nut wrench, then remove the wheel cylinder mounting bolts (arrows)

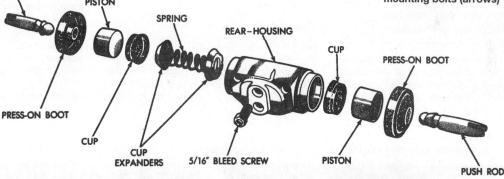

6.6a An exploded view of a typical wheel cylinder assembly (not all units are equipped with cup expanders or pushrods)

PUSH ROD • PISTON • SPRING • REAR-HOUSING • CUP • PRESS-ON BOOT

PRESS-ON BOOT • CUP • CUP EXPANDERS • 5/16" BLEED SCREW • PISTON • PUSH ROD

9

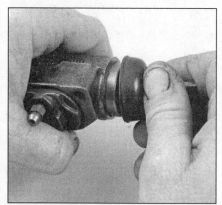

6.6b Remove the wheel cylinder dust seals

6.11 Install one piston cup with its open end facing in, and the flat side facing the piston

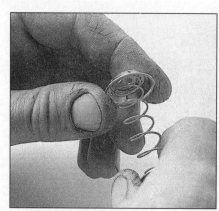

6.12a Attach the new cup expanders (if equipped) to the spring . . .

6.12b . . . then insert the expander/spring assembly (expanders are used only on 1974 and later models; on units without an expander, simply install the spring)

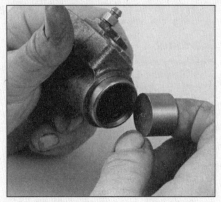

6.14 Install a piston into each end of the cylinder with the flat side toward the cup

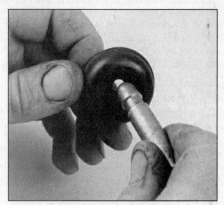

6.15 Insert the wheel cylinder pushrods into the new dust boots, then install the dust boots and pushrods onto both ends of the wheel cylinder

Installation

16   Installation is the reverse of removal. Attach the brake line to the wheel cylinder before installing the mounting bolts and tighten the line fitting after the wheel cylinder mounting bolts have been tightened. If available, use a flare-nut wrench to tighten the line fitting. **Note:** *It's a good idea to apply some RTV sealant around the opening in the backing plate before bolting in the wheel cylinder. This will keep out water and dust.*

17   Bleed the brakes (see Section 10). Don't drive the vehicle in traffic until the operation of the brakes has been thoroughly tested.

---

7   Master cylinder - removal, overhaul and installation

1   Before removing the master cylinder, read the detailed description of the various types of master cylinders used on the vehicles covered by this manual. Note that some units can be rebuilt, while others cannot. Also be advised that kits may no longer be available even for some rebuildable master cylinders, in which case you will have to install a new or rebuilt unit. So explore your options - take a few minutes to make some phone calls and find out what's available before proceeding. If you have an aluminum master cylinder, it cannot be rebuilt - it must be replaced. Although Chrysler originally supplied rebuildable aluminum master cylinder units on 1979 through 1982 models, rebuild kits are no longer available. Some parts departments may be able to direct you to an "old parts supplier" near you, but because properly rebuilt brake components are a safety issue, we feel that your best bet is to simply replace one of these older aluminum master cylinders with a new or rebuilt unit. For further information regarding buying a master cylinder unit, refer to Section 1.

Removal

Refer to illustrations 7.3 and 7.5

2   Place rags under the brake line fittings and prepare caps or plastic bags to cover the ends of the lines once they're disconnected. **Caution:** *Brake fluid will damage paint. Cover all painted surfaces and avoid spilling fluid during this procedure. Brake fluid can be siphoned out of the reservoir using a squeeze bulb, but wear safety goggles.*

3   Loosen the tube nuts at the ends of the brake lines where they enter the master cylinder. To prevent rounding off the flats on these nuts, a flare-nut wrench, which wraps around the nut, should be used **(see illustration)**.

4   Pull the brake lines away from the master cylinder slightly and plug the ends to prevent contamination.

5   Remove the two master cylinder mounting nuts. Remove the master cylinder from the booster, taking care not to kink the hydraulic lines **(see illustration)**.

6   Remove the reservoir cap(s), then discard any fluid remaining in the reservoir.

Overhaul

Refer to illustrations 7.7a, 7.7b, 7.7c, 7.9, 7.10, 7.11, 7.13, 7.14a and 7.14b

7   The following procedure applies only to one-piece cast-iron master cylinders **(see illustrations)**. With the exception of 1979 through 1982 units, aluminum master cylinders cannot be rebuilt - they must be replaced with a rebuilt or new unit. However, we have provided an exploded view of one of these units, in the event that you're able to locate a suitable rebuild kit **(see illustration)**.

8   Remove the caps or the cover and gas-

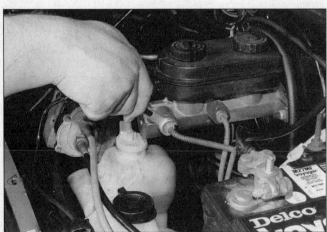

7.3 Use a flare-nut wrench, if available, to loosen and tighten the brake line fittings

7.5 To detach the master cylinder from the power brake booster, remove the two nuts and pull the master cylinder assembly straight off

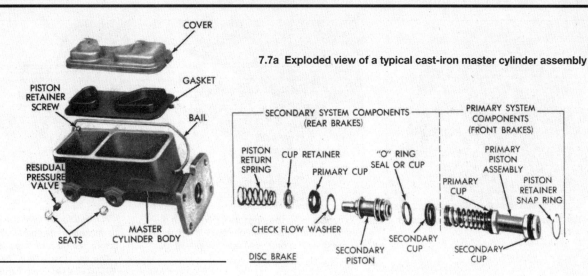

7.7a Exploded view of a typical cast-iron master cylinder assembly

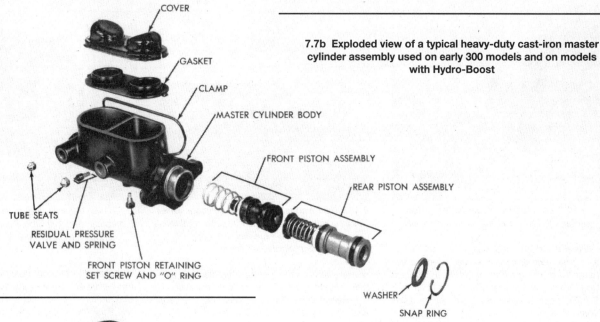

7.7b Exploded view of a typical heavy-duty cast-iron master cylinder assembly used on early 300 models and on models with Hydro-Boost

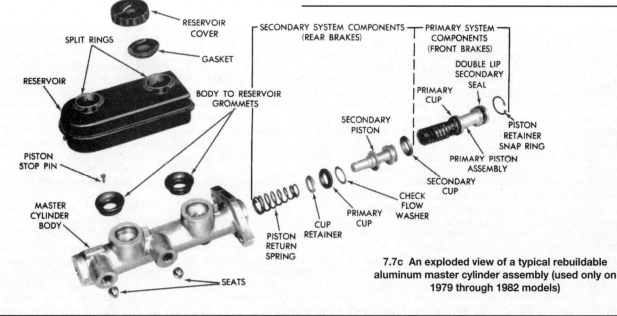

7.7c An exploded view of a typical rebuildable aluminum master cylinder assembly (used only on 1979 through 1982 models)

9

7.9 Remove the piston retainer screw

7.10 Depress the primary piston with a Phillips screwdriver and remove the snap-ring with a pair of snap-ring pliers

7.11 As you remove the primary piston assembly, note the relationship of the piston, spring and cups to one another - the new parts must be assembled EXACTLY like this during installation

7.13 Watching the order of the parts, remove the secondary piston assembly

7.14a To remove the master cylinder tube seats, tap a metal screw into them; this will give you something to pry on during removal

7.14b Using a prybar as a fulcrum, remove the old tube seats by pulling them out with a pair of pliers

ket and drain the brake fluid into a container, if you haven't already done so. Remove the plugs from the master cylinder outlets and drain the remaining brake fluid.

9   Remove the piston retainer screw **(see illustration)**.

10   Push in on the primary piston and remove the piston snap-ring **(see illustration)**.

11   Remove the primary piston assembly from the master cylinder bore **(see illustration)**.

12   Tap the open end of the master cylinder on the workbench to dislodge the secondary piston assembly.

13   Remove the secondary piston and spring from the master cylinder bore **(see illustration)**.

14   Examine the tube seats carefully with a small flashlight; unless they exhibit deep scoring from the old fittings or shown evidence that they have been leaking, do not remove them. If they have been leaking, install a self-tapping screw in the tube seats as shown and pry them out **(see illustrations)**. Remove the residual pressure valves and springs, if equipped. **Note:** *Make sure the master-cylinder rebuilding kit has new tube seats in it before removing the old ones.*

15   Remove the rubber cups from the ends of the pistons. However, don't remove the

primary cup of the primary piston. The primary piston must be replaced as an assembly. Remove the O-ring or seal cup from the second land of the secondary piston.

16   Clean the master cylinder with fresh brake fluid or brake system cleaner and blow dry it with compressed air. **Warning:** *DO NOT, under any circumstances, use petroleum-based solvents to clean brake parts.*

17   Clean the cylinder bore with fresh brake fluid and inspect the bore for scoring and pitting. Light scratches and corrosion on the cylinder bore walls can be usually be removed with crocus cloth or with a hone. However, deep scratches or score marks mean the cylinder must be replaced with a new unit. If the pistons or bore are severely corroded, replace them. Always use new piston cups and seals (or piston assemblies) when overhauling a master cylinder.

18   All components should be assembled wet after dipping them in clean brake fluid.

19   Carefully insert the complete secondary piston and return spring assembly into the master cylinder bore, easing the seals into the bore. Push the assembly all the way in.

20   Insert the primary piston assembly into the master cylinder bore.

21   Depress the primary piston assembly and install and tighten the secondary piston retainer screw or retainer pin.

22   Push in on the piston again and install the snap-ring.

23   Install the residual pressure valves and springs, if equipped, in the outlet ports.

24   If the tube seats had been removed, drive the new seats into place with a spare section of brake line (a flared end) with a tube fitting. Tap the fitting squarely into the master cylinder with the tubing section and a small hammer. Once it's in far enough, slide the tube nut into place and thread it into the outlet. Tighten the fitting to push the seat into place. **Warning:** *Be careful not to score or dent the sealing surface of the tube seat, or leaks may occur. Use a magnifying glass to get a close look inside after installing the tube seats.*

Installation

Refer to illustration 7.28

25   Whenever the master cylinder is removed, the entire hydraulic system must be bled. The time required to bleed the system can be reduced if the master cylinder is filled with fluid and bench bled before the master cylinder is installed on the vehicle.

26   Fill the reservoirs with brake fluid. The

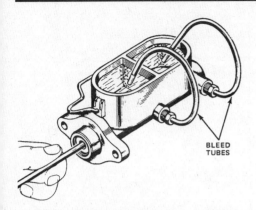

7.28 The quickest and cleanest method of bleeding the master cylinder is with a pair of bleed tubes and fittings, commonly available at auto parts stores

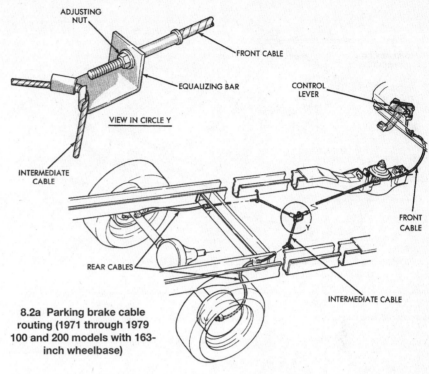

8.2a Parking brake cable routing (1971 through 1979 100 and 200 models with 163-inch wheelbase)

master cylinder should be held in a vise so the brake fluid won't spill during the bench bleeding procedure.

27   Hold your fingers tightly over the holes where the brake lines normally connect to the master cylinder to prevent air from being drawn back into the master cylinder.

28   Stroke the piston several times to ensure all air has been expelled. A large Phillips screwdriver can be used to push on the piston assembly. Wait several seconds each time for brake fluid to be drawn from the reservoir into the piston bore, then depress the piston again, removing your finger as brake fluid is expelled. Be sure to put your fingers back over the holes each time before releasing the piston. When the bleeding procedure is complete, temporarily install plugs in the holes. **Note:** *Another method for bench-bleeding master cylinders is to use a kit available in most auto parts stores. The kit contains plastic fittings that attach to the brake-line-fitting holes in the master cylinder and short hoses that connect to these fittings. With the hoses aimed inside the reservoir (held below the fluid level), you have a "closed" system and you simply stroke the piston slowly until no more air bubbles escape from the plastic tubing* **(see illustration).**

29   Carefully install the master cylinder by reversing the removal steps. **Caution:** *When reinstalling aluminum master cylinders, be very careful not to cross-thread the fittings when reinstalling the brake lines. It may help to start the two fittings into the master cylinder before it is fully bolted down.*

30   Bleed the brake system as described in Section 10.

8   Parking brake cables - replacement

1   Release the parking brake. Block the front wheels to keep the vehicle from rolling, then raise the vehicle and support it securely on jackstands.

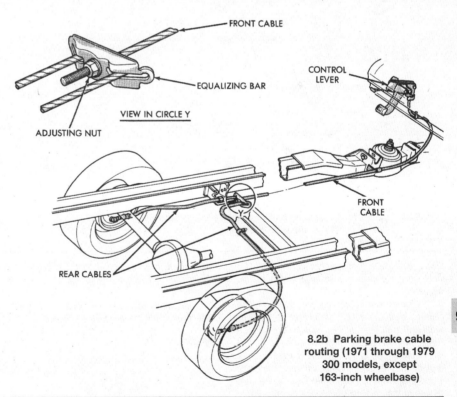

8.2b Parking brake cable routing (1971 through 1979 300 models, except 163-inch wheelbase)

Front cable

Refer to illustrations 8.2a, 8.2b, 8.2c, 8.3a and 8.3b

2   Remove the adjuster nut at the equalizing bar (1971 through 1979 models), or at the adjuster (1980 through 1989 models) or at the equalizer (1990 and later models) **(see illustrations).** Remove any cable retaining clips which attach the cable to the frame.

3   Remove the parking brake lever assembly **(see illustrations)** from its bracket, loosen the hex screw on the cable retaining

9

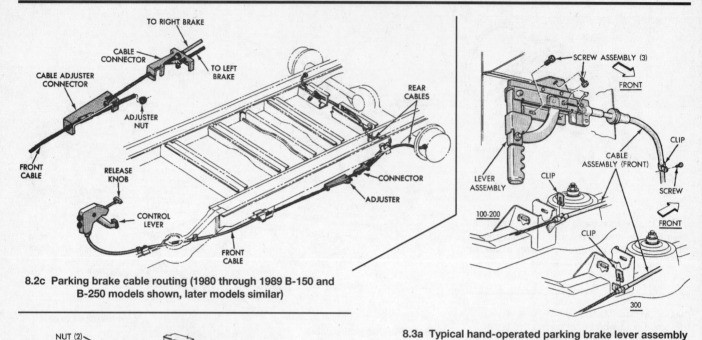

8.2c   Parking brake cable routing (1980 through 1989 B-150 and B-250 models shown, later models similar)

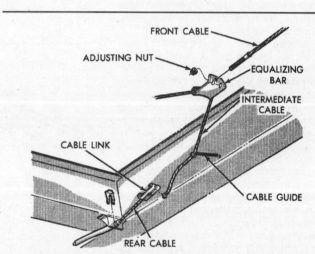

8.3a   Typical hand-operated parking brake lever assembly (1971 through 1977 models)

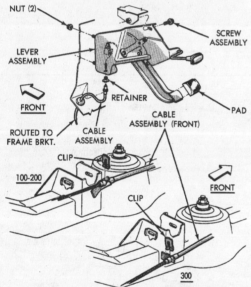

8.3b   Typical foot-operated parking brake lever assembly (1978 and later models)

8.6   Intermediate cable installation details (1971 through 1979 100 and 200 models)

clamp and, with a small screwdriver, pry the plug on the end of the cable out of the lever assembly.

4   Using a small screwdriver, push out the cable weatherproofing grommet at the firewall and pull the cable through the firewall from the engine compartment side.

5   Installation is the reverse of removal. It's a good idea to coat the firewall grommet with silicone sealant to keep out moisture. Be sure to adjust the parking brake (see Section 9).

Intermediate cable (1971 through 1979 100 and 200 models)

Refer to illustration 8.6

6   Unscrew and remove the adjusting nut **(see illustration)**. Disconnect the intermediate cable from the cable guide and the rear cable links.

7   Installation is the reverse of removal. Adjust the parking brake (see Section 9).

Rear cable

Refer to illustrations 8.9 and 8.10

Note: *The following procedure applies to either the left or right rear cable.*

8   Remove the rear brake shoes and disconnect the parking brake cable from the parking brake lever (see Section 5).

9   Compress the tangs of the cable retainer **(see illustration)** and withdraw the cable assembly from the brake backing plate.

10   On 1971 through 1979 models, disconnect the front end of the cable from the inter-mediate cable **(see illustration 8.6)**. On all other models, disconnect the front end of the cable from the equalizer **(see illustration)**.

11   Installation is the reverse of removal. Adjust the parking brake (see Section 9).

9   Parking brake - adjustment

Refer to illustration 9.3

1   The parking brake is pedal operated and is normally self-adjusting through the automatic adjusters in the rear brake drums. However, supplementary adjustment may be needed in the event of cable stretch, wear in the linkage or after installation of new components. **Note:** *Before making any adjustment to the parking brake assembly, make*

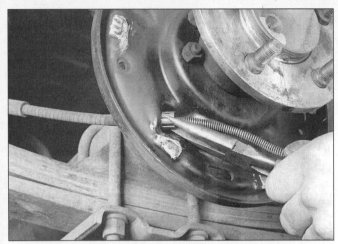

8.9 Compress the retainer tangs together with a small hose clamp (arrow), start the cable through, then remove the clamp and push the cable through the backing plate

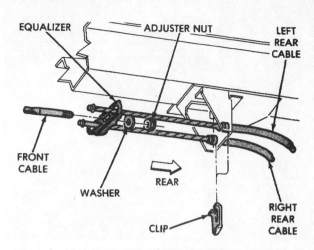

8.10 Rear cable installation details (typical)

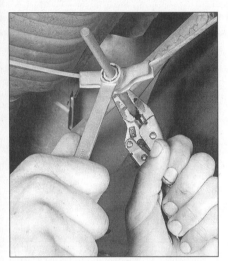

9.3 Hold the cable with locking pliers and turn the adjuster nut with a wrench

sure the rear drum brakes are properly adjusted first.

2   Raise the rear of the vehicle until the wheels are clear of the ground, support it securely on jackstands and block the front wheels. Release the parking brake pedal by pulling on the release lever.

3   The adjuster is located on the outside of the frame on the driver's side, though the location may vary over the years covered by this book. Hold the cable from turning with locking pliers, then loosen the locknut and tighten the adjuster nut **(see illustration)** until a slight drag is felt when the rear wheels are turned. **Note:** *If the threads appear rusty, apply penetrating oil before attempting adjustment.*

4   Loosen the parking brake adjustment until there's no longer any drag when the rear wheels are turned, then loosen the cable adjusting nut two additional turns.

5   Tighten the locknut, then apply and

release the parking brake several times. Confirm that the brake does not drag.

6   Lower the vehicle to the ground.

---

10   Brake system bleeding

Refer to illustration 10.11

Warning: *Wear eye protection when bleeding the brake system. If the fluid comes in contact with your eyes, immediately rinse them with water and seek medical attention.*

Note: *Bleeding the brake system is necessary to remove any air that's trapped in the system when it's opened during removal and installation of a hose, line, caliper, wheel cylinder or master cylinder.*

1   It will be necessary to bleed the system at the master cylinder and at all four brakes (and any anti-lock brake system control valves, if equipped) if air has entered the system due to low fluid level, or if the brake lines have been disconnected at the master cylinder.

2   If a brake line was disconnected only at a wheel, then only that caliper or wheel cylinder must be bled.

3   If a brake line is disconnected at a fitting located between the master cylinder and any of the brakes, that part of the system served by the disconnected line must be bled.

4   Remove any residual vacuum from the brake power booster (if equipped) by applying the brake several times with the engine off.

5   Remove the master cylinder reservoir cover and fill the reservoir with brake fluid. Reinstall the cover. **Note:** *Check the fluid level often during the bleeding operation and add fluid as necessary to prevent the fluid level from falling low enough to allow air bubbles into the master cylinder.*

6   Have an assistant on hand, as well as a supply of new brake fluid, an empty clear plastic container, a length of clear tubing to fit over the bleeder valve and a wrench to open and close the bleeder valve.

Conventional (non-ABS or RWAL) brakes

7   Bleed the system, following the procedure described in Steps 10 through 17, in the following order:

Master cylinder
Combination valve
Right rear wheel
Left rear wheel
Right front wheel
Left front wheel

Rear-Wheel Anti-Lock (RWAL) systems

8   On these systems, begin the bleeding procedure at the RWAL valve (see Section 15 for valve location). Bleed all air from the valve the same way as you would for a caliper or wheel cylinder (see below). After the RWAL valve is bled, proceed to bleed the wheel brakes in the following order, as described in Steps 10 through 17:

Master cylinder
Combination valve
Rear anti-lock valve
Right rear wheel
Left rear wheel
Right front wheel
Left front wheel

Four-wheel Anti-lock Brake System (ABS)

9

9   On these systems, bleed the system components in the following order, just as you would for a caliper or wheel cylinder (see Section 16 for component locations):

Combination valve
Rear anti-lock valve
Front anti-lock valve
Left rear wheel
Right rear wheel
Right front wheel
Left front wheel

10.11 To bleed the brakes, attach a clear piece of tubing to the bleeder screw fitting and submerge the other end in brake fluid - you can easily see any air bubbles in the tube and container (when no more bubbles appear, the air has been purged from the caliper or wheel cylinder)

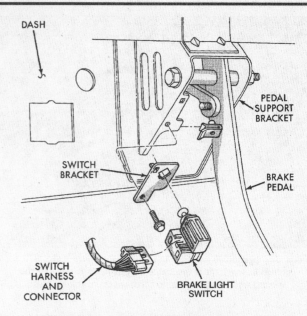

11.4 Brake light switch details (1995 and later models)

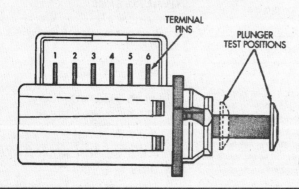

11.5 Brake light switch terminal identification (1995 and later models)

10   Beginning at the first component, loosen the bleeder screw slightly, then tighten it to a point where it's snug but can still be loosened quickly and easily.

11   Place one end of the tubing over the bleeder screw fitting and submerge the other end in brake fluid in the container (see illustration).

12   Have the assistant slowly depress the brake pedal, then hold the pedal firmly depressed.

13   While the pedal is held depressed, open the bleeder screw just enough to allow a flow of fluid to leave the valve. Watch for air bubbles to exit the submerged end of the tube. When the fluid flow slows after a couple of seconds, tighten the screw, then have your assistant release the pedal.

14   Repeat Steps 12 and 13 until no more air is seen leaving the tube, then tighten the bleeder screw and proceed to the next component in the bleeding sequence and perform the same procedure. Be sure to check the fluid in the master cylinder reservoir frequently.

15   Never use old brake fluid. It contains moisture which will allow the fluid to boil, rendering the brakes useless. When bleeding, make sure the fluid coming out of the bleeder

is not only free of bubbles, but clean also.

16   Refill the master cylinder with new fluid at the end of the operation.

17   Check the operation of the brakes. The pedal should feel solid when depressed, with no sponginess. If necessary, repeat the entire process. **Warning:** *Do not operate the vehicle if you are in doubt about the effectiveness of the brake system, or if the ABS warning light or brake warning light does not go off.*

11   Brake light switch - check, replacement and adjustment

Check

1   Apply the brake pedal - the brake lights should come on. If they don't, check the fuse (see Chapter 12). If the fuse is okay, perform the following test:

1971 through 1994 models

2   Adjust the switch (see below).

3   If, after the switch has been adjusted, the brake lights still don't come on when you apply the brake pedal:

a) *Verify that voltage is available at the switch.*

b) *If voltage is getting to the switch, verify that there's voltage on the other side of the switch when the pedal is applied.*

c) *If there's voltage on the switched side of the circuit when the pedal is applied, the switch is okay. Either the circuit to the brake lights is open, or there's something wrong with the lights themselves.*

d) *If there's no voltage on the switched side of the circuit when the pedal is applied, the switch is defective. Replace it (see below).*

1995 and later models

Refer to illustrations 11.4 and 11.5

4   Remove the steering column trim cover (see Chapter 11) and unplug the switch electrical connector **(see illustration)**.

5   Using an ohmmeter, check continuity between terminal pins 5 and 6 **(see illustration)** as follows:

a) *Pull the plunger all the way out to its fully extended position.*

b) *Attach test leads to pins 5 and 6 and note the ohmmeter reading.*

c) *If there's continuity, go to the next Step. If there's no continuity, replace the switch.*

6   Check continuity between terminal pins 1 and 2 and pins 3 and 4 as follows:

a) *Push the switch plunger inward to its fully retracted position.*

b) *Attach test leads to pins 1 and 2 and note the ohmmeter reading.*

c) *If continuity exists, the switch is okay. Replace the switch if the meter indicates a lack of continuity.*

Replacement

1971 through 1994 models

7   Remove the steering column trim cover (see Chapter 11). Unplug the switch electrical

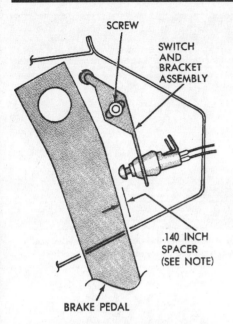

11.11 Typical brake light switch assembly (1974 through 1983 unit shown; all 1984 through 1986 units, and 1987 through 1993 units without cruise control, similar)

connector and unbolt the switch mounting bracket. On most units, this is simply a matter of removing the adjuster nut that's used to adjust the switch.

8   Installation is the reverse of removal. Be sure to adjust the switch after replacing it.

1995 and later models

9   Remove the steering column trim cover (see Chapter 11). Press and hold the brake in the applied position. Rotate the brake light switch counterclockwise about 1/16-turn to unlock the switch. Pull the switch straight out

of its mounting bracket and release the brake pedal. Unplug the electrical connector from the switch.

10   Pull the switch all the way out, then push it in three detent notches. Plug the electrical connector into the switch. Press and hold the brake pedal in the applied position. Align the switch locking collar index in the switch bracket, then insert the switch straight into the bracket. Turn the switch clockwise about 1/16-turn to lock it in position, then release the brake pedal. Lightly pull the brake pedal to the rear against the internal stop in the booster. This sets the switch plunger in its proper adjusted position.

Adjustment

All 1974 through 1986 models; 1987 through 1993 models without cruise control

Refer to illustration 11.11

11   Loosen the switch and bracket assembly **(see illustration)**.

12   With the pedal in its "free" position, i.e. not depressed, insert the proper spacer gauge (.140-inch [9/64-inch] for 1974 through 1983 models; .140-inch for 1984 through 1986 models without cruise control, .070-inch [approximately 5/64-inch] for 1984 through 1986 models with cruise control; .140-inch for 1987 through 1993 models without cruise control) between the brake pedal and the switch pushrod as shown. **Note:** *Do not use this procedure for 1987 through 1993 models with cruise control - a different procedure is needed to adjust them* (see below).

13   Push the switch bracket assembly toward the brake pedal until the pushrod is fully depressed against the spacer (but don't move the pedal!).

14   Tighten the switch bracket securely.

15   Remove the spacer, depress the pedal and verify that the brake lights now work okay. If they don't, readjust the switch!

1987 through 1993 models with cruise control

16   Push the switch through its clip in the mounting bracket until the switch is seated against the bracket - the brake pedal will move forward slightly.

17   Pull back gently on the brake pedal as far as it will go - the switch will ratchet backwards to the correct position and no further adjustment is required.

1994 models

Refer to illustrations 11.18 and 11.21

18   Remove the steering column knee bolster for access to the switch adjusting bolt **(see illustration)**.

19   Turn the adjusting bolt in a clockwise direction until the brake lights come on.

20   Turn the adjusting bolt in a counterclockwise direction until the brake lights go out.

21   Verify that the switch plunger is fully pressed into the switch barrel by the brake pedal striker **(see illustration)**.

22   Install the steering column knee bolster

1995 and later models

23   The switch automatically adjusts itself during installation (see Step 10).

12   Brake hoses and lines - check and replacement

Refer to illustrations 12.3 and 12.4

1   About every six months, with the vehicle raised and placed securely on jackstands, the flexible hoses which connect the steel

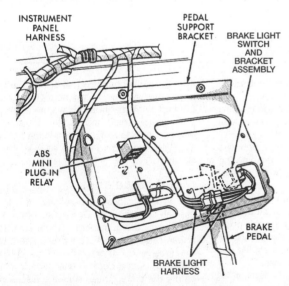

11.18 The brake light switch electrical connector is located on the knee bolster on 1994 models

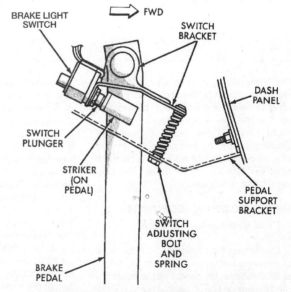

11.21 Brake light switch assembly (1994 models)

12.3 Disconnect the brake line from the hose fitting; although not necessary on this fitting, on some you will need a backup wrench

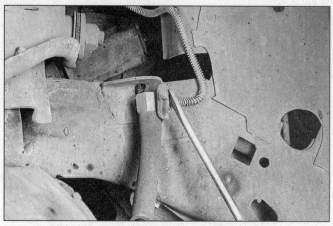

12.4 Before you can disconnect the hose from the line, you'll have to pry a U-clip off the bracket with a screwdriver

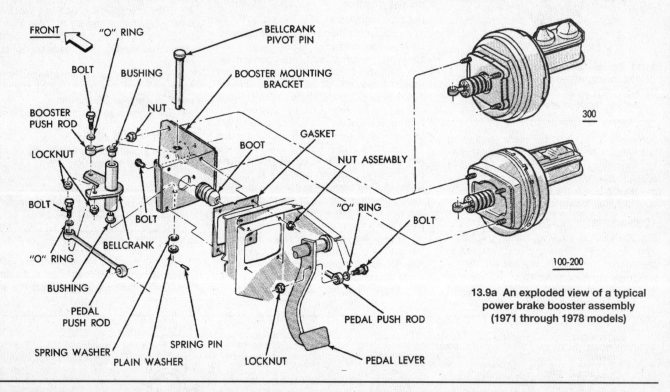

13.9a An exploded view of a typical power brake booster assembly (1971 through 1978 models)

brake lines with the front and rear brake assemblies should be inspected for cracks, chafing of the outer cover, leaks, blisters and other damage. These are important and vulnerable parts of the brake system and inspection should be complete. A light and mirror will be needed for a thorough check. If a hose exhibits any of the above defects, replace it with a new one.

Flexible hose replacement

2   Clean all dirt away from the ends of the hose.

3   Disconnect the brake line from the hose fitting using a back-up wrench on the fitting **(see illustration)**. Be careful not to bend the frame bracket or line. If necessary, soak the connections with penetrating oil.

4   Remove the U-clip (lock) from the

female fitting at the bracket **(see illustration)** and remove the hose from the bracket.

5   Disconnect the hose from the caliper, discarding the copper washers on either side of the fitting.

6   Using new copper washers, attach the new brake hose to the caliper, tightening the banjo bolt to the torque listed in this Chapter's Specifications.

7   Pass the female fitting through the frame or frame bracket. With the least amount of twist in the hose, install the fitting in this position. **Note:** *The weight of the vehicle must be on the suspension, so the vehicle should not be raised while positioning the hose.*

8   Install the U-clip (lock) in the female fitting at the frame bracket.

9   Attach the brake line to the hose fitting using a back-up wrench on the fitting.

10   Carefully check to make sure the suspension or steering components don't make contact with the hose. Have an assistant push on the vehicle and also turn the steering wheel lock-to-lock during inspection.

11   Bleed the brake system as described in Section 10.

Metal brake lines

12   When replacing brake lines, be sure to use the correct parts. Don't use copper tubing for any brake system components. Purchase prefabricated steel brake lines, with the tube ends already flared and fittings installed, from a dealer parts department or auto parts store. These lines are also sometimes bent to the proper shapes, but if you purchase straight steel tubing, be sure to use a bending tool to make kink-free bends.

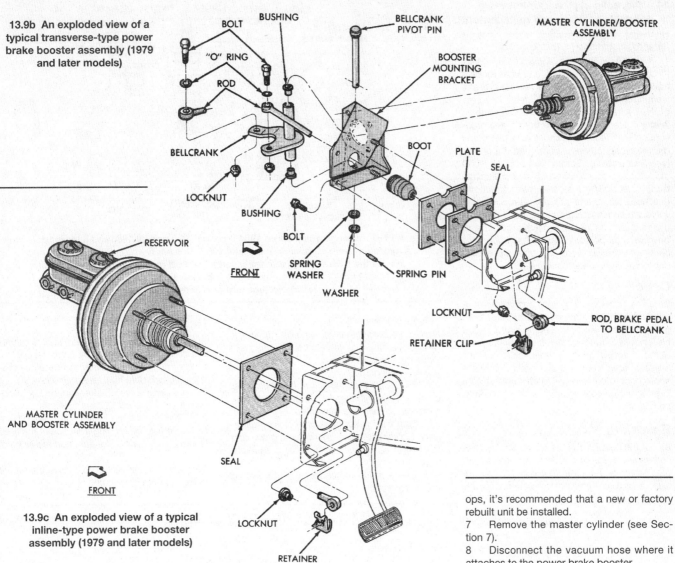

13.9b An exploded view of a typical transverse-type power brake booster assembly (1979 and later models)

13.9c An exploded view of a typical inline-type power brake booster assembly (1979 and later models)

13   When installing the new line make sure it's well supported in the brackets and has plenty of clearance between moving or hot components.
14   After installation, check the master cylinder fluid level and add fluid as necessary. Bleed the brake system as outlined in Section 10 and test the brakes carefully before placing the vehicle into normal operation.

13 Power brake booster - check, removal and installation

Operating check

1   Depress the brake pedal several times with the engine off and make sure that there is no change in the pedal reserve distance.
2   Depress the pedal and start the engine. If the pedal goes down slightly, operation is normal.

Airtightness check

3   Start the engine and turn it off after one or two minutes.
Depress the brake pedal several times slowly. If the pedal goes down farther the first time but gradually rises after the second or third depression, the booster is airtight.
4   Depress the brake pedal while the engine is running, then stop the engine with the pedal depressed. If there is no change in the pedal reserve travel after holding the pedal for 30 seconds, the booster is airtight.

Removal

Refer to illustrations 13.9a, 13.9b and 13.9c
5   The power brake booster unit requires no special maintenance apart from periodic inspection of the vacuum hose and the case.
6   Disassembly of the power unit requires special tools and is not ordinarily performed by the home mechanic. If a problem devel-

ops, it's recommended that a new or factory rebuilt unit be installed.
7   Remove the master cylinder (see Section 7).
8   Disconnect the vacuum hose where it attaches to the power brake booster.
9   Working under the dash, disconnect the brake pedal pushrod from the top of the brake pedal by removing the locknut (1971 through 1978 models) or by prying off the retainer clip (see illustrations).
10   Remove the nuts attaching the booster to the firewall.
11   Carefully lift the booster unit away from the firewall and out of the engine compartment.

Installation

12   To install the booster, place it into position and tighten the retaining nuts to the torque listed in this Chapter's Specifications. Connect the brake pedal. **Warning:** Use a new clip.
13   Install the master cylinder and vacuum hose. If the brake lines had been removed from the master cylinder, bleed the brake system (see Section 10).
14   Carefully test the operation of the brakes before placing the vehicle in normal operation.

9

14 Hydraulic system control valve - check, resetting and replacement

Refer to illustrations 14.1a and 14.1b

1   All 1971 through 1983, some 1984 through 1990 and all 1991 and later models have a hydraulic control valve with a brake warning switch which is activated if one side of the hydraulic system fails, and a hold-off (metering) valve to limit the front brake pressure until the rear brake shoes have overcome the return springs and contacted the drums **(see illustration)**. Some 1984 through 1990 models use a combination valve **(see illustration)** that performs the two functions described above and includes a proportioning valve which transmits full input pressure to the rear brakes up to a certain point, and beyond that point reduces the amount of pressure increase to the rear brakes in accordance with a predetermined ratio.

Check

2   The brake warning light normally comes on during the engine start sequence, then goes off again when the engine starts. It can be checked functionally by opening one bleeder screw slightly while an assistant presses down on the brake pedal. The light should come on with the ignition switch On. Tighten the bleeder again as soon as the light comes on.

Resetting

3   If the light for the brake warning system comes on, a leak or problem has occurred in the system and must be corrected. Once the leak or problem has been repaired or if the hydraulic system has been opened up for brake cylinder overhaul or a similar repair, the pressure differential valve must be centered. Once the valve is centered, the brake light on the dash will go out.

4   To center the valve, first fill the master cylinder reservoir and make sure the hydraulic system has been bled.

5   Turn the ignition switch to On or Accessory. Slowly press the brake pedal down and the piston will center itself, causing the light to go out.

6   Check the brake pedal for firmness and proper operation.

Replacement

7   If a pressure differential valve is defective or if it is leaking, it must be replaced. It is a non-serviceable unit and no repair operations are possible by the home mechanic.

8   Disconnect the brake warning light connector from the warning light switch.

9   Disconnect the brake line unions from the valve assembly. Plug the ends of the lines to prevent loss of brake fluid and the entry of dirt.

10   Remove the bolts and nuts securing the valve assembly to the chassis.

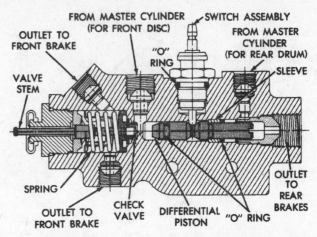

14.1a All 1971 through 1983, some 1984 through 1990 and all 1991 and later models have a hydraulic control valve with a brake warning switch which is activated if one side of the hydraulic system fails, and a hold-off (metering) valve to limit the front brake pressure until the rear brake shoes have overcome the return springs and contacted the drums

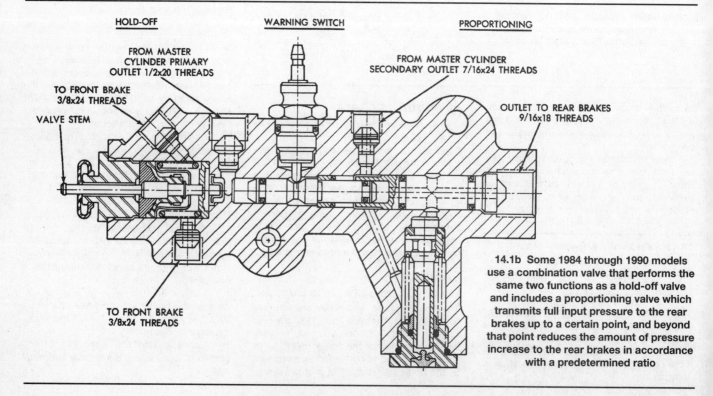

14.1b Some 1984 through 1990 models use a combination valve that performs the same two functions as a hold-off valve and includes a proportioning valve which transmits full input pressure to the rear brakes up to a certain point, and beyond that point reduces the amount of pressure increase to the rear brakes in accordance with a predetermined ratio

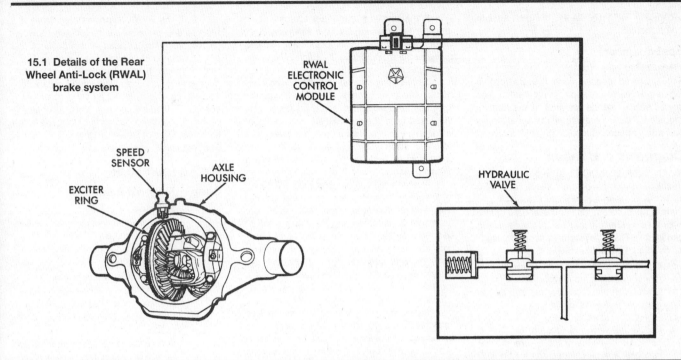

15.1 Details of the Rear Wheel Anti-Lock (RWAL) brake system

11   Remove the valve assembly and bracket.
12   Installation is the reverse of removal.
13   Refer to Section 10 and bleed the system after the replacement valve has been installed.
14   Center the valve as described in Steps 3 through 6.

15   Rear Wheel Anti-Lock (RWAL) brake system - general information

Refer to illustration 15.1
1   The Rear Wheel Anti-Lock (RWAL) brake system (see illustration) was introduced on 1990 models. It is designed to maintain vehicle maneuverability, directional stability and optimum deceleration under severe braking conditions on most road surfaces. RWAL does so by monitoring the rotational speed of the rear wheels and controlling the brake line pressure to the rear wheels while braking. This prevents the rear wheels from locking up during hard braking, regardless of the payload.

Components
Hydraulic valve
Refer to illustration 15.2
2   The hydraulic valve (see illustration), which consists of a dump valve and an isolation valve, is located on the driver's side frame rail above the rear axle. The valve operates by changing the rear brake fluid pressure in response to signals from the control module.

Control module
Refer to illustration 15.3
3   The RWAL electronic control module (see illustration) is mounted under the dash on the passenger's side (pull the glove box open past its stops for access). The function of the control module is to accept and process information received from the speed sensor and signal the hydraulic valve to control the hydraulic line pressure, avoiding wheel lock-up. The control module also constantly monitors the system, even under normal driving conditions, to find faults within the system.

Warning light
4   If a problem develops within the system, the ANTI-LOCK warning light will glow on the dashboard. A diagnostic code will also be stored, which, when retrieved by a service

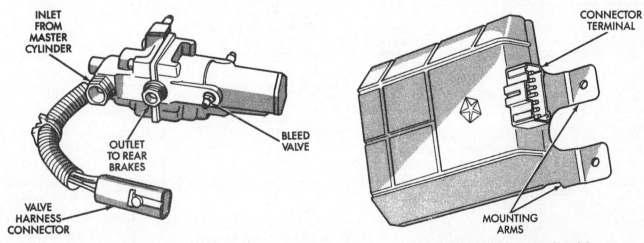

15.2 RWAL hydraulic valve

15.3 RWAL electronic control module

9

technician, will indicate the problem area or component.

Speed sensor

Refer to illustration 15.5

5   A speed sensor **(see illustration)** is located at the top of the rear differential. The speed sensor sends a signal to the control module indicating rear wheel rotational speed and rate of deceleration.

Diagnosis and repair

6   If the ANTI-LOCK warning light on the dashboard comes on and stays on, make sure the parking brake is not applied and there's no problem with the standard brake hydraulic system. If neither of these is the cause, the RWAL system is probably malfunctioning. Check the following:

a) *Make sure the brakes, calipers and wheel cylinders are in good condition.*
b) *Check the electrical connectors at the control module assembly.*
c) *Check the fuses.*
d) *Follow the wiring harness to the speed sensor and valve and make sure all connections are secure and the wiring isn't damaged.*

7   If the above preliminary checks don't identify the problem, find the RWAL Service Diagnostic connector, usually in the harness going to the module (see the *Wiring Diagrams* at the end of Chapter 12).

8   Temporarily ground the diagnostic connector and watch the amber RWAL warning light on the dashboard. Fault codes will be displayed (and the red BRAKE warning light will also go on) by a long flash, followed by a series of short flashes. Count the long and short flashes and compare to the following chart for diagnosis.

9   After the malfunction has been identified and repaired (some codes will require dealership repairs), the codes can be cleared from memory by disconnecting the battery for 30 seconds. **Note:** *Two different control modules are used on the 1992 RWAL system. Early models use the Type 1 module; All remaining production models use the type II module. The two modules can be distinguished by the manner in which they process fault codes 9 and 11. If the vehicle has a Type I module, these two codes will not be cleared by turning off the ignition; if the vehicle has a Type II module, they will be cleared when the ignition is turned off.*

10   Pull the RWAL harness plug from the module and measure the resistance between pin 13 (white/violet speed sensor ground) and pin 14 (red/violet speed sensor signal). If the resistance is greater than 2400 ohms, disconnect the harness at the speed sensor.

11   Check the resistance of the speed sensor itself. If it's less than 2400 ohms, the problem could be spreading of the pins in the harness. Do not attempt to bend the pins, see your dealer for a replacement harness.

RWAL fault code

Code	Failure detected
1	Not used
2	Open isolation valve wiring or bad control module
3	Open dump valve wiring or bad control module
4	Closed RWAL valve switch
5	Over 16 dump pulses generated in 2WD vehicles (disabled for 4WD)
6	Erratic speed sensor reading while rolling
7	Electronic control module fuse pellet open, isolation output missing, or valve wiring shorted to ground
8	Dump output missing or valve wiring shorted to ground
9	Speed sensor wiring/resistance (usually high reading)
10	Sensor wiring/resistance (usually low reading)
11	Brake switch always on, RWAL light comes on when speed exceeds 40 mph
12	Not used
13	Electronic control module phase lock loop failure
14	Electronic control module program check failure
15	Electronic control module RAM failure

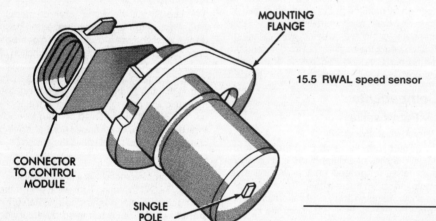

15.5 RWAL speed sensor

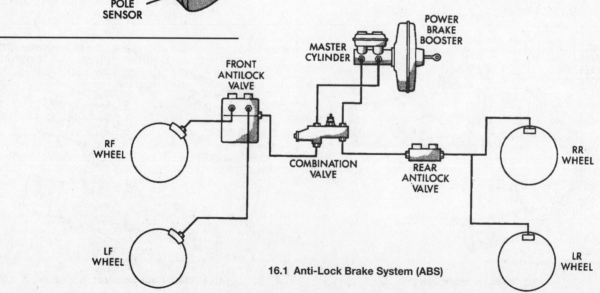

16.1 Anti-Lock Brake System (ABS)

16 Anti-lock Brake System (ABS) - general information

Description

Refer to illustration 16.1

Some 1994 and later models have a four-wheel Anti-lock Brake System (ABS) **(see illustration)** which maintains vehicle maneuverability, directional stability, and optimum deceleration under severe braking conditions on most road surfaces. It does so by monitoring the rotational speed of the wheels and controlling the brake line pressure to the wheels during braking. This prevents the wheels from locking up during hard braking.

Components

Hydraulic control unit

Refer to illustration 16.2

The hydraulic control unit **(see illustration)**, which contains the front anti-lock valve assembly and the pump/motor unit, is mounted on a bracket and plate on the driver's side inner fender panel just below the battery tray.

The front brake anti-lock valve, which consists of a solenoid valve body, provides two-channel pressure control of the front brakes. Each front brake is independently controlled. One channel controls the left front brake; the other channel controls the right.

When the ABS system is in operation, the solenoid valves are cycled open and closed quickly and continuously to modulate brake fluid pressure and control wheel lock-up and deceleration.

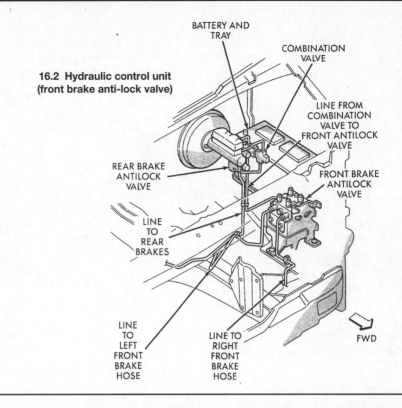

16.2 Hydraulic control unit (front brake anti-lock valve)

Controller Anti-lock Brake (CAB)

The CAB (the module) is also under the battery and battery tray, right next to the front anti-lock valve. The function of the CAB is to monitor the ABS system and control the anti-lock valve solenoids. It accepts and processes information received from the brake switch and wheel speed sensors to control the hydraulic line pressure and avoid wheel lock up. It also monitors the system and stores fault codes which indicate specific problems.

Pump/motor

The pump is operated by a DC-type motor controlled by the CAB. The pump supplies extra hydraulic fluid needed during anti-lock braking.

Wheel Speed Sensor (WSS)

Refer to illustrations 16.6a and 16.6b

A wheel speed sensor is mounted on the steering knuckle of each front wheel **(see illustration)**. A third sensor is mounted on

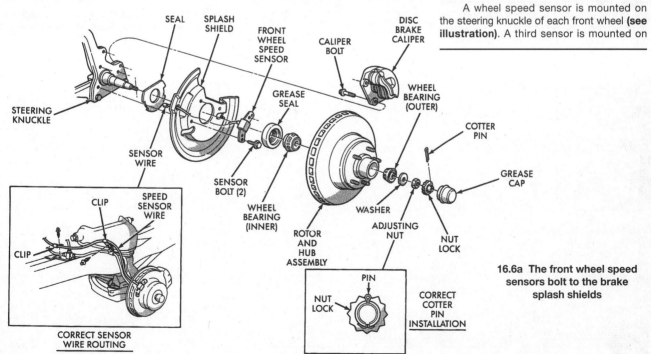

16.6a The front wheel speed sensors bolt to the brake splash shields

9

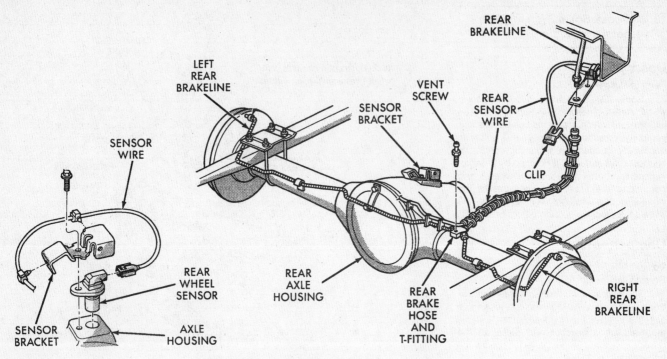

16.6b   Rear wheel speed sensor assembly details

top of the rear differential housing (see illustration). Each sensor assembly consists of a magnetic coil mounted adjacent to a "tone wheel" (front wheels) or "exciter ring" (rear) with an air gap between them. A sensor measures wheel speed by monitoring the rotation of the tone wheel/exciter ring. As the teeth of the tone wheel/exciter ring move through the magnetic field of the sensor, an AC voltage is generated. This signal frequency increases or decreases in proportion to the speed of the wheel. The CAB monitors these three signals for changes in wheel speed; if it detects the sudden deceleration of a wheel, i.e. wheel lockup, the CAB activates the ABS system.

Diagnosis and repair

The ABS system has self-diagnostic capabilities. Each time the vehicle is started, the system runs a self-test. The amber ABS warning light comes on during the test for about three seconds. If the amber warning light comes on and stays on during vehicle operation, or if it comes on intermittently, there may be a fault in the ABS system. However, the main brake system will function normally. Faults in the main brake system are indicated by a red warning light.

Although a special electronic tester is necessary to properly diagnose the system, the home mechanic can perform a few preliminary checks before taking the vehicle to a

dealer service department which is equipped with this tester.

a) *Make sure the brake calipers are in good condition.*
b) *Check the 60-way electrical connector at the controller.*
c) *Check the fuses.*
d) *Follow the wiring harness to the speed sensors and brake light switch and make sure all connections are secure and the wiring isn't damaged.*

If the above preliminary checks don't rectify the problem, the vehicle should be diagnosed by a dealer service department.

Chapter 10
Suspension and steering systems

Contents

Specifications

Torque Specifications

Front Suspension	Ft-lbs (unless otherwise indicated)
Lower balljoint nut	
11/16-18	135
3/4-16	175
Lower control arm bolt/nut	175
Shock absorber	
Upper nut	25
Lower bolts	200 in-lbs
Stabilizer bar	
Stabilizer-to-link nut	100 in-lbs
Stabilizer bushing clamp bolts	200 in-lbs
Upper control arm bolt/nut	210
Strut rod	
Rear nut	52
Front nuts/bolts	100
Upper balljoint	125
Upper balljoint nut	135

Rear suspension	
Shock absorber bolts/nuts	55
Leaf springs	
Center bolt nuts	15
Front eye-bolt nut*	
1971 through 1973	125
1974 through 1978	150
1979 through 1981	125
1982 on	100
Rear shackle-bolt nuts	
1971 through 1981	
Regular springs	40
Heavy-duty springs	155
1982 on	
Regular springs/two-piece shackle	35
Heavy-duty springs/one-piece shackle	155
U-bolt nuts	
1971 through 1973	65
1974 on	
100/150/200/250 models	45
300/350 models	110

*Referred to as a pivot bolt/nut on later models.

10

Steering

Drag link-to-center link nut	55
Idler arm-to-frame nut/bolt	
1971 through 1978	90
1979 through 1991	70
1992 and 1993	105
1994 on	115
Idler arm-to-center link nuts	47
Pitman arm-to-drag link nut	55
Pitman arm-to-steering gear nut	
1971 through 1991	175
1992 and 1993	225
1994 on	185
Tie-rod end-to-steering knuckle and tie-rod end-to-center link nuts	
1971 through 1976	
1/2 X 20	45
9/16 X 18	55
1977 and 1978	
1/2 X 20	47
9/16 X 20	55
5/8 X 18	75
1979 through 1989	
9/16 X 20	55
5/8 X 18	75
1990 on	
9/16 X 18	55
5/8 X 18	75
Tie-rod adjuster clamp nuts	
1971 through 1981	
Heavy duty	26
Except heavy duty	156 in-lbs
1982 on	
Heavy duty	26
Except heavy duty	200 in-lbs
Steering gear mounting bolts	100
Steering wheel-to-steering shaft nut	
1971 through 1977	24
1978 through 1981	60
1982 through 1997	45
1998 on	35
Wheel lug nuts	See Chapter 1

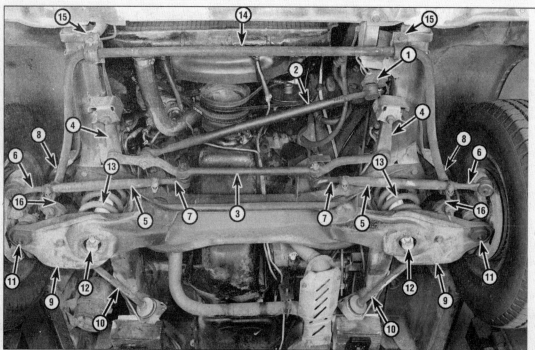

1.1 Front suspension and steering components

1. Pitman arm
2. Drag link
3. Center link
4. Idler arm
5. Tie-rod adjuster tube
6. Outer tie-rod
7. Inner tie-rod
8. Upper control arm
9. Lower control arm
10. Strut rod
11. Lower balljoint
12. Shock absorber
13. Coil spring
14. Stabilizer bar
15. Stabilizer bar bracket
16. Stabilizer bar link

1.2 Rear suspension components

1   *Rear axle assembly*
2   *Leaf spring*
3   *Shock absorber-to-axle mounting bracket*
4   *Spring-to-axle tube U-bolt*

1   General information

Front suspension

Refer to illustration 1.1

The models covered in this manual utilize an independent front suspension system with upper and lower control arms, coil springs and shock absorbers **(see illustration)**. The upper control arms are attached to longitudinal frame rails and the lower control arms are attached to a removable crossmember. Both control arms have replaceable bushings on their inner ends and replaceable balljoints on the outer ends. The lower control arms are positioned fore-and-aft by strut rods. On 1971 through 1978 models, an eccentric pivot bolt and two slotted brackets allow the pivot axis of each upper control arm to be changed to adjust caster and camber. On 1979 and later models, the eccentric pivot bolt is replaced by a "pivot bar" with a pair of retaining bolts installed vertically through the bar into slotted brackets; camber and caster are adjusted by repositioning the pivot bar to the left or right in relation to the centerline of the vehicle. Each steering knuckle is located by a pair of balljoints pressed into the upper and lower ends of the control arms. Some 1979 and later models are equipped with a stabilizer bar which is attached by a pair of brackets to the vehicle frame and by link rods to the lower control arms. On such models, body roll during cornering is reduced.

Rear suspension

Refer to illustration 1.2

The rear axle is suspended by two leaf springs **(see illustration)**. The rear axle itself consists of a cast-iron center housing for the differential with a pair of steel tubes for the axleshafts welded to either side; for information on servicing the rear axle, see Chapter 8.

Steering

The power steering system consists of a power steering pump, a recirculating-ball type steering gearbox, the steering linkage and a pair of steering knuckles. A Pitman arm transmits turning force from the steering box to the steering linkage, which consists of a drag link, a center link, a pair of idler arms and two tie-rods. The tie-rods connect the center link to the steering knuckle arms. A small U-joint connects the steering column to the steering gearbox. The steering column is designed to collapse in the event of an accident.

Frequently, when working on the suspension or steering system components, you may come across fasteners which seem impossible to loosen. These fasteners on the underside of the vehicle are continually subjected to water, road grime, mud, etc., and can become rusted or "frozen," making them extremely difficult to remove. In order to unscrew these stubborn fasteners without damaging them (or other components), be sure to use lots of penetrating oil and allow it to soak in for a while. Using a wire brush to clean exposed threads will also ease removal of the nut or bolt and prevent damage to the threads. Sometimes a sharp blow with a hammer and punch is effective in breaking the bond between a nut and bolt threads, but care must be taken to prevent the punch from slipping off the fastener and ruining the threads. Heating the stuck fastener and surrounding area with a torch sometimes helps too, but isn't recommended because of the obvious dangers associated with fire. Long breaker bars and extension, or "cheater," pipes will increase leverage, but never use an extension pipe on a ratchet - the ratcheting mechanism could be damaged. Sometimes, turning the nut or bolt in the tightening (clockwise) direction first will help to break it loose. Fasteners that require drastic measures to unscrew should always be replaced with new ones.

Since most of the procedures that are dealt with in this chapter involve jacking up the vehicle and working underneath it, a good pair of jackstands will be needed. A hydraulic floor jack is the preferred type of jack to lift the vehicle, and it can also be used to support certain components during various operation. **Warning 1:** *Never, under any circumstances, rely on a jack to support the vehicle while working on it.* **Warning 2:** *Whenever any of the suspension or steering fasteners are loosened or removed they must be inspected and, if necessary, be replaced with new ones of the same part number or of original equipment quality and design. Torque specifications must be followed for proper reassembly and component retention. Never attempt to heat or straighten any suspension or steering components. Instead, replace any bent or damaged part with a new one.*

2   Stabilizer bar and bushings - removal and installation

Removal

Refer to illustrations 2.2 and 2.3

1   Apply the parking brake. Raise the front of the vehicle and support it securely on jackstands.

2   Remove the nuts that attach the stabilizer bar to the link rods **(see illustration)**. If it is necessary to remove the links themselves (you're going to replace the lower control arm, for instance), remove the nuts from the lower end of the links instead.

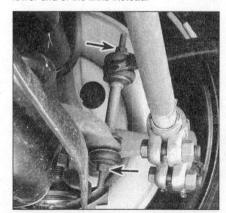

2.2 To disconnect the stabilizer bar from the link rods, remove the upper nuts (arrow); to disconnect the links themselves from the lower control arm, remove the lower nuts (arrow) (left link assembly shown)

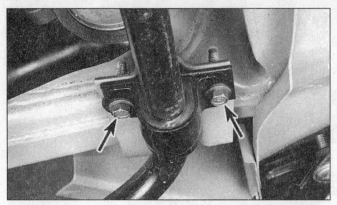

2.3 To detach the stabilizer bar from the frame, remove these bolts (arrows) from both brackets (left bracket shown)

3.3 To disconnect the lower end of the shock absorber from the lower control arm, remove these two bolts (arrows)

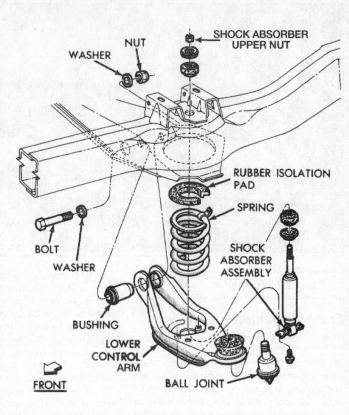

3.2 An exploded view of the shock absorber, coil spring and lower control arm assemblies

3   Remove the stabilizer bar-to-frame mounting brackets **(see illustration)**.
4   Remove the stabilizer bar from the vehicle.
5   Pull the bushings off the stabilizer bar and inspect them for cracks, hardness and other signs of deterioration. If the bushings are damaged, replace them. Inspect the bushings in the lower ends of the links, replacing them if necessary.

Installation

6   Position the stabilizer bar bushings on the bar.
7   Push the brackets over the bushings and raise the bar up to the frame. Install the bracket bolts but don't tighten them completely at this time.
8   Attach the stabilizer to the link bolts (or the links to the control arms), and tighten the nuts to the torque listed in this Chapter's Specifications.
9   Tighten the bracket bolts to the torque listed in this Chapter's Specifications.

3   Shock absorber (front) - removal and installation

Removal

Refer to illustrations 3.2 and 3.3
1   Loosen the wheel lug nuts, raise the

vehicle and support it securely on jackstands. Apply the parking brake. Remove the wheel.
2   Remove the shock absorber upper nut **(see illustration)**. Use an open end wrench to keep the stem from turning. If the nut won't loosen because of rust, squirt some penetrating oil on the stem threads and allow it to soak in for awhile. It may be necessary to keep the stem from turning with a pair of locking pliers, since the flats provided for a wrench are quite small.
3   Remove the two shock absorber-to-lower control arm mounting bolts **(see illustration)** and pull the shock absorber out from the wheel well. Remove the washers and the rubber grommets from the top of the shock absorber.

Installation

4   Extend the new shock absorber as far as possible. Position a new washer and rubber grommet on the stem and guide the shock up into the upper mount.
5   Install the upper rubber grommet and washer and wiggle the stem back-and-forth to ensure the grommets are centered in the mount. Tighten the stem nut to the torque listed in this Chapter's Specifications.
6   Install the lower mounting bolts and tighten them to the torque listed in this Chapter's Specifications.

4   Balljoints - check and replacement

Check

Upper balljoint

1   Loosen the wheel lug nuts, raise the front of the vehicle and support it securely on jackstands. Remove the wheel.
2   Place a floor jack under the lower control arm and raise it slightly. Using a large screwdriver or prybar, pry up on the upper control arm and watch for movement at the balljoint. Any movement indicates a worn balljoint. Now grasp the steering knuckle and attempt to move the top of it in-and-out - if any play is felt, the balljoint will have to be replaced.

Lower balljoint

3   Remove the floor jack from under the lower control arm.
4   Grasp the the bottom of the tire and attempt to move it in-and-out - if any play is felt, the balljoint will have to be replaced.

Replacement

5   Loosen the wheel lug nuts, raise the front of the vehicle and support it securely on jackstands. Remove the wheel.

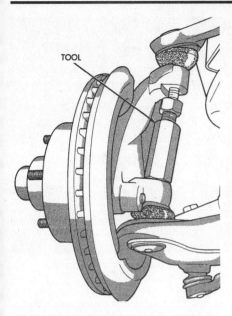

TOOL

4.9a The best way to separate a balljoint stud from the steering knuckle is with this special tool - this illustration shows the proper setup for upper balljoint separation (the tool must be inverted for lower balljoint separation)

Upper balljoint

Refer to illustrations 4.9a, 4.9b, 4.9c and 4.9d

6   Remove the brake caliper and hang it out of the way with a piece of wire (see Chapter 9).

7   Remove the cotter pin from the upper balljoint stud and back off the nut several turns.

8   Place a floor jack under the outer end of the lower control arm. **Warning:** *The jack must remain under the control arm during removal and installation of the balljoint to hold the spring and control arm in position.*

9   Separate the balljoint from the steering knuckle as follows:

a) *Remove the cotter pin and loosen the nut on the upper balljoint stud a few turns, but don't remove it.*

b) *Position a balljoint stud remover over the lower balljoint stud, allowing the tool to rest on the knuckle arm* **(see illustrations)**. *Set the tool nut securely against the upper stud. Tighten the tool to apply pressure to the upper stud and strike the knuckle sharply with a hammer to loosen the stud. Don't try to force out the stud with the tool alone.*

c) *If you're not worried about damaging the balljoint boot, i.e. you're planning to replace the balljoint anyway, it's okay to use a picklefork type balljoint separator* **(see illustration)**.

d) *You can also loosen the balljoint stud nut and give the knuckle or steering arm a few sharp raps with a hammer* **(see illustration)**, *but we don't recommend this method unless you're working on a vehicle which makes the use of any other means difficult because of the shape of the knuckle, its close proximity to the ballstud nut, etc.*

10   If the upper balljoint has a hex on top, unscrew it from the control arm. After you remove the balljoint, clean the balljoint bore in the control arm. If the balljoint is pressed in, i.e. there is no hex on top with which to unscrew it, remove the upper control arm (see Section 8) and have the old balljoint pressed out and a new unit pressed in at an automotive machine shop. **Note:** *If you can rent or borrow a balljoint removal/installation press, the control arm doesn't have to be removed from the vehicle. Follow the tool manufacturer's instructions.*

Lower balljoint

Refer to illustrations 4.12

11   Remove the cotter pin from the lower balljoint stud and back off the nut several turns.

12   Place a floor jack under the lower control arm. Remove the shock absorber and attach a spring compressor to the coil spring.

Separate the lower balljoint stud from the steering knuckle **(see illustration 4.9a through 4.9d)** and lower the control arm slowly to prevent the coil spring from flying out. Remove the spring and set it aside.

13   Remove the lower control arm (see Section 8), then take the arm to a dealer service department or other repair shop to have the old balljoint pressed out and the new one installed. **Note:** *If you can rent or borrow a balljoint removal/installation press, the coil spring and control arm don't have to be removed from the vehicle. Follow the tool manufacturer's instructions.*

All balljoints

14   While the balljoint stud is disconnected from the steering knuckle, inspect the tapered holes in the steering knuckle. Remove any accumulated dirt. If out-of-roundness, deformation or other damage is noted, the steering knuckle must be replaced with a new one.

15   Inspect the balljoint boot(s) for cracks and tears. If a boot is damaged, replace it. You can install a new balljoint boot and retainer using a large socket and a hammer.

16   Installation is the reverse of removal. After the suspension has been reassembled, raise the lower control arm with a floor jack to simulate normal ride height, then tighten all fasteners to the torque listed in this Chapter's Specifications. **Note:** *Raising the suspension prior to tightening the suspension fasteners is only necessary if a control arm has been removed.*

17   If a cotter pin doesn't line up with an opening in the castle nut, tighten (never loosen) the nut just enough to allow installation of the cotter pin.

18   Install the grease fittings and lubricate the new balljoints (see Chapter 1).

19   Install the wheels and lower the vehicle. Tighten the lug nuts to the torque listed in the Chapter 1 Specifications.

20   The front end alignment should be checked, and if necessary, adjusted, by a dealer service department or alignment shop.

10

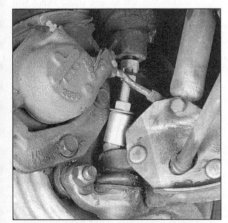

4.9b You can make your own version of the special tool with a large bolt and nut, a large washer and a big socket

4.9c If necessary, use a picklefork to separate the balljoint stud from the steering knuckle (but remember, it will damage the balljoint boot)

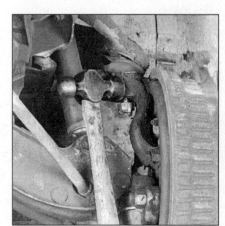

4.9d If you don't have a picklefork tool, you can also rap the knuckle sharply with a hammer to loosen the upper balljoint stud

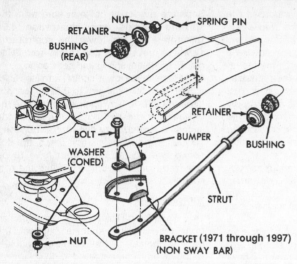

6.2a An exploded view of the strut rod assembly

6.2b Remove the cotter pin and nut (arrow) from the rear end of the strut rod

5 Steering knuckle - removal and installation

Removal

1　Loosen the wheel lug nuts, raise the vehicle and support it securely on jackstands. Remove the wheel.
2　Remove the disc brake caliper and disc (see Chapter 9).
3　Disconnect the tie-rod end from the steering knuckle (see Section 12).
4　Support the lower control arm with a floor jack to prevent the coil spring from extending. **Warning:** *The jack must remain in this position throughout the entire procedure.*
5　Disconnect the upper and lower balljoints from the steering knuckle (see Section 4).
6　Carefully check the steering knuckle for cracks, especially around the steering arm and spindle mounting area. Check for elongated balljoint stud holes. Replace the steering knuckle if any of these conditions are found.

Installation

7　Position the steering knuckle on the upper and lower balljoints and push the

balljoint studs into the holes in the knuckle. Install the nuts and tighten them to the torque listed in this Chapter's Specifications. Use new cotter pins to secure the nuts.
8　Install the brake disc and caliper (see Chapter 9).
9　Connect the tie-rod end to the steering knuckle (see Section 12).
10　Install the wheel and lug nuts. Lower the vehicle and tighten the lug nuts to the torque listed in the Chapter 1 Specifications.

6 Strut rod - removal and installation

Refer to illustrations 6.2a, 6.2b and 6.3

1　Loosen the wheel lug nuts, raise the front of the vehicle and support it securely on jackstands. Remove the wheel.
2　Remove the cotter pin and nut from the rear end of the strut rod **(see illustrations)**.
3　Remove the two nuts and bolts that attach the forward end of the strut to the lower control arm **(see illustration)**, then remove the strut rod.
4　Inspect the bushings and retainers for wear. Replace as necessary.
5　Installation is the reverse of removal. Be sure to tighten all fasteners to the torque listed in this Chapter's Specifications.

7 Coil springs - removal and installation

1　Loosen the front wheel lug nuts, raise the front of the vehicle and support it securely on jackstands. Remove the wheels.
2　Using a wood block, support the brake pedal in its full "up" position.
3　Remove the brake caliper assembly (see Chapter 9) and hang it out of the way with a piece of wire. Don't disconnect the brake hose.
4　Remove the shock absorber (see Section 3).
5　Remove the strut rod (see Section 6).

6　Attach a coil spring compressor to the coil spring. These can be obtained at equipment rental yards or at some auto parts stores.
7　Loosen - but don't remove - the lower balljoint nut. Place a floor jack under the lower control arm to control the movement of the coil spring. Carefully separate the balljoint from the steering knuckle (see Section 4) and slowly lower the floor jack until the spring compressor is bearing all of the force of the spring.
8　Remove the coil spring and set it in a safe location.
9　Check the spring for deep nicks and corrosion, which will cause premature failure of the spring. Replace the spring if these or any other questionable conditions are evident.
10　Installation is the reverse of removal. Be sure to tighten all fasteners to the torque listed in this Chapter's Specifications (the wheel lug nut specifications are in Chapter 1).

8 Control arms - removal and installation

1　Loosen the wheel lug nuts, raise the front of the vehicle and support it securely on jackstands. Remove the wheel.

Upper control arm

Refer to illustrations 8.4a, 8.4b and 8.4c

2　Disconnect the upper end of the shock absorber from the upper control arm (see Section 3).
3　Remove the coil spring assembly (see Section 7).
4　On 1971 through 1978 models, remove the eccentric pivot bolts from the control arm and frame **(see illustrations)**. On 1979 and later models, remove the two retaining bolts from the pivot bar **(see illustration)**.
5　Remove the control arm.
6　Check the control arm for distortion and cracks. If the arm is damaged, replace it.
7　Inspect the bushing in the upper control arm for cracking, hardness and general deterioration. If the bushing needs to be replaced,

6.3 Remove these nuts (arrows), pull the bolts up through the top of the lower control arm and remove the strut rod

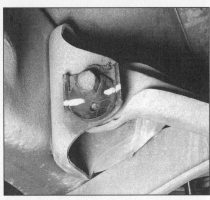

8.4a On 1971 through 1978 models, mark the position of the eccentric adjusters before removing the bolts

take the control arm to a dealer service department or an automotive machine shop and have the old bushing pressed out and a

new bushing pressed in.

8   Installation is the reverse of removal. Be sure to tighten the fasteners to the torque values listed in this Chapter's Specifications.

Lower control arm

Refer to illustration 8.11

9   Disconnect the shock absorber from the lower control arm (see Section 3).

10   Remove the coil spring assembly (see Section 7).

11   Remove the nut and pivot bolt from the inner end of the lower control arm **(see illustration)**.

12   Remove the lower control arm.

13   Inspect the bushing in the lower control arm for cracking, hardness and general deterioration. If the bushing needs to be replaced, take the control arm to a dealer service department or an automotive machine shop and have the old bushing pressed out and a new bushing pressed in.

14   Installation is the reverse of removal. Tighten all fasteners to the torque values listed in this Chapter's Specifications. Tighten the lug nuts to the torque listed in the Chapter 1 Specifications.

9   Shock absorber (rear) - removal and installation

Refer to illustration 9.2
Note: *Remove only one shock absorber at a time*

1   Loosen the wheel lug nuts, raise the vehicle and support it securely on jackstands positioned under the frame rails.

2   Remove the nut and bolt from the upper end of the shock absorber **(see illustration)**. Use a back-up wrench to keep the bolt from turning. If the nut won't loosen because of rust, squirt some penetrating oil on the threads and allow it to soak in for awhile.

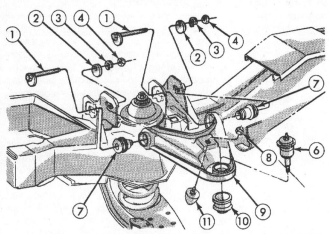

8.4b An exploded view of the upper control arm assembly (1971 through 1978 models)

1	Eccentric cam and bolt assembly	6	Upper balljoint
2	Eccentric cam	7	Bushing
3	Lock washer	8	Locknut
4	Nut	9	Upper control arm
5	not used	10	Upper balljoint seal
		11	Bumper

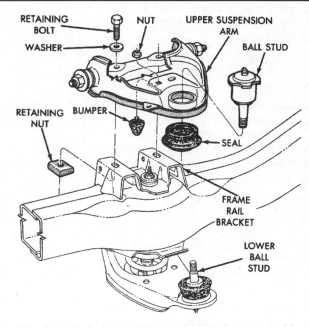

8.4c An exploded view of the upper control arm assembly (1979 and later models)

8.11 To detach the lower control arm from the frame, remove this nut and pull out the bolt

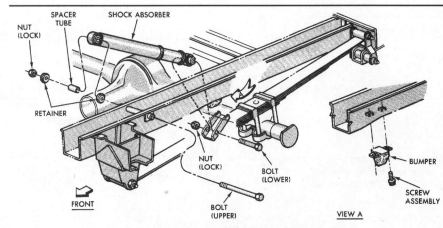

9.2 Rear shock absorber mounting details

10

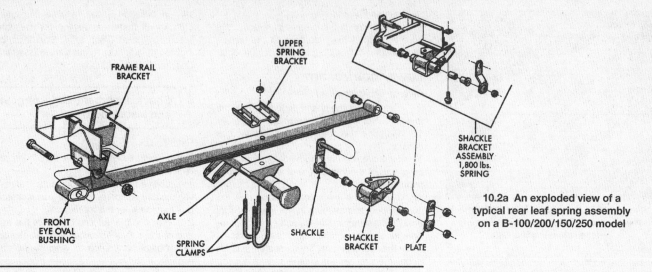

10.2a An exploded view of a typical rear leaf spring assembly on a B-100/200/150/250 model

3   Remove the lower mounting nut and bolt and remove the shock absorber.
4   Extend the new shock absorber as far as possible.
5   Installation is the reverse of removal. Be sure to tighten the upper and lower shock fasteners to the torque listed in this Chapter's Specifications.

10   Leaf springs and bushings - removal and installation

Removal

Refer to illustrations 10.2a and 10.2b

1   Loosen the wheel lug nuts, raise the rear of the vehicle and support it securely on jackstands placed under the frame rails. Remove the wheel and support the rear axle with a floor jack placed under the axle tube on the side being worked on. Don't raise the axle - just support its weight.
2   Remove the nuts, lock washers and U-bolts that attach the leaf spring to the axle **(see illustrations)**.
3   Remove the nut and bolt attaching the rear of the spring to the shackle.
4   Remove the front nut and bolt and

remove the leaf spring assembly.
5   Inspect the bushings at the ends of the spring. If they're cracked or torn, remove them with a bushing driver and install new ones, or take the spring to an automotive machine shop and have the job done there.

Installation

6   Position the spring on the axle tube. Make sure that the spring center bolt engages the locating hole on the axle tube.
7   Line up the spring front eye with the mounting bracket and install the front bolt and nut.
8   Install the rear shackle assembly.
9   Tighten the shackle bolt and the front bolt until all slack is taken up.
10   Install the U-bolts, new lock washers and nuts. Tighten the nuts until they push the lock washers against the spring bracket.
11   Install the wheel and lug nuts. Remove the jackstands and lower the vehicle. Tighten the lug nuts to the torque listed in this Chapter's Specifications.
12   Tighten the U-bolts, pivot bolts and shackle bolts to the torque listed in this Chapter's Specifications.

11   Steering wheel - removal and installation

Warning: *1995 and later models covered by this manual are equipped with a Supplemental Restraint System (SRS), more commonly known as an airbag(s). All 1995 and later models are equipped with a driver side airbag and all 1998 and later models are equipped with a passenger side airbag, located in the instrument panel. Always disconnect the negative battery cable and wait at least two minutes before working in the vicinity of any airbag system component to avoid the possibility of accidental deployment of the airbag, which could cause personal injury (see Chapter 12, Section 28). Do not use electrical test equipment on the airbag system wiring or tamper with it in any way.*

Removal

Refer to illustrations 11.2a, 11.2b, 11.3a, 11.3b, 11.4a, 11.4b, 11.4c, 11.4d, 11.5 and 11.6

1   Disconnect the cable from the negative battery terminal and, if equipped with an airbag, wait two minutes.

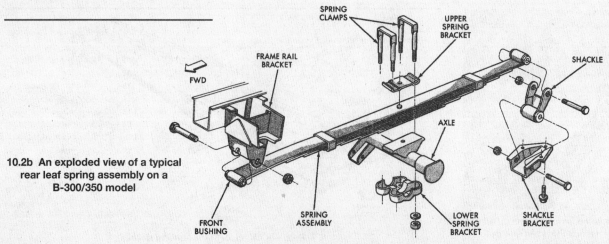

10.2b An exploded view of a typical rear leaf spring assembly on a B-300/350 model

11.2a An exploded view of a typical early steering wheel assembly; to detach the horn pad on one of these models, carefully pry it off with a small screwdriver

SCREW–(2)

PLATE, PAD RETAINER

STEERING SHAFT SPLINES

SCREW–(3)

NUT

SCREW–(3)

HORN PAD

SWITCH ASSEMBLY

WASHER, PLAIN

HORN BUTTON ASSEMBLY

STEERING WHEEL ASSEMBLY

BUTTON RETAINER

COVER (WITH SPEED CONTROL SWITCHES)

STEERING WHEEL

HORN RELAY GROUND WIRE

STEERING SHAFT

NUT & WASHER

SOURCE FOR HORN RELAY GROUND

SPEED CONTROL WIRE HARNESS

COVER RETAINING SCREW

11.2b An exploded view of a typical later-model steering wheel assembly; to detach the horn pad on one of these models, remove the pad retaining screws from the side of the wheel facing the instruments

BRACE

ENGINE COVER

STEERING COLUMN COVER

11.3a On airbag-equipped models this under-dash trim panel must be removed to get to the main electrical connector for the system (early model shown)

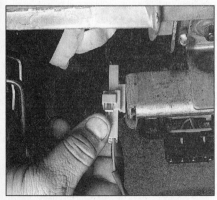

11.3b After removing the under-dash trim panel, unplug the big yellow electrical connector - the airbag system is now disabled

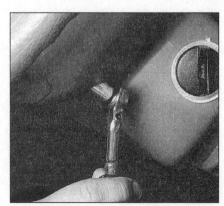

11.4a To remove the airbag module, remove the four screws (two on each side) from the backside of the steering wheel

2   On some earlier models without an airbag, you can simply pry off the horn pad **(see illustration)**. On these models, remove the horn button by turning it counterclockwise. On other older models, remove the horn pad from the retainer by taking out the two screws from underneath **(see illustration)**.

3   On models with an airbag, remove the trim panel from the lower left side of the dash, and unplug the yellow electrical connector for the airbag module **(see illustrations)**.

4   On models with an airbag, remove the four bolts that attach the module to the steering wheel **(see illustration)**. Lift off the module and unplug the electrical connector for the module and the horn switch wire **(see illustrations)**. **Warning:** Set the airbag module down with the upholstered side facing up.

10

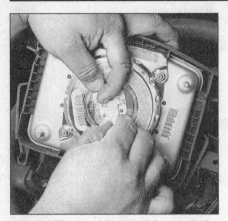

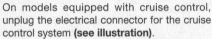

11.4b Flip the airbag module over and unplug the module electrical connector

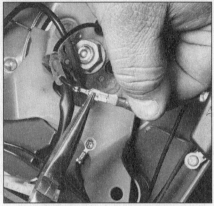

11.4c Disconnect the horn switch wire

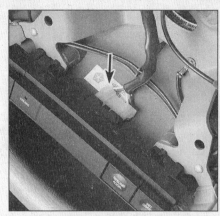

11.4d If the vehicle is equipped with cruise control, unplug this electrical connector (arrow)

On models equipped with cruise control, unplug the electrical connector for the cruise control system (**see illustration**).

5   On all models, remove the steering wheel retaining nut and mark the position of the steering wheel to the shaft (**see illustration**), unless marks already exist.

6   On all models, use a puller to detach the steering wheel from the shaft (**see illustration**). Don't hammer on the shaft to dislodge the wheel.

Installation

Refer to illustration 11.7

7   On models with an airbag, make sure the clockspring is centered properly. **Warning:** *Failure to center the clockspring could prevent the airbag from deploying in an accident.* To center the clockspring:

   a) *Place the front wheels in the straight-ahead position.*

   b) *Depress the two locking tabs to disengage the locking mechanism* (**see illustration**).

   c) *With the mechanism disengaged, rotate the clockspring rotor in a clockwise direction until it reaches the end of its travel (but don't try to wind it up beyond the point at which it takes up all the slack, or you will damage the mechanism).*

   d) *From the end of its travel, rotate the rotor 2-1/2 full turns back (counterclockwise). The horn wire should end up at the top and the airbag module wire at the bottom.*

8   To install the wheel, align the mark on the steering wheel hub with the mark on the shaft and slide the wheel onto the shaft. Install the nut and tighten it to the torque listed in this Chapter's Specifications.

9   Reattach the horn wire, plug in the electrical connector for the cruise control (if equipped) and, on models with an airbag, plug in the electrical connector to the airbag module.

10   Install the horn pad or airbag module.

11   On models with an airbag, plug in the yellow electrical connector under the dash and install the under-dash trim panel.

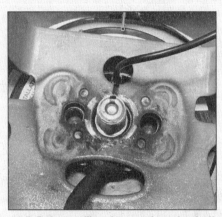

11.5 Before pulling the steering wheel off the steering shaft, mark the alignment of the wheel to the shaft to insure proper realignment during reassembly

12   Connect the cable to the negative terminal of the battery.

13   When the engine is started on airbag-equipped models, the airbag system warning light should come on only momentarily. If the light comes on and remains on after the engine is started, take the vehicle to a dealer to have the system checked and reset. Because of the special electronic tester required, this job is impossible at home. **Warning:** *Failure to have the airbag system reset could result in the failure of the airbag to deploy in an accident.*

12   Steering linkage - inspection, removal and installation

Inspection

Refer to illustration 12.1

1   The steering linkage (**see illustration**) connects the steering gearbox to the front wheels and keeps the wheels in proper relation to each other. The linkage begins with the Pitman arm, which connects the steering gear shaft to the drag link, which in turn is connected to the center link. The center link, suspended between two idler arms, is connected

11.6 Use a steering wheel puller to remove the wheel - DON'T hammer on the shaft in an attempt to loosen it!

to a pair of tie-rod assemblies, which in turn are connected to the steering knuckles. When the steering wheel is turned, the drag link moves the center link which transmits this motion to the steering knuckles via the tie-rods. Each tie-rod consists of an adjuster tube, an inner and outer tie-rod end and a pair of clamps to hold the three pieces together.

2   Set the wheels in the straight-ahead position and lock the steering wheel.

3   Raise one side of the vehicle until the tire is approximately one inch off the ground. Grasp the front and rear of the tire and using light pressure, wiggle the wheel back-and-forth. There should be no noticeable play (if there is, make sure the wheel bearings are properly adjusted). If the play in the steering system is noticeable, inspect each steering linkage pivot point and ballstud for looseness and replace parts as necessary.

4   Raise the front of the vehicle and support it on jackstands. Push up, then pull down on the center link near each idler arm with a force of about 25 pounds each way. If there is any looseness at either idler arm, replace that arm.

5   Check for torn ballstud boots, frozen balljoints and bent or damaged linkage components.

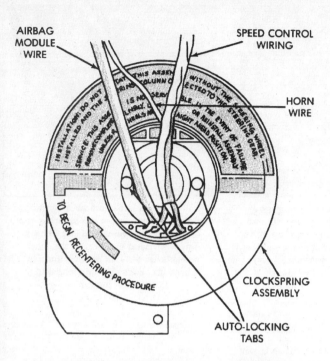

11.7 Details of the airbag system clockspring

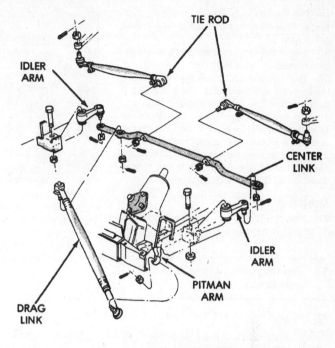

12.1 Steering linkage details

Removal and installation

Tie-rod

Refer to illustration 12.8

6   Loosen the wheel lug nuts, raise the vehicle and support it securely on jackstands. Apply the parking brake. Remove the wheel.

7   Remove the cotter pin and loosen, but do not remove, the castle nut that secures the ballstud to the steering knuckle.

8   Using a small puller, separate the tie-rod end from the steering knuckle **(see illustration)**. Remove the castle nut and pull the tie-rod from the knuckle. **Caution:** *The use of a picklefork-type balljoint separator most likely will cause damage to the balljoint boot.*

9   If a tie-rod end must be replaced, count the number of threads showing and jot down this number to maintain correct toe-in during reassembly. Loosen the adjuster tube clamp and unscrew the tie-rod end.

10   Lubricate the threaded portion of the tie-rod end with chassis grease. Screw the new tie-rod end into the adjuster tube and adjust the distance from the tube to the ballstud by threading the tie-rod into the adjuster tube until the same number of threads are showing as before (the number of threads showing on both sides of the adjuster tube should be within three threads of each other). Don't tighten the adjuster tube clamp yet.

11   To install the tie-rod, insert the ballstud into the steering knuckle. Make sure both the ballstud is fully seated. Install the nut and tighten it to the torque listed in this Chapter's Specifications. If a ballstud spins when attempting to tighten the nut, force it into the tapered hole with a large pair of pliers.

12   Install a new cotter pin. If necessary, tighten the nut slightly to align a slot in the

12.8 Install a small puller and press the ballstud out of the steering knuckle

nut with the hole in the ballstud.

13   Tighten the adjuster tube clamp nut to the torque listed in this Chapter's Specifications. The adjuster tube clamp bolts should be nearly horizontal.

14   Install the wheel and lug nuts, lower the vehicle and tighten the lug nuts to the torque listed in the Chapter 1 Specifications. Drive the vehicle to an alignment shop to have the front end alignment checked and, if necessary, adjusted.

Center link

Refer to illustration 12.16

15   Raise the front of the vehicle and support it securely on jackstands. Apply the parking brake.

16   Remove the cotter pins and loosen, but do not remove, the nuts securing the drag link, inner tie-rod ends and idler arms to the

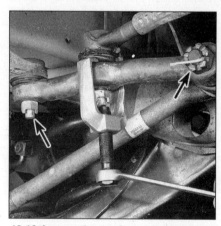

12.16 Loosen the castle nut and separate the drag link (shown) from the center link with a small puller; separate the idler arms and tie-rods from the center link the same way

center link. Separate the center link from the drag link, the inner tie-rod ends and the idler arms with a two-jaw puller **(see illustration)**, then remove the nuts.

17   Installation is the reverse of the removal procedure. If the ballstuds spin when attempting to tighten the nuts, force them into the tapered holes with a large pair of pliers. Be sure to tighten all fasteners to the torque listed in this Chapter's Specifications.

Pitman arm

18   Detach the drag link from the Pitman arm using the same technique described in Step 16. To remove the Pitman arm from the steering gear, refer to Section 13.

10

Idler arms

Refer to illustration 12.20

Note: *The following procedure applies to either idler arm.*

19   Raise the front of the vehicle and support it securely on jackstands.

20   Remove the cotter pin and loosen the castle nut that secures the idler arm ballstud to the center link **(see illustration)**. Attach a small puller (see Step 16) and separate the idler arm from the center link. Remove the nut and bolt from the other end of the idler arm.

21   Installation is the reverse of removal. Be sure to tighten the fasteners to the torque listed in this Chapter's Specifications.

Drag link

22   Use the technique described in Steps 16 and 17 to remove and install the drag link.

---

13   Steering gear - removal and installation

Removal

Refer to illustrations 13.2, 13.3a, 13.3b, 13.4, 13.5 and 13.6

Warning: *Don't allow the steering shaft to turn with the steering wheel removed. If the shaft turns, the airbag coil assembly (the mechanism which protects the airbag wiring when the steering wheel is turned) will become uncentered, which will cause the airbag harness to break when the vehicle is returned to service. To prevent the shaft from turning, turn the ignition key to the Lock position before beginning work or run the seat belt through the steering wheel and clip the seat belt into place.*

1   Raise the front of the vehicle and support it securely on jackstands. Apply the parking brake.

2   On models with power steering, place a drain pan under the steering gear and disconnect the hose (low-pressure side) and metal line (high-pressure side) **(see illustration)**. Use a flare-nut wrench if you have one to unscrew the hose and line fittings. Cap the hose and line to prevent excessive fluid loss and contamination.

3   Locate the small universal joint that couples the steering shaft or intermediate shaft to

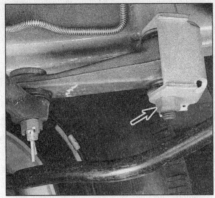

12.20 To remove the idler arm, remove the cotter pin, loosen the castle nut and install a small puller (see illustration 12.16) and press the ballstud out of the center link, then remove the nut (arrow) and bolt that attach the other end of the arm to the frame bracket

the steering gear input shaft. On some models, this coupling is protected by a fabric or rubber dust cover. Remove the dust cover, if equipped. Mark the relationship of the U-joint to the input shaft. Remove the coupler spring pin or lower pinch bolt **(see illustrations)**.

4   Loosen the Pitman arm nut. Mark the relationship of the Pitman arm to the shaft so it can be installed in the same position **(see illustration)**.

5   Separate the Pitman arm from the shaft with a Pitman arm puller **(see illustration)**. Remove the nut and detach the Pitman arm.

6   Support the steering gear and remove the mounting bolts **(see illustrations)**. Lower the unit, separate the intermediate shaft from the steering gear input shaft and remove the steering gear from the vehicle.

Installation

7   Raise the steering gear into position and connect the steering column U-joint coupler; be sure to match up the alignment marks you made before disassembly.

8   Install the mounting bolts and washers and tighten them securely.

9   Slide the Pitman arm onto the shaft. Again, make sure the alignment marks match up. Install the washer and nut and tighten the nut to the torque listed in this Chapter's

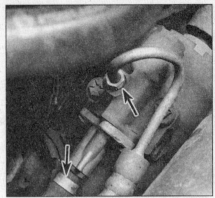

13.2 If the vehicle has power steering, place a drain pan under the steering gear, disconnect the hose and line fittings (arrows), hang them up so the fluid doesn't run out and cap the ends to prevent excessive fluid loss and contamination

Specifications.

10   Install the U-joint coupler spring pin or the lower pinch bolt. Tighten the bolt securely.

11   Connect the power steering hose and line to the steering gear and fill the power steering pump reservoir with the recommended fluid (see Chapter 1).

12   Lower the vehicle and bleed the steering system (see Section 15).

---

14   Power steering pump - removal and installation

Refer to illustration 14.3, 14.4a, 14.4b, 14.5a and 14.5b

1   Loosen the pump drivebelt and slip the belt over the pulley (see Chapter 1).

2   Using a suction gun, suck out as much power steering fluid from the reservoir as possible. Position a drain pan under the pump and disconnect the high pressure line and fluid return hose. It may be necessary to remove the air cleaner housing for access to the return hose (if so, see Chapter 4). Cap the ends of the lines to prevent excessive fluid leakage and the entry of contaminants.

3   Remove the pump and bracket from the engine as a single assembly **(see illustra-**

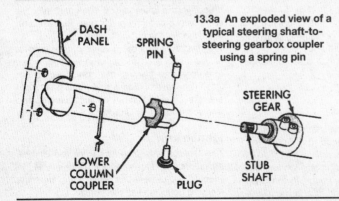

13.3a An exploded view of a typical steering shaft-to-steering gearbox coupler using a spring pin

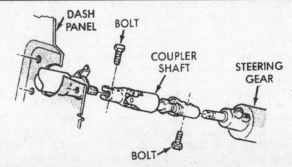

13.3b An exploded view of a typical steering shaft-to-steering gearbox coupler using pinch bolts

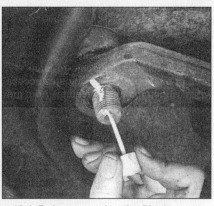

13.4 Before removing the Pitman arm, mark its relationship to the steering gear shaft

13.5 Use a Pitman arm puller to separate the Pitman arm from the steering gear shaft

tion). (You can't separate the pump from the bracket while they're bolted to the engine.)

4   If it is necessary to remove the pulley from the pump, first measure how far the pump shaft protrudes from the face of the pulley hub. Remove the pulley from the shaft with a special power steering pump pulley removal tool **(see illustration)**. This tool can be purchased at most auto parts stores. Then remove the pump from the bracket **(see illustration)**.

5   Install the new pump onto the bracket and tighten the pump-to-bracket bolts securely. A special pulley installation tool is available at most auto parts stores for pressing the pulley back onto the pump shaft **(see illustration)**, but a suitable alternative can be fabricated from a long bolt, nut, washer and a socket of the same diameter as the pulley

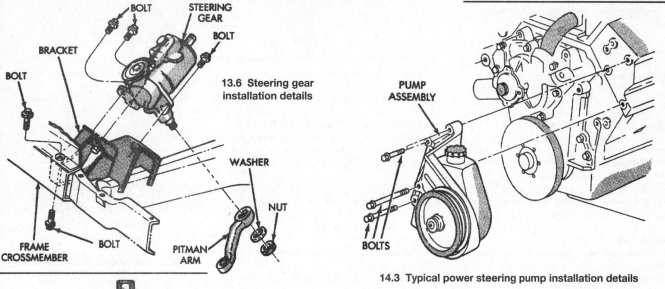

13.6 Steering gear installation details

14.3 Typical power steering pump installation details

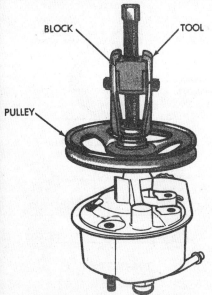

14.4a You'll need this special tool, or one like it, to remove the pulley from the power steering pump; suitable tools are available at most auto parts stores

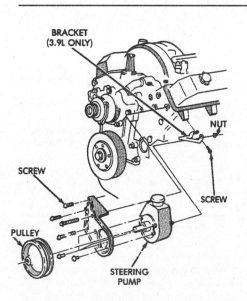

14.4b Typical power steering pump bracket installation details

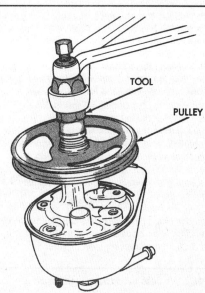

14.5a There's also a special tool available for installing the power steering pump pulley, but . . .

10

14.5b . . . a long bolt with the same thread pitch as the internal threads of the power steering pump shaft, a nut, washer and a socket that is the same diameter as the pulley hub can also be used to install the pulley on the shaft

hub **(see illustration)**. Push the pulley onto the shaft until the shaft protrudes from the hub the previously recorded amount.

6   Installation of the pump/bracket assembly is the reverse of removal. Be sure to bleed the power steering system when you're done (see Section 15).

15   Power steering system - bleeding

1   Following any operation in which the power steering fluid lines have been disconnected, the power steering system must be bled to remove all air and obtain proper steering performance.

2   With the front wheels in the straight ahead position, check the power steering fluid level and, if low, add fluid until it reaches the Cold (C) mark on the dipstick.

3   Start the engine and allow it to run at fast idle. Recheck the fluid level and add more if necessary to reach the Cold (C) mark on the dipstick.

4   Bleed the system by turning the wheels from side-to-side, without hitting the stops. This will work the air out of the system. Keep the reservoir full of fluid as this is done.

5   When the air is worked out of the system, return the wheels to the straight ahead position and leave the vehicle running for several more minutes before shutting it off.

6   Road test the vehicle to be sure the steering system is functioning normally and noise-free.

7   Recheck the fluid level to be sure it is up to the Hot (H) mark on the dipstick while the engine is at normal operating temperature. Add fluid if necessary (see Chapter 1).

16   Wheels and tires - general information

Refer to illustration 16.1

Most vehicles covered by this manual are equipped with metric-sized fiberglass or

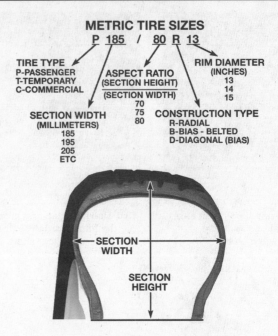

16.1 Metric tire size code

steel-belted radial tires **(see illustration)**. Use of other size or type of tires may affect the ride and handling of the vehicle. Don't mix different types of tires, such as radials and bias belted, on the same vehicle as handling may be seriously affected. It's recommended that tires be replaced in pairs on the same axle, but if only one tire is being replaced, be sure it's the same size, structure and tread design as the other.

Because tire pressure has a substantial effect on handling and wear, the pressure on all tires should be checked at least once a month or before any extended trips (see Chapter 1).

Wheels must be replaced if they are bent, dented, leak air, have elongated bolt holes, are heavily rusted, out of vertical symmetry or if the lug nuts won't stay tight. Wheel repairs that use welding or peening are not recommended.

Tire and wheel balance is important to the overall handling, braking and performance of the vehicle. Unbalanced wheels can adversely affect handling and ride characteristics as well as tire life. Whenever a tire is installed on a wheel, the tire and wheel should be balanced by a shop with the proper equipment.

17   Front end alignment - general information

Refer to illustration 17.1

A front end alignment refers to the adjustments made to the front wheels so they are in proper angular relationship to the suspension and the ground **(see illustration)**. Front wheels that are out of proper alignment not only affect steering control, but also

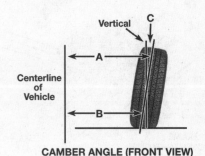

CAMBER ANGLE (FRONT VIEW)

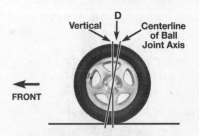

CASTER ANGLE (SIDE VIEW)

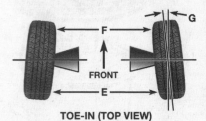

TOE-IN (TOP VIEW)

17.1 Front end alignment details

A minus B = C (degrees camber)
D = degrees caster
E minus F = toe-in (measured in inches)
G - toe-in (expressed in degrees)

increase tire wear. The front end adjustments possible on these vehicles are camber, caster and toe-in.

Getting the proper front wheel alignment is a very exacting process, one in which complicated and expensive machines are necessary to perform the job properly. Because of this, you should have a technician with the proper equipment perform these tasks. We will, however, use this space to give you a basic idea of what is involved with front end alignment so you can better understand the process and deal intelligently with the shop that does the work.

Toe-in is the turning in of the front wheels. The purpose of a toe specification is to ensure parallel rolling of the front wheels. In a vehicle with zero toe-in, the distance between the front edges of the wheels will be the same as the distance between the rear edges of the wheels. The actual amount of toe-in is normally only a fraction of an inch. Toe-in adjustment is controlled by the tie-rod end position on the tie-rod. Incorrect toe-in will cause the tires to wear improperly by making them scrub against the road surface.

Caster is the tilting of the top of the front steering axis from the vertical. A tilt toward the rear is positive caster and a tilt toward the front is negative caster. This angle is adjusted by altering the position of the upper control arm. On 1978 and earlier models it's adjusted by turning the eccentric bolts. On 1979 and later models its done by loosening the bolts attaching the upper control arm pivot bar to the frame, then moving the pivot bar as necessary.

Camber is the tilting of the front wheels from vertical when viewed from the front of the vehicle). It is adjusted using the same technique as for caster adjustment.

Notes

Chapter 11 Body

Contents

1 General information

The Dodge and Plymouth vans covered by this manual are built with a body-on-frame construction. The frame is a ladder-type, consisting of two C-section steel side rails joined by crossmembers. The number of crossmembers depends on the vehicle wheelbase and load rating.

Van body styles are of two types: the *van*, which has front seats only, no side windows, and is generally built for commercial use or the basis for aftermarket van conversions; and the *wagon*, which is designed for passenger use, with side windows and rear seats, the number of which varies with the van length and options.

Certain components are particularly vulnerable to accident damage and can be unbolted and repaired or replaced. Among these parts are the doors, seats, bumpers and door glass. The fenders are welded to the body and require professional repair to replace.

Only general body maintenance practices and body panel repair procedures within the scope of the do-it-yourselfer are included in this Chapter.

2 Body - maintenance

1 The condition of your vehicle's body is very important, because the resale value depends a great deal on it. It's much more difficult to repair a neglected or damaged body than it is to repair mechanical components. The hidden areas of the body, such as the wheel wells, the frame and the engine compartment, are equally important, although they don't require as frequent attention as the rest of the body.

2 Once a year, or every 12,000 miles, it's a good idea to have the underside of the body steam-cleaned. All traces of dirt and oil will be removed and the area can then be inspected carefully for rust, damaged brake lines, frayed electrical wires, damaged cables and other problems. The front suspension components should be greased after completion of this job.

3 At the same time, clean the engine and the engine compartment with a steam cleaner or water soluble degreaser. Do not steam clean the engine with the engine cover removed, or water and cleaning solution may get on interior components.

4 The wheel wells should be given close attention, since undercoating can peel away and stones and dirt thrown up by the tires can cause the paint to chip and flake, allowing rust to set in. If rust is found, clean down to the bare metal and apply an anti-rust paint.

5 The body should be washed about once a week. Wet the vehicle thoroughly to soften the dirt, then wash it down with a soft sponge and plenty of clean soapy water. If the surplus dirt is not washed off very carefully, it can wear down the paint. After scrubbing an area with the sponge, rinse the sponge in a separate bucket of clear water before soaping it for use on the body again, or loosed dirt can be dragged around on the paint with the sponge.

6 Spots of tar or asphalt thrown up from the road should be removed with a cloth soaked in solvent.

7 Once every six months, wax the body and chrome trim. If a chrome cleaner is used to remove rust from any of the vehicle's plated parts, remember that the cleaner also removes part of the chrome, so use it sparingly, and after using chrome cleaner, apply paste wax to the chrome for further protection against the elements.

11

3 Vinyl trim - maintenance

Don't clean vinyl trim with detergents, caustic soap or petroleum-based cleaners. Plain car-washing soap and water works just fine, with a soft brush to clean dirt that may be ingrained. Wash the vinyl as frequently as the rest of the vehicle.

After cleaning, application of a high quality rubber and vinyl protectant will help prevent oxidation and cracks. The protectant can also be applied to weather-stripping, vacuum lines and rubber hoses, which often fail as a result of chemical degradation, and to the tires. **Caution:** *Do not use protectant on vinyl-covered steering wheels.*

4 Upholstery and carpets - maintenance

1   Every three months remove the floor-mats and clean the interior of the vehicle (more frequently if necessary). Use a stiff whisk broom to brush the carpeting and loosen dirt and dust, then vacuum the upholstery and carpets thoroughly, especially along seams and crevices.

2   Dirt and stains can be removed from carpeting with basic household or automotive carpet shampoos available in spray cans. Follow the directions and vacuum again, then use a stiff brush to bring back the "nap" of the carpet.

3   Most interiors have cloth or vinyl upholstery, either of which can be cleaned and maintained with a number of material-specific cleaners or shampoos available in auto supply stores. Follow the directions on the product for usage, and always spot-test any upholstery cleaner on an inconspicuous area (bottom edge of a back seat cushion) to ensure that it doesn't cause a color shift in the material.

4   After cleaning, vinyl upholstery should be treated with a protectant. **Note:** *Make sure the protectant container indicates the product can be used on seats - some products may make a seat too slippery.* **Caution:** *Do not use protectant on vinyl-covered steering wheels.*

5 Body repair - minor damage

See photo sequence

Repair of minor scratches

1   If the scratch is superficial and does not penetrate to the metal of the body, repair is very simple. Lightly rub the scratched area with a fine rubbing compound to remove loose paint and built up wax. Rinse the area with clean water.

2   Apply touch-up paint to the scratch, using a small brush. Continue to apply thin

layers of paint until the surface of the paint in the scratch is level with the surrounding paint. Allow the new paint at least two weeks to harden, then blend it into the surrounding paint by rubbing with a very fine rubbing compound. Finally, apply a coat of wax to the scratch area.

3   If the scratch has penetrated the paint and exposed the metal of the body, causing the metal to rust, a different repair technique is required. Remove all loose rust from the bottom of the scratch with a pocket knife, then apply rust-inhibiting paint to prevent the formation of rust in the future. Using a rubber or nylon applicator, coat the scratched area with glaze-type filler. If required, the filler can be mixed with thinner to provide a very thin paste, which is ideal for filling narrow scratches. Before the glaze filler in the scratch hardens, wrap a piece of smooth cotton cloth around the tip of a finger. Dip the cloth in thinner and then quickly wipe it along the surface of the scratch. This will ensure that the surface of the filler is slightly hollow. The scratch can now be painted over as described earlier in this Section.

Repair of dents

4   When repairing dents, the first job is to pull the dent out until the affected area is as close as possible to its original shape. There is no point in trying to restore the original shape completely as the metal in the damaged area will have stretched on impact and cannot be restored to its original contours. It is better to bring the level of the dent up to a point which is about 1/8-inch below the level of the surrounding metal. In cases where the dent is very shallow, it is not worth trying to pull it out at all.

5   If the back side of the dent is accessible, it can be hammered out gently from behind using a soft-face hammer. While doing this, hold a block of wood firmly against the opposite side of the metal to absorb the hammer blows and prevent the metal from being stretched.

6   If the dent is in a section of the body which has double layers, or some other factor makes it inaccessible from behind, a different technique is required. Drill several small holes through the metal inside the damaged area, particularly in the deeper sections. Screw long, self tapping screws into the holes just enough for them to get a good grip in the metal. Now the dent can be pulled out by pulling on the protruding heads of the screws with locking pliers.

7   The next stage of repair is the removal of paint from the damaged area and from an inch or so of the surrounding metal. This is easily done with a wire brush or sanding disk in a drill motor, although it can be done just as effectively by hand with sandpaper. To complete the preparation for filling, score the surface of the bare metal with a screwdriver or the tang of a file or drill small holes in the affected area. This will provide a good grip

for the filler material. To complete the repair, see the Section on filling and painting.

Repair of rust holes or gashes

8   Remove all paint from the affected area and from an inch or so of the surrounding metal using a sanding disk or wire brush mounted in a drill motor. If these are not available, a few sheets of sandpaper will do the job just as effectively.

9   With the paint removed, you will be able to determine the severity of the corrosion and decide whether to replace the whole panel, if possible, or repair the affected area. New body panels are not as expensive as most people think and it is often quicker to install a new panel than to repair large areas of rust.

10   Remove all trim pieces from the affected area except those which will act as a guide to the original shape of the damaged body, such as headlight shells, etc. Using metal snips or a hacksaw blade, remove all loose metal and any other metal that is badly affected by rust. Hammer the edges of the hole in to create a slight depression for the filler material.

11   Wire brush the affected area to remove the powdery rust from the surface of the metal. If the back of the rusted area is accessible, treat it with rust inhibiting paint.

12   Before filling is done, block the hole in some way. This can be done with sheet metal riveted or screwed into place, or by stuffing the hole with wire mesh. At a professional body shop, rusted out areas are repaired by welding new metal in to replace damaged areas that are cut out.

13   Once the hole is blocked off, the affected area can be filled and painted. See the following subsection on filling and painting.

Filling and painting

14   Many types of body fillers are available, but generally speaking, body repair kits which contain filler paste and a tube of resin hardener are best for this type of repair work. A wide, flexible plastic or nylon applicator will be necessary for imparting a smooth and contoured finish to the surface of the filler material. Mix up a small amount of filler on a clean piece of wood or cardboard (use the hardener sparingly). Follow the manufacturer's instructions on the package, otherwise the filler will set incorrectly.

15   Using the applicator, apply the filler paste to the prepared area. Draw the applicator across the surface of the filler to achieve the desired contour and to level the filler surface. As soon as a contour that approximates the original one is achieved, stop working the paste. If you continue, the paste will begin to stick to the applicator. Continue to add thin layers of paste at 20-minute intervals until the level of the filler is just above the surrounding metal.

16   Once the filler has hardened, the excess can be removed with a body file. From then

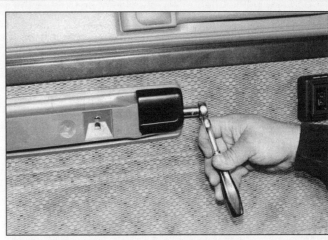

9.2a The armrest is secured by two
Phillips screws (1971 through 1997)

9.2b Remove the interior door handle by taking out one bolt on
the end (1971 through 1997)

on, progressively finer grades of sandpaper should be used, starting with a 180-grit paper and finishing with 600-grit wet-or-dry paper. Always wrap the sandpaper around a flat rubber or wooden block, otherwise the surface of the filler will not be completely flat. During the final sanding of the filler surface, the wet-or-dry paper should be periodically rinsed in water. This will ensure that a very smooth finish is produced in the final stage.

17   At this point, the repair area should be surrounded by a ring of bare metal, which in turn should be encircled by the finely feathered edge of good paint. Rinse the repair area with clean water until all of the dust produced by the sanding operation is gone.

18   Spray the entire area with a light coat of primer. This will reveal any imperfections in the surface of the filler. Repair the imperfections with fresh filler paste or glaze filler and once more smooth the surface with sandpaper. Repeat this spray-and-repair procedure until you are satisfied that the surface of the filler and the feathered edge of the paint are perfect. Rinse the area with clean water and allow it to dry completely.

19   The repair area is now ready for painting. Spray painting must be carried out in a warm, dry, windless and dust-free atmosphere. These conditions can be created if you have access to a large indoor work area, but if you are forced to work in the open, you will have to pick the day very carefully. If you are working indoors, dousing the floor in the work area with water will help settle the dust which would otherwise be in the air. If the repair area is confined to one body panel, mask off the surrounding panels. This will help minimize the effects of a slight mismatch in paint color. Trim pieces such as chrome strips, door handles, etc., will also need to be masked off or removed. Use masking tape and several thicknesses of newspaper for the masking operations.

20   Before spraying, shake the paint can thoroughly, then spray a test area until the spray painting technique is mastered. Cover

the repair area with a thick coat of primer. The thickness should be built up using several thin layers of primer rather than one thick one. Using 600-grit wet-or-dry sandpaper, rub down the surface of the primer until it is very smooth. While doing this, the work area should be thoroughly rinsed with water and the wet-or-dry sandpaper periodically rinsed as well. Allow the primer to dry before spraying additional coats.

21   Spray on the top coat, again building up the thickness by using several thin layers of paint. Begin spraying in the center of the repair area and then, using a circular motion, work out until the whole repair area and about two inches of the surrounding original paint is covered. Remove all masking material 10 to 15 minutes after spraying on the final coat of paint. Allow the new paint at least two weeks to harden, then use a very fine rubbing compound to blend the edges of the new paint into the existing paint. Finally, apply a coat of wax.

6   Body repair - major damage

1   Major damage must be repaired by an auto body shop specifically equipped to perform body and frame repairs. These shops have the specialized equipment required to do the job properly.

2   If the damage is extensive, the body and frame must be checked for proper alignment or the vehicle's handling characteristics may be adversely affected and other components may wear at an accelerated rate.

3   Due to the fact that some of the major body components (hood, doors) are separate and replaceable units, any seriously damaged components should be replaced rather than repaired. Sometimes the components can be found in a wrecking yard that specializes in used vehicle components, often at considerable savings over the cost of new parts.

7   Hinges and locks - maintenance

Once every 3000 miles, or every three months, the hinges and latch assemblies on the doors, hood and trunk should be given a few drops of light oil or lock lubricant. The door latch strikers should also be lubricated with a thin coat of grease to reduce wear and ensure free movement. Lubricate the door locks with spray-on graphite lubricant.

8   Windshield and fixed glass - replacement

Replacement of the windshield and fixed glass requires the use of special fast-setting adhesive/caulk materials and some specialized tools and techniques. These operations should be left to a dealer service department or a shop specializing in glass work.

9   Door trim panel - removal and installation

Refer to illustrations 9.2a, 9.2b, 9.4a, 9.4b, 9.5, 9.6, 9.7 and 9.8

Removal

1   Disconnect the negative cable from the battery.

2   On 1971 through 1997 models, remove the two Phillips screws from the arm rest and remove the armrest (see illustration). Unbolt the one bolt securing the interior door handle and remove the handle (see illustration).

3   On models equipped with manual window regulators, remove the screw retaining the window crank, then remove the handle. On power regulator models, pry out the control switch assembly and disconnect the electrical connector, or leave the switch in the panel and disconnect the electrical con-

11

These photos illustrate a method of repairing simple dents. They are intended to supplement *Body repair - minor damage* in this Chapter and should not be used as the sole instructions for body repair on these vehicles.

1   If you can't access the backside of the body panel to hammer out the dent, pull it out with a slide-hammer-type dent puller. In the deepest portion of the dent or along the crease line, drill or punch hole(s) at least one inch apart . . .

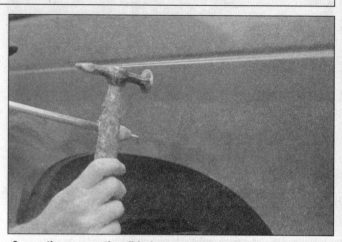

2   . . . then screw the slide-hammer into the hole and operate it. Tap with a hammer near the edge of the dent to help 'pop' the metal back to its original shape. When you're finished, the dent area should be close to its original contour and about 1/8-inch below the surface of the surrounding metal

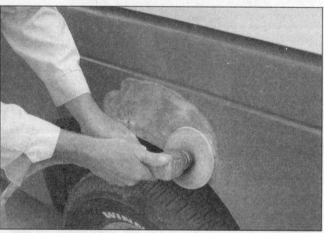

3   Using coarse-grit sandpaper, remove the paint down to the bare metal. Hand sanding works fine, but the disc sander shown here makes the job faster. Use finer (about 320-grit) sandpaper to feather-edge the paint at least one inch around the dent area

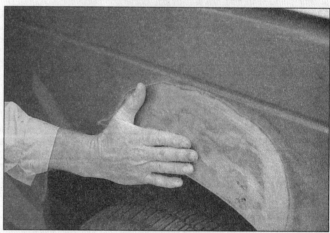

4   When the paint is removed, touch will probably be more helpful than sight for telling if the metal is straight. Hammer down the high spots or raise the low spots as necessary. Clean the repair area with wax/silicone remover

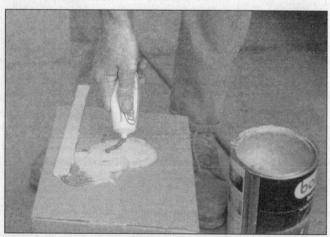

5   Following label instructions, mix up a batch of plastic filler and hardener. The ratio of filler to hardener is critical, and, if you mix it incorrectly, it will either not cure properly or cure too quickly (you won't have time to file and sand it into shape)

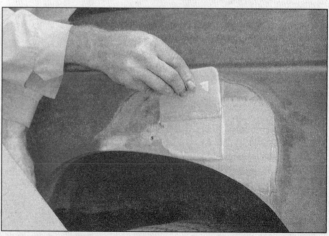

6   Working quickly so the filler doesn't harden, use a plastic applicator to press the body filler firmly into the metal, assuring it bonds completely. Work the filler until it matches the original contour and is slightly above the surrounding metal

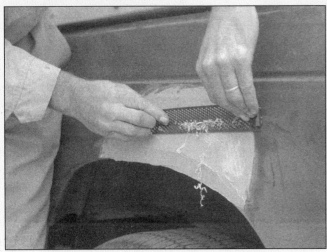

7   Let the filler harden until you can just dent it with your fingernail. Use a body file or Surform tool (shown here) to rough-shape the filler

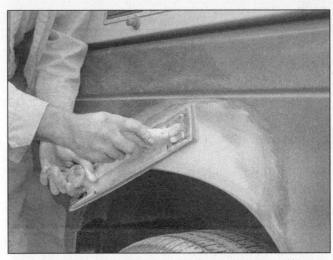

8   Use coarse-grit sandpaper and a sanding board or block to work the filler down until it's smooth and even. Work down to finer grits of sandpaper - always using a board or block - ending up with 360 or 400 grit

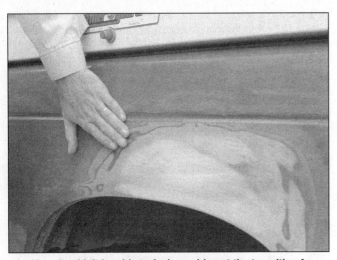

9   You shouldn't be able to feel any ridge at the transition from the filler to the bare metal or from the bare metal to the old paint. As soon as the repair is flat and uniform, remove the dust and mask off the adjacent panels or trim pieces

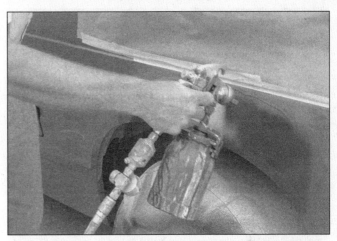

10   Apply several layers of primer to the area. Don't spray the primer on too heavy, so it sags or runs, and make sure each coat is dry before you spray on the next one. A professional-type spray gun is being used here, but aerosol spray primer is available inexpensively from auto parts stores

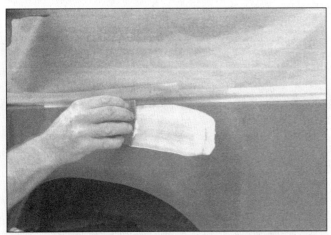

11   The primer will help reveal imperfections or scratches. Fill these with glazing compound. Follow the label instructions and sand it with 360 or 400-grit sandpaper until it's smooth. Repeat the glazing, sanding and respraying until the primer reveals a perfectly smooth surface

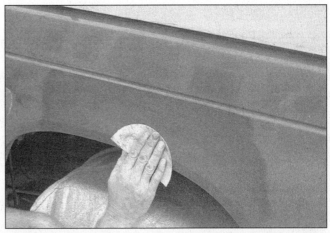

12   Finish sand the primer with very fine sandpaper (400 or 600-grit) to remove the primer overspray. Clean the area with water and allow it to dry. Use a tack rag to remove any dust, then apply the finish coat. Don't attempt to rub out or wax the repair area until the paint has dried completely (at least two weeks)

9.4a Remove the screw from the pull cup, then remove the pull cup from the door panel

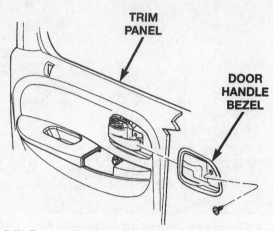

9.4b Remove the screw from the door handle bezel and remove bezel

9.5 Pry carefully, working around the door trim panel until all of the clips are disengaged (1991 through 1997)

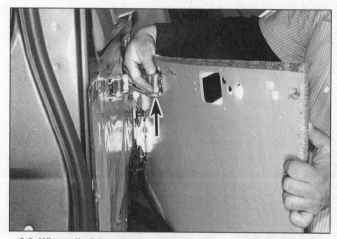

9.6 When all of the clips are disengaged, pull the panel out far enough to disconnect any electrical connectors

nector after the panel is removed **(see illustration 9.5)**.

4   On 1998 and later models, remove the screw from the pull cup and remove the pull cup **(see illustration)**. Remove the screw from the door handle bezel and remove bezel **(see illustration)**.

5   Insert a putty knife or wide-bladed, flat screwdriver between the trim panel and the door and disengage the retaining clips one at a time **(see illustration)**. Work around the outer edge until the panel is free.

6   Once all of the clips are disengaged, detach the trim panel, disconnect the wiring plug from the electric window switch and any other electrical connectors and remove the trim panel from the vehicle **(see illustration)**.

7   For access to the inner door, carefully peel back the plastic watershield **(see illustration)**.

8   Some late-model vans have separate door pull straps above the door panel. If the strap is to be removed, peel up the end flap and remove the retaining screws **(see illustration)**.

Installation

9   Prior to installation of the door panel, be sure to reinstall any clips in the panel which may have come out of the panel during the removal procedure and remained in the door itself.

10   Plug in the electrical connectors and place the panel in position on the door. Press the door panel into place until the clips are seated and install the armrest/door handle. Install the manual regulator window crank or power window switch assembly (if it was removed).

9.7 Carefully peel back the watershield - it will be reused on reassembly

9.8 To remove the door pull strap, pull up each end to expose the two screws

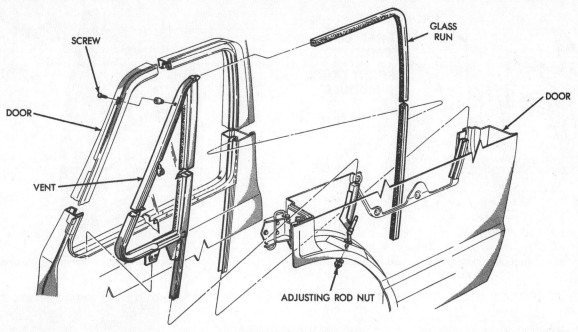

10.2 Vent window glass details

11.3 On vans with factory-mounted door speakers, remove the four screws and drop the speaker down to allow more working room inside the door

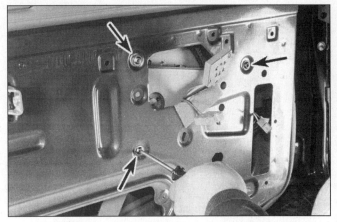

11.4 With the glass removed from the regulator track, remove the three screws to remove the regulator

10 Door window glass - removal and installation

Refer to illustration 10.2

1   Remove the door trim panel and plastic watershield (see Section 9).

Vent window

2   Peel back the weatherstripping to remove the upper retaining screw holding the vent frame to the door (see illustration).

3   Remove one beltline weatherstrip seal and the adjusting rod nut under the door (see illustration 10.2).

4   Lower the window glass all the way and tilt the vent glass and channel assembly toward the rear and lift it out of the door.

5   Installation is the reverse of removal.

Glass

6   Lower the glass and remove the vent window (steps 2 through 4).

7   Detach the glass by sliding it forward until the channel separates from the regulator arm roller.

8   Installation is the reverse of removal.

11 Window regulator - removal and installation

Manual windows

Early models

Refer to illustrations 11.3 and 11.4

1   Remove the door trim panel and water

shield as described in Section 9.

2   Remove the vent window assembly and slide the door glass forward until it comes off the channel rollers, then lower the glass to the bottom of the door.

3   If equipped, remove the door-mounted stereo speaker (see illustration).

4   Remove the regulator bolts and lift the regulator out of the access hole (see illustration).

5   Installation is the reverse of removal.

1998 and later models

Refer to illustration 11.8

6   Remove the door trim panel and water shield as described in Section 9.

7   Remove the door glass.

8   Remove the front door module screws

11

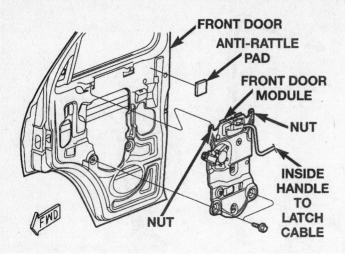

**FRONT DOOR
ANTI-RATTLE
PAD**

**FRONT DOOR
MODULE**

NUT

**INSIDE
HANDLE
TO
LATCH
CABLE**

NUT

FWD

**11.8 Remove the front door module screws and nuts and remove
the module**

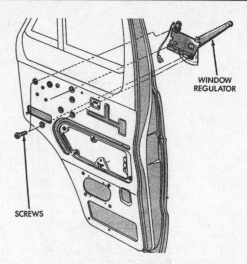

**WINDOW
REGULATOR**

SCREWS

11.14 Remove the regulator mounting screws

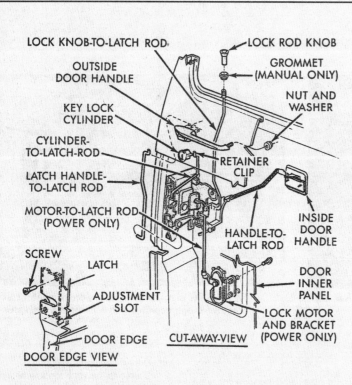

LOCK KNOB-TO-LATCH ROD

LOCK ROD KNOB

**OUTSIDE
DOOR HANDLE**

**GROMMET
(MANUAL ONLY)**

**NUT AND
WASHER**

**KEY LOCK
CYLINDER**

**CYLINDER-
TO-LATCH-ROD**

**RETAINER
CLIP**

**LATCH HANDLE-
TO-LATCH ROD**

**MOTOR-TO-LATCH ROD
(POWER ONLY)**

**INSIDE
DOOR
HANDLE**

**HANDLE-TO-
LATCH ROD**

SCREW

LATCH

**ADJUSTMENT
SLOT**

**DOOR
INNER
PANEL**

**LOCK MOTOR
AND BRACKET
(POWER ONLY)**

DOOR EDGE

CUT-AWAY-VIEW

DOOR EDGE VIEW

12.2 Door latch and linkage details

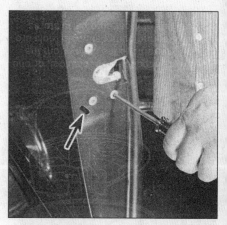

**12.3 Unbolt the latch assembly from the
end of the door - arrow indicates access
hole for outside handle-to-latch
adjustment**

18   Disconnect the electrical connector.
19   Remove the front door module screws
and nuts (see illustration 11.8) and lift the
module out of the door.
20   Installation is the reverse of removal.

12   Front door latch, lock cylinder and handles - removal and installation

Refer to illustrations 12.2, 12.3, 12.5 and 12.8
1   Raise the glass fully and remove the
door trim panel and plastic watershield (see
Section 9).

Latch

2   Disconnect the operating links from the
latch (see illustration).
3   Remove the three retaining screws and
lift the latch out of the door through the inside
panel opening (see illustration).
4   Installation is the reverse of removal.

and nuts (see illustration) and lift the module
out of the door.
9   Installation is the reverse of removal.

Power windows

Early models

Refer to illustration 11.14
10   Remove the door trim panel and water
shield as described in Section 9.
11   Raise the door glass to the top.
12   Remove the vent window frame.
13   Slide the glass forward until the channel

separates from the window regulator. Lower
the glass and channel to the bottom of the
door.
14   Remove the regulator screws (see illustration) and lift the regulator part way out
and disconnect the electrical connector.
Remove the regulator from the door.
15   Installation is the reverse of removal.

1998 and later models

16   Remove the door trim panel and water
shield as described in Section 9.
17   Remove the door glass.

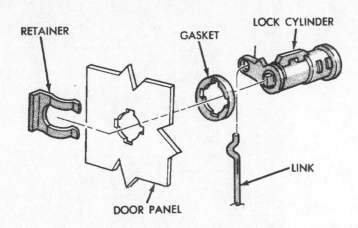

RETAINER   GASKET   LOCK CYLINDER

DOOR PANEL   LINK

12.5 Remove the retainer, disconnect the link and detach the lock cylinder from the door

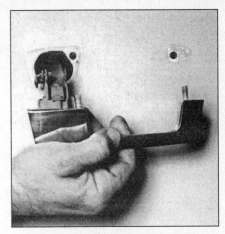

12.8 Remove the nuts, withdraw the handle from the door and disconnect the link

13.2a Remove the screws along the lower edge of the bezel, then the screw inside the glove box and one at the driver's side

13.2b Pull the instrument cluster bezel up and away, until the wiring to the radio can be disconnected

Lock cylinder

5   Disconnect the operating link at the lock cylinder. Use a screwdriver to push the retaining clip off, then withdraw the lock cylinder from the door **(see illustration)**.
6   Installation is the reverse of removal.

Outside handle

7   Disconnect the operating links from the outside handle.
8   Remove the nuts and lift the handle from the door **(see illustration)**.
9   Installation is the reverse of removal.

Inside handle

10   Remove the bolt and rotate the handle away from the door **(see illustration 9.3)**. Disconnect the operating links from the handle.
11   Installation is the reverse of removal.

Latch-to-outside-handle adjustment

12   If the outside handle is not releasing the latch properly, remove the inner door panel

and watershield (see Section 9).
13   Depress the outer door handle button several times, making sure it comes back out all the way when released.
14   Insert a 4mm hex wrench through the adjustment hole in the door jamb and loosen the screw **(see illustration 12.2)**. Reaching inside the door, push up on the latch release lever to eliminate any play.
15   Tighten the adjustment screw and check for proper operation of the outside handle.

13 Instrument cluster bezel - removal and installation

Refer to illustrations 13.2a, 13.2b and 13.3.
Warning: *1995 and later models covered by this manual are equipped with a Supplemental Restraint System (SRS), more commonly known as an airbag(s). All 1995 and later models are equipped with a driver side airbag and all 1998 and later models are equipped with a passenger side airbag, located in the instrument panel. Always disconnect the neg-*

ative battery cable and wait at least two minutes before working in the vicinity of any airbag system component to avoid the possibility of accidental deployment of the airbag, which could cause personal injury (see Chapter 12, Section 28). Do not use electrical test equipment on the airbag system wiring or tamper with it in any way.
1   Disconnect the cable from the negative battery terminal.
2   On 1971 through 1997 models, remove the instrument cluster retaining screws. There are several underneath, and one at each end (open the driver's door for access to the left screw; open the glove box for access to the right screw). Grasp the bezel securely, then pull it out and detach it from the instrument panel **(see illustrations)**.
3   On 1998 and later models, using a trim stick or wide flat-bladed tool, gently pry around the perimeter of the cluster to disengage the snap clip retainers securing the bezel to the instrument panel. Carefully pull the driver's side of the bezel away sufficiently to clear the trip odometer rest button, then

11

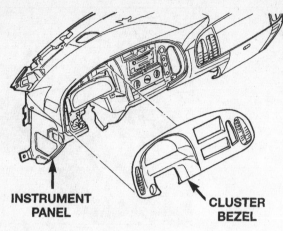

13.3 Using a trim stick or wide flat-bladed tool, gently pry around the perimeter of the cluster and disengage the snap clip retainers securing the bezel to the instrument panel

14.2 Use a marking pen to draw a line (arrow) around the hood hinge

remove the bezel from the instrument panel **(see illustration)**.

4   Installation is the reverse of removal.

14   Hood - removal, installation and adjustment

Refer to illustrations 14.2, 14.10 and 14.11

Note: *The hood is heavy and somewhat awkward to remove and install - at least two people should perform this procedure.*

Removal and installation

1   Use blankets or pads to cover the cowl area of the body and the fenders. This will protect the body and paint as the hood is lifted off.

2   Scribe or draw alignment marks around the hinge to insure proper alignment during installation **(see illustration)**.

3   Disconnect the wiring for the underhood light, if equipped.

4   Have an assistant support the weight of the hood. Remove the hinge-to-hood bolts.

5   Lift off the hood.

6   Installation is the reverse of removal.

Adjustment

7   Fore-and-aft and up-and-down adjustment of the hood is done by moving the hood after loosening the bolts.

8   Scribe a line around the entire hinge plate so you can judge the amount of movement **(see illustration 14.2)**.

9   Loosen the bolts and move the hood into correct alignment. Move it only a little at a time. Tighten the hinge bolts and carefully lower the hood to check the alignment.

10   If necessary after installation, the entire hood latch assembly can be adjusted up-and-down as well as from side-to-side on the radiator support so the hood closes securely and is flush with the fenders. To do this, scribe a line around the hood latch and striker mounting bolts to provide a reference point. Then loosen the bolts and reposition the latch assembly and striker assemblies as necessary **(see illustration)**. Following

adjustment, retighten the mounting bolts.

11   Finally, adjust the hood bumpers on the radiator support so the hood, when closed, is flush with the fenders **(see illustration)**.

12   The hood latch assembly, as well as the hinges, should be periodically lubricated with white lithium-base grease to prevent sticking and wear.

15   Hood latch release cable - removal and installation

1   Open the hood and disconnect the hood latch release cable from the latch assembly **(see illustration 14.10)**.

2   Remove the cable from the routing clips above the grille and along the fender cowl.

3   From inside the vehicle, remove the grommet from the dashboard.

4   Remove the release handle mounting screws from the instrument panel

5   Guide the hood release cable out of the dashboard and remove the cable from the vehicle.

6   Installation is the reverse of removal.

16   Radiator grille - removal and installation

Refer to illustrations 16.1, 16.3 and 16.4

Early models

1   Remove both headlight bezels **(see illustration)**.

2   Remove the three plastic buttons holding the top of the grille to the radiator support.

3   Remove the two screws on each side exposed by removal of the headlight bezels **(see illustration)**.

4   Remove the mounting screws from the turn signal lights **(see illustration 16.1)**. Remove the turn signal lens/housings, allowing access to the grille screws underneath

14.10 Typical hood latch mechanism - loosen the two bolts to align the latch for proper hood engagement, then tighten

14.11 Screwing the threaded hood bumpers in or out aligns the front of the hood with the front fenders

16.1 Remove the four screws to detach the headlight bezel - arrows indicate two screws for turn signal light removal

16.3 Remove these two screws (arrows) exposed when the bezel is removed

16.4 Remove the turn signal light to access a screw like this at each end of the grille

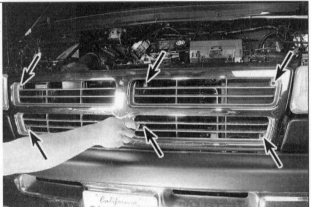

16.6 On late-model vans, remove these six screws (arrows) and the grille comes out

17.3a On older vans, remove this bolt underneath each side of the front bumper . . .

17.3b . . . then in the fenderwell, use an extension and socket through these two holes (arrows) to remove the two remaining front bumper nuts on each side

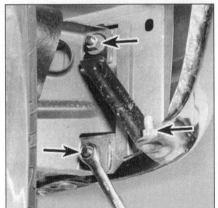

17.3c On later-model vans, remove these three nuts on each side to remove the front bumper

17.3d To remove the rear bumper, remove these four bolts on each side holding the bumper bracket to the frame

(one under each light) **(see illustration)**. Remove the grille.

5   Installation is the reverse of removal.

Later models

Refer to illustration 16.6

6   Remove the six screws, one at each corner and two in the center, and detach the grille **(see illustration)**.

7   Installation is the reverse of removal.

17 Bumpers - removal and installation

Refer to illustrations 17.3a, 17.3b, 17.3c and 17.3d

1   Disconnect any wiring harnesses or any other components that would interfere with bumper removal.

2   Support the bumper with a floor jack. Scribe or draw alignment marks around the

bolt heads, then loosen the retaining bolts. It's a good idea to have an assistant balance the bumper as the bolts are loosened.

3   Remove the retaining bolts and detach the bumper **(see illustrations)**.

4   Installation is the reverse of removal. Tighten the retaining bolts securely.

11

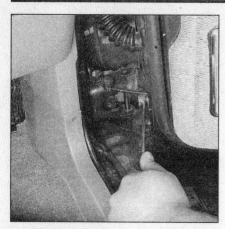

18.4 With the door supported on a padded jack, remove the door-to-hinge bolts

5   Before tightening the mounting bolts, check for proper bumper alignment with the body sheet metal. Adjust it as necessary and tighten the mounting bolts.

---

18   Front door - removal, installation and adjustment

Refer to illustrations 18.4 and 18.6

Removal and installation

1   Remove the door trim panel (see Section 9). Disconnect any electrical connectors and push them through the door opening so they won't interfere with door removal.
2   Place a floor jack under the door and have an assistant on hand to support the door when the hinge bolts are removed. **Note:** *Place a wood block and rag between the head of the jack and the door to protect the door's painted surfaces.*
3   Scribe around the door hinges.
4   Remove the hinge-to-door bolts and carefully lift off the door **(see illustration)**.
5   Installation is the reverse of removal.

Adjustment

6   Following installation of the door, check the alignment and adjust, if necessary, as follows:

a) *Up/down and forward/backward adjustments are made by loosening the hinge-to-body bolts and moving the door as necessary to gain proper alignment with the surrounding body panels (fender, roof, sill plate, etc.). The lower body-to-hinge bolts are accessed by removing the lower interior kick panels, and the upper bolts through the glovebox (passenger side) and through an access hole up behind the dash on the driver's side (it is easier to access these by removing the left-hand dash panel).*
b) *In-and-out adjustments of the hinge end of the doors is done by adjusting the hinge-to-door bolts. Loosen the bolts, move the door to align with the front fender, then tighten the bolts.*

18.6 Use a wrench (some models require a Torx tool, some a box-end wrench as shown) to loosen the door striker, then move it as necessary to adjust the door closed position

b) *The door lock striker can also be adjusted both up-and-down and sideways to provide positive engagement with the lock mechanism. This is done by loosening the striker, moving it as necessary, then retightening.* **Note:** *Make a reference mark around the striker before loosening it* **(see illustration)**.

---

19   Rear door(s) handle, latch and linkage - removal and installation

Refer to illustrations 19.2, 19.3, and 19.7
1   The latch mechanisms are virtually the same for front doors and single rear doors **(see illustration 12.2)**.
2   The latches for double side doors are the same for double rear doors. The right-hand door (opens first) has the handle and latch, while the left door has the striker (for the right door), interior handle, and its own latches to the body **(see illustration)**.

Early-models

3   On 1971 through 1990 vans, the top and bottom latches of the left door are of a "deadbolt" design, where operating the central handle pulls top and bottom rods to release the "bolts" that fit into plates on the body **(see illustration)**.
4   The left and right door latches may be removed by taking off the interior panel or metal access panel.
5   For the right door latch, disconnect the rod to the exterior handle (right door), unbolt the latch and remove it through the access hole.
6   For the left door latch assembly, remove the lower "slam bolt" **(see illustration 19.3)** through the access hole, pull up on the handle and detach the upper slam bolt. Unbolt the handle, remove the three bolts and remove the latch through the access hole.

19.2 The left door of a pair (viewed from the outside) has the striker for the right door (lower arrow) and an interior handle (upper arrow) only

Later models

7   1991 and later models have latches at the top and bottom of the left door, fitting onto strikers located on the body at the top and bottom of the door opening. Removing an extension panel retained by four screws allows access to the handle, striker plate and upper and lower latches **(see illustration)**.
8   To remove the left door latches, unbolt them from the door, detach the rods from the interior handle, separate the rods from the latches, and remove the latches through the access hole.
9   Installation is the reverse of removal. On models with extension panels **(see illustration 19.7)**, make sure you reuse all spacers or grommets.

---

20   Hinged side and rear door(s) - removal and installation

Refer to illustrations 20.2a and 20.2b
1   The removal and installation procedure for dual side doors, dual rear doors, and single rear doors is basically the same as in Section 18 for front doors, except that the hinges for these doors are exposed.
2   The hinge-to-door bolts are accessed by opening the door, and a door can be removed by supporting it with a jack and a helper, removing the hinge-to-door bolts and slipping the door off the hinges **(see illustration)**. **Note:** *Most side and rear doors have a metal check strap. Push the pin out to release the strap from the body* **(see illustration)**.
3   The doors can also be removed with the hinges attached, if necessary. The hinges are bolted to the body. Pulling back the interior panels allows access to the hinge bolts inside the van. **Note:** *Mark a reference mark around the hinges before removing them.*
4   Installation is the reverse of removal. Reuse the sealing washers on the hinge-to-body bolts and align the doors with the surrounding sheetmetal before fully tightening the hinge bolts.

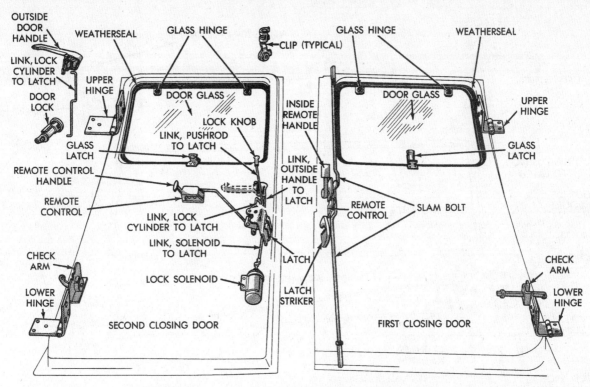

19.3 Side and rear paired door details (1990 and later models)

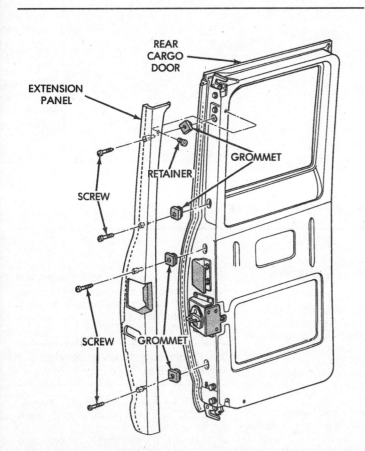

19.7 Later vans have an extension panel on the right door that allows access to the upper and lower latches

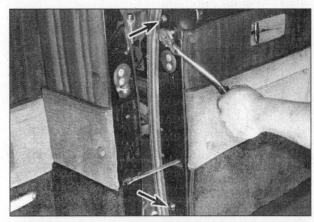

20.2a Upper and lower hinge bolts (arrows) on a typical rear door - there are three at each hinge

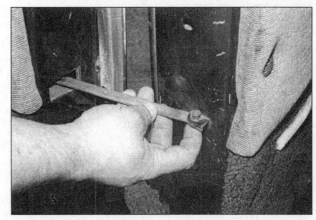

20.2b Push this pin out to release the door check strap

11

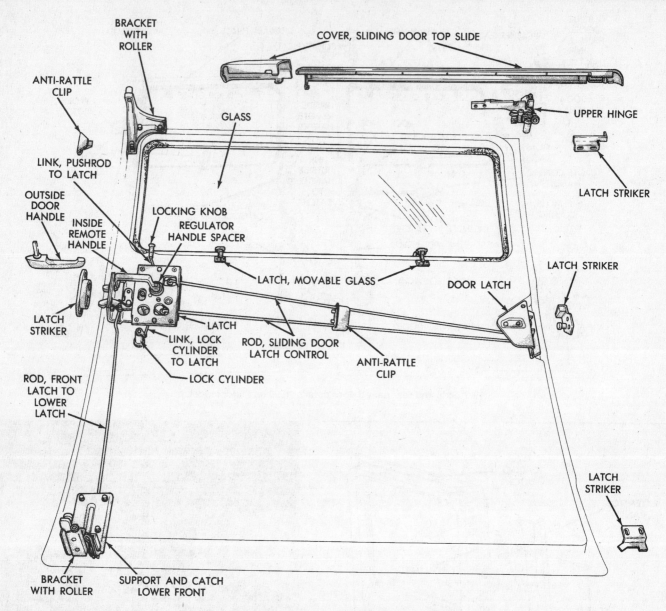

BRACKET
WITH
ROLLER

COVER, SLIDING DOOR TOP SLIDE

ANTI-RATTLE
CLIP

GLASS

UPPER HINGE

LINK, PUSHROD
TO LATCH

LATCH STRIKER

OUTSIDE
DOOR
HANDLE

LOCKING KNOB

INSIDE
REMOTE
HANDLE

REGULATOR
HANDLE SPACER

LATCH STRIKER

LATCH, MOVABLE GLASS

DOOR LATCH

LATCH
STRIKER

LATCH

LINK, LOCK
CYLINDER
TO LATCH

ROD, SLIDING DOOR
LATCH CONTROL

ANTI-RATTLE
CLIP

ROD, FRONT
LATCH TO
LOWER
LATCH

LOCK CYLINDER

LATCH
STRIKER

BRACKET
WITH ROLLER

SUPPORT AND CATCH
LOWER FRONT

21.1 Sliding side door mounting and latching details

21 Sliding side door latch and linkage - removal and installation

Refer to illustration 21.1

Latch

1   Remove the interior door handle (one Allen screw), the knob from the lock-pull, and the screws from the door latch access panel **(see illustration)**. Remove the access panel.
2   Pull the outside door handle and its shaft out of the door with the escutcheon and seals.
3   Disconnect the operating links from the forward latch **(see illustration 21.1)**.
4   Remove the retaining screws from the inside of the door and the front edge of the door and remove the latch.
5   Installation is the reverse of removal.

Lock cylinder

6   Remove the access panel as in Step 1.
7   Disconnect the operating link at the lock cylinder. Use a screwdriver to push the retaining clip off, then withdraw the lock cylinder from the door.
8   Installation is the reverse of removal.

Outside handle

9   Remove the one screw retaining the inside door handle to the shaft of the exterior handle.
10   Pull the outside handle and shaft out of the door with the escutcheon and seals.
11   Installation is the reverse of removal.

Rear latch

12   To remove the rear latch, detach the anti-rattle bracket from the tabs on the door, disconnect the rods, and unbolt the latch from the rear of the door.

22 Sliding side door - removal and installation

1   With the door closed, remove the lower rear screw from the roller track outside cover (on the exterior of the van).
2   Remove the rear upper hinge cover and the rear upper roller hinge plate. Pull the plate from the hinge.
3   Remove the front lower roller bracket

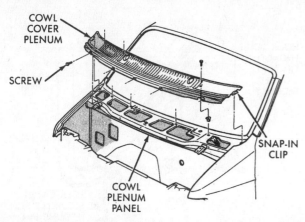

23.3 Cowl grille removal - early models do not have snap-in clips

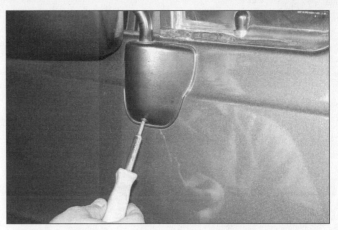

24.2 On some mirrors you will have to remove this screw and slide the plastic bracket cover up

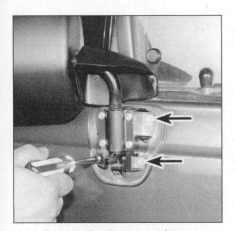

24.3 Remove the three mounting screws (arrows) to remove the mirror

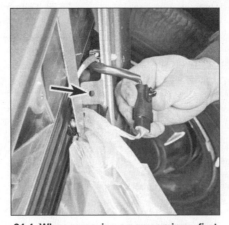

24.4 When removing a power mirror, first detach this electrical connector from its bracket (arrow) then unplug it

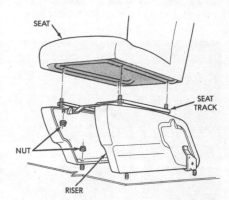

25.1 Typical front seat mounting details

screws and separate the roller bracket from the support bracket.

4   With a helper, remove the door by sliding it to the rear and guiding the rollers out of the rear of each track.

5   Installation is the reverse of removal. During installation, clean each of the rollers and tracks and lubricate with multi-purpose grease.

23 Cowl grille - removal and installation

Refer to illustration 23.3

1   Mark the position of the windshield wiper blades on the windshield with a wax marking pen or pieces of tape.

2   Remove the wiper arms.

3   Raise the hood and remove the cowl retaining screws, disconnect the windshield washer hoses and detach the cowl from the vehicle **(see illustration)**. **Note:** *Late-model vans have a snap-in clip at each end of the cowl grille.*

4   Installation is the reverse of removal. Make sure to align the wiper blades with the marks made during removal.

24 Outside rear view mirror - removal and installation

Refer to illustrations 24.2, 24.3 and 24.4

1   On some models, the mirrors are secured by screws from the outside. On later models, the screws are covered by a plastic housing.

2   Remove the screw at the bottom of the plastic cover and slide the cover up the mirror arm **(see illustration)**. **Note:** *Apply some silicone spray to the mirror arm before sliding the cover up, or the rubber grommet on the cover can be dislodged.*

3   With the cover out of the way, remove the four mounting screws and remove the mirror **(see illustration)**.

4   For power outside mirrors, first remove the interior door panel (see Section 9). Pull the electrical connector out of the door bracket and disconnect the harness **(see illustration)**. Then remove the mirror from the outside as in Steps 2 and 3, feeding the wiring out through the door as the mirror is removed.

5   Installation is the reverse of removal.

25 Seats - removal and installation

Refer to illustration 25.1

Warning: *1995 and later models covered by this manual are equipped with a Supplemental Restraint System (SRS), more commonly known as an airbag(s). All 1995 and later models are equipped with a driver side airbag and all 1998 and later models are equipped with a passenger side airbag, located in the instrument panel. Always disconnect the negative battery cable and wait at least two minutes before working in the vicinity of any airbag system component to avoid the possibility of accidental deployment of the airbag, which could cause personal injury (see Chapter 12, Section 28). Do not use electrical test equipment on the airbag system wiring or tamper with it in any way.*

Front

1   Although the shape of the risers have changed over the years, all van front seats are of the bucket type, mounted on sheet-metal risers **(see illustration)**.

2   Working inside the vehicle, remove the seat nuts, and separate the seat track assembly from the riser.

11

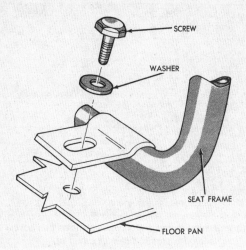

25.6a Most rear seats are mounted to the floor with clamps over the tubular frame

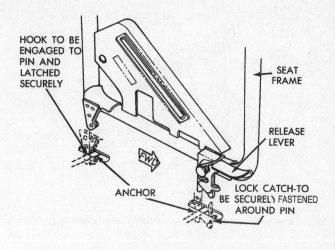

25.6b Optional "latch-in" rear seats are usually easily removed by releasing the lever

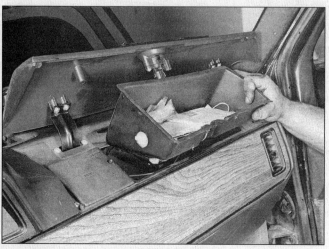

26.4 Depress the top and bottom of the glove box and lift it out

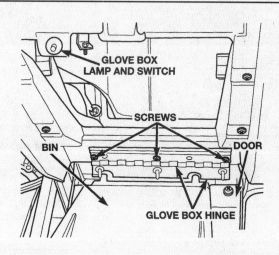

26.7 Remove the three screws securing the glove box hinge to the instrument panel and remove the glove box

3   If desired, the seat and riser assembly can be removed as a unit after removing the riser-to-floor bolts (some models are mounted with nuts on studs from the floor).

4   Installation is the reverse of removal.

Rear

Refer to illustrations 25.6a and 25.6b

5   The type and number of rear seats varies with the year and model of van. Some have no rear seats (cargo vans), while some have convert-a-bed seats or as many as three passenger bench seats.

6   Seat mounting varies, from clamps bolted to the floor securing the tubular seat framework on early models, to quick-release latches on seats designed to be easily removed to suit the vehicle's usage (**see illustrations**). Release the catch or unbolt the

clamps to remove the rear seats. **Note:** *Some models with optional third seats may have storage compartments underneath. On these models, remove the storage trays and unbolt the seat frame from the floor.*

26   Glove box - removal and installation

Refer to illustrations 26.4 and 26.7

Early models

1   Remove the screws and detach the glove box door.

2   Remove the glove box by depressing and withdrawing it from the instrument panel opening.

3   Installation is the reverse of removal.

Later models (through 1997)

4   Open the glove box door. Squeeze and remove the plastic glove box (**see illustration**).

5   Installation is the reverse of removal.

Later models (1998 on)

6   Open the glove box. Depress each side of the glove box inward far enough to clear the stop bumper on each side. Keep the sides depressed and hinge the glove box down until it stops and release the sides.

7   Hold onto the glove box and remove the three screws securing the glove box hinge to the instrument panel (**see illustration**). Remove the glove box.

8   Installation is the reverse of removal.

Chapter 12
Chassis electrical system

Contents

Specifications

Torque specifications
Steering column upper mounting nuts/bolts
1977 and earlier ... 30 ft-lbs
1978 to 1990 ... 110 in-lbs
1991 on ... 105 in-lbs

1 General information

The electrical system is a 12-volt, negative ground type. Power for the lights and all electrical accessories is supplied by a lead/acid-type battery which is charged by the alternator.

This Chapter covers repair and service procedures for the various electrical components not associated with the engine. Information on the battery, alternator, distributor and starter motor can be found in Chapter 5.

It should be noted that when portions of the electrical system are serviced, the negative cable should be disconnected from the battery to prevent electrical shorts and/or fires.

2 Electrical troubleshooting - general information

A typical electrical circuit consists of an electrical component, any switches, relays, motors, fuses, fusible links or circuit breakers related to that component and the wiring and connectors that link the component to both the battery and the chassis. To help you pinpoint an electrical circuit problem, wiring diagrams are included at the end of this book.

Before tackling any troublesome electrical circuit, first study the appropriate wiring diagrams to get a complete understanding of what makes up that individual circuit. Trouble spots, for instance, can often be narrowed down by noting if other components related to the circuit are operating properly. If several components or circuits fail at one time, chances are the problem is in a fuse or ground connection, because several circuits are often routed through the same fuse and ground connections.

Electrical problems usually stem from simple causes, such as loose or corroded connections, a blown fuse, a melted fusible link or a bad relay. Visually inspect the condition of all fuses, wires and connections in a problem circuit before troubleshooting it.

If testing instruments are going to be utilized, use the diagrams to plan ahead of time where you will make the necessary connections in order to accurately pinpoint the trouble spot.

The basic tools needed for electrical troubleshooting include a circuit tester or voltmeter (a 12-volt bulb with a set of test leads can also be used), a continuity tester, which includes a bulb, battery and set of test leads, and a jumper wire, preferably with a circuit breaker incorporated, which can be used to bypass electrical components. Before attempting to locate a problem with test instruments, use the wiring diagram(s) to decide where to make the connections.

Voltage checks

Voltage checks should be performed if a circuit is not functioning properly. Connect one lead of a circuit tester to either the negative battery terminal or a known good ground. Connect the other lead to a connector in the circuit being tested, preferably nearest to the battery or fuse. If the bulb of the tester lights, voltage is present, which means that the part of the circuit between the connector and the battery is problem free. Continue checking the rest of the circuit in the same fashion. When you reach a point at which no voltage is present, the problem lies between that point and the last test point with voltage.

3.1a Typical early model fuse box

3.1b On later models, the fuse block is located under the glove box door (1985 model shown)

3.1c Late models use blade type fuses (1996 model shown)

Most of the time the problem can be traced to a loose connection. **Note:** *Keep in mind that some circuits receive voltage only when the ignition key is in the Accessory or Run position.*

Finding a short

One method of finding shorts in a circuit is to remove the fuse and connect a test light or voltmeter in its place to the fuse terminals. There should be no voltage present in the circuit. Move the wiring harness from side-to-side while watching the test light. If the bulb goes on, there is a short to ground somewhere in that area, probably where the insulation has rubbed through. The same test can be performed on each component in the circuit, even a switch.

Ground check

Perform a ground test to check whether a component is properly grounded. Disconnect the battery and connect one lead of a self-powered test light, known as a continuity tester, to a known good ground. Connect the other lead to the wire or ground connection being tested. If the bulb goes on, the ground is good. If the bulb does not go on, the ground is not good.

Continuity check

A continuity check is done to determine if there are any breaks in a circuit - if it is passing electricity properly. With the circuit off (no power in the circuit), a self-powered continuity tester can be used to check the circuit. Connect the test leads to both ends of the circuit (or to the "power" end and a good ground), and if the test light comes on the circuit is passing current properly. If the light doesn't come on, there is a break somewhere in the circuit. The same procedure can be used to test a switch, by connecting the continuity tester to the switch terminals. With the switch turned On, the test light should come on.

Finding an open circuit

When diagnosing for possible open circuits, it is often difficult to locate them by sight because oxidation or terminal misalignment are hidden by the connectors. Merely

wiggling a connector on a sensor or in the wiring harness may correct the open circuit condition. Remember this when an open circuit is indicated when troubleshooting a circuit. Intermittent problems may also be caused by oxidized or loose connections.

Electrical troubleshooting is simple if you keep in mind that all electrical circuits are basically electricity running from the battery, through the wires, switches, relays, fuses and fusible links to each electrical component (light bulb, motor, etc.) and to ground, from which it is passed back to the battery. Any electrical problem is an interruption in the flow of electricity to and from the battery.

3   Fuses - general information

Refer to illustrations 3.1a, 3.1b, 3.1c, 3.1d and 3.3

The electrical circuits of the vehicle are protected by a combination of fuses, circuit breakers and fusible links. On early models the fuse block is found under the left side of the instrument panel below the steering column, under a cover **(see illustration)** while on later models, the fuse block is located under the glove box door **(see illustrations)**. On 1995 and later models there's an additional fuse box in the engine compartment. Some van conversion models may also have additional fuse boxes throughout the vehicle, but usually they're mounted under the dash **(see illustration)**.

Each of the fuses is designed to protect a specific circuit, and the various circuits are identified on the fuse panel itself.

On early models the fuses are tubular glass body types and on later models plastic miniaturized fuses are used. If an electrical component fails, always check the fuse first. A blown fuse is easily detected with a test light - if voltage is available on one side of the fuse but not the other, it's blown (the circuit must be energized for this check). You can also visually inspect the element for evidence of damage **(see illustration)**.

Be sure to replace blown fuses with the

3.1d Van conversions may be equipped with a separate fuse block that protects the circuits for the passenger compartment accessories

correct type. Fuses of different ratings are physically interchangeable, but only fuses of the proper rating should be used. Replacing a fuse with one of a higher or lower value than specified is not recommended. Each electrical circuit needs a specific amount of protection. The amperage value of each fuse is stamped on the metal end cap (glass type) or molded into the fuse body (blade type).

If the replacement fuse immediately fails, don't replace it again until the cause of the problem is isolated and corrected. In most cases, the cause will be a short circuit in the wiring caused by a broken or deteriorated wire.

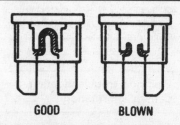

GOOD   BLOWN

3.3 The fuses can be checked visually to determine if they are blown (later model fuse shown)

FUSIBLE LINK CHART

Wire Gauge	Color Code	Color
12 Ga.	BK	Black
14 Ga.	RD	Red
16 Ga.	DB	Dark Blue
18 Ga.	GY	Gray
20 Ga.	OR	Orange
30 Ga.	LG	Light Green

4.2 Always replace a burned out fusible link with one of the same color code

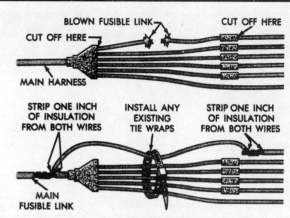

4.3 Fusible link repair details

4   Fusible links - general information

Refer to illustrations 4.2 and 4.3

Note: *1971 through 1992 models are equipped with fusible links. 1993 and later models are not equipped with fusible links but instead use fuse and relay control centers located in the in the engine compartment.*

Some circuits are protected by fusible links. The links are used in circuits which are not ordinarily fused, such as the ignition circuit.

Although the fusible links appear to be a heavier gauge than the wire they are protecting, the appearance is due to the thick insulation. All fusible links are several wire gauges smaller than the wire they are designed to protect. The fusible links are color coded and a link should always be replaced with one of the same color obtained from a dealer parts department or auto parts store **(see illustration)**. On most models the fusible links are located in the wiring harness along the firewall, behind or near the power brake booster.

To repair a fusible link, the burned link must be removed and a new link of the same gauge put in its place. The procedure is as follows:

a) Disconnect the negative cable from the battery.

b) Disconnect the fusible link from the wiring harness.

c) Cut the damaged fusible link out of the wiring just behind the connector.

d) Strip the insulation back approximately 1-inch **(see illustration)**.

e) Position the connector on the new fusible link and twist or crimp it into place.

f) Use rosin core solder at each end of the new link to obtain a good solder joint.

g) Use plenty of electrical tape around the soldered joint. No wires should be exposed.

h) Connect the battery ground cable. Test the circuit for proper operation.

5   Circuit breakers - general information

Refer to illustration 5.1

Circuit breakers protect components such as power windows, power door locks **(see illustration)** and headlights. Some circuit breakers are located in the fuse box.

On some models the circuit breaker resets itself automatically, so an electrical overload in a circuit breaker protected system will cause the circuit to fail momentarily, then come back on. If the circuit does not come back on, check it immediately. Once the condition is corrected, the circuit breaker will resume its normal function.

6   Relays - general information

Several electrical accessories in the vehicle use relays to transmit the electrical signal to the component. If the relay is defective, that component will not operate properly.

The various relays are spread out throughout the engine compartment and under the instrument panel.

If a faulty relay is suspected, it can be removed and tested by a dealer service department or a repair shop. Defective relays must be replaced as a unit.

7   Turn signal and hazard flashers - check and replacement

Turn signal flasher

Refer to illustrations 7.4a and 7.4b

1   The turn signal flasher, a small canister-shaped unit located near the fuse block, flashes the turn signals.

2   When the flasher unit is functioning properly, a clicking sound can be heard during its operation. If the turn signals fail on one side or the other and the flasher unit does not make its characteristic clicking sound, a faulty turn signal bulb is indicated.

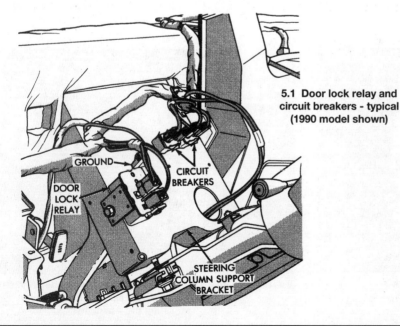

5.1 Door lock relay and circuit breakers - typical (1990 model shown)

12

7.4a The turn signal flasher is located behind the glove box near the fuses on some models (1985 model shown)

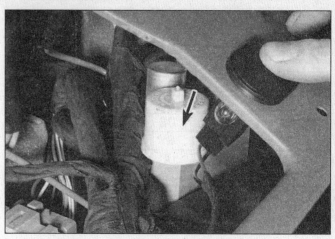

7.4b On later models, the turn signal/hazard flasher is also located behind the glove box, but on the right end (1996 model shown)

3   If both turn signals fail to blink, the problem may be due to a blown fuse, a faulty flasher unit, a broken switch or a loose or open connection. If a quick check of the fuse box indicates that the turn signal fuse has blown, check the wiring for a short before installing a new fuse.

4   To remove the flasher, simply pull it out of its electrical connector or flasher/relay module **(see illustrations)**.

5   Make sure that the replacement unit is identical to the original. Compare the old one to the new one before installing it.

6   Installation is the reverse of removal.

Hazard flasher

Note: *On some models the hazard flasher and turn signal flasher is combined into one unit.*

7   The hazard flasher, a small canister-shaped unit located in the fuse block, flashes all four turn signals simultaneously when activated.

8   The hazard flasher is checked in a fashion similar to the turn signal flasher (see Steps 2 and 3).

9   To replace the hazard flasher, pull it from the back of fuse block or from the flasher/relay module.

10   Make sure the replacement unit is identical to the one it replaces. Compare the old one to the new one before installing it.

11   Installation is the reverse of removal.

8   Headlights - replacement

Sealed beam headlights (1971 through 1993)

Refer to illustrations 8.2 and 8.3

1   Disconnect the negative cable from the battery.

2   Remove the retaining screws and detach the headlight bezel **(see illustration)**.

3   Remove the retainer screws, taking care not to disturb the adjusting screws **(see illustration)**. **Note:** *On models with round headlights it isn't necessary to remove the screws completely - just loosen them a couple of*

turns, then rotate the retainer clockwise to allow the screw heads to pass through the larger openings in the retainer flanges.

4   Remove the retainer and pull the headlight out far enough to allow the connector to be unplugged.

5   Remove the headlight.

6   To install the headlight, plug the connector in, place the headlight in position and install the retainer and screws. Tighten the screws securely.

7   Place the bezel in position and install the retaining screws.

Halogen bulb headlights (1994 and later models)

Refer to illustrations 8.8 and 8.10

8   Lift the lock/release lever and slide one end of the headlight assembly forward **(see illustration)**.

9   Remove the opposite lock/release lever and separate the headlight assembly from the body. Do not pull on the wiring harness and damage the electrical connections.

8.2 Remove the headlight bezel screws

8.3 Remove only the retainer screws (arrows) - do not disturb the headlight aim adjusting screws unless you're adjusting the headlights

8.8 Lift the lock/release lever and slide one end of the headlight assembly forward

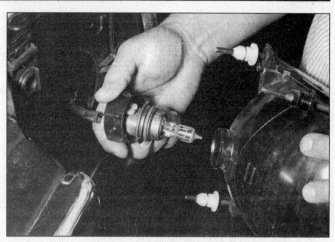

8.10 Turn the socket holder counterclockwise

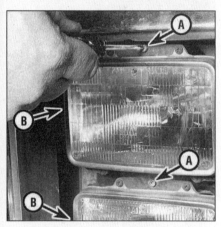

9.1 The vertical (A) and horizontal (B) adjusting screws are located at the top and side of the headlights

10   Turn the socket holder counterclockwise and remove the bulb assembly from the headlight assembly **(see illustration)**.
11   Pull the old bulb out of its socket and replace it with a new one. Be careful to not touch the new bulb - oil from your skin can cause the bulb to overheat and burn out. If you do touch the bulb, clean it with rubbing alcohol.
12   Installation is the reverse of removal.

9   Headlights - adjustment

Refer to illustrations 9.1 and 9.2
Note: *The headlights must be aimed correctly. If adjusted incorrectly they could blind the driver of an oncoming vehicle and cause a serious accident or seriously reduce your ability to see the road. The headlights should be checked for proper aim every 12 months and any time a new headlight is installed or front end body work is performed. It should be emphasized that the following procedure is only an interim step which will provide temporary adjustment until the headlights can be adjusted by a properly equipped shop.*

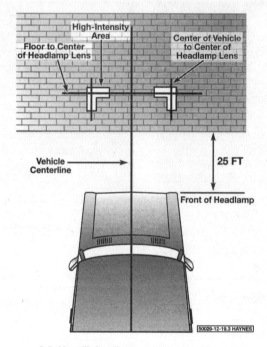

9.2 Headlight alignment screen dimensions

1   Headlights have two spring loaded adjusting screws, one on the top **(see illustration)** controlling up-and-down movement and one on the side controlling left-and-right movement.
2   There are several methods of adjusting the headlights. The simplest method requires a blank wall 25 feet in front of the vehicle and a level floor **(see illustration)**.
3   Position masking tape vertically on the wall in reference to the vehicle centerline and the centerlines of both headlights.
4   Position a horizontal tape line in reference to the centerline of all the headlights. **Note:** *It may be easier to position the tape on the wall with the vehicle parked only a few inches away.*
5   Adjustment should be made with the vehicle sitting level, the gas tank half-full and no unusually heavy load in the vehicle.
6   Starting with the low beam adjustment,

position the high intensity zone so it is two inches below the horizontal line and two inches to the side of the headlight vertical line away from oncoming traffic. Adjustment is made by turning the top adjusting screw clockwise to raise the beam and counterclockwise to lower the beam. The ad-justing screw on the side should be used in the same manner to move the beam left or right.
7   With the high beams on, the high intensity zone should be vertically centered with the exact center just below the horizontal line.
Note: *It may not be possible to position the headlight aim exactly for both high and low beams. If a compromise must be made, keep in mind that the low beams are the most used and have the greatest effect on driver safety.*
8   Have the headlights adjusted by a dealer service department or service station at the earliest opportunity.

12

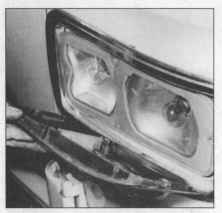

10.1a On early models, the turn signal bulb is accessible after removing the screws and detaching the lens

10.1b Push in on the bulb and rotate it counterclockwise to remove it

10.1c On later models, remove the screws (arrows) that retain the lens cover to the body (1985 model shown)

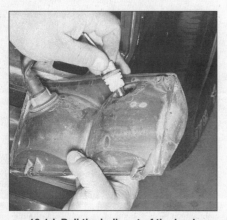

10.1d Pull the bulb out of the back of the housing

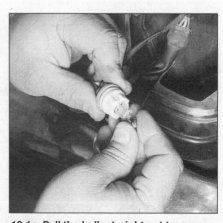

10.1e Pull the bulb straight out to remove it from the holder

10.2a Remove the screws (arrows) from the side marker lens

10   Bulb replacement

Turn signal/parking/side marker lights

Refer to illustrations 10.1a through 10.1e, 10.2a and 10.2b

1   On early models, remove the screws that secure the turn signal/parking light lens, rotate the bulb holder counterclockwise and pull the bulb out **(see illustrations)**. On later models, remove the screws and pull the housing out, then remove the bulb and holder from the rear of the housing **(see illustration)**.

2   Remove the screws and detach the side marker housing for access to the bulb **(see illustrations)**.

3   Installation is the reverse of removal.

Tail light

Refer to illustrations 10.4a, 10.4b, 10.5a, 10.5b and 10.5c

4   On 1993 and earlier models, remove the screws and detach the tail light lens **(see illustration)**. To remove a bulb, push it in and turn it counterclockwise, then pull it out **(see illustration)**.

5   On 1994 and later models, remove the tail light housing screws and detach the

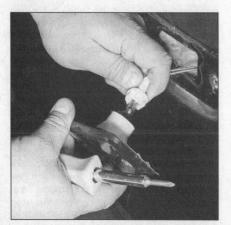

10.2b Unlock the bulb holder and remove the assembly from the lens cover

10.4a Remove the tail light lens screws (arrows) (1985 model shown)

10.4b Push in on the bulb and rotate it counterclockwise to remove it

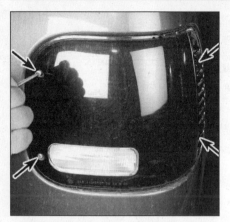

10.5a Location of the tail light housing screws (arrows) on a 1994 or later model

10.5b Turn the bulb holder counterclockwise and pull it out

10.5c The bulb slides straight out of the holder

10.8 After removing the instrument cluster, turn the bulb holders counterclockwise, lift them out and pull out the bulbs

10.10a Remove the screws (arrows) and pull the license plate lamp assembly out for access to the bulb (1985 model shown)

housing from the body **(see illustration)**. Turn the bulb holder counterclockwise and remove it from the back of the housing **(see illustration)**. Pull the bulb out of the holder and replace it with a new one **(see illustration)**.

6   Installation is the reverse of removal.

Instrument cluster lights

Refer to illustration 10.8

7   To gain access to the instrument cluster illumination lights, the cluster will have to be removed (see Section 20). The bulbs can then be removed and replaced from the rear of the cluster.

8   Rotate the bulb holder counterclockwise to remove it **(see illustration)**.

9   Installation is the reverse of removal.

License plate light

Refer to illustrations 10.10a and 10.10b

10   Remove the screws, detach the license plate lamp and remove the bulb from the back **(see illustrations)**.

11   Installation is the reverse of removal.

10.10b Remove the bulb holder from the assembly

Interior lights

Refer to illustration 10.12

12   Pry the interior lens off and detach the bulb from the terminals **(see illustration)**. It may be necessary to pry the bulb out - if this

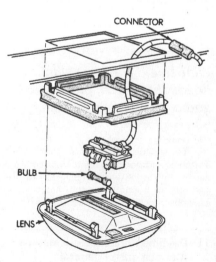

10.12 Exploded view of the dome light assembly

12

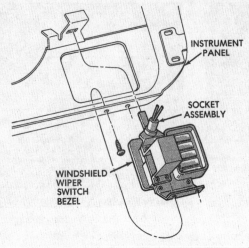

10.15a Windshield wiper/washer switch lamp

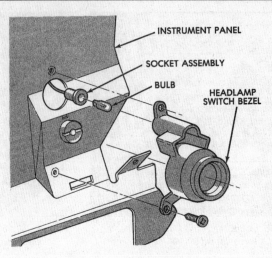

10.15b Headlight switch lamp

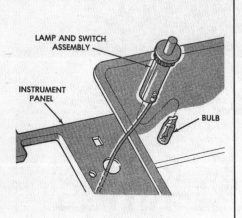

10.15c Glovebox lamp

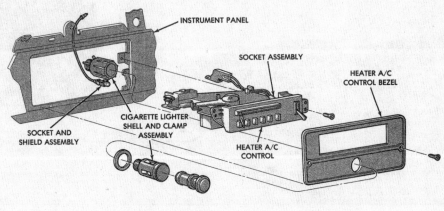

10.15d Cigarette lighter and heater control lamp

is the case, pry only on the ends of the bulb (otherwise the glass may shatter).

13   On early models, remove the screws, detach the lens and push in and rotate the bulb counterclockwise to replace it.

14   Installation is the reverse of removal.

Windshield washer switch, headlight switch, glove box and cigarette lighter bulbs

Refer to illustrations 10.15a, 10.15b, 10.15c and 10.15d

15   To replace the bulbs in the various switches and compartments will require a certain amount of disassembly. Refer to the appropriate Sections for the removal and assembly procedures **(see illustrations)**.

11   Daytime Running Light Module (Canada only) - general information and replacement

Refer to illustration 11.2

1   Canadian models are required to illuminate the headlights at less than 50-percent during daylight, and are equipped with a daytime running light module that controls the intensity of the headlights during daytime hours. The module is mounted in the engine compartment on the bottom of the cowl near the windshield wiper motor.

2   Remove the bolts retaining the module to the vehicle body **(see illustration)**.

3   Disconnect the electrical connector and remove the module from the engine compartment.

4   Installation is the reverse of removal.

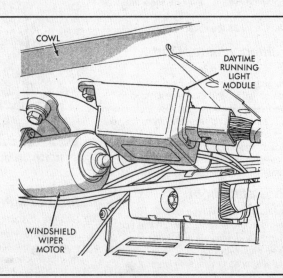

11.2 The daytime running light module is located near the windshield wiper motor

12 Radio and amplifier - removal and installation

Warning: *1995 and later models covered by this manual are equipped with a Supplemental Restraint System (SRS), more commonly known as an airbag(s). All 1995 and later models are equipped with a driver side airbag and all 1998 and later models are equipped with a passenger side airbag, located in the instrument panel. Always disconnect the negative battery cable and wait at least two minutes before working in the vicinity of any airbag system component to avoid the possibility of accidental deployment of the airbag, which could cause personal injury (see Chapter 12, Section 28). Do not use electrical test equipment on the airbag system wiring or tamper with it in any way.*

1   Disconnect the negative cable at the battery.

Radio

Early models
Refer to illustration 12.3

2   Remove the glove box (see Chapter 11).
3   Remove the radio bezel mounting screws and remove the bezel **(see illustration).**
4   Remove the radio to cluster mounting bolts and carefully push the radio away from the cluster toward the front of the vehicle far enough to release the radio mounting tabs from their openings in the panel.

5   Reach into the glove box opening and unplug the antenna, power and speaker electrical connectors
6   If the vehicle is equipped with air conditioning, remove the right defroster distribution duct (see Chapter 3).
7   Remove the radio-to-rear support bracket nuts, tilt the radio up and then pull toward the rear of the vehicle to release the radio from the rear support bracket.
8   Remove the radio through the glove box opening.
9   Installation is the reverse of removal.

Later models
Refer to illustration 12.11

10   Remove the instrument cluster bezel (see Chapter 11).
11   Remove the radio mounting screws **(see illustration).**
12   Remove the ground strap screw.
13   Slide the radio out and support it, disconnect the antenna and electrical connectors, then remove the assembly from the instrument panel.
14   Installation is the reverse of removal.

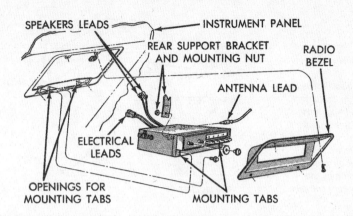

12.3 Radio mounting details on an early model (1974 model shown)

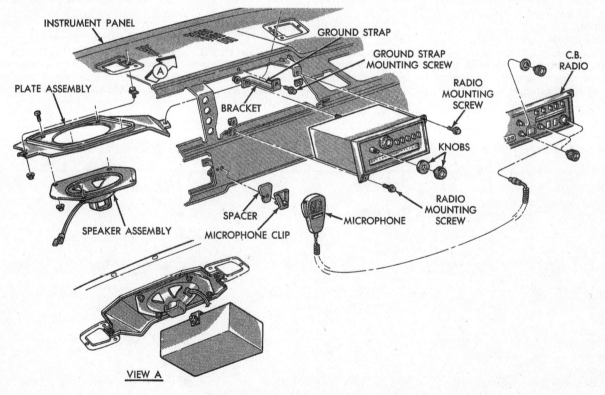

12.11 Radio mounting details on a 1978 model

12

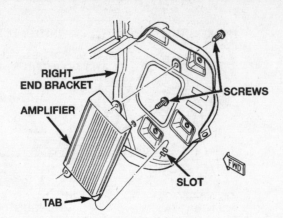

12.19 Reach through the glove box case right side opening, support the amplifier, then remove the two screws from the passenger side end cap opening in the instrument panel

13.2 The antenna is screwed onto its base

Amplifier (1998 and later models)

Refer to illustration 12.19

15   Disconnect the cable from the negative battery terminal.
16   Open the door and remove the passenger side end cap from the instrument panel.
17   Open the glove box. Depress each side of the glove box inward far enough to clear the stop bumper on each side. Keep the sides depressed and hinge the glove box down until it stops and release the sides.
18   Reach through the glove box case left side opening and disconnect the two electrical connectors from the amplifier.
19   Reach through the glove box case right side opening and support the amplifier, then remove the two screws from the passenger side end cap opening in the instrument panel **(see illustration)**.
20   Lift the amplifier upward enough to disengage the mounting tab from the slot in the lower rear corner of the instrument panel bracket. Remove the amplifier through the glove box case left side opening.
21   Installation is the reverse of removal.

13   Antenna - removal and installation

Antenna mast

Refer to illustration 13.2

1   The antenna mast can be unscrewed and replaced with a new one in the event it is damaged, following the procedure below. Follow the procedure beginning with Step 4 to replace the antenna and cable as an assembly.
2   Use a small wrench to unscrew the mast **(see illustration)**.
3   Install the new antenna mast finger tight and tighten it securely with the wrench.

Antenna and cable

Refer to illustrations 13.7a, 13.7b, 13.8a and 13.8b

4   Disconnect the negative cable at the battery.
5   On earlier models reach around behind the radio and unplug the antenna lead. On later models it will probably be necessary to remove the radio and unplug the antenna.
6   Remove the glove box (Chapter 11).

7   Working through the glove box opening, detach the antenna cable from the clips, then open the right door and detach the cable grommet from the door pillar **(see illustrations)**. Connect a long piece of string or thin wire to the antenna cable.
8   Remove the antenna cap nut, using needle-nose pliers, then lift off the adapter **(see illustrations)**. Be very careful - the pliers could slip and scratch the fender. It's a good idea to surround the base of the antenna with rags to prevent against scratching.
9   Install the antenna mast temporarily and push the antenna down far enough so that you can grasp the cable end through the fender opening.
10   Remove the antenna mast, then pull the cable through the fender and into the fenderwell.
11   Push the new cable end up through the fender, then install the adapter, cap nut and antenna mast.
12   Connect the string or wire to the radio end of the new cable and pull it through into the passenger compartment.

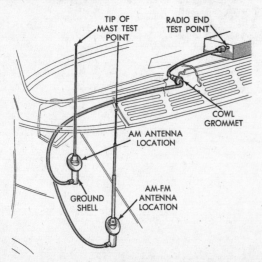

13.7a Mounting location of the AM and AM/FM antenna on a 1978 model

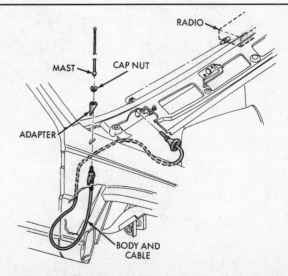

13.7b Mounting location of the antenna on a 1996 model

13   Connect the cable to the radio, secure it with the clips, then push the grommet into place in the door pillar.

14   Install the glove box and connect the battery negative cable.

14 Multi-function switch (1991 and later models) - check and replacement

Warning: *1995 and later models covered by this manual are equipped with a Supplemental Restraint System (SRS), more commonly known as an airbag(s). All 1995 and later models are equipped with a driver side airbag and all 1998 and later models are equipped with a passenger side airbag, located in the instrument panel. Always disconnect the negative battery cable and wait at least two minutes before working in the vicinity of any airbag system component to avoid the possibility of accidental deployment of the airbag, which could cause personal injury (see Chapter 12, Section 28). Do not use electrical test equipment on the airbag system wiring or tamper with it in any way.*

1   The multi-function switch is located on the left side of the steering column. It incorporates the turn signal, hazard flasher, headlight dimmer and windshield wiper/washer functions into one switch.

Check

Refer to illustrations 14.2a through 14.2d and 14.3a through 14.3e

2   Remove the steering column covers **(see illustrations)** and unplug the multi-function switch connector **(see illustrations)**.

3   Use an ohmmeter or self-powered test light and the accompanying diagrams to

13.8a Unscrew the cap nut using needle-nose pliers - use care so the pliers don't slip and scratch the fender

13.8b Lift the adapter off the fender

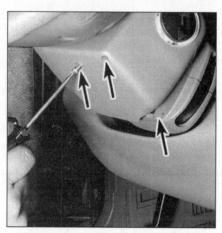

14.2a Remove the screws (arrows) from the lower steering column cover

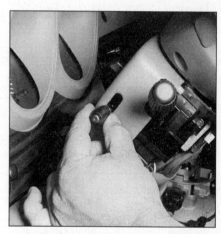

14.2b Unscrew the steering column tilt lever

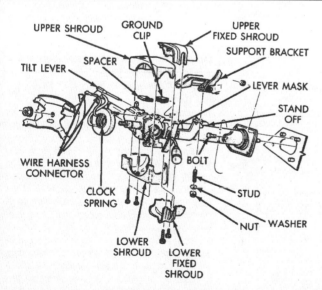

14.2c Exploded view of the late model steering column equipped with the multi-function switch

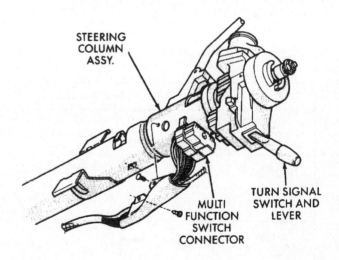

14.2d Unplug the electrical connector from the switch

12

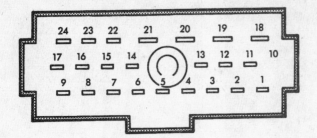

14.3a Multi-function switch terminal details (1991 through 1997)

SWITCH POSITION	CONTINUITY BETWEEN
LOW BEAM	18 AND 19
HIGH BEAM	19 AND 20
OPTICAL HORN	20 AND 21

14.3c Headlight dimmer switch continuity details

SWITCH POSITIONS		CONTINUITY BETWEEN
TURN SIGNAL	HAZARD WARNING	
NEUTRAL	OFF	12 AND 14 AND 15
LEFT	OFF	15 AND 16 AND 17
LEFT	OFF	12 AND 14
LEFT	OFF	22 AND 23 WITH OPTIONAL CORNER LAMPS
RIGHT	OFF	11 AND 12 AND 17
RIGHT	OFF	14 AND 15
RIGHT	OFF	23 AND 24 WITH OPTIONAL CORNER LAMPS
NEUTRAL	ON	11 AND 12 AND 13 AND 15 AND 16

14.3b Turn signal and hazard flasher continuity details - the switch must be in the indicated position for each check

INTERMITTENT WIPE SWITCH CONTINUITY CHART

SWITCH POSITION	CONTINUITY BETWEEN
OFF	PIN 6 & PIN 7
DELAY	PIN 8 & PIN 9 PIN 2 & PIN 4 PIN 1 & PIN 2
LOW	PIN 4 & PIN 6
HIGH	PIN 4 & PIN 5

*RESISTANCE AT MAXIMUM DELAY POSITION SHOULD BE BETWEEN 270,000 OHMS AND 300,000 OHMS.
*RESISTANCE AT MINIMUM DELAY POSITION SHOULD BE ZERO WITH OHMMETER SET ON HIGH OHM SCALE.

14.3d Make the continuity checks for the windshield wiper/intermittent wiper with the multi-function switch in the indicated positions (1991 through 1997)

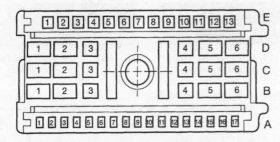

STEERING COLUMN CONNECTOR
VIEWED FROM TERMINAL (ENGAGEMENT) END

MULTI-FUNCTION SWITCH

SWITCH POSITION		CONTINUITY BETWEEN
TURN SIGNAL	HAZARD WARNING	
NEUTRAL	OFF	A1 & E2, A1 & E6, A1 & E7, A2 & A3, E2 & E6, E2 & E7
LEFT	OFF	A1 & E2, A1 & E6, A2 & A3, A7 & E9, E2 & E6, E7 & E9
RIGHT	OFF	A1 & E2, A1 & E7, A2 & A3, A6 & E9, E2 & E7, E6 & E9
NEUTRAL	ON	A6 & E9, A7 & E9, E1 & E8, E2 & E6, E2 & E7, E2 & E9, E6 & E9, E7 & E9

14.3e Multi-function switch terminal details and continuity chart (1998 and later models)

check the multi-function switch for continuity between the terminals with the switch in each position (see illustrations).

Replacement

Refer to illustrations 14.4a, 14.4b and 14.4c
4   Unplug the electrical connector, remove the screws, then detach the switch from the steering column (see illustrations).
5   Installation is the reverse of removal.

15 Headlight switch - replacement

Refer to illustrations 15.3, 15.4, 15.5a and 15.5b
Warning: *1995 and later models covered by this manual are equipped with a Supplemental Restraint System (SRS), more commonly known as an airbag(s). All 1995 and later models are equipped with a driver side airbag and all 1998 and later models are equipped with a passenger side airbag, located in the instrument panel. Always disconnect the negative battery cable and wait at least two minutes before working in the vicinity of any airbag system component to avoid the possibility of accidental deployment of the airbag, which could cause personal injury (see Chapter 12, Section 28). Do not use electrical test equipment on the airbag system wiring or tamper with it in any way.*
1   Disconnect the negative cable at the battery.

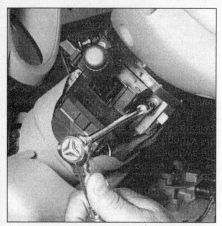

14.4a Use a special tamperproof Torx-head tool to remove the screws and lift the switch off the steering column (1991 through 1997)

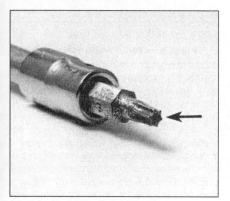

14.4b The tamperproof T-20 Torx socket is recessed (arrow) to slip over the post on the fastener (1991 through 1997)

2   On later models, if necessary, remove the instrument cluster bezel (see Chapter 11).
3   Remove the screws from the outer headlight switch assembly **(see illustration)**.
4   Reach under the dash and press the release button on the bottom **(see illustration)** and withdraw the switch knob and shaft.
5   Remove the retaining nut and lower the switch from the back of the instrument panel **(see illustrations)**.
6   Installation is the reverse of removal.

16   Headlight dimmer switch - check and replacement

Note: *On 1991 and later models, the head-light dimmer switch is incorporated into the multi-function switch (see Section 14).*

1988 and earlier floor-mounted switch

Check
Refer to illustrations 16.1 and 16.2

1   Pull back the carpet, remove the bolts, then unplug the connector and detach the

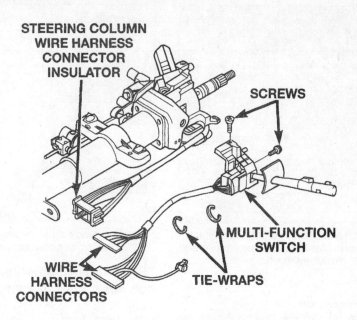

14.4c Multi-function switch mounting details (1998 and later models)

STEERING COLUMN
WIRE HARNESS
CONNECTOR
INSULATOR

SCREWS

MULTI-FUNCTION
SWITCH

WIRE
HARNESS
CONNECTORS

TIE-WRAPS

15.3 Remove the screws from the outer headlight switch assembly

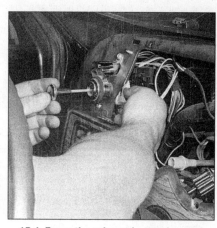

15.4 Press the release button located under the switch and pull the headlight selector knob straight out

15.5a Using a wide blade screwdriver, remove the headlight switch fastener

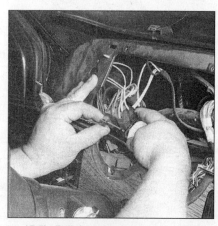

15.5b Pull the tab out to release the electrical connector from the headlight switch

12

16.1 Remove the two bolts and detach the connector (arrows), then remove the headlight dimmer switch

16.2 Checking a floor-mounted dimmer switch with an ohmmeter

switch **(see illustration)**.

2   Use an ohmmeter to check the continuity between the switch terminals. Connect one lead to the center terminal and the other lead to the low beam terminal. There should be continuity with the switch in either position **(see illustration)**. Repeat the test with the high beam terminal - it should only have continuity in one position. Replace the switch if it fails the test.

Replacement

3   Plug the electrical connector into the new switch, place the switch in position and install the bolts. Tighten the bolts securely.

1989 and 1990 steering column-mounted switch

Warning: *Some models have airbags. The airbag is armed and can deploy (inflate) anytime the battery is connected. To prevent accidental deployment (and possible injury), disconnect the negative battery cable whenever working near airbag components. After the battery is disconnected, wait two minutes before beginning work (the system has a backup capacitor that must fully discharge). For more information see Section 28.*

Check

Refer to illustrations 16.5, 16.6 and 16.9

4   Remove the screws and detach the steering column cover.

5   Locate the electrical connector and leave it plugged in **(see illustration)**.
6   Use a voltmeter to check for battery voltage at the switch terminals. There should be voltage on terminals A and D with the switch on LOW beam. Now switch to HIGH beam and there should be voltage on terminals B and C **(see illustration)**. If there isn't, replace the switch.

Replacement

7   Remove the two bolts and lower the switch from the steering column.
8   Place the new switch in position and install the mounting bolts finger tight.
9   Insert an adjustment pin (fabricated from a piece of wire) into the adjusting hole in the switch **(see illustration)**. Push the switch to the rear to take up the slack in the actuator control rod, then tighten the bolts securely.
10   Remove the pin, plug in the connector and install the steering column cover.

17 Turn signal/hazard warning switch - check and replacement

Warning: *1995 and later models covered by this manual are equipped with a Supplemental Restraint System (SRS), more commonly known as an airbag(s). All 1995 and later models are equipped with a driver side airbag and all 1998 and later models are equipped with a passenger side airbag, located in the instrument panel. Always disconnect the negative battery cable and wait at least two minutes before working in the vicinity of any airbag system component to avoid the possi-*

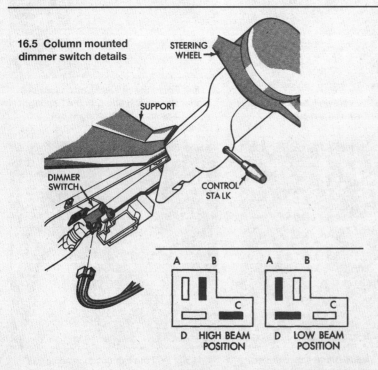

16.5 Column mounted dimmer switch details

16.6 Dimmer switch terminal designations

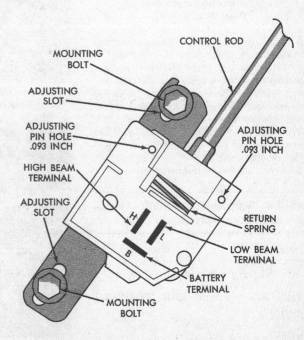

16.9 Dimmer switch adjustment details - you can fabricate your own adjustment pin with a paper clip

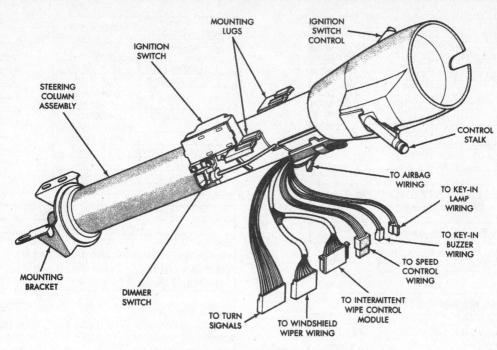

17.8 Steering column electrical connector details

bility of accidental deployment of the airbag, which could cause personal injury (see Chapter 12, Section 28). *Do not use electrical test equipment on the airbag system wiring or tamper with it in any way.*

1   The turn signal switch is located at the top end of the steering column. The hazard warning switch is mounted under the turn signal switch on most later models.

Check

1971 through 1977 models

2   Both the turn signal and hazard flashers are operated by a small canister-shaped relay located near the glove box (see Section 2). When the flasher unit is functioning properly, an audible click can be heard during its operation. If the turn signal light fails on one side, and the flasher unit does not make its characteristic clicking sound, a faulty turn signal bulb is indicated.

3   If both turn signal lights fail to blink, the problem may be due to a blown fuse, a faulty flasher unit, a broken switch or a loose or open connection. If a quick check of the fuse box indicates that the turn signal fuse has blown, check the wiring for a short before installing a new fuse (see *Wiring diagrams* at the end of this Chapter).

4   The flasher relay is located near the glove box. Make sure that the replacement relay is identical to the original.

5   Installation is the reverse of removal.

1978 through 1996 models

Refer to illustrations 17.8, 17.9a, 17.9b, 17.9c and 17.9d

6   Disconnect the negative cable from the battery.

7   Remove the screws and detach the steering column lower cover.

8   Unplug the turn signal switch electrical connector (see illustration).

9   Use an ohmmeter or self-powered test

light to check for continuity between the indicated switch connector terminal (see illustrations).

10   Replace the turn signal or hazard switch if the continuity is not as specified.

WIRE CAVITY	STANDARD COLUMN SWITCH HARNESS WIRE COLOR	TILT COLUMN SWITCH HARNESS WIRE COLOR	FUNCTION
H	WHITE	WHITE	STOP LIGHT SWITCH
G	DARK GREEN	DARK BROWN/RED TRACER	RIGHT REAR
F	YELLOW/BLACK TRACER	DARK GREEN/RED TRACER	LEFT REAR
E	RED	RED	TURN SIGNAL FLASHER
D	DARK BLUE	PINK	HAZARD WARNING FLASHER
C	LIGHT GREEN	TAN	RIGHT FRONT
B	YELLOW	LIGHT GREEN	LEFT FRONT
A	DARK GREEN/WHITE TRACER	BLACK/RED TRACER	HORN

SWITCH CONTINUITY CHART (Fig. 1)

Turn Signal Switch

Switch Position:	Left	Neutral	Right
Continuity Between:	E and B	H and G	E and C
Continuity Between:	E and F	H and F	E and G
Continuity Between:	H and G		H and F

Hazard Warning Switch

Switch Position:	Off	On
Continuity Between:	H and G	D and B
Continuity Between:	H and F	D and C
Continuity Between:		D and F
Continuity Between:		D and G

17.9a Terminal guide and continuity chart for the turn signal/hazard flasher switch on standard and tilt steering columns from 1978 through 1986 (except 1979 through 1981 tilt steering columns)

12

WIRE CAVITY	WIRE COLOR	APPLICATION
8	BLACK	HORN
7	LIGHT BLUE	LEFT FRONT
6	DARK BLUE	RIGHT FRONT
5	BROWN	HAZARD WARNING FLASHER
4	PURPLE	TURN SIGNAL FLASHER
3	YELLOW	LEFT REAR
2	DARK GREEN	RIGHT REAR
1	WHITE	STOP LIGHT SWITCH

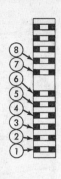

SWITCH CONTINUITY CHART (Fig. 2) With Tilt Column

Turn Signal Switch

Switch Position:	Left	Neutral	Right
Continuity Between:	4 and 3	1 and 2	4 and 2
Continuity Between:	4 and 7	1 and 3	4 and 6
Continuity Between:	1 and 2		1 and 3

Hazard Warning Switch

Switch Position:	Off	On
Continuity Between:	1 and 2	5 and 2
Continuity Between:	1 and 3	5 and 3
Continuity Between:		5 and 6
Continuity Between:		5 and 7

17.9b Terminal guide and continuity chart for the turn signal/hazard flasher switch on 1979 through 1981 tilt steering columns

WIRE CAVITY	WIRE COLOR	APPLICATION
10	WHITE/TAN	STOP LIGHT SWITCH
9	BROWN/RED	RIGHT REAR
8	DK GRN/RED	LEFT REAR
7	RED/BLACK	TURN SIGNAL FLASHER
6	PINK	HAZARD WARNING FLASHER
5		RIGHT FRONT
4	LIGHT GREEN	LEFT FRONT
3	BLACK/RED	HORN
2	BLACK	HORN GROUND
1		

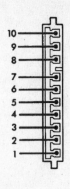

SWITCH CONTINUITY CHART

Turn Signal Switch

Switch Position:	Left	Neutral	Right
Continuity Between:	7 and 4	10 and 9	7 and 5
Continuity Between:	7 and 8	10 and 8	7 and 9
Continuity Between:	10 and 9	—	10 and 8

Hazard Warning Switch

Switch Position:	Off	On
Continuity Between:	10 and 9	6 and 4
Continuity Between:	10 and 8	6 and 5
Continuity Between:	—	6 and 8
Continuity Between:	—	6 and 9

17.9c Turn signal and hazard flasher terminal guide and continuity chart for 1987 through 1990 models

VIEW FROM TERMINAL CASE

SWITCH POSITIONS		CONTINUITY BETWEEN
TURN SIGNAL	**HAZARD WARNING**	
NEUTRAL	OFF	12 AND 14 AND 15
LEFT	OFF	15 AND 16 AND 17
LEFT	OFF	12 AND 14
LEFT	OFF	22 AND 23 WITH OPTIONAL CORNER LAMPS
RIGHT	OFF	11 AND 12 AND 17
RIGHT	OFF	14 AND 15
RIGHT	OFF	23 AND 24 WITH OPTIONAL CORNER LAMPS
NEUTRAL	ON	11 AND 12 AND 13 AND 15 AND 16

17.9d Turn signal and hazard flasher terminal guide and continuity chart for 1991 and later models

Replacement

Note: *On 1991 and later models the turn signal and hazard flasher switch function is incorporated into the multi-function switch. Refer to Section 14 for the replacement procedure.*

11   Remove the steering wheel (see Chapter 10).

Standard (non-tilt) column

Refer to illustration 17.13

12   Remove the screw and detach the lever from the switch.

13   Remove the retaining screws and detach the turn signal switch **(see illustration)**. Carefully pull the wiring harness out through the top of the steering column. **Note:**

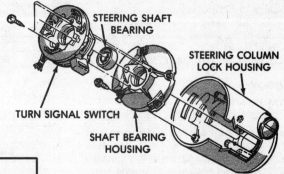

17.13 Turn signal switch mounting details (standard steering column)

It may be necessary to unbolt the steering column from the underside of the dash to allow harness removal.

14   Installation is the reverse of removal. If you unbolted the column from the dash, tighten the mounting nuts to the torque listed in this Chapter's Specifications.

Tilt column

Refer to illustrations 17.15a, 17.15b, 17.16, 17.17a, 17.17b and 17.18

15   Depress the lock plate using a lock plate depressor (available at most auto parts stores), then use a small screwdriver or pick to pry out the retaining ring **(see illustrations)**. If the recommended tool is not available, you might be able to depress the lock

17.15a Install a lock plate depressor tool on the steering column

17.15b Depress the lock plate and pry the retaining ring out of its groove

17.16 Remove the lockplate

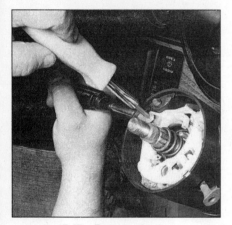

17.17a Remove the turn signal stalk screw

17.17b Remove the three turn signal switch screws

17.18 Remove the hazard flasher button retaining screw

plate by hand far enough so you can get at the retaining ring (if you are forced to take this route, an assistant would be very helpful).

16   Remove the lock plate and cancelling cam **(see illustration)**.

17   Remove the screw and detach the turn signal stalk, then remove the three turn signal switch mounting screws **(see illustrations)**.

18   Unscrew the hazard switch knob **(see illustration)**.

19   Remove the switch and guide the wiring harness and electrical connector up through the steering column. **Note:** *It may be necessary to unbolt the steering column from the dash to allow the electrical connector to pass through.*

20   Installation is the reverse of removal. If you unbolted the column from the dash, tighten the mounting nuts to the torque listed in this Chapter's Specifications.

18 Wiper/washer switch - replacement

Dash-mounted switches

Early style (knob-actuated)

Refer to illustration 18.2

Warning: *1995 and later models covered by this manual are equipped with a Supplemen-*

tal Restraint System (SRS), more commonly known as an airbag(s). All 1995 and later models are equipped with a driver side airbag and all 1998 and later models are equipped with a passenger side airbag, located in the instrument panel. Always disconnect the negative battery cable and wait at least two minutes before working in the vicinity of any airbag system component to avoid the possibility of accidental deployment of the airbag,

which could cause personal injury (see Chapter 12, Section 28). *Do not use electrical test equipment on the airbag system wiring or tamper with it in any way.*

1   Remove the instrument cluster (see Section 20).

2   Loosen the set screw from the underside of the switch knob, then pull the knob off the switch **(see illustration)**.

3   Unscrew the switch mounting nut from

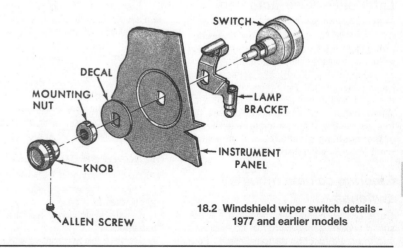

18.2 Windshield wiper switch details - 1977 and earlier models

18.6a Remove the two windshield wiper switch mounting screws (1978 through 1986 models) . . .

18.6b . . . and lower the switch from the dash

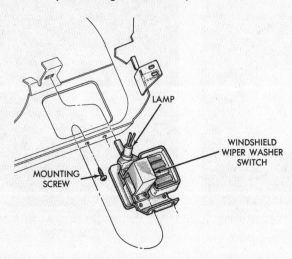

18.8 Windshield wiper switch installation details - 1978 and later dash-mounted switches

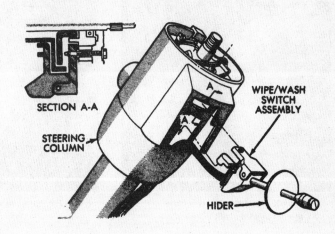

18.14 Wiper/washer switch removal details - 1990 and earlier steering column mounted switch

the front of the dash.

4   Reach through the dash and disconnect the electrical connector from the switch, then remove the switch from the dash.

5   Installation is the reverse of removal.

Later style (lever-actuated)

Refer to illustrations 18.6a, 18.6b and 18.8

6   Remove the screws from the lower edge of the instrument panel and lower the switch from the panel **(see illustrations)**.

7   Unplug the electrical connector and detach the light bulb from the switch.

8   Installation is the reverse of removal. When installing the switch, make sure the tang on the top of the switch fits into the slot in the mounting bracket behind the instrument panel **(see illustration)**.

Steering column mounted switches

Note: *On 1991 and later models the windshield wiper/washer switch is incorporated into the multi-function switch. Refer to Sec-*

tion 14 for the replacement procedure.

9   These models are equipped with a multi-function lever located on the left side of the steering column which controls the wiper/washer, turn signal and dimmer switches.

10   Disconnect the cable from the negative terminal of the battery and remove the steering wheel (see Chapter 10). If your vehicle is equipped with an airbag, be sure to read the airbag warning in the steering wheel removal and installation procedure.

11   Remove screws and detach the steering column lower cover.

12   Detach the wiring harness trough from the steering column (some later models) and unplug the electrical connector.

Standard steering column

Refer to illustrations 18.14, 18.15 and 18.16

13   Remove the lock housing cover screws and pull the cover off.

14   Pull the switch hider out for access **(see illustration)**.

15   Remove the two screws and detach the

switch from the stalk, then remove the wiper knob from the end of the stalk **(see illustration)**.

16   Rotate the control stalk fully clockwise, align the slot and pin, then pull the stalk out **(see illustration)**.

17   Installation is the reverse of removal.

Tilt steering column

Refer to illustrations 18.25 and 18.27

18   Compress the lock plate and remove the retaining ring (see Section 17).

19   Remove the lock plate, cancelling cam and upper bearing spring.

20   Remove the multi-function stalk retaining screw.

21   Remove the hazard flasher knob by pushing it in and unscrewing it.

22   Unplug the wiper/washer switch electrical connector and remove the turn signal switch (see Section 17).

23   Remove the ignition key lamp.

24   Remove the lock cylinder (see Section 23).

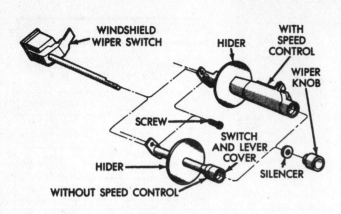

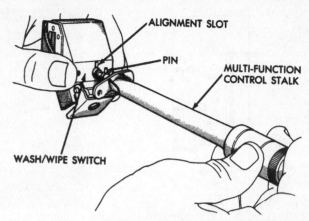

18.15 Wiper/washer stalk details

18.16 Line up the pin with the alignment slot and withdraw the multi-function stalk

25   Remove the ignition buzzer switch and wedge spring **(see illustration)**.
26   Remove the three screws and lift off the column housing cover.
27   Press the wiper switch pivot pin out with a punch **(see illustration)**.
28   Remove the wiper/washer switch.
29   Use tape to secure the dimmer switch rod, pull the switch hider up the control stalk, then remove the two screws.
30   Pull the knob off the end of the multi-function stalk, turn the stalk fully clockwise, align the pin with the slot and pull the stalk straight out.
31   Installation is the reverse of removal.

19 Ignition switch - check and replacement

Warning: *1995 and later models covered by this manual are equipped with a Supplemental Restraint System (SRS), more commonly known as an airbag(s). All 1995 and later models are equipped with a driver side airbag and all 1998 and later models are equipped with a passenger side airbag, located in the instrument panel. Always disconnect the negative battery cable and wait at least two minutes before working in the vicinity of any airbag system component to avoid the possibility of accidental deployment of the airbag, which could cause personal injury (see Chapter 12, Section 28). Do not use electrical test equipment on the airbag system wiring or tamper with it in any way.*

1977 and earlier models
Check

1   Remove the switch from the instrument panel for access to the terminals (see Step 6).
2   Turn the switch to the Run position. Connect the probes of an ohmmeter to the B (battery +) terminal and either of the I (ignition) terminals - there should be continuity.
3   Leave one of the probes connected to the B terminal and attach the other probe to the S (start) terminal. Turn the key to the Start

position - the meter should indicate continuity.
4   Remove the probe from the S terminal and attach it to the A (accessory) terminal. Turn the key to the Accessory position and verify that the meter indicates continuity.
5   If the switch fails any of these tests, replace it.

18.25 Use needle-nose pliers to lift out the buzzer switch and wedge spring

Replacement
Refer to illustration 19.7
6   Remove the instrument cluster (see Section 20).
7   Unscrew the ignition switch retaining nut from the front of the dash **(see illustration)**.

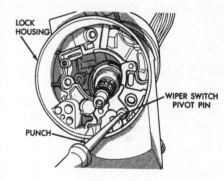

18.27 Use a thin punch or screwdriver to press the pivot pin out

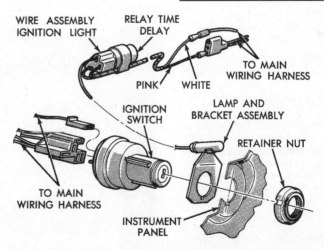

19.7 Ignition switch installation details - 1977 and earlier models

12

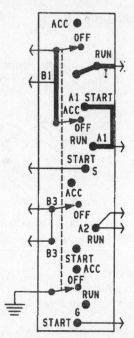

19.12 Schematic of a typical 1978 through 1990 ignition switch

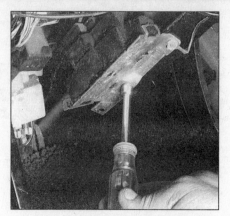

19.17 Remove the ignition switch mounting screws

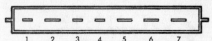

**IGNITION SWITCH CONNECTOR
LOOKING INTO SWITCH**

WIRE CAVITY	WIRE COLOR	APPLICATION
1	YELLOW	STARTER RELAY
2	DARK BLUE	IGNITION RUN/START
3	GRAY/BLACK	BRAKE WARNING LAMP
4	PINK/BLACK	IGNITION SWITCH BATTERY FEED
	PINK OR PINK/WHITE	
5	BLACK/ORANGE OR BLACK/TAN	RUN ACCESSORY
6	BLACK OR BLACK/WHITE	ACCESSORY
7	RED	IGNITION SWITCH BATTERY FEED

19. 21 Typical later model ignition switch terminal guide

8   Remove the switch from the instrument panel and detach the electrical connector and the light bracket (if equipped).
9   Installation is the reverse of removal.

1978 through 1990 models

Refer to illustrations 19.12 and 19.17
10   The ignition switch is located on the steering column and is actuated by a rod attached to the key lock cylinder.

Check

11   Remove the switch (see Steps 14 through 17).
12   Use an ohmmeter or self-powered test light and the accompanying diagram to check for continuity between the switch terminals with the switch in the various positions **(see illustration)**.
13   If the switch does not have correct continuity, replace it.

Replacement

14   Disconnect the negative cable from the battery.
15   Insert the key into the lock cylinder and turn it to the ACC position.
16   Remove the steering column cover. On some models it may be necessary to remove the bolts or nuts and lower the steering column for access to the switch.
17   Remove the screws **(see illustration)**, then detach the switch from the actuator rod, lower it from the steering column and unplug the electrical connector.
18   Installation is the reverse of removal. With the switch in the LOCK position engage it to the actuator rod, then push up on the switch to remove any slack from the rod before fully tightening the screws.

1991 through 1997 models

Refer to illustrations 19.21 and 19.25
19   The ignition switch is located on right side of the steering column and is held in place by three Torx T-20 tamper proof screws which require a special tool for removal (available at auto parts stores).

Check

20   Remove the switch (see Steps 14 through 17).
21   Use an ohmmeter or self-powered test light and check for continuity between the ignition switch battery feed terminal and the and the Starter relay, Run and Accessory terminals with the ignition key in the appropriate positions **(see illustration)**.
22   If the switch does not have correct continuity, replace it.

Replacement

23   Disconnect the negative cable from the battery.
24   Remove the steering column cover.
25   Remove the tamper proof screws, detach the switch from column, then unplug the electrical connector and remove switch from the steering column **(see illustration)**.
26   Installation is the reverse of removal.

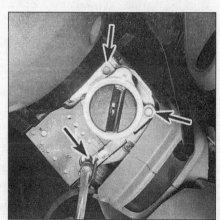

19.25 The ignition switch on 1991 and later models is held in place by three tamper-proof Torx-head screws (arrows)

1998 and later models

Check

Note: *A procedure for checking the ignition switch is not available at the time of publication.*

Replacement

Refer to illustrations 19.29, 19.30, 19.32, 19.33a and 19.33b
27   Disconnect the cable from the battery negative terminal.
28   Remove the steering column cover and knee bolster.
29   Using a small flat-bladed screwdriver, press on the retaining clip and rotate the alarm switch 1/4 turn **(see illustration)**. Remove the alarm switch.
30   **Note:** *Note the position of the wiring harness and all tie wraps between the ignition switch and the 48-way electrical connector at the lower portion of the steering column. The harness must be correctly routed and secured with the tie wraps during installation to prevent damage to the harness. Carefully cut and remove all tie wraps securing the wiring harness to the steering column* **(see illustration)**.
31   If equipped with an overdrive (OD) transmission, unclip the OD wire harness at the ignition switch.
32   Using a T4 Torx wrench, remove the screws securing the ignition switch to the steering column **(see illustration)**.
33   Remove the screw securing the 48-way electrical connector at the base of the steering column and separate the switch **(see illustration)**. **Note:** *The 48-way connector contains three different switch connectors: one 17-way connector for the multi-function switch, one 13-way connector for the multi-function switch and an 18-way connector for the ignition switch* **(see illustration)**.

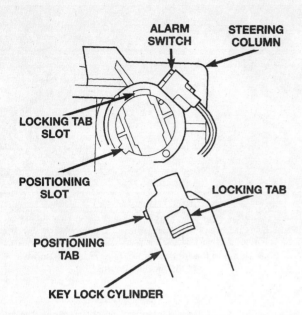

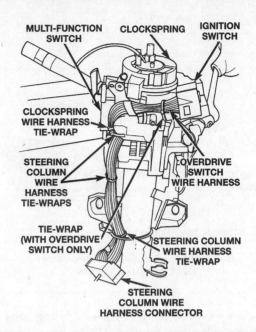

19.29 Using a small flat-bladed screwdriver, press on the retaining clip and rotate the alarm switch 1/4 turn, then remove the alarm switch

19.30 The harness must be correctly routed and secured with the tie wraps during installation to prevent damage to the harness

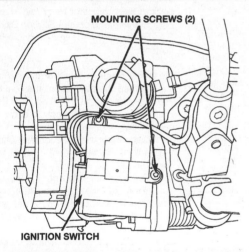

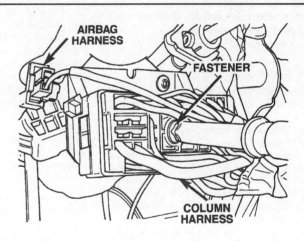

19.32 Remove the screws with a T4 Torx wrench

19.33a Remove the screw securing the 48-way electrical connector at the base of the steering column and separate the switch . . .

34   Separate the 17-way and the 13-way electrical connectors from the 18-way electrical connector.

35   Separate the multi-function switch wiring harness from the ignition switch harness. Remove the ignition switch and harness from the steering column and vehicle.

36   Installation is the reverse of removal.

20 Instrument cluster - removal and installation

1977 and earlier models

Refer to illustration 20.2

1   Disconnect the cable from the negative terminal of the battery.

19.33b . . . then disconnect the 17-way and 13-way connectors from the 18-way connector

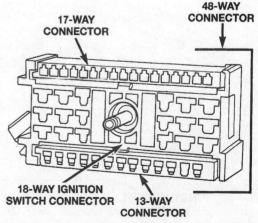

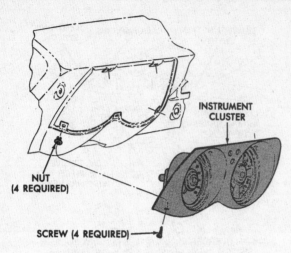

20.2 Instrument cluster mounting details (1977 and earlier models)

20.9 Remove the instrument cluster retaining screws (1985 model shown)

2   Remove the screws securing the instrument cluster to the instrument panel **(see illustration)**.
3   Pull the cluster out of the instrument panel far enough to disconnect the speedometer cable and any electrical connectors, then remove the cluster.
5   Installation is the reverse of removal.

1978 through 1997 models

Refer to illustration 20.9
6   Disconnect the negative cable from the battery.
7   Remove the instrument cluster bezel (see Chapter 11).
8   On some later automatic transmission models, it will be necessary to detach the gearshift cable from the steering column lever.
9   Remove the screws securing the cluster to the instrument panel **(see illustration)**. Pull the cluster out, reach behind the cluster and disconnect the speedometer cable then unplug the electrical connectors and remove the cluster.
10   Installation is the reverse of removal.

1998 and later models

Refer to illustration 20.13
11   Disconnect the cable from the battery negative terminal.
12   Remove the instrument cluster bezel (see Chapter 11).
13   Remove the four screws securing the instrument cluster to the instrument panel **(see illustration)**. Carefully pull the instrument cluster straight out and disengage the two self-docking electrical connectors attached to the instrument panel.
14   Installation is the reverse of removal.
Note: *The two self-docking electrical connectors will connect automatically when the instrument cluster is reinstalled into the instrument panel.*

21   Speedometer cable - check and replacement

Warning: *Some models have airbags. The airbag is armed and can deploy (inflate) anytime the battery is connected. To prevent*

accidental deployment (and possible injury), disconnect the negative battery cable whenever working near airbag components. After the battery is disconnected, wait two minutes before beginning work (the system has a backup capacitor that must fully discharge). For more information see Section 28.
1   Disconnect the negative cable from the battery.
2   Disconnect the speedometer cable from the cruise control adapter or transmission.
3   Detach the cable from the routing clips in the engine compartment and pull the cable up to provide enough slack to allow disconnection from the speedometer.
4   Disconnect the speedometer cable from the back of the instrument cluster (see Section 20).
5   Remove the cable from the vehicle.
6   Prior to installation, lubricate the speedometer end of the cable with spray-on speedometer cable lubricant (available at auto parts stores).
7   Installation is the reverse of removal. Make sure the grommet in the firewall is properly seated.

22   Horn - check and replacement

Check

Early models (horns with two terminals)

1   On early models each horn has two terminals - one for power and one for ground. The horn receives power from the fuse box and is grounded through the horn switch at the steering wheel. Before performing any checks on the system be sure to check the fuses.
2   Using a jumper wire connected to a good ground, ground the negative terminal of the horn. If the horn sounds, the horn switch or the wire between the switch and the horn is bad (open).

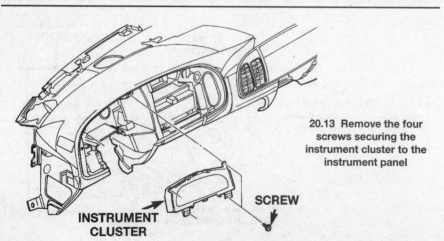

20.13 Remove the four screws securing the instrument cluster to the instrument panel

23.8 Remove the bolts from the windshield wiper motor

24.5 Insert a thin screwdriver into the lock release slot and press the spring latch while pulling the lock cylinder out

3   Using a jumper wire connected to a battery voltage source, apply power to the positive terminal of the horn and push on the horn switch. If the horn sounds, the wire from the fuse box is shorted or open.

4   If the horn didn't sound in either test, try turning the adjusting screw on the horn. If the horn still doesn't work, replace it.

Later models (horns with one terminal)

5   On later models each horn has only one terminal. The horn switch on the steering wheel grounds the horn relay which in turn sends voltage to the horn. Before performing any checks on the system be sure to check the fuses.

6   Disconnect the wire from the horn and attach a test light to the wire. Have an assistant push on the horn pad - if the test light illuminates, reconnect the wire and try adjusting the horn with the adjusting screw. If the horn still makes no sound, replace it.

7   If the test light did not light up, the problem lies in the horn relay, the horn switch or the wiring connecting the components.

Replacement

8   Unplug the electrical connector(s) and remove the mounting bolt.

9   Installation is the reverse of removal.

23 Windshield wiper motor - check and replacement

Refer to illustration 23.8

Check

1   If the wiper motor doesn't run at all, first check for a blown fuse (see Section 3).

2   Check the wiper switch (see Section 14 or Section 18).

3   Turn the ignition switch and wiper switch on.

4   Connect a jumper wire between the wiper motor and ground. If the motor works

now, check for a bad ground connection.

5   If the wiper motor still doesn't work, turn the wipers on and check for voltage at the motor connector. If there is voltage, remove the motor and check it's operation off the vehicle with fused wires from the battery. If the motor works, check for a binding linkage. If the motor still doesn't work, replace it.

6   If there is no voltage at the motor, the problem lies in the switch or wiring.

Replacement

7   Disconnect the negative cable from the battery.

8   Unplug the wiper motor electrical connector and remove the mounting bolts **(see illustration)**.

9   Lower the motor far enough to gain access to the crank arm drive link retainer bushing.

10   Separate the crank arm from the drive link by prying the retainer bushing from the crank arm pin with a large screwdriver.

11   Remove the motor.

12   Installation is the reverse of removal.

24 Ignition lock cylinder - removal and installation

Warning: *Some models have airbags. The airbag is armed and can deploy (inflate) anytime the battery is connected. To prevent accidental deployment (and possible injury), disconnect the negative battery cable whenever working near airbag components. After the battery is disconnected, wait two minutes before beginning work (the system has a backup capacitor that must fully discharge). For more information see Section 28.*

1990 and earlier models

Refer to illustration 24.5

1   Disconnect the negative cable from the battery.

2   Remove the turn signal/hazard warning switch (see Section 17).

3   Remove the ignition key lamp.

4   Place the lock cylinder in the LOCK position.

5   Insert a thin screwdriver or punch into the lock release slot located next to the switch mounting screw boss, press down on the spring latch and withdraw the cylinder from the steering column **(see illustration)**.

6   Insert the new lock cylinder (without the key) into position until it contacts the ignition switch actuating rod. Move the rod up and down to remove any slack and align the components. Insert the key and push the lock cylinder in. The lock cylinder will snap into place automatically when the components align themselves, locking it in the housing.

7   The remainder of installation is the reverse of removal.

1991 and later models

Refer to illustrations 24.8, 24.9 and 24.10

8   Insert the key and turn it to the LOCK position. Press the retaining pin in with a small screwdriver until it is flush with the surface **(see illustration)**.

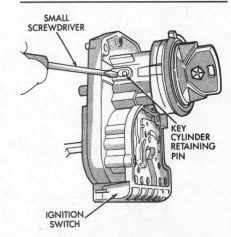

24.8 Push the retaining pin in to unseat the lock cylinder

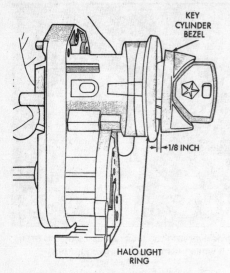

24.9 The lock cylinder will protrude about 1/8-inch from the switch once it's unseated - don't try to remove it until you rotate it to the Lock position and remove the key

9   Turn the key clockwise to the OFF position, which will unseat the lock cylinder, but don't try to remove it yet **(see illustration)**. With the cylinder unseated, rotate the key counterclockwise to the LOCK position, remove the key, then remove the lock cylinder from the ignition switch.

10   Installation is the reverse of removal, making sure that on column shift models the shifter is in Park. As the switch is engaged to the column shift park slider linkage, make sure the column lock flag on the switch is parallel with the electrical connectors **(see illustration)**.

11   Insert the lock cylinder in the LOCK position until it bottoms in the switch. While pushing the lock cylinder in, insert the key and turn it clockwise to the RUN position.

25 Cruise control system - description and check

Refer to illustration 25.6

1   The cruise control system maintains vehicle speed with a vacuum actuated servo motor located in the engine compartment, which is connected to the throttle linkage by a cable. The system consists of the vacuum servo motor, brake switch, control switches, a relay and associated vacuum hoses. Later models also incorporate the Vehicle Speed Sensor (VSS) and the Powertrain Control Module (PCM). Cruise controls all work by the same basic principles; however, the hardware used varies considerably depending on model and year of manufacture. Some later systems require special testers and diagnostic procedures which are beyond the scope of the home mechanic. Listed below are some general procedures that may be used to locate common problems.

2   Locate and check the fuse (see Section 3).

3   Have an assistant operate the brake lights while you check their operation (voltage from the brake light switch deactivates the cruise control).

4   If the brake lights don't come on or don't shut off, correct the problem and retest the cruise control.

5   Inspect the control linkage between the cruise control servo (or actuator) and the throttle linkage. This will consist of either a cable, chain or metal rod. The cruise control servo is usually fist-sized or slightly larger and is located near the carburetor or throttle body.

6   Visually inspect the vacuum hose(s) and wires connected to the cruise control servo and transducer (if equipped) and replace as necessary **(see illustration)**.

7   Test drive the vehicle to determine if the cruise control is now working. If it isn't, take it to a dealer service department or an automotive electrical specialist for further diagnosis and repair.

26 Power door lock system - description and check

1   The power door lock system operates the door lock actuators mounted in each door. The system consists of the switches, actuators, relays and associated wiring. Diagnosis can usually be limited to simple checks of the wiring connections and actuators for minor faults which can be easily repaired.

2   Power door lock systems are operated by bi-directional solenoids located in the doors. The lock switches have two operating positions: Lock and Unlock. These switches activate a relay which in turn connects voltage to the door lock solenoids. Depending on which way the relay is activated, it reverses polarity, allowing the two sides of the circuit to be used alternately as the feed (positive) and ground side.

3   Some vehicles may have anti-theft systems incorporated into the power locks. If you are unable to locate the trouble using the following general steps, consult a dealer service department or other qualified repair shop.

4   Always check the circuit protection first (see Section 3).

5   Operate the door lock switches in both directions (Lock and Unlock) with the engine off. Listen for the faint click of the relay operating.

6   If there's no click, check for voltage at the switches. If no voltage is present, check the wiring between the fuse block and the switches for shorts and opens.

7   If voltage is present but no click is heard, test the switch for continuity. Replace it if there's no continuity in both switch positions.

8   If the switch has continuity but the relay doesn't click, check the wiring between the switch and relay for continuity. Repair the wiring if there's no continuity.

9   If the relay is receiving voltage from the

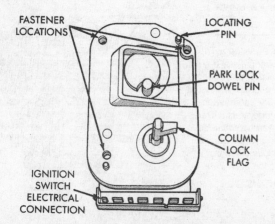

24.10 Make sure the switch column lock flag is parallel with the electrical connection before installation

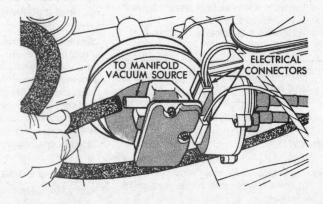

25.6 Check the cruise control servo and vacuum and electrical connections for damage

switch but is not sending voltage to the solenoids, check for a bad ground at the relay case. If the relay case is grounding properly, replace the relay.

10   If all but one lock solenoids operate, remove the trim panel from the affected door (see Chapter 11) and check for voltage at the solenoid while the lock switch is operated. One of the wires should have voltage in the Lock position; the other should have voltage in the Unlock position.

11   If the inoperative solenoid is receiving voltage, replace the solenoid.

12   If the inoperative solenoid isn't receiving voltage, check for an open or short in the wire between the lock solenoid and the relay. **Note:** *It's common for wires to break in the portion of the harness between the body and door (opening and closing the door fatigues and eventually breaks the wires).*

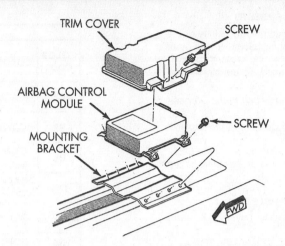

28.5 The airbag control module is mounted underneath the driver's seat

27   Power window system - description and check

1   The power window system operates the electric motors mounted in the doors which lower and raise the windows. The system consists of the control switches, the motors, glass mechanisms (regulators) and associated wiring.

2   The windows can be lowered and raised from the master control switch by the driver or by remote switches located at the individual windows. Each window has a separate motor which is reversible. The position of the control switch determines the polarity and therefore the direction of operation. Some systems are equipped with relays that control current flow to the motors.

3   Each motor is equipped with a separate circuit breaker in addition to the fuse or circuit breaker protecting the whole circuit. This prevents one stuck window from disabling the whole system.

4   The power window system will only operate when the ignition switch is ON. In addition, when activated the window lockout switch at the master control switch disables the switches at the passenger's window also. Always check these items before troubleshooting a window problem.

5   These procedures are general in nature, so if you can't find the problem using them, take the vehicle to a dealer service department.

6   If the power windows don't work at all, check the fuse or circuit breaker.

7   If the windows only operate from the master control switch, check the window lockout switch for continuity in the unlocked position. Replace it if it doesn't have continuity.

8   Check the wiring between the switches and fuse panel for continuity. Repair the wiring, if necessary.

9   If only one window is inoperative from the master control switch, try the other control switch at the window. **Note:** *This doesn't apply to the drivers door window.*

10   If the same window works from one

switch, but not the other, check the switch for continuity.

11   If the switch tests OK, check for a short or open in the wiring between the affected switch and the window motor.

12   If one window is inoperative from both switches, remove the trim panel from the affected door and check for voltage at the motor while the switch is operated.

13   If voltage is reaching the motor, disconnect the glass from the regulator (see Chapter 11). Move the window up and down by hand while checking for binding and damage. Also check for binding and damage to the regulator. If the regulator is not damaged and the window moves up and down smoothly, replace the motor. If there's binding or damage, lubricate, repair or replace parts, as necessary.

14   If voltage isn't reaching the motor, check the wiring in the circuit for continuity between the switches and motors.

15   Test the windows after you are done to confirm proper repairs.

28   Airbag - general information

Refer to illustration 28.5

1995 and later models covered by this manual are equipped with a Supplemental Restraint Systems (SRS), more commonly known as airbags. All 1995 and later models are equipped with a driver side airbag and all 1998 and later models are equipped with a passenger side airbag, located in the instrument panel. This system is designed to protect the driver, and on 1998 and later models, the front seat passenger from head-on or frontal collision. It consists of an airbag module in the center of the steering wheel and an airbag module in the instrument panel. The airbag control module, which also contains an impact sensor, is mounted under the driver's seat.

Note: *1996 and later models are also*

equipped with seat belt pre-tensioner which uses an explosive charge (similar to that of the airbag) to tighten the seat belt during a frontal collision. This system is also controlled by the airbag control module and is activated at the same time as the airbag is deployed.

Airbag module

All airbag modules contain a housing incorporating the cushion (airbag) and inflator unit. The inflator assembly is mounted on the back of the housing over a hole through which gas is expelled, inflating the airbag almost instantaneously when an electrical signal is sent from the control module.

On the steering wheel airbag module, the specially wound wire that carries this signal to the module is called a clockspring. This clockspring is a flat, ribbon-like electrically conductive tape which winds and unwinds as the steering wheel is turned so it can transmit an electrical signal regardless of steering wheel position.

The passenger side airbag module is wired to the control module with normal wiring since it is mounted directly to the instrument panel and does not move.

Airbag control module

The airbag control module contains the impact sensor and an on-board microprocessor which monitors the operation of the system **(see illustration)**. It checks this system every time the vehicle is started, causing the AIRBAG light to go on, then off, if the system is operating properly. If there is a fault in the system, the light will go on and stay on and the airbag control module will store fault codes indicating the nature of the fault. If the AIRBAG light does go on and stay on, the vehicle should be taken to your dealer immediately for service.

The control module also contains a back-up capacitor which is capable of deploying the airbag if the battery becomes disconnected.

12

Disabling the system

Whenever working in the vicinity of the steering wheel, steering column or near other components of the airbag system, the system should be disarmed. To do this perform the following steps:

a) *Turn the ignition switch to the Off position.*

b) *Disconnect the cable from the negative battery terminal.*

c) *Wait at least two minutes for the back-up power supply to be depleted before beginning work.*

Enabling the system

To enable the airbag system, perform the following steps:

a) *Turn the ignition switch to the Off position.*

b *Connect the cable to the negative battery terminal.*

c) *Turn the ignition switch to the On position. Confirm that the airbag warning light glows for 6 to 8 seconds, then goes out, indicating the system is functioning properly.*

29   Wiring diagrams - general information

Refer to illustration 29.4

Since it isn't possible to include all wiring diagrams for every year and model covered by this manual, the following diagrams are those that are typical and most commonly needed.

Prior to troubleshooting any circuits, check the fuse and circuit breakers (if equipped) to make sure they're in good condition. Make sure the battery is properly charged and check the cable connections (see Chapter 1).

When checking a circuit, make sure that all connectors are clean, with no broken or loose terminals. When unplugging a connector, do not pull on the wires. Pull only on the connector housings themselves.

Refer to the accompanying chart **(see illustration)** for the wire color codes applicable to your vehicle.

WIRE COLOR CODE CHART					
COLOR CODE	**COLOR**	**STANDARD TRACER COLOR**	**COLOR CODE**	**COLOR**	**STANDARD TRACER CODE**
BK	BLACK	WT	PK	PINK	BK OR WH
BR	BROWN	WT	RD	RED	WT
DB	DARK BLUE	WT	TN	TAN	WT
DG	DARK GREEN	WT	VT	VIOLET	WT
GY	GRAY	BK	WT	WHITE	BK
LB	LIGHT BLUE	BK	YL	YELLOW	BK
LG	LIGHT GREEN	BK	*	WITH TRACER	
OR	ORANGE	BK			

29.4 Wiring diagram color codes

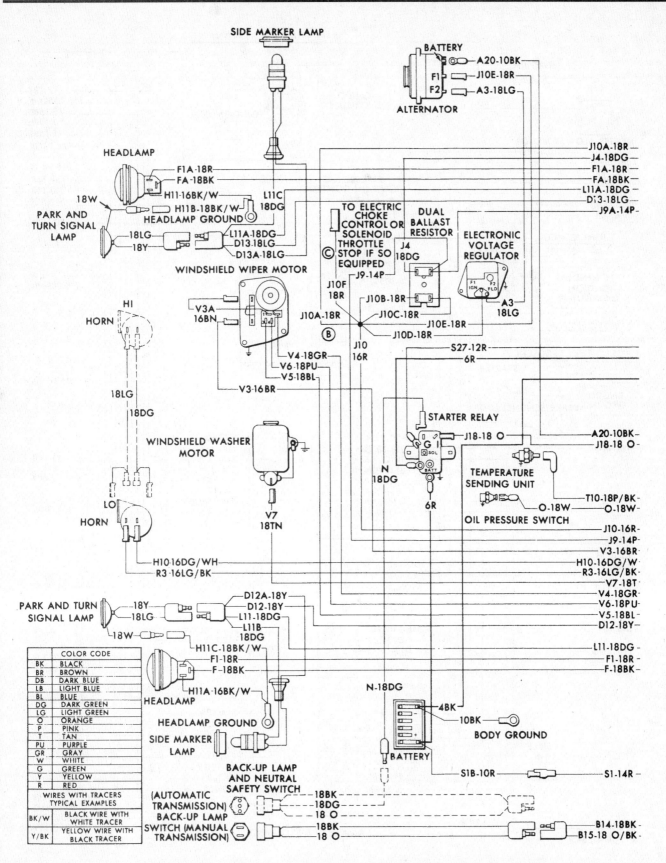

Wiring diagram - 1971 through 1976 models (1977 and 1978 models similar) (1 of 6)
Front lights, charging system, starter relay, engine switches

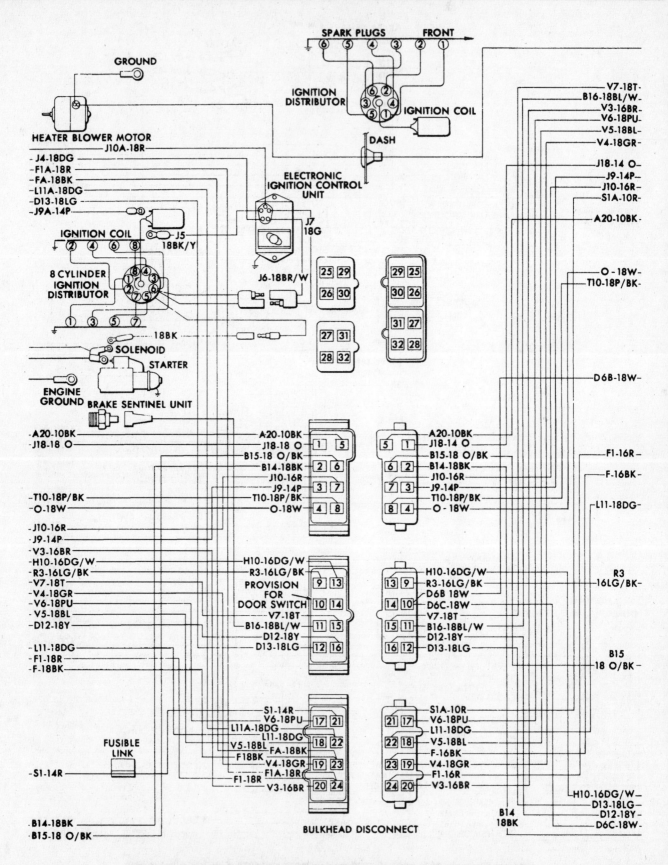

Wiring diagram - 1971 through 1976 models (1977 and 1978 models similar) (2 of 6)
Heater, ignition distributor, bulkhead disconnect block

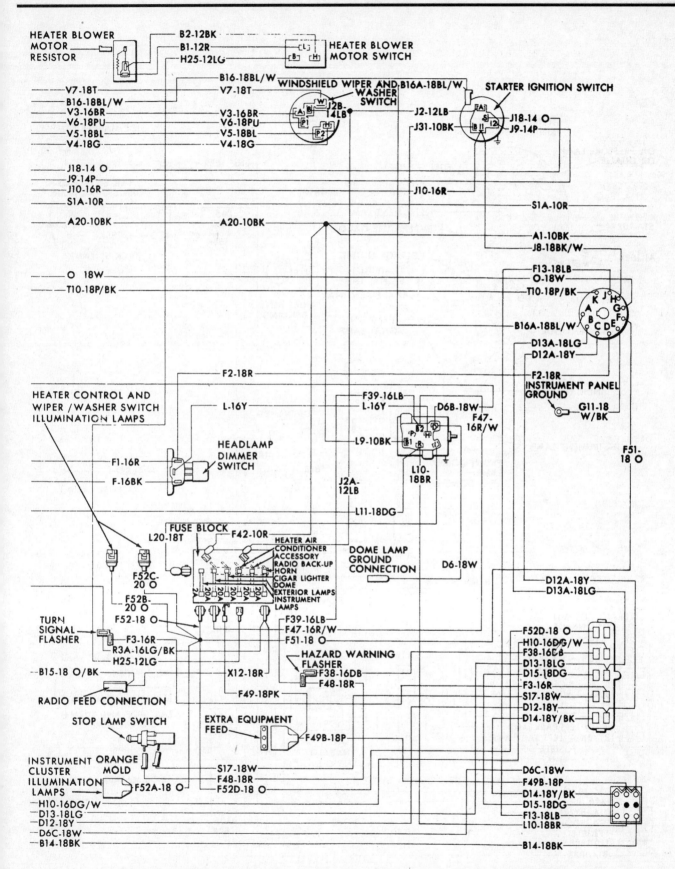

Wiring diagram - 1971 through 1976 models (1977 and 1978 models similar) (3 of 6)
Heater, wiper/washer, ignition switch, instrument lights

12

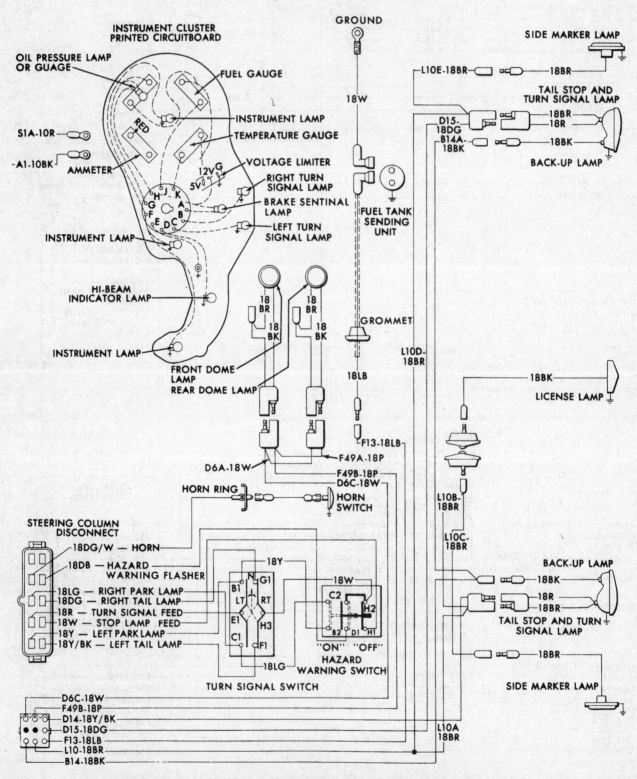

Wiring diagram - 1971 through 1976 models (1977 and 1978 models similar) (4 of 6)
Instrument cluster, steering column, rear lights

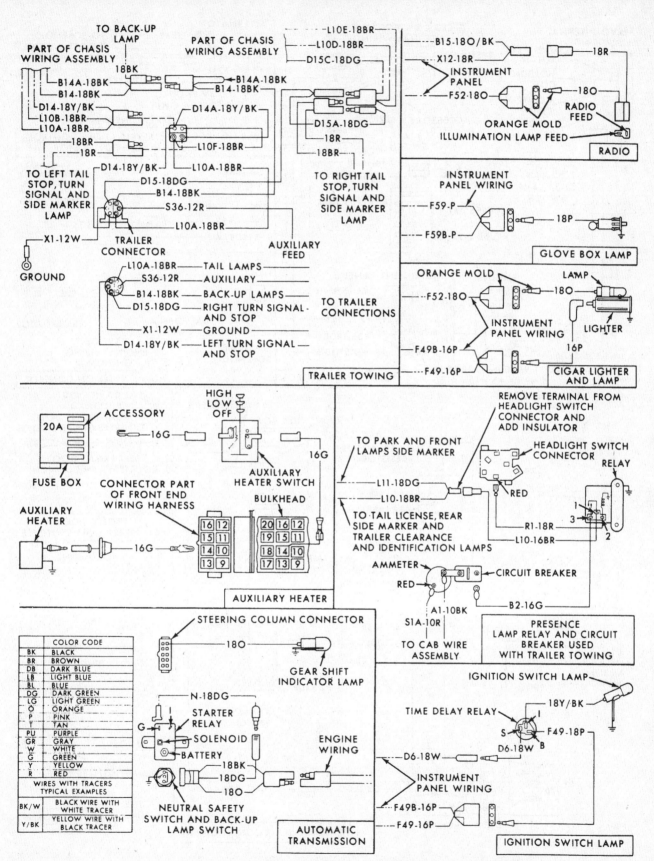

Wiring diagram - 1971 through 1976 models (1977 and 1978 models similar) (5 of 6)
Auxiliary wiring systems

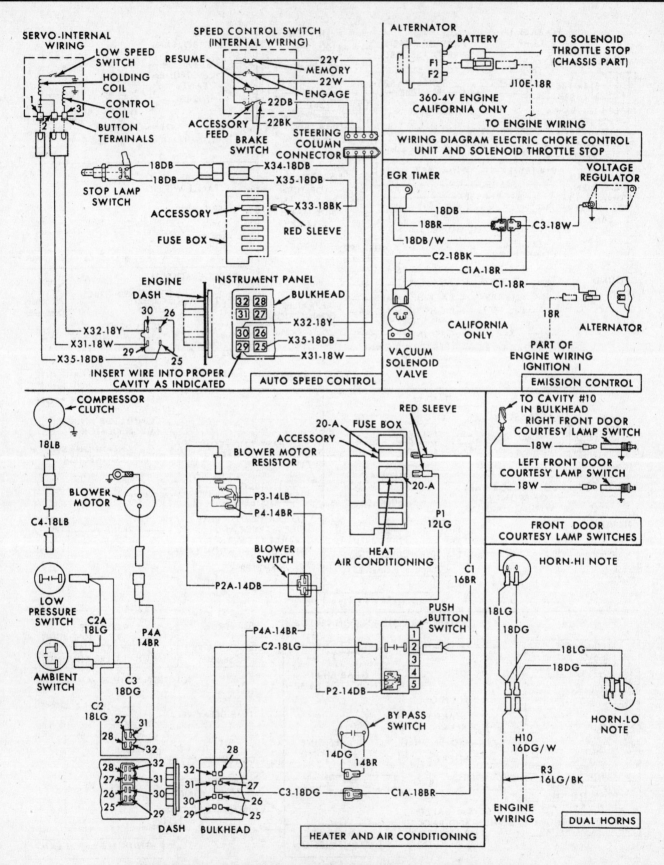

Wiring diagram - 1971 through 1976 models (1977 and 1978 models similar) (6 of 6)
Auxiliary wiring systems

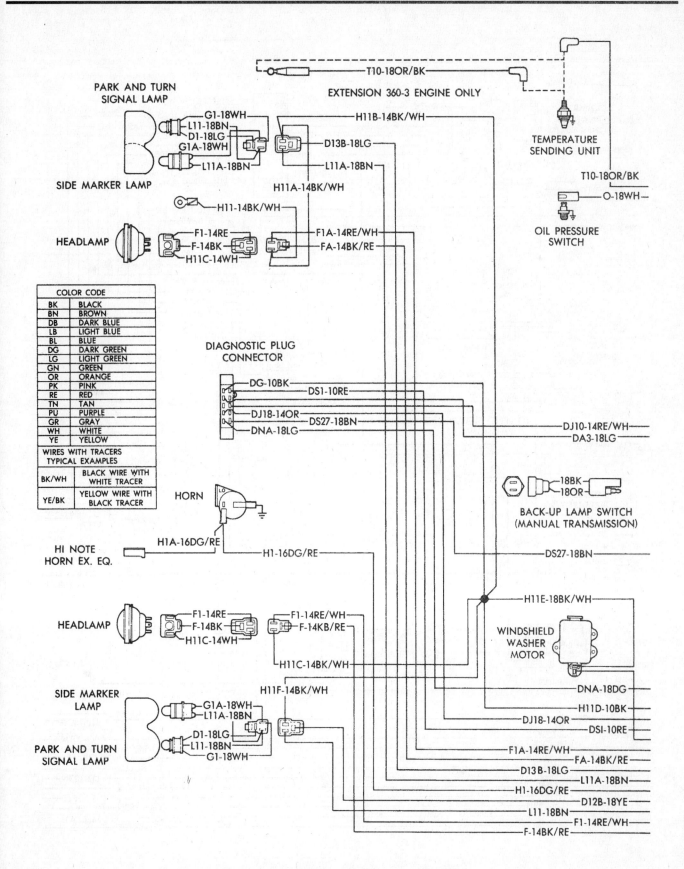

PARK AND TURN SIGNAL LAMP

- G1-18WH
- L11-18BN
- D1-18LG
- G1A-18WH
- L11A-18BN

SIDE MARKER LAMP

HEADLAMP
- F1-14RE
- F-14BK
- H11C-14WH

H11-14BK/WH

H11A-14BK/WH

H11B-14BK/WH

D13B-18LG

L11A-18BN

F1A-14RE/WH

FA-14BK/RE

EXTENSION 360-3 ENGINE ONLY

T10-18OR/BK

TEMPERATURE SENDING UNIT

T10-18OR/BK

O-18WH

OIL PRESSURE SWITCH

COLOR CODE	
BK	BLACK
BN	BROWN
DB	DARK BLUE
LB	LIGHT BLUE
BL	BLUE
DG	DARK GREEN
LG	LIGHT GREEN
GN	GREEN
OR	ORANGE
PK	PINK
RE	RED
TN	TAN
PU	PURPLE
GR	GRAY
WH	WHITE
YE	YELLOW

WIRES WITH TRACERS TYPICAL EXAMPLES	
BK/WH	BLACK WIRE WITH WHITE TRACER
YE/BK	YELLOW WIRE WITH BLACK TRACER

DIAGNOSTIC PLUG CONNECTOR

- DG-10BK
- DS1-10RE
- DJ18-14OR
- DS27-18BN
- DNA-18LG

DJ10-14RE/WH

DA3-18LG

18BK
18OR

BACK-UP LAMP SWITCH (MANUAL TRANSMISSION)

HORN

LO

H1A-16DG/RE

HI NOTE HORN EX. EQ.

H1-16DG/RE

DS27-18BN

H11E-18BK/WH

WINDSHIELD WASHER MOTOR

HEADLAMP
- F1-14RE
- F-14BK
- H11C-14WH

F1-14RE/WH

F-14KB/RE

H11C-14BK/WH

SIDE MARKER LAMP

H11F-14BK/WH

- G1A-18WH
- L11A-18BN
- D1-18LG
- L11-18BN
- G1-18WH

PARK AND TURN SIGNAL LAMP

- DNA-18DG
- H11D-10BK
- DJ18-14OR
- DSI-10RE
- F1A-14RE/WH
- FA-14BK/RE
- D13B-18LG
- L11A-18BN
- H1-16DG/RE
- D12B-18YE
- L11-18BN
- F1-14RE/WH
- F-14BK/RE

12

Wiring diagram - 1979 through 1984 models (1 of 6)
Front lights, horn, engine switches

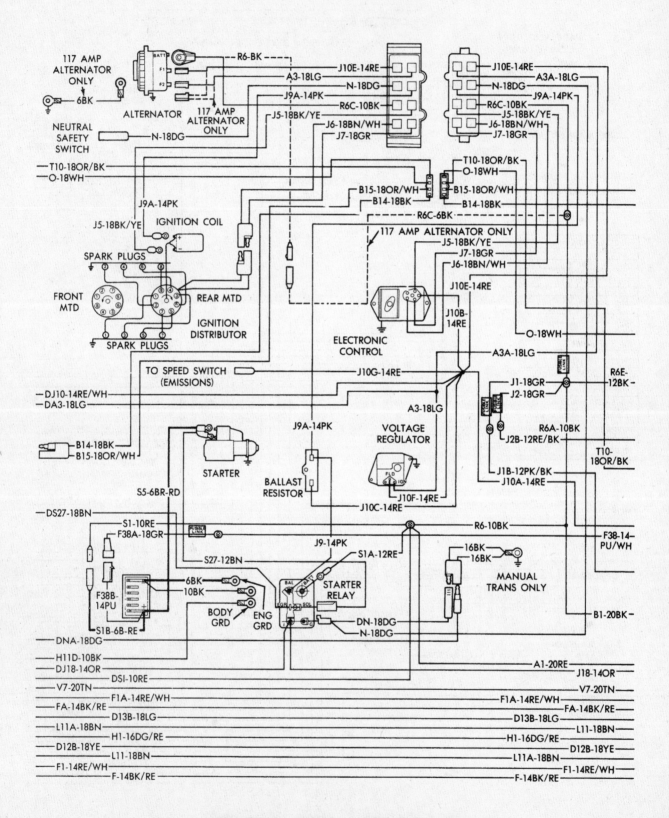

Wiring diagram - 1979 through 1984 models (2 of 6)
Charging system, ignition distributor, starting system

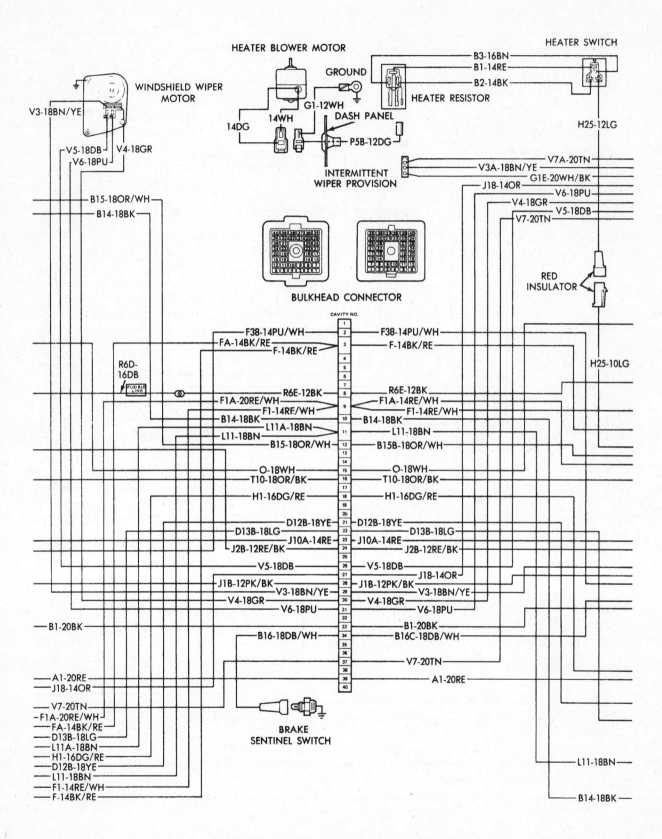

Wiring diagram - 1979 through 1984 models (3 of 6)
Heater, wiper motor, bulkhead disconnect block

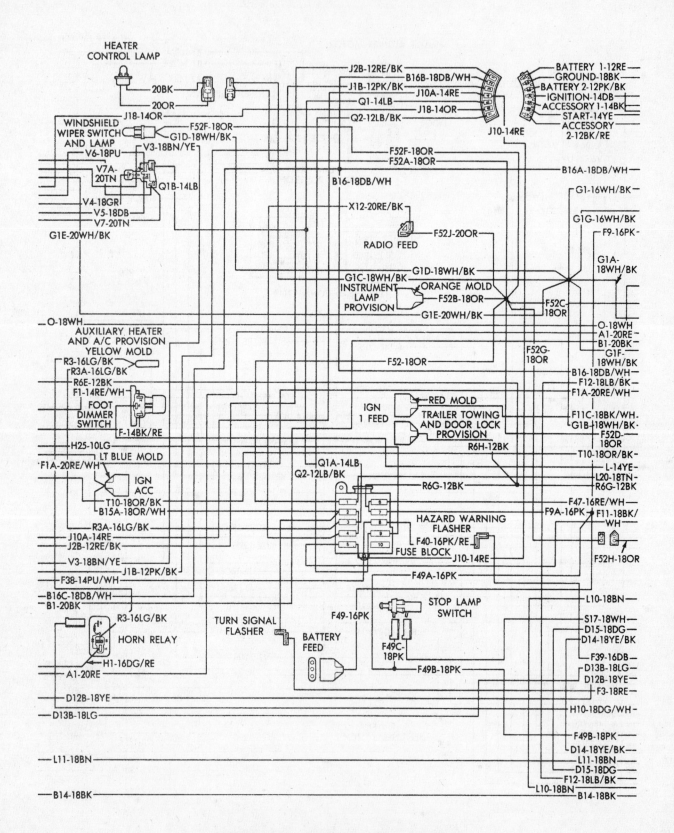

Wiring diagram - 1979 through 1984 models (4 of 6)
Fuse block, ignition switch, auxiliary wiring systems

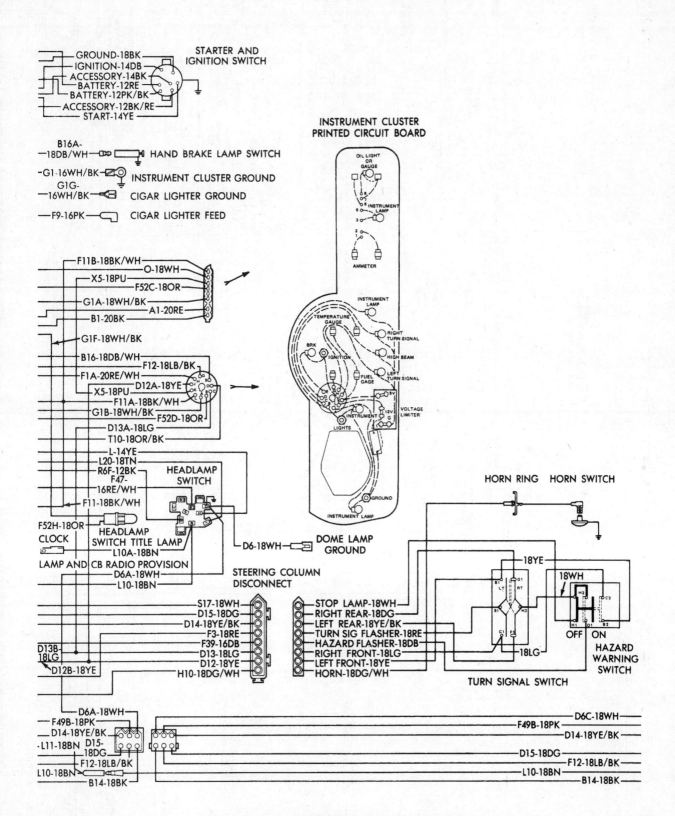

Wiring diagram - 1979 through 1984 models (5 of 6)
Instrument cluster and switches

12

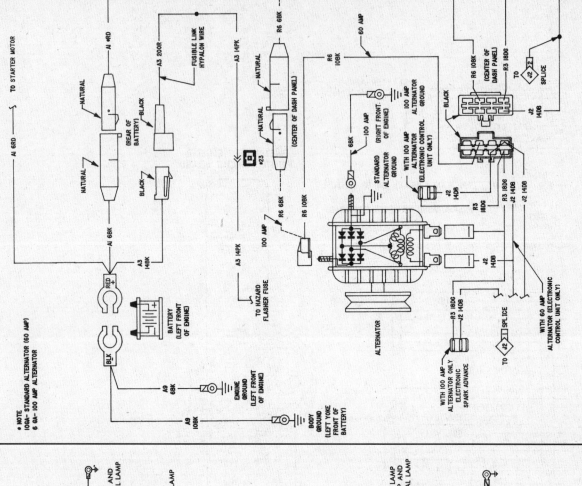

Wiring diagram - 1985 through 1987 models (1 of 11)

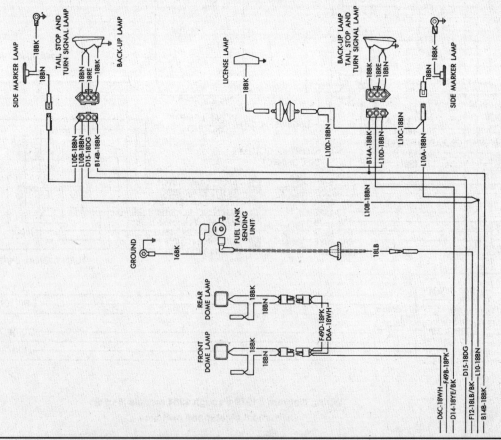

Wiring diagram - 1979 through 1984 models (6 of 6)

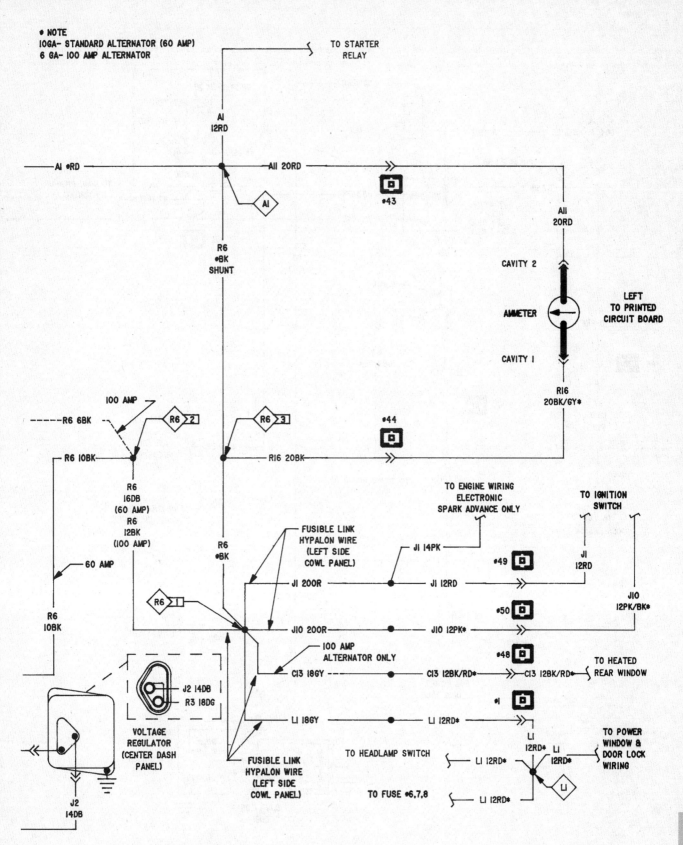

Wiring diagram - 1985 through 1987 models (2 of 11)
Charging system

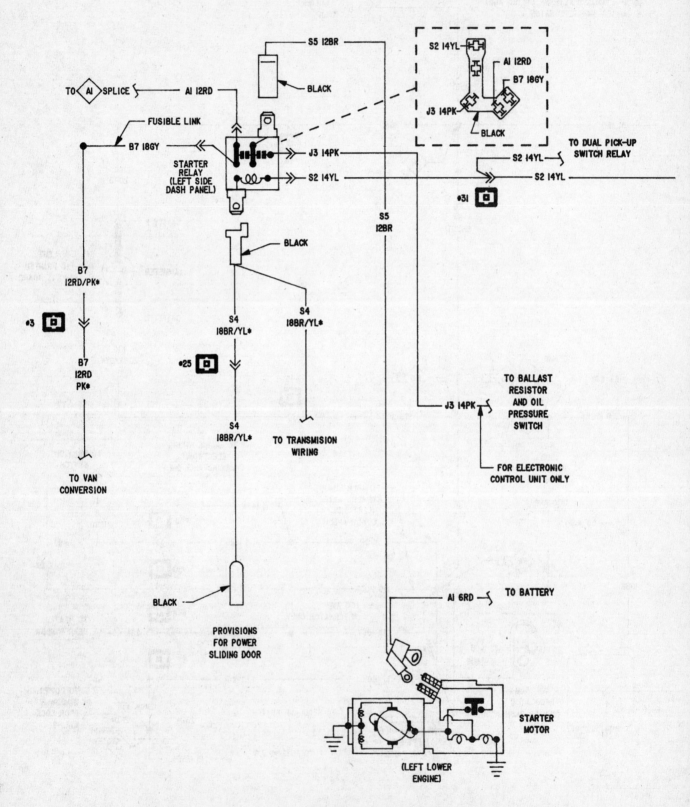

Wiring diagram - 1985 through 1987 models (3 of 11)
Starting system

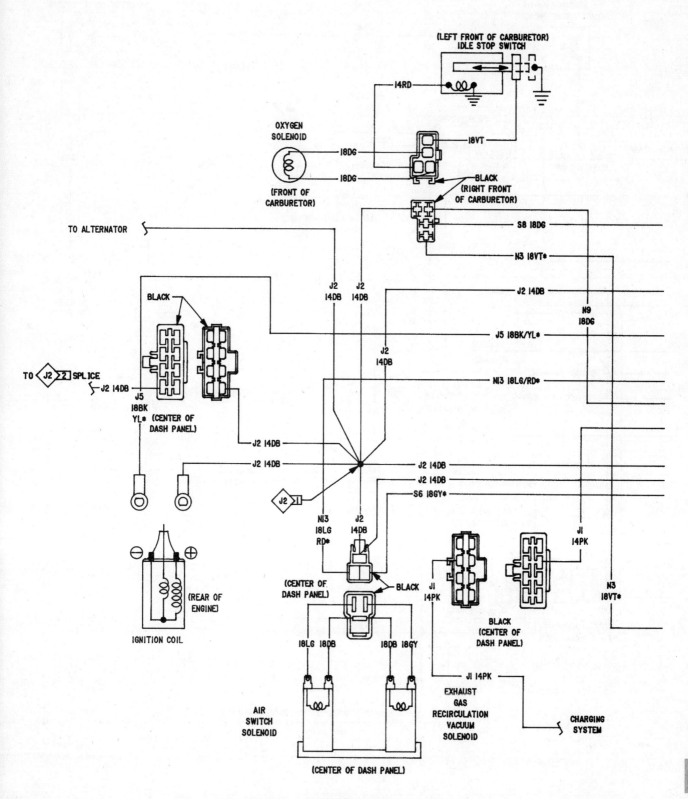

Wiring diagram - 1985 through 1987 models (4 of 11)
Ignition system - 318 V8

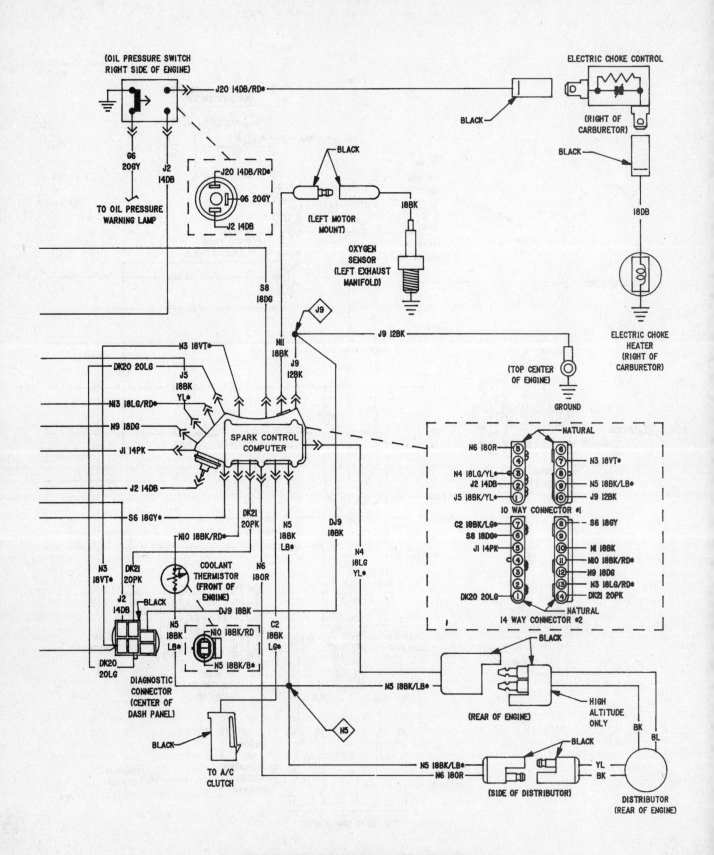

Wiring diagram - 1985 through 1987 models (5 of 11)
Ignition system - 318 V8

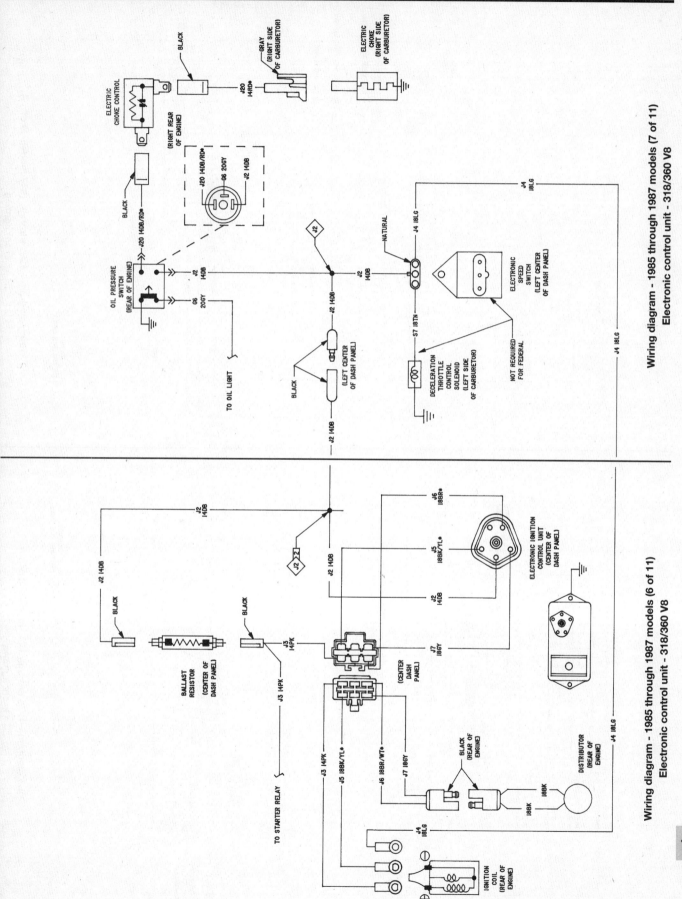

Wiring diagram - 1985 through 1987 models (7 of 11)
Electronic control unit - 318/360 V8

Wiring diagram - 1985 through 1987 models (6 of 11)
Electronic control unit - 318/360 V8

12

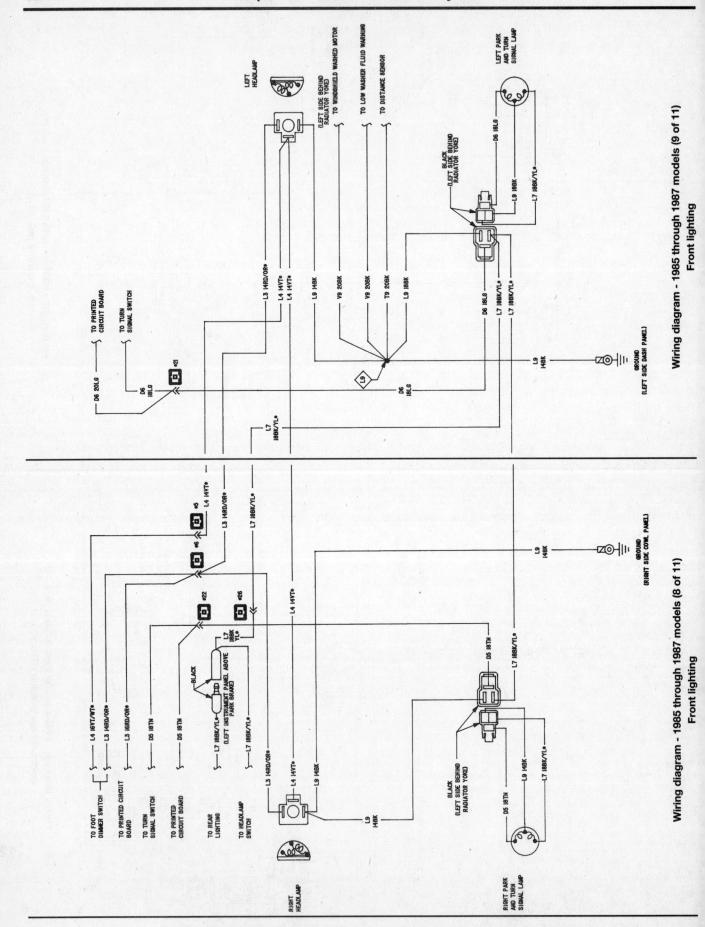

Wiring diagram - 1985 through 1987 models (9 of 11)
Front lighting

Wiring diagram - 1985 through 1987 models (8 of 11)
Front lighting

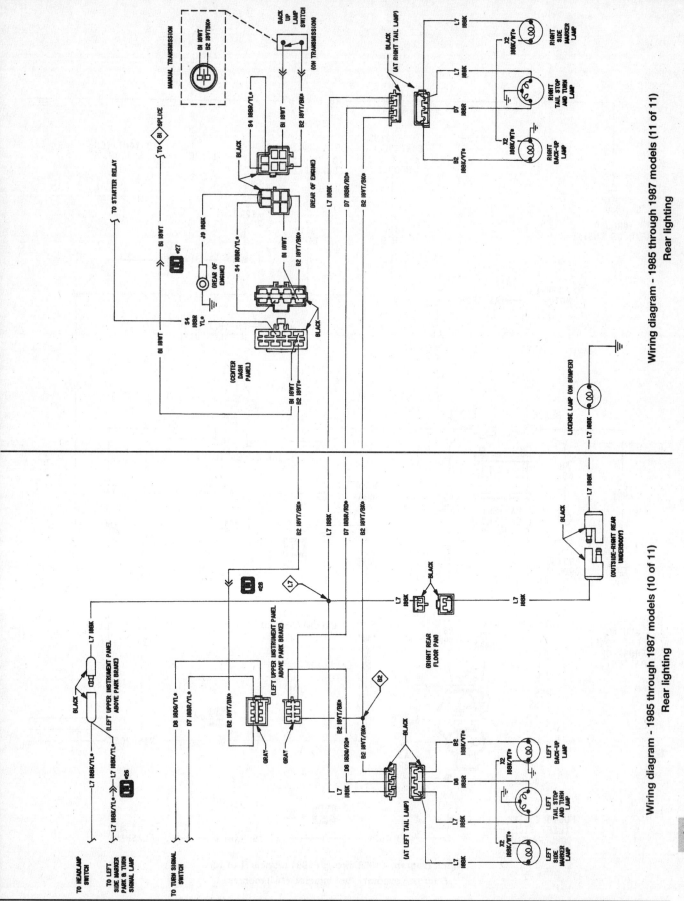

Wiring diagram - 1985 through 1987 models (11 of 11)
Rear lighting

Wiring diagram - 1985 through 1987 models (10 of 11)
Rear lighting

12

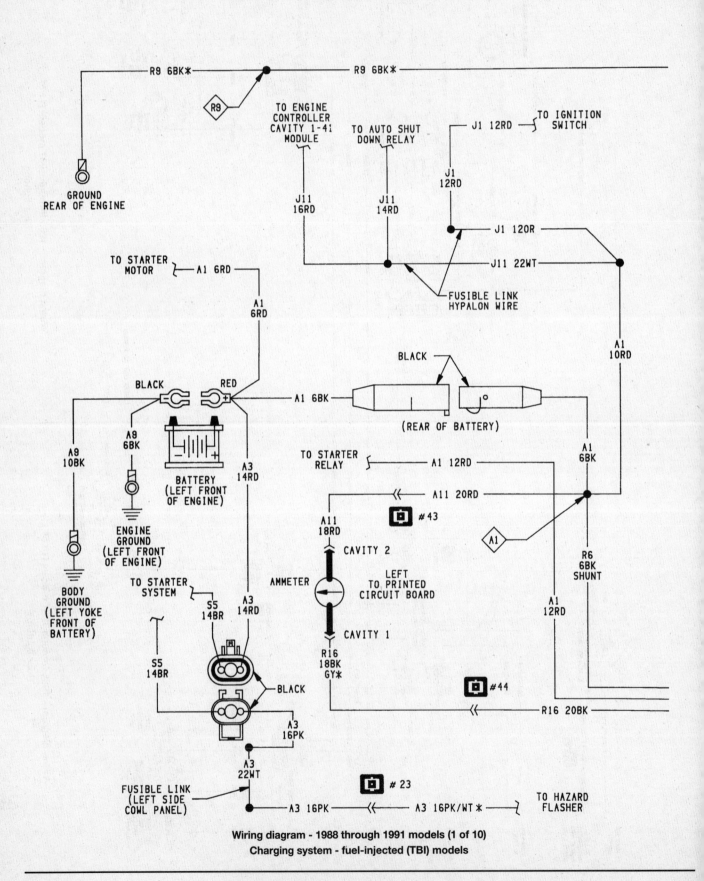

Wiring diagram - 1988 through 1991 models (1 of 10)

Charging system - fuel-injected (TBI) models

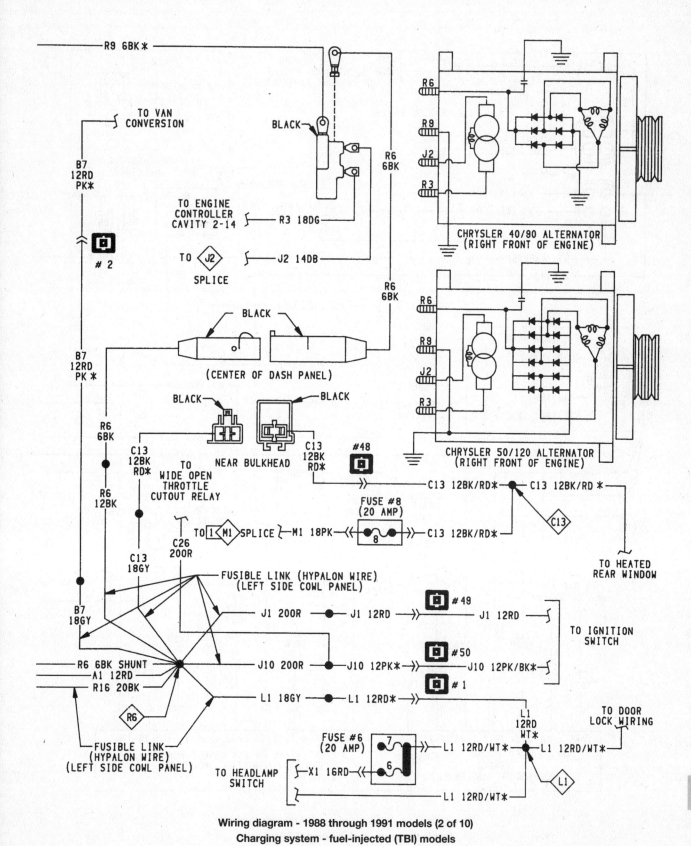

Wiring diagram - 1988 through 1991 models (2 of 10)
Charging system - fuel-injected (TBI) models

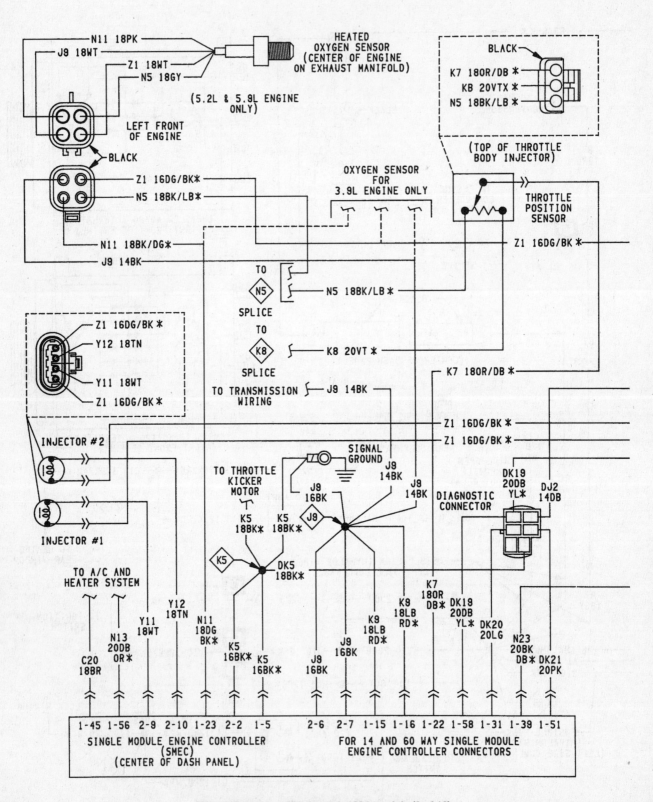

Wiring diagram - 1988 through 1991 models (3 of 10)
Ignition system - fuel-injected (TBI) models

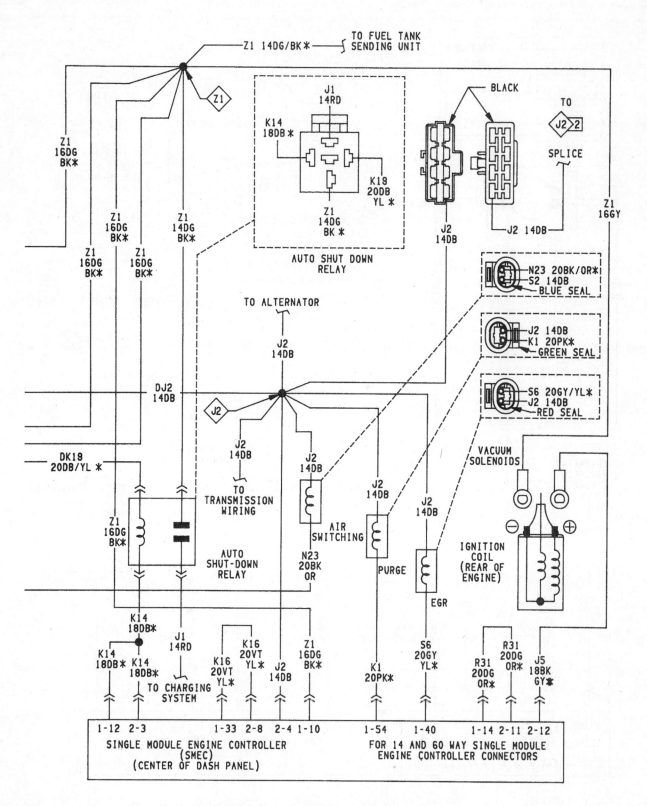

Wiring diagram - 1988 through 1991 models (4 of 10)
Ignition system - fuel-injected (TBI) models

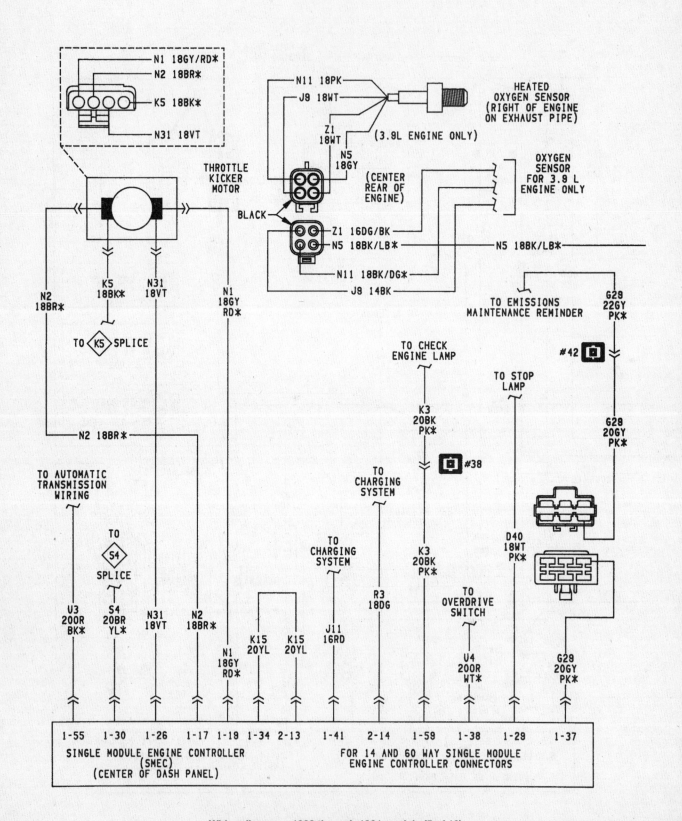

N1 18GY/RD✳
N2 18BR✳
K5 18BK✳
N31 18VT

N11 18PK
J9 18WT
Z1 18WT
N5 18GY

HEATED
OXYGEN SENSOR
(RIGHT OF ENGINE
ON EXHAUST PIPE)

(3.9L ENGINE ONLY)

(CENTER
REAR OF
ENGINE)

OXYGEN
SENSOR
FOR 3.9 L
ENGINE ONLY

THROTTLE
KICKER
MOTOR

BLACK

Z1 16DG/BK
N5 18BK/LB✳   N5 18BK/LB✳
N11 18BK/DG✳
J9 14BK

N2
18BR✳

K5
18BK✳

N31
18VT

N1
18GY
RD✳

TO EMISSIONS
MAINTENANCE REMINDER

G29
22GY
PK✳

TO K5 SPLICE

TO CHECK
ENGINE LAMP

#42

K3
20BK
PK✳

G29
20GY
PK✳

N2 18BR✳

#38

TO AUTOMATIC
TRANSMISSION
WIRING

TO
S4
SPLICE

TO
CHARGING
SYSTEM

TO
CHARGING
SYSTEM

TO STOP
LAMP

D40
18WT
PK✳

K3
20BK
PK✳

TO
OVERDRIVE
SWITCH

V3
200R
BK✳

S4
20BR
YL✳

N31
18VT

N2
18BR✳

N1
18GY
RD✳

K15
20YL

K15
20YL

J11
16RD

R3
18DG

U4
200R
WT✳

G29
20GY
PK✳

1-55   1-30   1-26   1-17   1-19   1-34   2-13   1-41   2-14   1-59   1-38   1-29   1-37

SINGLE MODULE ENGINE CONTROLLER
(SMEC)
(CENTER OF DASH PANEL)

FOR 14 AND 60 WAY SINGLE MODULE
ENGINE CONTROLLER CONNECTORS

Wiring diagram - 1988 through 1991 models (5 of 10)
Ignition system - fuel-injected (TBI) models

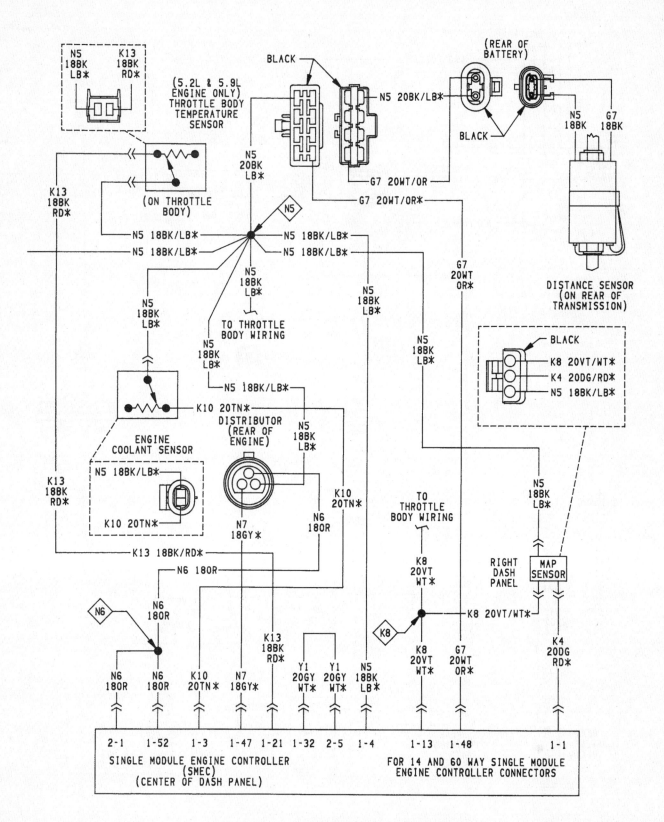

Wiring diagram - 1988 through 1991 models (6 of 10)
Ignition system - fuel-injected (TBI) models

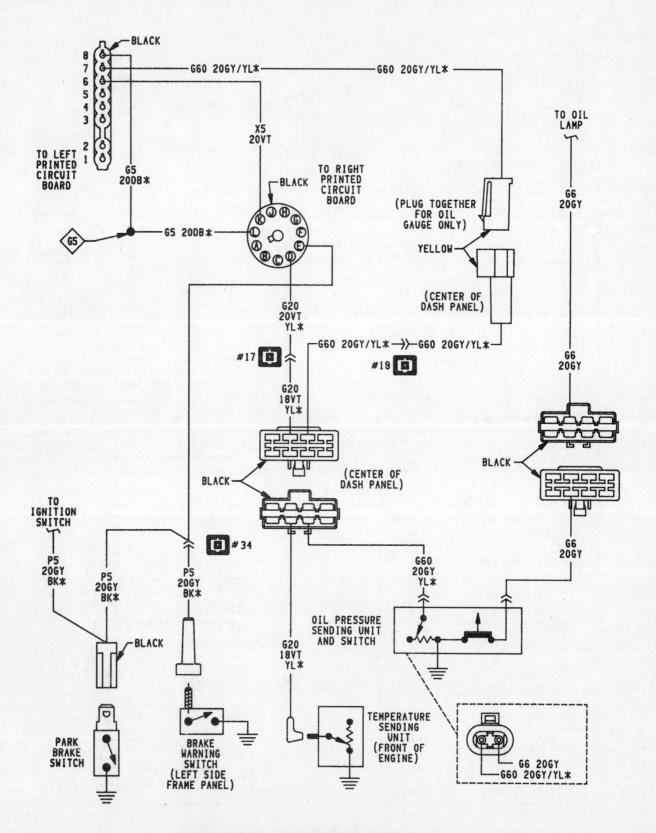

Wiring diagram - 1988 through 1991 models (7 of 10)
Oil pressure, coolant temperature and brake warning light

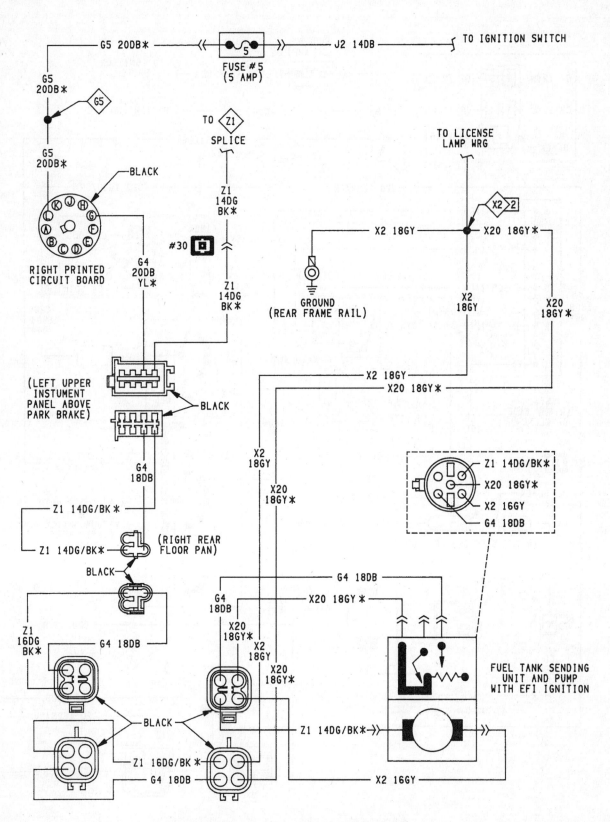

Wiring diagram - 1988 through 1991 models (8 of 10)
Fuel pump and gauge - fuel-injected (TBI) models

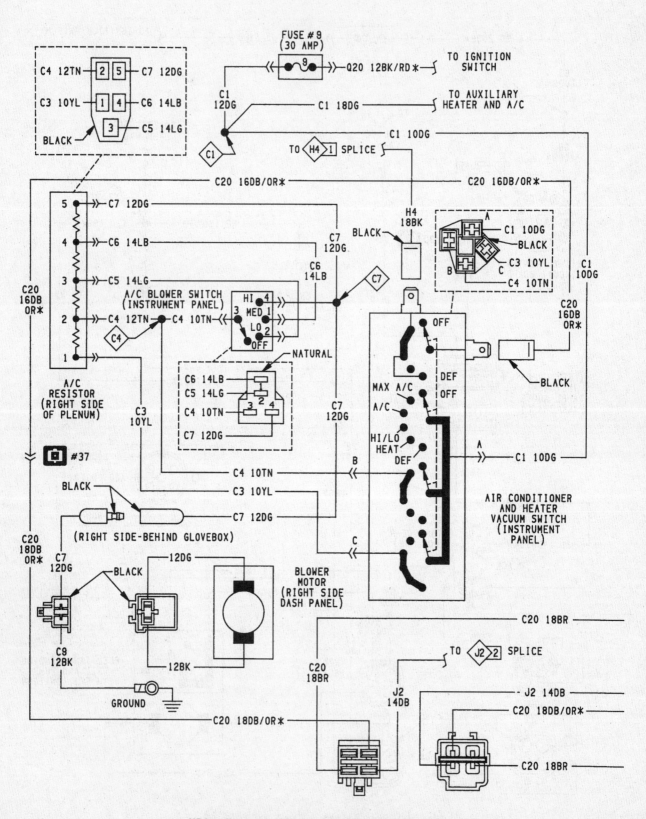

Wiring diagram - 1988 through 1991 models (9 of 10)
Heater and air conditioning system

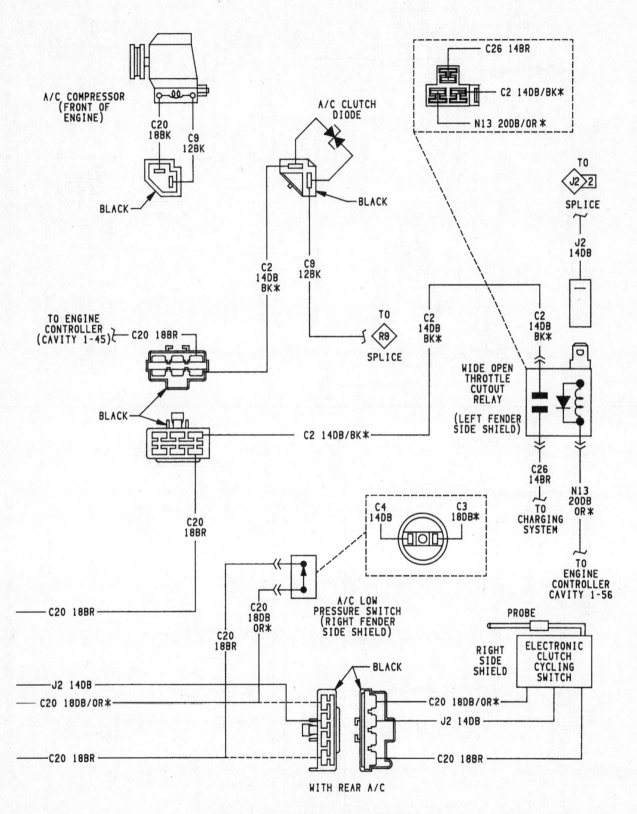

Wiring diagram - 1988 through 1991 models (10 of 10)
Heater and air conditioning system

12

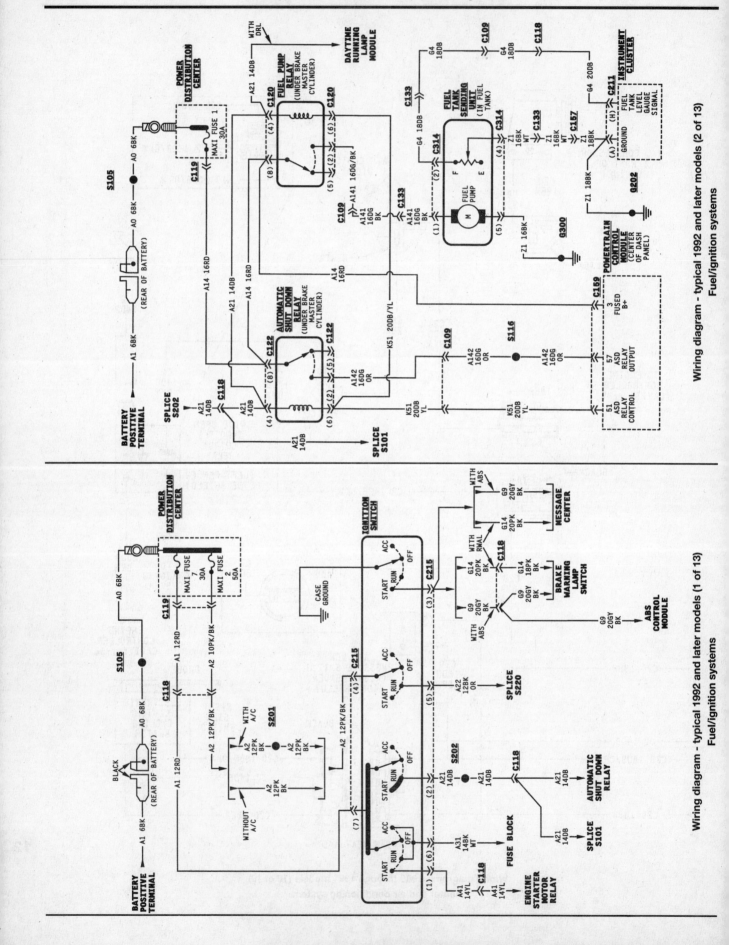

Wiring diagram - typical 1992 and later models (2 of 13)
Fuel/ignition systems

Wiring diagram - typical 1992 and later models (1 of 13)
Fuel/ignition systems

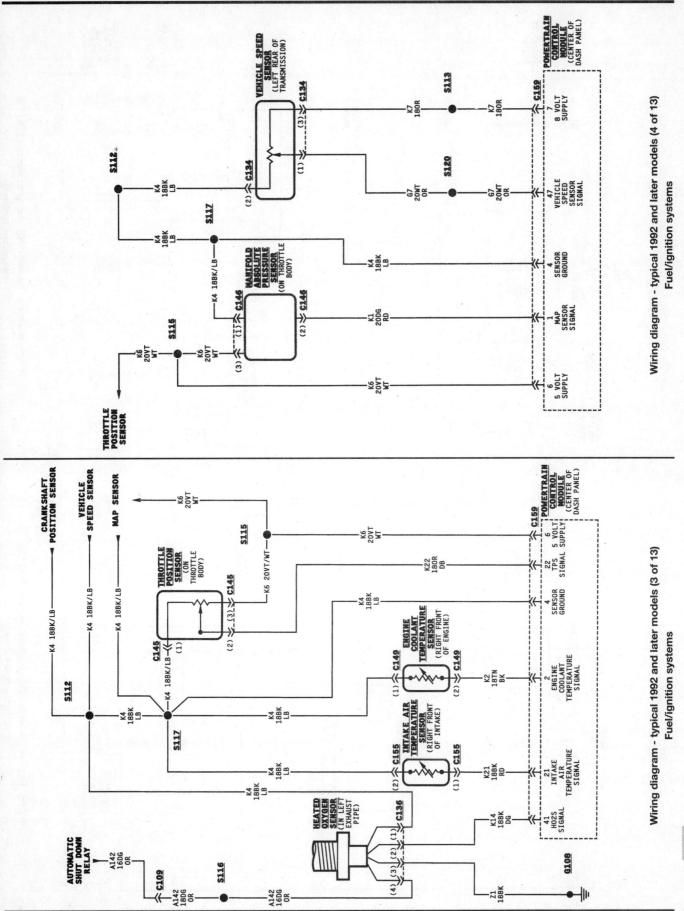

Wiring diagram - typical 1992 and later models (4 of 13)
Fuel/ignition systems

Wiring diagram - typical 1992 and later models (3 of 13)
Fuel/ignition systems

12

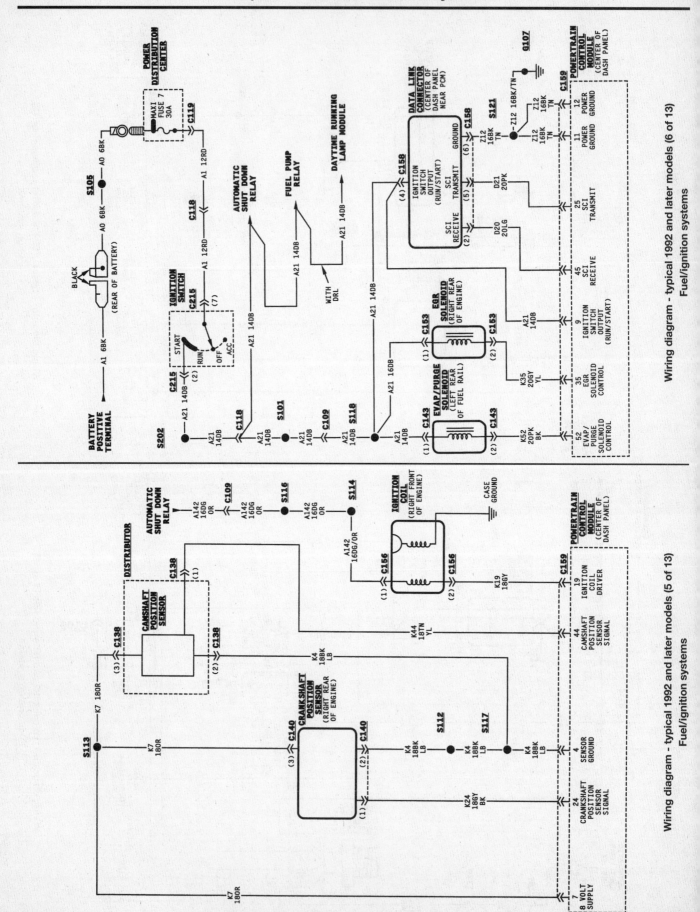

Wiring diagram - typical 1992 and later models (6 of 13) Fuel/ignition systems

Wiring diagram - typical 1992 and later models (5 of 13) Fuel/ignition systems

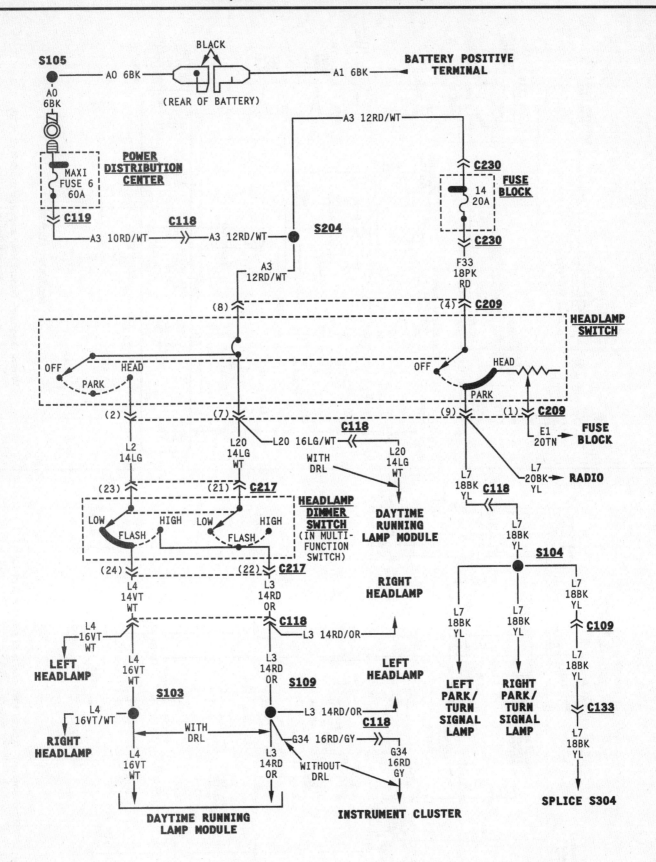

Wiring diagram - typical 1992 and later models (7 of 13)
Front lighting

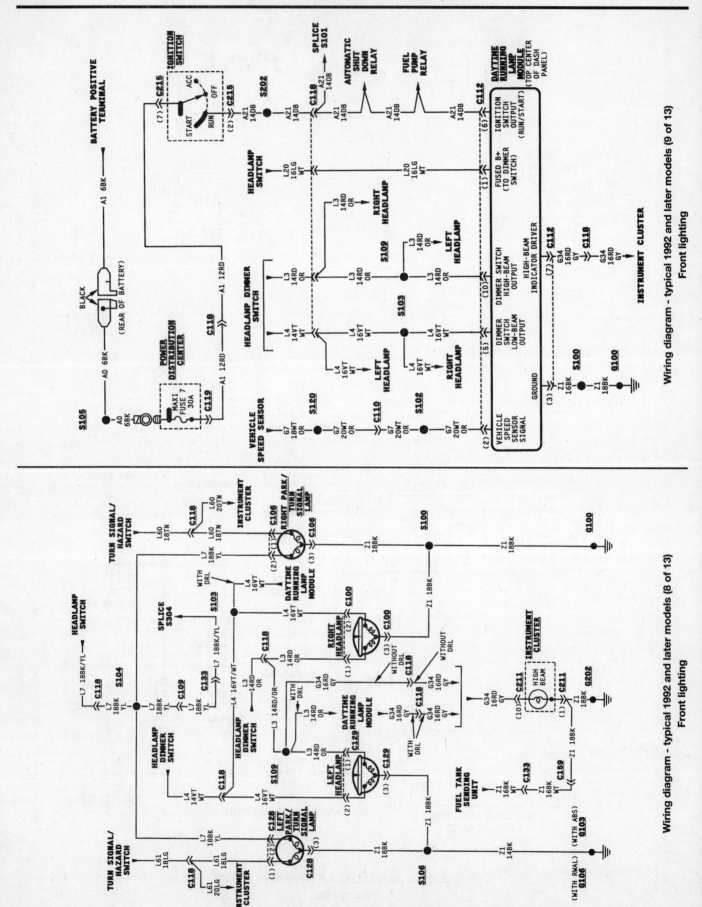

Wiring diagram - typical 1992 and later models (9 of 13)
Front lighting

Wiring diagram - typical 1992 and later models (8 of 13)
Front lighting

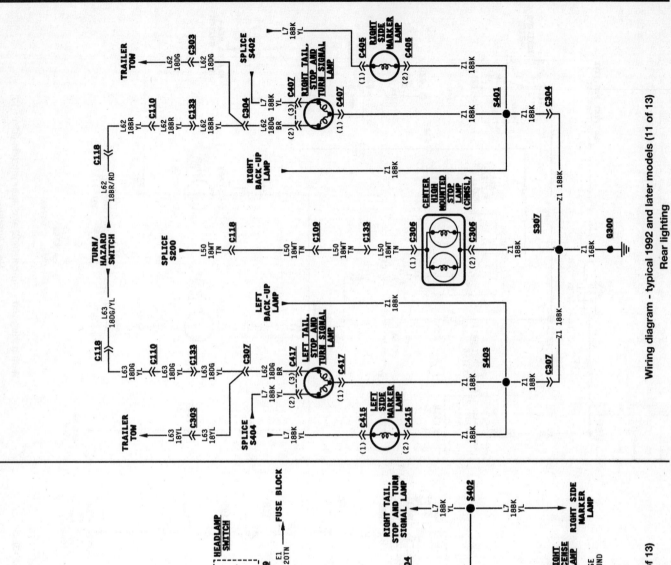

Wiring diagram - typical 1992 and later models (11 of 13)
Rear lighting

Wiring diagram - typical 1992 and later models (10 of 13)
Rear lighting

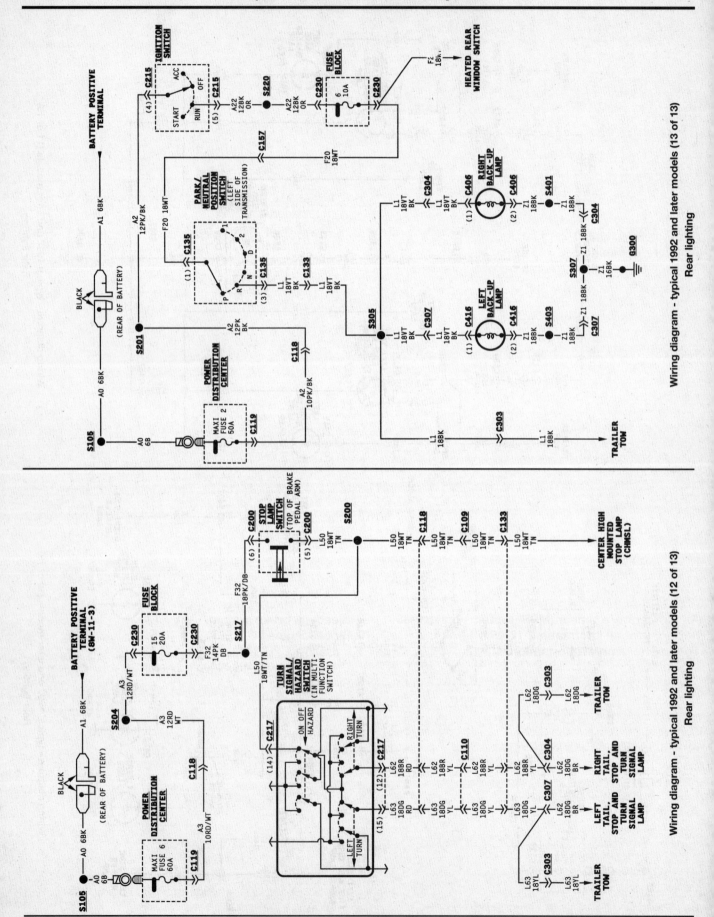

Wiring diagram - typical 1992 and later models (13 of 13)
Rear lighting

Wiring diagram - typical 1992 and later models (12 of 13)
Rear lighting

Index

Notes

Haynes Automotive Manuals

NOTE: New manuals are added to this list on a periodic basis. If you do not see a listing for your vehicle, consult your local Haynes dealer for the latest product information.

ACURA
*12020 **Integra** '86 thru '89 **& Legend** '86 thru '90

AMC
Jeep CJ - see JEEP (50020)
14020 **Mid-size models,** Concord, Hornet, Gremlin & Spirit '70 thru '83
14025 **(Renault) Alliance & Encore** '83 thru '87

AUDI
15020 **4000** all models '80 thru '87
15025 **5000** all models '77 thru '83
15026 **5000** all models '84 thru '88

AUSTIN-HEALEY
Sprite - see MG Midget (66015)

BMW
*18020 **3/5 Series** not including diesel or all-wheel drive models '82 thru '92
*18021 **3 Series** except 325iX models '92 thru '97
18025 **320i** all 4 cyl models '75 thru '83
18035 **528i & 530i** all models '75 thru '80
18050 **1500 thru 2002** except Turbo '59 thru '77

BUICK
Century (front wheel drive) - see GM (829)
*19020 **Buick, Oldsmobile & Pontiac Full-size (Front wheel drive)** all models '85 thru '98
Buick Electra, LeSabre and Park Avenue; **Oldsmobile** Delta 88 Royale, Ninety Eight and Regency; **Pontiac** Bonneville
19025 **Buick Oldsmobile & Pontiac Full-size (Rear wheel drive)**
Buick Estate '70 thru '90, Electra '70 thru '84, LeSabre '70 thru '85, Limited '74 thru '79
Oldsmobile Custom Cruiser '70 thru '90, Delta 88 '70 thru '85, Ninety-eight '70 thru '84
Pontiac Bonneville '70 thru '81, Catalina '70 thru '81, Grandville '70 thru '75, Parisienne '83 thru '86
19030 **Mid-size Regal & Century** all rear-drive models with V6, V8 and Turbo '74 thru '87
Regal - see GENERAL MOTORS (38010)
Riviera - see GENERAL MOTORS (38030)
Roadmaster - see CHEVROLET (24046)
Skyhawk - see GENERAL MOTORS (38015)
Skylark '80 thru '85 - see GM (38020)
Skylark '86 on - see GM (38025)
Somerset - see GENERAL MOTORS (38025)

CADILLAC
*21030 **Cadillac Rear Wheel Drive** all gasoline models '70 thru '93
Cimarron - see GENERAL MOTORS (38015)
Eldorado - see GENERAL MOTORS (38030)
Seville '80 thru '85 - see GM (38030)

CHEVROLET
*24010 **Astro & GMC Safari Mini-vans** '85 thru '93
24015 **Camaro V8** all models '70 thru '81
24016 **Camaro** all models '82 thru '92
Cavalier - see GENERAL MOTORS (38015)
Celebrity - see GENERAL MOTORS (38005)
24017 **Camaro & Firebird** '93 thru '97
24020 **Chevelle, Malibu & El Camino** '69 thru '87
24024 **Chevette & Pontiac T1000** '76 thru '87
Citation - see GENERAL MOTORS (38020)
*24032 **Corsica/Beretta** all models '87 thru '96
24040 **Corvette** all V8 models '68 thru '82
*24041 **Corvette** all models '84 thru '96
10305 **Chevrolet Engine Overhaul Manual**
24045 **Full-size Sedans** Caprice, Impala, Biscayne, Bel Air & Wagons '69 thru '90
24046 **Impala SS & Caprice and Buick Roadmaster** '91 thru '96
Lumina - see GENERAL MOTORS (38010)

24048 **Lumina & Monte Carlo** '95 thru '98
Lumina APV - see GM (38035)
24050 **Luv Pick-up** all 2WD & 4WD '72 thru '82
*24055 **Monte Carlo** all models '70 thru '88
Monte Carlo '95 thru '98 - see LUMINA (24048)
24059 **Nova** all V8 models '69 thru '79
*24060 **Nova and Geo Prizm** '85 thru '92
24064 **Pick-ups '67 thru '87** - Chevrolet & GMC, all V8 & in-line 6 cyl, 2WD & 4WD '67 thru '87; Suburbans, Blazers & Jimmys '67 thru '91
*24065 **Pick-ups '88 thru '98** - Chevrolet & GMC, all full-size pick-ups, '88 thru '98; Blazer & Jimmy '92 thru '94; Suburban '92 thru '98; Tahoe & Yukon '98
24070 **S-10 & S-15 Pick-ups** '82 thru '93, **Blazer & Jimmy** '83 thru '94,
*24071 **S-10 & S-15 Pick-ups** '94 thru '96 **Blazer & Jimmy** '95 thru '96
*24075 **Sprint & Geo Metro** '85 thru '94
*24080 **Vans - Chevrolet & GMC,** V8 & in-line 6 cylinder models '68 thru '96

CHRYSLER
25015 **Chrysler Cirrus, Dodge Stratus, Plymouth Breeze** '95 thru '98
25025 **Chrysler Concorde, New Yorker & LHS, Dodge Intrepid, Eagle Vision,** '93 thru '97
10310 **Chrysler Engine Overhaul Manual**
*25020 **Full-size Front-Wheel Drive** '88 thru '93
K-Cars - see DODGE Aries (30008)
Laser - see DODGE Daytona (30030)
*25030 **Chrysler & Plymouth Mid-size** front wheel drive '82 thru '95
Rear-wheel Drive - see Dodge (30050)

DATSUN
28005 **200SX** all models '80 thru '83
28007 **B-210** all models '73 thru '78
28009 **210** all models '79 thru '82
28012 **240Z, 260Z & 280Z** Coupe '70 thru '78
28014 **280ZX** Coupe & 2+2 '79 thru '83
300ZX - see NISSAN (72010)
28016 **310** all models '78 thru '82
28018 **510 & PL521 Pick-up** '68 thru '73
28020 **510** all models '78 thru '81
28022 **620 Series Pick-up** all models '73 thru '79
720 Series Pick-up - see NISSAN (72030)
28025 **810/Maxima** all gasoline models, '77 thru '84

DODGE
400 & 600 - see CHRYSLER (25030)
*30008 **Aries & Plymouth Reliant** '81 thru '89
30010 **Caravan & Plymouth Voyager Mini-Vans** all models '84 thru '95
*30011 **Caravan & Plymouth Voyager Mini-Vans** all models '96 thru '98
30012 **Challenger/Plymouth Saporro** '78 thru '83
30016 **Colt & Plymouth Champ** (front wheel drive) all models '78 thru '87
*30020 **Dakota Pick-ups** all models '87 thru '96
30025 **Dart, Demon, Plymouth Barracuda, Duster & Valiant** 6 cyl models '67 thru '76
*30030 **Daytona & Chrysler Laser** '84 thru '89
Intrepid - see CHRYSLER (25025)
*30034 **Neon** all models '95 thru '99
*30035 **Omni & Plymouth Horizon** '78 thru '90
*30040 **Pick-ups** all full-size models '74 thru '93
*30041 **Pick-ups** all full-size models '94 thru '96
*30045 **Ram 50/D50 Pick-ups & Raider and Plymouth Arrow Pick-ups** '79 thru '93
30050 **Dodge/Plymouth/Chrysler** rear wheel drive '71 thru '89
*30055 **Shadow & Plymouth Sundance** '87 thru '94
*30060 **Spirit & Plymouth Acclaim** '89 thru '95
*30065 **Vans - Dodge & Plymouth** '71 thru '96

EAGLE
Talon - see Mitsubishi Eclipse (68030)
Vision - see CHRYSLER (25025)

FIAT
34010 **124 Sport Coupe & Spider** '68 thru '78
34025 **X1/9** all models '74 thru '80

FORD
10355 **Ford Automatic Transmission Overhaul**
*36004 **Aerostar Mini-vans** all models '86 thru '96
*36006 **Contour & Mercury Mystique** '95 thru '98
36008 **Courier Pick-up** all models '72 thru '82
36012 **Crown Victoria & Mercury Grand Marquis** '88 thru '96
10320 **Ford Engine Overhaul Manual**
36016 **Escort/Mercury Lynx** all models '81 thru '90
*36020 **Escort/Mercury Tracer** '91 thru '96
*36024 **Explorer & Mazda Navajo** '91 thru '95
36028 **Fairmont & Mercury Zephyr** '78 thru '83
36030 **Festiva & Aspire** '88 thru '97
36032 **Fiesta** all models '77 thru '80
36036 **Ford & Mercury Full-size,** Ford LTD & Mercury Marquis ('75 thru '82); Ford Custom 500, Country Squire, Crown Victoria & Mercury Colony Park ('75 thru '87); Ford LTD Crown Victoria & Mercury Gran Marquis ('83 thru '87)
36040 **Granada & Mercury Monarch** '75 thru '80
36044 **Ford & Mercury Mid-size,** Ford Thunderbird & Mercury Cougar ('75 thru '82); Ford LTD & Mercury Marquis ('83 thru '86); Ford Torino, Gran Torino, Elite, Ranchero pick-up, LTD II, Mercury Montego, Comet, XR-7 & Lincoln Versailles ('75 thru '86)
36048 **Mustang V8** all models '64-1/2 thru '73
36049 **Mustang II** 4 cyl, V6 & V8 models '74 thru '78
36050 **Mustang & Mercury Capri** all models Mustang, '79 thru '93; Capri, '79 thru '86
*36051 **Mustang** all models '94 thru '97
36054 **Pick-ups & Bronco** '73 thru '79
36058 **Pick-ups & Bronco** '80 thru '96
36059 **Pick-ups, Expedition & Mercury Navigator** '97 thru '98
36062 **Pinto & Mercury Bobcat** '75 thru '80
36066 **Probe** all models '89 thru '92
36070 **Ranger/Bronco II** gasoline models '83 thru '92
*36071 **Ranger** '93 thru '97 & **Mazda Pick-ups** '94 thru '97
36074 **Taurus & Mercury Sable** '86 thru '95
*36075 **Taurus & Mercury Sable** '96 thru '98
*36078 **Tempo & Mercury Topaz** '84 thru '94
36082 **Thunderbird/Mercury Cougar** '83 thru '88
*36086 **Thunderbird/Mercury Cougar** '89 and '97
36090 **Vans** all V8 Econoline models '69 thru '91
*36094 **Vans** full size '92-'95
*36097 **Windstar Mini-van** '95-'98

GENERAL MOTORS
*10360 **GM Automatic Transmission Overhaul**
*38005 **Buick Century, Chevrolet Celebrity, Oldsmobile Cutlass Ciera & Pontiac 6000** all models '82 thru '96
*38010 **Buick Regal, Chevrolet Lumina, Oldsmobile Cutlass Supreme & Pontiac Grand Prix** front-wheel drive models '88 thru '95
*38015 **Buick Skyhawk, Cadillac Cimarron, Chevrolet Cavalier, Oldsmobile Firenza & Pontiac J-2000 & Sunbird** '82 thru '94
*38016 **Chevrolet Cavalier & Pontiac Sunfire** '95 thru '98
38020 **Buick Skylark, Chevrolet Citation, Olds Omega, Pontiac Phoenix** '80 thru '85
38025 **Buick Skylark & Somerset, Oldsmobile Achieva & Calais and Pontiac Grand Am** all models '85 thru '95
*38030 **Cadillac Eldorado** '71 thru '85, **Seville** '80 thru '85, **Oldsmobile Toronado** '71 thru '85 **& Buick Riviera** '79 thru '85
*38035 **Chevrolet Lumina APV, Olds Silhouette & Pontiac Trans Sport** all models '90 thru '95
General Motors Full-size Rear-wheel Drive - see BUICK (19025)

(Continued on other side)

---

* Listings shown with an asterisk (*) indicate model coverage as of this printing. These titles will be periodically updated to include later model years - consult your Haynes dealer for more information.

Haynes North America, Inc., 861 Lawrence Drive, Newbury Park, CA 91320-1514 • (805) 498-6703

Haynes Automotive Manuals (continued)

NOTE: New manuals are added to this list on a periodic basis. If you do not see a listing for your vehicle, consult your local Haynes dealer for the latest product information.

GEO
Metro - see CHEVROLET Sprint (24075)
Prizm - '85 thru '92 see CHEVY (24060), '93 thru '96 see TOYOTA Corolla (92036)
*40030 Storm all models '90 thru '93
Tracker - see SUZUKI Samurai (90010)

GMC
Safari - see CHEVROLET ASTRO (24010)
Vans & Pick-ups - see CHEVROLET

HONDA
42010 Accord CVCC all models '76 thru '83
42011 Accord all models '84 thru '89
42012 Accord all models '90 thru '93
42013 Accord all models '94 thru '95
42020 Civic 1200 all models '73 thru '79
42021 Civic 1300 & 1500 CVCC '80 thru '83
42022 Civic 1500 CVCC all models '75 thru '79
42023 Civic all models '84 thru '91
*42024 Civic & del Sol '92 thru '95
*42040 Prelude CVCC all models '79 thru '89

HYUNDAI
*43015 Excel all models '86 thru '94

ISUZU
Hombre - see CHEVROLET S-10 (24071)
*47017 Rodeo '91 thru '97; Amigo '89 thru '94; Honda Passport '95 thru '97
*47020 Trooper & Pick-up, all gasoline models Pick-up, '81 thru '93; Trooper, '84 thru '91

JAGUAR
*49010 XJ6 all 6 cyl models '68 thru '86
*49011 XJ6 all models '88 thru '94
*49015 XJ12 & XJS all 12 cyl models '72 thru '85

JEEP
*50010 Cherokee, Comanche & Wagoneer Limited all models '84 thru '96
50020 CJ all models '49 thru '86
*50025 Grand Cherokee all models '93 thru '98
50029 Grand Wagoneer & Pick-up '72 thru '91 Grand Wagoneer '84 thru '91, Cherokee & Wagoneer '72 thru '83, Pick-up '72 thru '88
*50030 Wrangler all models '87 thru '95

LINCOLN
Navigator - see FORD Pick-up (36059)
59010 Rear Wheel Drive all models '70 thru '96

MAZDA
61010 GLC Hatchback (rear wheel drive) '77 thru '83
61011 GLC (front wheel drive) '81 thru '85
*61015 323 & Protegé '90 thru '97
*61016 MX-5 Miata '90 thru '97
*61020 MPV all models '89 thru '94
Navajo - see Ford Explorer (36024)
61030 Pick-ups '72 thru '93
Pick-ups '94 thru '96 - see Ford Ranger (36071)
61035 RX-7 all models '79 thru '85
*61036 RX-7 all models '86 thru '91
61040 626 (rear wheel drive) all models '79 thru '82
*61041 626/MX-6 (front wheel drive) '83 thru '91

MERCEDES-BENZ
63012 123 Series Diesel '76 thru '85
*63015 190 Series four-cyl gas models, '84 thru '88
63020 230/250/280 6 cyl sohc models '68 thru '72
63025 280 123 Series gasoline models '77 thru '81
63030 350 & 450 all models '71 thru '80

MERCURY
See FORD Listing.

MG
66010 MGB Roadster & GT Coupe '62 thru '80
66015 MG Midget, Austin Healey Sprite '58 thru '80

MITSUBISHI
*68020 Cordia, Tredia, Galant, Precis & Mirage '83 thru '93
*68030 Eclipse, Eagle Talon & Ply. Laser '90 thru '94
*68040 Pick-up '83 thru '96 & Montero '83 thru '93

NISSAN
72010 300ZX all models including Turbo '84 thru '89
*72015 Altima '93 thru '97
*72020 Maxima all models '85 thru '91
*72030 Pick-ups '80 thru '96 Pathfinder '87 thru '95
72040 Pulsar all models '83 thru '86
*72050 Sentra all models '82 thru '94
*72051 Sentra & 200SX all models '95 thru '98
*72060 Stanza all models '82 thru '90

OLDSMOBILE
*73015 Cutlass V6 & V8 gas models '74 thru '88
For other OLDSMOBILE titles, see BUICK, CHEVROLET or GENERAL MOTORS listing.

PLYMOUTH
For PLYMOUTH titles, see DODGE listing.

PONTIAC
79008 Fiero all models '84 thru '88
79018 Firebird V8 models except Turbo '70 thru '81
79019 Firebird all models '82 thru '92
For other PONTIAC titles, see BUICK, CHEVROLET or GENERAL MOTORS listing.

PORSCHE
*80020 911 except Turbo & Carrera 4 '65 thru '89
80025 914 all 4 cyl models '69 thru '76
80030 924 all models including Turbo '76 thru '82
*80035 944 all models including Turbo '83 thru '89

RENAULT
Alliance & Encore - see AMC (14020)

SAAB
*84010 900 all models including Turbo '79 thru '88

SATURN
87010 Saturn all models '91 thru '96

SUBARU
89002 1100, 1300, 1400 & 1600 '71 thru '79
*89003 1600 & 1800 2WD & 4WD '80 thru '94

SUZUKI
*90010 Samurai/Sidekick & Geo Tracker '86 thru '96

TOYOTA
92005 Camry all models '83 thru '91
92006 Camry all models '92 thru '96
92015 Celica Rear Wheel Drive '71 thru '85
*92020 Celica Front Wheel Drive '86 thru '93
92025 Celica Supra all models '79 thru '92
92030 Corolla all models '75 thru '79
92032 Corolla all rear wheel drive models '80 thru '87
92035 Corolla all front wheel drive models '84 thru '92
*92036 Corolla & Geo Prizm '93 thru '97
92040 Corolla Tercel all models '80 thru '82
92045 Corona all models '74 thru '82
92050 Cressida all models '78 thru '82
92055 Land Cruiser FJ40, 43, 45, 55 '68 thru '82
92056 Land Cruiser FJ60, 62, 80, FZJ80 '80 thru '96
*92065 MR2 all models '85 thru '87
92070 Pick-up all models '69 thru '78
*92075 Pick-up all models '79 thru '95
*92076 Tacoma '95 thru '98, 4Runner '96 thru '98, & T100 '93 thru '98
*92080 Previa all models '91 thru '95
92085 Tercel all models '87 thru '94

TRIUMPH
94007 Spitfire all models '62 thru '81
94010 TR7 all models '75 thru '81

VW
96008 Beetle & Karmann Ghia '54 thru '79
96012 Dasher all gasoline models '74 thru '81
*96016 Rabbit, Jetta, Scirocco, & Pick-up gas models '74 thru '91 & Convertible '80 thru '92
96017 Golf & Jetta all models '93 thru '97
96020 Rabbit, Jetta & Pick-up diesel '77 thru '84
96030 Transporter 1600 all models '68 thru '79
96035 Transporter 1700, 1800 & 2000 '72 thru '79
96040 Type 3 1500 & 1600 all models '63 thru '73
96045 Vanagon all air-cooled models '80 thru '83

VOLVO
97010 120, 130 Series & 1800 Sports '61 thru '73
97015 140 Series all models '66 thru '74
*97020 240 Series all models '76 thru '93
97025 260 Series all models '75 thru '82
*97040 740 & 760 Series all models '82 thru '88

TECHBOOK MANUALS
10205 Automotive Computer Codes
10210 Automotive Emissions Control Manual
10215 Fuel Injection Manual, 1978 thru 1985
10220 Fuel Injection Manual, 1986 thru 1996
10225 Holley Carburetor Manual
10230 Rochester Carburetor Manual
10240 Weber/Zenith/Stromberg/SU Carburetors
10305 Chevrolet Engine Overhaul Manual
10310 Chrysler Engine Overhaul Manual
10320 Ford Engine Overhaul Manual
10330 GM and Ford Diesel Engine Repair Manual
10340 Small Engine Repair Manual
10345 Suspension, Steering & Driveline Manual
10355 Ford Automatic Transmission Overhaul
10360 GM Automatic Transmission Overhaul
10405 Automotive Body Repair & Painting
10410 Automotive Brake Manual
10415 Automotive Detailing Manual
10420 Automotive Eelectrical Manual
10425 Automotive Heating & Air Conditioning
10430 Automotive Reference Manual & Dictionary
10435 Automotive Tools Manual
10440 Used Car Buying Guide
10445 Welding Manual
10450 ATV Basics

SPANISH MANUALS
98903 Reparación de Carrocería & Pintura
98905 Códigos Automotrices de la Computadora
98910 Frenos Automotriz
98915 Inyección de Combustible 1986 al 1994
99040 Chevrolet & GMC Camionetas '67 al '87 Incluye Suburban, Blazer & Jimmy '67 al '91
99041 Chevrolet & GMC Camionetas '88 al '95 Incluye Suburban '92 al '95, Blazer & Jimmy '92 al '94, Tahoe y Yukon '95
99042 Chevrolet & GMC Camionetas Cerradas '68 al '95
99055 Dodge Caravan & Plymouth Voyager '84 al '95
99075 Ford Camionetas y Bronco '80 al '94
99077 Ford Camionetas Cerradas '69 al '91
99083 Ford Modelos de Tamaño Grande '75 al '87
99088 Ford Modelos de Tamaño Mediano '75 al '86
99091 Ford Taurus & Mercury Sable '86 al '95
99095 GM Modelos de Tamaño Grande '70 al '90
99100 GM Modelos de Tamaño Mediano '70 al '88
99110 Nissan Camionetas '80 al '96, Pathfinder '87 al '95
99118 Nissan Sentra '82 al '94
99125 Toyota Camionetas y 4Runner '79 al '95

* Listings shown with an asterisk (*) indicate model coverage as of this printing. These titles will be periodically updated to include later model years - consult your Haynes dealer for more information.

Over 100 Haynes motorcycle manuals also available

5-98

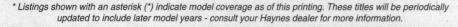

Haynes North America, Inc., 861 Lawrence Drive, Newbury Park, CA 91320-1514 • (805) 498-6703